FRAMINGHAM STATE COLLEGE

3 3014 00057 1127

D1218825

Optical Methods in Biology

Optical Methods in Biology

ELIZABETH M. SLAYTER

HENRY WHITTEMORE
LIBRARY

STATE COLLEGE
FRAMINGHAM, MASS.

WILEY-INTERSCIENCE
A DIVISION OF JOHN WILEY & SONS
New York · London · Sydney · Toronto

Copyright © 1970 by John Wiley & Sons, Inc.

All Rights reserved. No part of this book may be reproduced
by any means, nor transmitted, nor translated into a machine
language without the written permission of the publisher.

10 9 8 7 6 5 4 3 2 1

Library of Congress Catalog Card Number: 76–77833

SBN 471   79670   0

Printed in the United States of America

QH
205
S5

Preface

Current research in the life sciences relies heavily on microscopy, diffraction, spectroscopy, and related optical techniques. Although many optical instruments have been developed which can be used essentially as black boxes—that is, with little knowledge of the optical principles on which the instruments are based—the complete exploitation of instrumental methods still requires a general understanding of optical concepts. Unfortunately, such understanding is difficult for the biologist to acquire unless he is fortunate enough to have had unusually extensive training in physics or some related discipline. Many accounts of optical instruments are so descriptive as to be trivial; more serious treatises may be too mathematical in approach to be accessible to the nonspecialist. Although discussions of most topics of interest at a suitable level of presentation do exist, this literature is widely scattered and often difficult to locate. The problem is compounded by the fact that a variety of conventions and terminologies is used and by the fact that the extent of physical and mathematical knowledge assumed is not always the same. Thus there has been a need for a single and unified account of the optical principles which are important in basic biological research; it is that need which this volume endeavors to fulfill.

It is intended that this book could be used either in a graduate course in optical instrumentation or as a reference for the working scientist. The level of presentation assumes that the student reader (a) has at some time taken a course in general physics in which, as is usually the case, the subject of classical optics has been dealt with hastily and (b) has some knowledge of integral and differential calculus but little idea of how these mathematical methods might be relevant to biology or biochemistry. Thus the first chapters are devoted to an account of optical principles *as they apply to biological optical instrumentation*. Emphasis and even content are thus somewhat different from that of treatises on optics per se. Wherever possible, these accounts are essentially verbal rather than mathematical in nature. In particular, the use of calculus is avoided when arithmetical or algebraic methods can suffice. However, certain topics cannot be discussed meaningfully without the use of more advanced mathematical methods. Thus the discussions of resolving power and of the interpretation of x-ray diffraction data (in Chapters 10 and 31, respectively) have required some explanation of Fourier integral methods. Even there, however, a basically verbal description has been attempted.

It should be made clear that this book is not a treatise on the design of optical instruments nor on the state of the art of instrumentation. Few biological scientists can divide their efforts between the design of instruments and the application of instrumental methods; the former is, quite properly, a separate discipline. Neither is the book a handbook of techniques nor a review of what has been learned by applying optical methods to biological problems, since these aspects of optical instrumentation are dealt with in literature which discusses the respective biological systems. The book is, in fact, an account of the *principles* on which optical methods used in biological research are based.

Chapters 29 through 31 are intended to provide the reader with an understanding of the nature of the information which can be obtained in x-ray diffraction studies. They should not be considered more than an introduction to the very specialized field of biological x-ray crystallography; in this respect they are more superficial in nature than the remainder of the book.

Finally, the emphasis placed on high resolution electron microscopy in Chapter 20 and in other parts of the book reflects not only my particular concern but also the widespread interest that exists in this subject at the present time. It is felt that a specific discussion of the subject is of particular importance here, since the factors that affect ultimate resolution in the electron image have rarely been described fully or clearly.

I wish to thank Bernard Talbot for reading the entire text during its formative stages and for contributing Chapter 27. I also thank Jean Bennett, Hannah Gay, Ian Gay, William Longley, Susan Lowey, Richard Morgan, and Henry Slayter for reviewing portions of the text in their respective areas of interest and Katherine Bowler for help in preparing the manuscript.

Lincoln Center, Massachusetts   ELIZABETH M. SLAYTER

Contents

Optical Methods in Biology

CHAPTER 1

The Nature of Light: Geometrical, Wave, and Quantum Optics

Optical phenomena may be considered in three frames of reference: *geometrical* optics, which utilizes the concept of the "ray"; *wave* or *physical* optics, which treats light as a wave motion; and *quantum* optics, which treats light as a stream of particles of zero mass (i.e., zero rest mass) termed *photons*. Despite certain appearances to the contrary, these three points of view are not mutually inconsistent. The reconciliation of the wave and particle properties of light are discussed later in this chapter.

The user of optical instruments fortunately need not ponder the question of the nature of light other than for his own intellectual satisfaction. He may in fact on various occasions speak of light as rays, waves, *or* photons, switching concepts and mixing terminologies with apparent abandon. It is important nonetheless to be familiar with the vocabulary and basic concepts of each of these approaches to optical effects and to understand their range of application and limitations. A number of optical effects can be described and/or explained in terms that may seem quite unrelated; this fact may cause the optical literature to seem extremely confusing to the uninitiated. This chapter and this book as a whole, endeavor to provide the vocabulary and other bases of understanding that will render the optical literature effectively accessible to the biologist. In this chapter the disciplines of geometrical, physical, and quantum optics are defined, and certain basic concepts of each are presented.

GEOMETRICAL OPTICS

Geometrical optics deals with the behavior of light as directly observed macroscopically; it describes the paths traveled by light beams and predicts the location and magnification of images. The physical nature of light itself is not considered; hence geometrical optics makes no predictions concerning the interactions of light beams with matter nor with each other. The laws of geometrical optics can be regarded as an extrapolation of the observations of wave optics to the case of infinitesimal wavelength.

In this book, a geometrical treatment is used to explain the focusing action of lenses and lens systems and the third order abberations which occur in lens images, as well as to describe light paths through instruments (microscopes, spectrophotometers, etc.) which may, however, require a discussion of wave or photon properties for explanation of their specific action.

WAVE OPTICS

Wave or physical optics treats light as a periodic oscillation of electric and magnetic fields in space. These oscillations are analogous in many respects to the mechanical waves observed in fluids and other forms of matter.

A wave-optical treatment must be used in explaining such instruments as the phase, interference, or polarization microscopes and in accounting for such effects as the formation of interference colors and diffraction patterns. In particular, the wave theory of light clarifies the relationship between diffraction patterns and microscope images and accounts for the limited resolution of the latter.

QUANTUM OPTICS

The quantum approach to optics depicts a light beam as consisting of a stream of "chunks" of energy of defined sizes, designated photons. The photon may be described as a fundamental particle of zero mass. (More precisely, the photon must be described as a particle of zero rest mass, since moving photons, like other moving particles, are associated with a relativistic mass determined by their velocity.) The emission of light and its interaction with matter can only be explained from this point of view. Quantum optics provides the most consistent means of considering optical effects, since, at least in principle, the approach can be applied to the description of any and all optical phenomena.

The quantum optical treatment is required for the explanation of absorption, fluorescence and related effects, and of such devices as the laser and fluorophotometer.

LIGHT RAYS

A *ray*, which is a line representing the direction in which light travels, seems, superficially, to be some sort of unit component of a light beam. Attempts to isolate individual rays, however, produce increasingly anomalous effects as smaller and smaller portions of light beams are examined, and it becomes evident that rays as units of light have no physical reality.

Rays, in fact, represent the *directions of energy transfer* by light beams; so viewed, they offer a physically valid and useful concept.

The path of light, that is, the directions of energy transfer, in any optical system may be represented by a *ray diagram* in which each arrow symbolizes one of an infinite number of possible directions of energy flow. A few characteristic rays are plotted in such diagrams. In general, rays proceed outward from a source in all possible directions, traveling in straight lines until they intercept some device (such as a lens or prism) which acts as a ray course modifier. Figure 1-1 is a ray diagram of an optical instrument in its most general form. Note that such diagrams provide no information concerning the relative intensities of energy flow in different directions (i.e., rays, though drawn as arrows, are not vectors; they represent directions without specifying magnitudes).

Groups of a few rays, that is, narrow beams of light, are variously referred to as *bundles* of rays or *pencils* of light.

LAWS OF GEOMETRICAL OPTICS

The behavior of light rays is described by three basic laws: the principle of reversibility, the law of reflection, and the law of refraction (Snell's law).

The *principle of reversibility* states that, if the direction of a ray is reversed in any system, the ray exactly retraces its path through the system. It is important to note that the reversibility principle applies only to the location of light paths, and not to the intensity of the light.

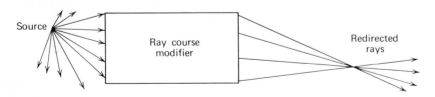

Fig. 1-1 Ray diagram.

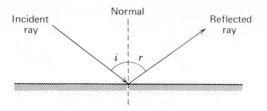

Fig. 1-2 Reflection.

The *law of reflection*, illustrated by Fig. 1-2, states that, when light is reflected from a surface, *the angle of incidence is equal to the angle of reflection* and that *the reflected ray lies in the plane of incidence.** The *plane of incidence* is that plane which contains the incident ray and the normal (perpendicular) to the reflecting surface at the point of incidence. Thus in Fig. 1-2 the plane of incidence is the plane of the page. The angles of incidence and reflection are sometimes defined as those made by the respective rays with the reflecting surface; this convention is unfortunate, since, in the case of refraction, angles must be defined as those between the rays and the normal to the surface at the point of incidence. The choice of convention does not alter the form of the law of reflection; it determines the numerical values of the specified angles.

REFRACTION AND REFRACTIVE INDEX

Refractive index (designated n), a property of the medium through which light travels, is defined as the ratio of the velocity of light in vacuum (designated c, and equal approximately to 3×10^{10} cm/sec) to that in the medium; that is,

(1-a) $$n_{\text{med}} = \frac{v_{\text{vac}}}{v_{\text{med}}} = \frac{c}{v_{\text{med}}}.$$

The refractive index of a vacuum is thus exactly unity; that of air is only very slightly greater. The refractive indices of substances generally lie between 1 and 2, although a few substances have indices somewhat greater than 2. Some representative values are included in Table 1-1.

---

*It is interesting to note the following quote from Canto XV of *Purgatory* in Dante's *Divine Comedy*, written ca. 1300, which makes plain that the law of reflection has been understood for many centuries (Longfellow's translation):

> "As when from off the water, or a mirror,
> The sunbeam leaps unto the opposite side,
> Ascending upwards in the self-same measure
> That it descends, and deviates as far
> From falling of a stone in line direct,
> (As demonstrate experiment and art) . . ."

Table 1-1   Refractive Indices of Certain Media

Substance	n
Vacuum	1.00000...
Air (15°C; 760 mm Hg)	1.000277
Water	1.33
Ethanol	1.36
Rock salt	1.54
Carbon disulfide	1.63
Various optical glasses	1.45–1.95
Diamond	2.42

Refractive indices of all substances are constant only for a particular wavelength (color) of light. The values given in Table 1-1 refer, as they usually do, to the refraction of the yellow D-line light from a sodium lamp. In all considerations to be discussed in this book the refractive index of air (not water!) for all wavelengths may be taken as unity. Means for measurement of refractive indices are described in Chapter 26, and the variation of refractive index with wavelength is discussed in Chapter 23.

Note that the refractive index of any medium is *inversely* proportional to the velocity of light in that medium.

The *law of refraction*, also known as *Snell's law*, is stated by

(1-b)
$$\boxed{\frac{\sin i}{\sin r} = \frac{n_2}{n_1}.}$$

in which i, the angle of incidence, is the angle between the incident ray and the normal to the surface at the point of incidence; r, the angle of refraction, is the angle between the refracted ray and the normal; and n_1 and n_2, respectively, are the refractive indices of the media in which incidence and refraction occur. The law of refraction further states that the refracted ray lies in the plane of incidence. The refraction of light is illustrated in Fig. 1-3.

Frequently light is incident upon an optically more dense medium (i.e., a medium of higher refractive index) from *air*. Then (1-b) becomes simply

(1-c)
$$\frac{\sin i}{\sin r} = n_{med},$$

where n_{med} is the refractive index of the optically more dense medium.

Note that here, in contrast to the case of reflection, the angles i and r *must* be expressed as those made by the rays with the normal.

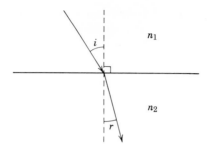

Fig. 1-3 Refraction.

It is convenient to remember that, according to the law of refraction, light entering a medium of higher index is bent *toward* the normal, and light entering an optically less dense medium is bent *away from* the normal. (Note that the common use of the terms "optically more dense" and "optically less dense" to refer to media of higher and lower refractive indices, respectively, should not be confused with the optical density, a measure of light absorption (discussed in Chapter 24).)

TOTAL INTERNAL REFLECTION

The phenomenon of total internal reflection is a consequence of the behavior of light as described by the law of refraction. Consider a ray a, as illustrated by Fig. 1-4, which enters, from air, a medium of refractive index n at a grazing angle; that is, at an angle approximating 90° to the normal. The ray is refracted in the medium, as shown, in accordance with (1-b) or (1-c). Rays entering the medium at any other angle of incidence, for example, ray b in Fig. 1-4, must be refracted within the angle α_c between ray a and the normal. If now the directions of these rays are reversed they retrace their paths according to the principle of reversibility.

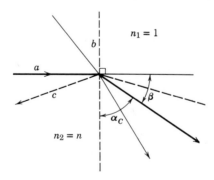

Fig. 1-4 Total internal reflection.

Rays a, b, and any other rays which fall within angle α reemerge into air along the paths shown. If, however, a ray such as c is incident within the angle β between ray a and the surface, *there is no corresponding path in air which it may take*. Ray c has, as it were, "no place to go" and is reflected back into the medium at an angle determined by the law of reflection. In general, rays incident upon a surface from the side of larger refractive index are totally reflected if the angle of incidence is greater than the so-called *critical angle*. The critical angle corresponding to any value of n may be found from (1-b); it must be borne in mind, that light incident at the critical angle is refracted into air at exactly 90°. Then

$$(1\text{-}d) \quad \sin \text{(critical angle)} = \frac{n_{\text{air}}}{n_{\text{med}}} \times \sin 90° = \frac{1}{n_{\text{med}}} \quad \text{or} \quad \alpha_C = \sin^{-1} \frac{1}{n_{\text{med}}}.$$

As the term "total internal reflection" implies all the intensity of the light beam is reflected back into the medium of higher index. In other cases in which light is incident upon an interface between media of different refractive indices the energy of the light beam is invariably divided between reflected and refracted rays (each of which travels in a direction dictated by the geometric laws). The resulting loss of light intensity from the main beam, whether reflected or refracted, may be an important consideration in the design of optical instruments. In practical systems, total internal reflection is often exploited to produce redirection of light beams without energy loss.

PROPERTIES OF MECHANICAL WAVES

An understanding of the nature of light as a wave requires an appreciation of the properties of wave motions in general. Some macroscopic properties of waves in fluids, which can be observed directly, are noted.

Waves, despite superficial appearances to the contrary, do not represent a net transport of matter; rather, they transport energy. Tidal fluctuations apart, a rough ocean is not continuously transported onto the shore. Shifts of rocky substratum may be detected on a seismograph hundreds or thousands of miles away within a few minutes. In fact, waves move as a propagation of *disturbance*, involving only local, *periodic* movements of matter.

A series of waves may be generated in a pan of water by tipping with some fixed frequency. Careful observation then shows that crests and troughs appear at any point in the pan with a frequency imparted by the periodic motion of the water. Replacement of the water by a medium of different viscosity changes the spacing between troughs and crests but not the frequency with which troughs and crests alternate at a given position.

When two or more wave motions are generated, as by throwing a pair of stones into a pond, complicated interactions can be observed in the regions where wave trains cross, but the forms of the original disturbances are unchanged when the individual disturbances pass beyond the region of overlap.

A point source of disturbance (such as the stone thrown in a pond) produces a circular wavefront, and a linear source of disturbance (such as the tipping of a rectangular pan) produces a linear wavefront. It is not difficult to realize that in three dimensions a point source of disturbance would produce a *spherical* wavefront whereas a linear source would produce a *planar* wavefront. A small portion of a spherical wavefront, at some distance from the source, approximates a planar wave.

In the vicinity of an obstacle, such as a breakwater in front of a harbor, wave motions are distorted. Beyond the obstacle, waves of altered amplitude, but unaltered frequency, are observed.

A SUGGESTION OF THE WAVELIKE CHARACTER OF LIGHT

The wavelike nature of light is suggested by an observation so simple as to be taken entirely for granted. Consider, as in Fig. 1-5, a window from the sides of which A and B observe, respectively, a bird and a flower placed beyond the window and on opposite sides of it. The light paths from the bird to A and from the flower to B must cross, yet the image of the bird seen by A is in no way distorted by the fact that B is looking at the flower, or even by the fact that the flower is there, whether B observes it or not. These facts suggest that there is no passage of matter or of charge between objects and viewers, because otherwise the two sets of rays would have deflected each other. In short, the light beams passing to A and B show no net interaction outside the region of overlap. Proof of the wavelike character of light is constituted by observation of interactions *within* the region of overlap, that is, by observation of the diffraction phenomena described in Chapter 3.

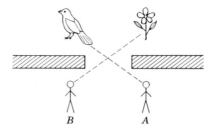

Fig. 1-5.

THE ETHER

Assuming then that light is in fact a wave motion, the question which immediately arises is: What vibrates as light waves pass through it? The medium of vibration obviously cannot be matter in the ordinary sense, since light travels at maximum velocity c, and without attenuation of intensity, through a vacuum, where there is no matter at all! Yet, to complete the analogy between light and mechanical waves, *something* that vibrates must be postulated. For this purpose, 19th century physicists proposed the existence of an "ether"—a massless medium permeating all space, matter and vacuum alike, which vibrates during the transmission of light. The ether was, and remains, a satisfactory concept in terms of attempts to pictorialize the nature of light waves, but it has absolutely no physical reality. When attempts were made to characterize the ether, it was found to possess some extraordinarily improbable qualities. For example, it was necessary to postulate that the viscosity of the ether was enormous, yet not the slightest evidence could be found that this "medium" hindered the motions of stars or planets. These difficulties failed to re-solve themselves, and eventually it became clear that a strict analogy between light and mechanical waves does not apply. Light waves can instead be described consistently as electromagnetic in character.

THE ELECTROMAGNETIC CHARACTER OF LIGHT

In 1864 Maxwell developed, without reference to the behavior of light, a series of equations describing the propagation of electric and magnetic fields. In these equations a quantity c, included originally as an appropri-ate constant without particular significance, has the units of distance per unit time; that is, of *velocity*. Subsequently the value of c was found from independent experimental measurements to be approximately 3×10^{10} cm/sec or the same, within experimental error, as values previously obtained for the velocity of light by direct measurements. The identity of light with the propagation of electromagnetic oscillations was thus strongly suggested. Further development of electromagnetic theory confirmed this hypothesis.

What is the physical picture of light waves which emerges from Max-well's treatment? It is one in which the periodic motions of matter that characterize mechanical waves are replaced by *periodic fluctuations of electric and magnetic field strength*. The electric field and, likewise, the magnetic field may be represented by a set of vectorial values of electric and magnetic force at all points in space. That is, at any point, a magni-tude and a direction of electric and of magnetic field strength may be specified. Fluctuations of electric field strength are always associated with fluctuations at right angles thereto of magnetic field strength. Conventional representations of electromagnetic waves, however, normally represent only the magnitude and direction of the vector representing the *electric* field strength, the associated magnetic vector being merely implied.

Figure 1-6 is a graph representing an electromagnetic wave. The reader should understand exactly what a graph of this sort represents. The ordinate (*y*-axis) of Fig. 1-6 is the electric field strength, a vectorial quantity. The abscissa (*z*-axis) may represent either time *or* position. If it represents *time*, the curve shows the variation of values of the electric field strength at a fixed point as a function of time. If, as is more usual, the *z*-axis represents *position*, the curve gives an instantaneous representation of the positions of the tip of the electric vector associated with the electromagnetic wave. The positive direction of the *z*-axis represents the direction of propagation of the wave.

A property of light waves predicted by Maxwell's equations is that they should be purely *transverse*; that is, all components of vibration must take place in directions at right angles to the direction of propagation. In this respect, light waves differ from mechanical waves, which are associated with longitudinal components of vibration (i.e., components of vibration which are parallel to the direction of propagation) and, specifically, from sound waves, which are *purely* longitudinal mechanical vibrations. Note that vibrations of a transverse wave may occur in *any* or *all* directions which are perpendicular to the direction of propagation of the wave. The vibration symbolized in Fig. 1-6, for example, oscillates in the plane of the page; another wave motion might vibrate in and out of the plane of the paper.

The student is almost certain to find the electromagnetic character of light waves difficult to visualize. Fortunately, it is not essential to do so. The practical worker who attempts to understand wave-optical effects is at liberty to imagine them in mechanical terms—by invoking an "ether"

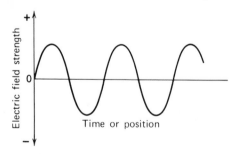

Fig. 1-6 An electromagnetic wave.

as it were. The interactions, if not the fundamental nature, of light waves can effectively be understood in this way. The visualization of light waves in terms of mechanical analogy is recommended to the reader with the caution that he should refrain from taking the pictorial representation too seriously.

DEVELOPMENT OF WAVE EQUATIONS

A quantitative treatment of the properties of waves is required in many optical considerations. Here a general equation for wave motions is derived; this relationship can be obtained with reference to mechanical vibrations, but the equation satisfactorily describes electromagnetic vibrations also. Characteristic properties of wave motions are then defined with reference to the wave equation.

Figure 1-7 is a plot of a wave motion; it can most simply be thought of as a section through a mechanical wave train, the displacements of a material medium in the y direction being plotted as a function of position in the z direction. The maximum value of y is $\pm A_0$, and the distance along the z-axis corresponding to one complete cycle is designated λ. This curve is described by the equation

(1-e)
$$y = A_0 \sin \frac{2\pi z}{\lambda},$$

if the distance z is expressed in terms of fractions of the distance λ. (For example, note that, when $z = 0$, $\lambda/2$, λ, or any integral multiple of $\lambda/2$, $\sin(2\pi mz/\lambda)$, and therefore y is equal to zero. If $z = \lambda/4$ or, in general, $z = (m + \frac{1}{4}\lambda)$, $y = A_0 \sin(\pi/2) = A_0$, while, if $z = 3\lambda/4$ or, in general, $(m + \frac{3}{4})\lambda$, $y = A_0 \sin(3\pi/2) = -A_0$. Inspection shows that these values fit the curve plotted in Fig. 1-7.) Wave motions which can be described by (1-e) are known as *sine waves* or *simple harmonic motions*.

A wave which is not static, but which propagates with velocity v is depicted in Fig. 1-8. An equation describing such a wave must have the property that at any specific position $y(t + \Delta t) = y(t)$, since the crests of

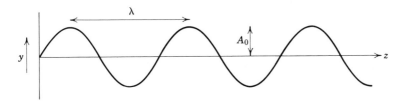

Fig. 1-7 A wave motion.

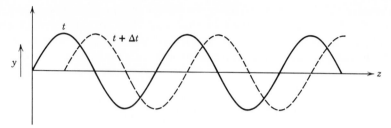

Fig. 1-8 Successive positions of a wave train.

the wave do not grow larger or smaller with time, but simply progress in the $+z$ direction. The general form of equation possessing this property is

(1-f) $$y = f(z - vt).$$

to establish that this is so, note that $\Delta z = v\Delta t$, so that

(1-g)
$$\begin{aligned}
y(t + \Delta t) &= f(z + \Delta z - vt - v\Delta t) \\
&= f(z + v\Delta t - vt - v\Delta t) \\
&= f(z - vt) = y(t).
\end{aligned}$$

Combination of the general equation (1-f) with the more specific equation (1-e) gives

(1-h) $$y = A_0 \sin\left[\frac{2\pi(z - vt)}{\lambda}\right].$$

Although equation (1-h) correctly describes sine wave motions, some re-arrangement is required in order to obtain a more useful form of wave equation. As already noted, the velocity of a wave depends upon the properties of the medium through which it travels (for mechanical and electromagnetic waves alike). It is desirable, however, to describe a wave motion in terms which are independent of the nature of the medium. This may be done by equating velocity to certain more fundamental properties, as given by

(1-i) $$v = \frac{\text{distance along } z \text{ axis representing one cycle } (= \text{wavelength})}{\text{time per cycle}}$$
$$= \text{wavelength} \times \text{number of cycles per unit time}$$
$$= \text{wavelength} \times \text{frequency}.$$

The product given by (1-i) is designated $\lambda\nu$, where λ is the wavelength, as depicted in Fig. 1-7, and ν is the frequency of vibration which is, as noted above, an inherent property of any wave motion imparted by the source. It is convenient to express frequency as the *angular velocity* $\omega = 2\pi\nu$ radians/sec. Application of these considerations to (1-h) gives

(1-j) $$y = A_0 \sin \left(\frac{2\pi z}{\lambda} - \omega t \right),$$

which, by routine trigonometric rearrangement,* may be expressed as

(1-k) $$y = A_0 \sin \left(\omega t - \frac{2\pi z}{\lambda - \pi} \right).$$

Note that the term $(2\pi z/\lambda - \pi)$ specifies what particular fraction of a cycle the wave has completed. Should some arbitrary fraction of a cycle be added to this quantity, the positions of the crests of the wave would be shifted, but the general form of the wave equation would be unchanged. Hence it is possible to write, more concisely

(1-l) $$\boxed{y = A_0 \sin (\omega t - \alpha)} ,$$

or, for that matter,

(1-m) $$y = A_0 \cos (\omega t - \alpha'),$$

in which $\alpha = 2\pi z/\lambda$ is now an arbitrary term which provides a reference point for the position of crests along the z direction. The quantity α designates the *phase* of the wave motion.

CHARACTERISTICS OF WAVES AS SPECIFIED BY THE WAVE EQUATION

Equation 1-l and equivalent expressions specify, either explicitly or by implication, the four fundamental properties of wave motions. They are *amplitude, frequency, phase*, and *state of polarization*.

The quantity A_0, representing the maximum displacement in the y direction of Fig. 1-7, clearly specifies the magnitude of the wave motion. A_0 is termed the *amplitude* of the wave; this quantity must be distinguished from the instantaneous amplitude, A $(=y)$. In terms of analogy to mechanical waves, A may also be referred to as the displacement.

If Fig. 1-7 represents the motions of a single particle executing simple harmonic motion in the y direction, the velocity of the particle is

(1-n) $$\frac{dy}{dt} = A_0 \omega \cos (\omega t - \alpha).$$

At $y = 0$, dy/dt is maximal and $\cos (\omega t - \alpha)$ assumes its maximal value of unity. Thus

(1-o) $$v_{max} = \left(\frac{dy}{dt} \right)_{y=0} = \omega A_0.$$

*That is, by making use of the identities $\sin (A - B) = \sin A \cos B - \cos A \sin B$ and $\sin (-A) = - \sin A$.

At $y = 0$, also, the potential energy of the particle is zero and thus

(1-p)   Total energy = kinetic energy = $\frac{1}{2}mv^2 = \frac{1}{2}m\omega^2 A_0^2$.

Equation (1-p) shows that, for a particle of constant mass and for a constant frequency of vibration, the *energy* of the wave motion varies as the *square* of its amplitude. The amplitude A_0 is the constant maximum amplitude of the wave, and it must be distinguished from its instantaneous amplitude, designated A. Thus the energy content of a wave of constant A_0 is constant. The quantity *intensity* is defined in electromagnetic theory as

(1-q) $I \equiv \dfrac{cn}{8\pi} A_0^2$.

Thus, for a wave motion of a given frequency, in a given medium,

(1-r)   Energy density = $KA_0^2 = KI$.

Intensity, a measure of the energy density of the wave, is one of the directly observable properties of light waves. It is an inherent property of the wave and must be distinguished from *brightness*, a quantity dependent upon the response characteristics of a detector such as, for example, the human eye.

Frequency, appearing in (1-l) in the form of ω, is the quantity which determines the "*kind*" or *color* of light and is thus also a quantity amenable to direct observation. Frequency is related to the wavelength λ by the expression

(1-s) $\nu = \dfrac{c}{\lambda_{vac}} = \dfrac{c}{n_{med}\lambda_{med}}$.

Frequency is an invariant property of a wave motion, whereas λ is a related quantity which varies according to the refractive index of the transmitting medium. Rationally, then, the *kind* of light should always be specified in terms of its *frequency*. But optical convention is irrational in this case; light, at least in the range of visible frequencies, is normally referred to in terms of its wavelength. Wavelength in this context in fact, implies the *in vacuum wavelength*. The value so specified may differ appreciably from the actual wavelength in a particular medium.

A third characteristic of wave motions is the quantity *phase* specified by α in (1-l). Phase refers to *relative* position in a cyclic motion; all crests of a wave are exactly in phase with each other and all crests are exactly out of phase with all troughs. In experimental terms it might seem that the concept of phase is a useless one in the case of light waves. Imagine that, in some unspecified way, a measurement of the position of a crest is made. Employing unusually precise instrumentation, the time of measurement is limited to only a microsecond $\sim 10^{-6}$ sec. During this time, however, the crest moves a distance $vt = 3 \times 10^{10}$ cm/sec $\times 10^{-6}$ sec $= 3 \times 10^{4}$ cm, or 30 meters. This distance is enormous in comparison with the wavelength (for visible light) of less than 10^{-4} cm. The experiment is worthless, for the *absolute* phase of a light wave is indeed immeasurable. The quantity α is nonetheless of interest, for, if (1-l) is applied to two wave motions, the phase difference between the two, that is, $(\alpha_1 - \alpha_2)$, can be measured with precision. Phase difference is *not* a directly observable quantity, but it is deduced by observing the changes in intensity which result from the interaction (*interference*) of the two beams. Phase difference and the related quantity, path difference, are discussed in Chapter 3, where a description of wave interactions is given.

The relative phases of waves are specified in terms of fractions of a wavelength $((\alpha_1 - \alpha_2) = 0, \lambda/4, \lambda/2, \lambda$, etc.) or as *phase angles*. One complete rotation, or cycle of vibration, represents 360° (i.e., 2π rad). Thus, for in-phase vibrations, $(\alpha_1 - \alpha_2) = 0°, 360°, \ldots, (360m)°$ (where m is any integer) or $0, 2\pi, \ldots, 2m\pi$ rad. For vibrations which are exactly *out* of phase, $(\alpha_1 - \alpha_2) = 180°, \ldots, (2m+1)180°$, or $\pi, \ldots, (2m+1)\pi$ rad. Any intermediate phase angle may be specified; for example, 90° or $\pi/2$ rad, etc.

A *wavefront* may be defined as a surface of equal phase of the wave. The direction of energy transfer by a wave motion (and thus the direction of rays; cf. p. 3) is *perpendicular* to the wavefront at any point and parallel to the direction of propagation of the light.

Equations 1-e, 1-h and 1-j include terms in z and y but not in x. This implies that the vibrations described by these equations lie in the yz plane; not only are they perpendicular to the direction of propagation, but also they lie in the plane of the page. The orientation of the plane of the page is arbitrary, however; thus motions might equally well occur, for example, at right angles to the plane of the page. This is the case represented in Fig. 1-9, where only those points at which the wave crosses the z-axis are indicated. The wave motion of Fig. 1-9 propagates along the plane of the page, but its direction of polarization is at right angles to that plane.

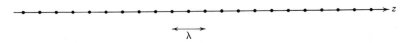

Fig. 1-9 A wave motion polarized at right angles to the plane of the page.

The wave motions represented in Figs. 1-7 and 1-9 can also be drawn as they would appear if observed by looking along the z-axis toward the source of the wave. Such a representation is the *projection* in the xy plane of the periodic motion. The xy projections of the wave motions of Figs. 1-7 and 1-9 are sketched in Figs. 1-10 and 1-11, respectively.

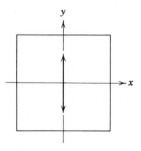

Fig. 1-10 Projection of a plane polarized motion.

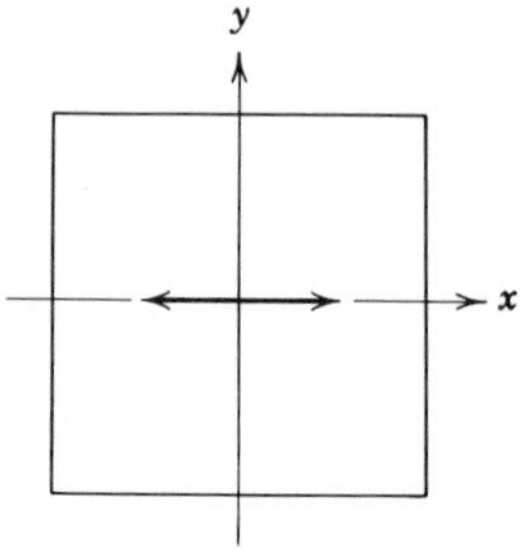

Fig. 1-11 Projection of a plane polarized motion.

Typical light beams do not consist of rays vibrating in any single arbitary direction but of vibrations in *all* directions perpendicular to the direction of propagation. The xy projection of such a beam is indicated symbolically in Fig. 1-12. This figure represents an unpolarized beam, whereas Figs. 1-10 and 1-11 represent two different *plane polarized* vibrations, that is, waves for which there is only a single component of vibration perpendicular to the z-axis.

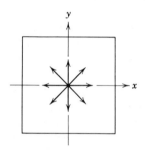

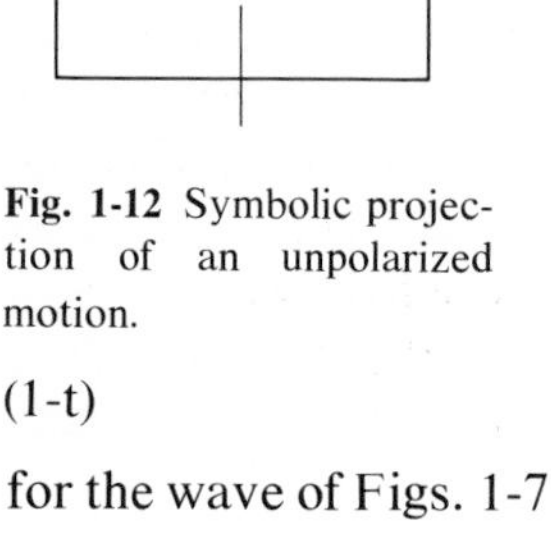

Fig. 1-12 Symbolic projection of an unpolarized motion.

The *plane of polarization* of a wave motion is defined as the plane which includes the direction of propagation and (in the case of electromagnetic radiation) the electric vector associated with the wave motion. Thus Fig. 1-10 represents a wave polarized in the yz plane, while the wave of Fig. 1-11 is polarized in the xz plane. The equations for these two waves are

(1-t)
$$y = A_0 \sin(\omega t - \alpha)$$

for the wave of Figs. 1-7 and 1-10, as already developed, and

(1-u)
$$z = A_0 \sin(\omega t - \alpha)$$

for the wave of Figs. 1-9 and 1-11.

A third case of a plane polarized motion is represented by Fig. 1-13 a, where the projection of the vibratory motion in the xy plane makes some angle, θ, with the y-axis. As is shown in Chapter 3, two motions acting at the same point combine to produce a net disturbance which may be quite different in form from the components, while, conversely, a single vibratory motion may be equivalent to the combination of two wave disturbances. The vibration represented in Fig. 1-13a may be considered

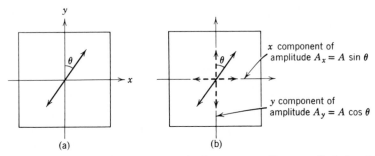

Fig. 1-13 Resolution of a plane polarized motion into two mutually perpendicularly polarized components.

to be the resultant of its projections on the x and y axes, as suggested by Fig. 1-13b. The wave motion of Fig. 1-13a is thus represented by the *pair* of equations

(1-v)
$$y = A_0 \cos \theta \sin (\omega t - \alpha),$$
$$x = A_0 \sin \theta \sin (\omega t - \alpha).$$

Partially polarized, as well as plane-polarized or unpolarized light beams may exist. Figure 1-14 shows the xy plane representation of such a wave. The projection of a *longitudinal* wave motion in the xy plane is illustrated by Fig. 1-15 for comparison.

The state of polarization of a light beam cannot be directly observed, but it can be deduced if the light beam is so treated as to produce changes in intensity which are dependent upon the state of polarization. The properties of polarized light beams are discussed in Chapters 4 through 6.

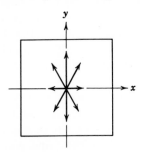

Fig. 1-14 Symbolic projection of a partially polarized motion.

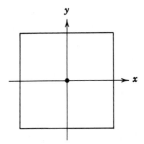

Fig. 1-15 Projection of a longitudinally polarized motion.

In summary, wave equations specify four important properties of wave motions in general, and of light beams in particular. They are: the *amplitude* (and thus the related property of *intensity*), the *frequency* (and thus the related quantity, *wavelength*), the *phase*, and the state of *polarization*. Two of these properties — intensity and wavelength — are subject to direct observation, whereas the relative phases of two or more beams and the state of polarization must be *deduced* after these effects have been exploited to produce observable changes in intensity.

EQUATION FOR A SPHERICAL WAVEFRONT

Equation 1-1 and equivalent expressions describe plane waves, in which the energy per unit area of the wavefront remains constant as the vibration propagates. For a spherical wavefront radiating outward from a point origin, the intensity diminishes as the square, and thus the amplitude as the first power, of the distance r from the origin. Thus

(1-w)
$$A_r = \frac{A_0}{r}.$$

Accordingly, the equation which describes a spherical wavefront is

(1-x)
$$y = \frac{A_0}{r} \sin (\omega t - \alpha).$$

MONOCHROMATICITY

In experimental practice there exists a fifth important property of light beams in addition to the four specified by the wave equation. This is the *monochromaticity*, a measure of the extent to which real light beams conform to the idealized sine function equations.

A sine equation such as (1-l) implies the existence of a sine wave extending from $z = -\infty$ to $z = +\infty$. This is equivalent to a pulse of radiation of infinite duration. In reality, as is discussed in Chapter 7, light is emitted in *finite* pulses as excited atoms lose excess energy. Idealized and real wave pulses are represented by Fig. 1-16a and Fig. 1-16b, respectively. (Note, however, that pulses of radiation, although finite, in fact include many more cycles of vibration than are shown in Fig. 1-16b). The pulse illustrated in Fig. 1-16b appears to consist of vibrations of only a single wavelength, λ, which rise in amplitude from zero to some value A_0 during the duration of the pulse and subsequently diminish again to zero. Mathematically this sequence of vibrations can be shown to be equivalent to the superposition of a number of idealized, infinite sine waves of different frequencies. Thus, *to the extent that emission occurs over a limited time, a light beam consists of a mixture of frequencies.*

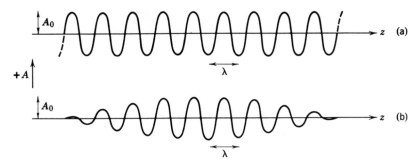

Fig. 1-16 (*a*) Portion of an infinite sine wave; (*b*) finite pulse of radiation.

Experimentally there is no such thing as a strictly monochromatic light beam. The term is used, however, in a relative sense to mean "a beam containing a very narrow range of wavelengths." A *polychromatic* beam is one which consists of a number of different wavelengths or a continuous range of wavelengths, while a *white* beam consists of a rather broad continuum of wavelengths. (Note, for example, the use of the term "white x-ray radiation" in Chapter 28.)

The nature of monochromaticity is considered further in Chapter 3.

PHASE AND GROUP VELOCITIES OF LIGHT

A consequence of the fact that light travels as pulses, representing a mixture of frequencies, is the distinction between the *phase velocity* (also called the wave velocity) of individual frequencies and the *group velocity* of the pulse as a whole.

In vacuuo, electromagnetic waves of all frequencies travel with identical velocities. Hence a pulse of radiation of different frequencies travels at the same velocity as the components; that is, phase and group velocities are equal. In all media, however, velocity varies with frequency and thus, as is discussed further in Chapter 3, phase and group velocities differ.

Although refractive indices specify ratios of phase velocities at the specified wavelength, the phase velocity of light cannot be measured directly (just as the absolute phase of a light wave cannot be determined). Direct measurements of the velocity of light are measurements of the transfer of *energy* by *pulses* of radiation as a whole; thus they yield *group* velocities. Alternatively, it may be said that information cannot be carried by a light wave without *modulation* of the simple harmonic motion of a single frequency (i.e., without alteration of the pure sine wave form of the vibration). The transmission of information thus represents the passage of a pulse of different frequencies and so occurs at a velocity character-istic of the *group* of wave forms. While phase velocities can be greater than c, group velocities can be at most c, as specified by the theory of relativity.

COMPLEX REPRESENTATION OF LIGHT WAVES

In many applications the *summation* of a large number of wave equa-tions must be computed. The process is most conveniently carried out when the wave equation (1-l) is expressed in an equivalent form which makes use of complex numbers (i.e., numbers containing terms including the "imaginary" quantity $i = \sqrt{-1}$). The elementary properties of complex numbers are reviewed in Appendix C, where a derivation and discussion of the complex form of the wave equation are also given. The latter yields, for the wave equation,

$$(1\text{-y}) \qquad\qquad A = A_0 e^{i\omega t} e^{-i\varphi},$$

in which A is the instantaneous amplitude and φ is the phase of the wave. This elegant but mathematically more sophisticated form of the wave equation is used in this book only in discussing the interference filter (Chapter 22) and the determination of electron density distribution by x-ray diffraction methods (Chapter 31).

PHOTOEMISSION

Certain surfaces possess the property of photoemissivity; that is, they emit electrons upon illumination. A certain energy input is required simply to overcome the forces within the material which keep electrons from leaving the surface. This energy, a characteristic of the material, is known as the *work function*. An electron which acquires energy in excess of the work function from a light beam leaves the surface with a velocity determined by the net energy remaining after escape.

Properties of a photoemissive process could be predicted on the basis of the wave theory of the character of light. The energy of a wave should be distributed *uniformly* over the wavefront; hence, when a photoemissive surface is illuminated by a beam of very low intensity, the energy density of radiation reaching a single electron in the surface should be extremely low. A finite time should thus elapse before *any* electron in the surface could accumulate sufficient energy for escape; a lag in photoemission should thus be observed at very low incident intensities, while doubling of the intensity should reduce the length of the lag and should increase the average velocity of the electrons emitted. The wavelength of the incident light should not be a critical factor in the process. In fact, these predictions are refuted by experiment. *No* lag period can be detected, no matter how low the incident intensity. An increase of the intensity level increases the *rate* of photoemission, but is without effect on the average velocity of the electrons. Electron velocity is, instead, a function of wavelength; the shorter the incident wavelength (i.e., the greater the frequency), the larger is the *maximum* obtainable electron velocity.

These results, so much at variance with predictions based upon the wave theory, can easily be accounted for if light is considered to be a stream of "concentrated chunks" of energy which are related in size to the wavelength of the light. According to this view, an electron may instantaneously absorb a finite amount of energy if it absorbs a "chunk" or *quantum* of energy from the light beam. Hence the failure to observe any lag in photoemission at low intensities. If the absorbed quantum corresponds to an energy less than the work function of the electrons, no emission results, whereas, if the quantum energy level is in excess of the work function, the surplus energy is converted into kinetic energy of motion of the ejected electron. Shorter wavelengths excite higher emission velocities; thus the energy per quantum must be directly proportional to the frequency of the radiation and inversely proportional to the wavelength. The appropriate quantitative relationship is

(1-z) $\boxed{E = \text{energy per quantum} = h\nu}$,

where h (Planck's constant) $= 6.6 \times 10^{-27}$ erg sec.

These considerations make it plain that the frequency, not the intensity, of the incident light determines whether photoemission occurs, and with what velocities photoelectrons can leave the surface. The intensity of the incident light, according to either the wave or photon point of view, merely determines how many photoelectrons are emitted per unit time.

The quantum properties of light, strikingly illustrated by the phenomena associated with photoemission, must also be invoked in considering the emission and absorption of light by matter. These processes are discussed in Chapter 7. It is also possible, though relatively inconvenient, to explain other optical phenomena which are compatible with a wavelike nature of light in terms of the quantum concept of the nature of light beams.

APPLICATION OF RAY WAVE AND QUANTUM TREATMENTS OF LIGHT

Rather different sets of terms and concepts apply to the treatment of light as rays, waves, or photons. How is the frame of reference to choose in studying an optical effect or instrument selected? Since all three treatments are in fact consistent, any treatment may be selected which is adequate for describing the type of phenomenon studied. *Any* optical effect can be described in terms of quantum optics, although the complexity of the description may be formidable. Wave optics does not adequately describe interactions of light with *matter*, such as the photoelectric effect, but can be applied to the study of interactions between light beams (phenomena of interference and diffraction) or to the determination of directions of transmission of light beams. Geometrical optics, which does not consider the nature of light, is adequate only for the determination of directions and magnifications. Very often the simplest method of treatment is most suitable. Thus, for example, the refraction of light by a lens can be explained in terms of the propagation of wavefronts, but it is more easily represented by the refraction of light rays.

The variety of possible approaches to the discussion of optical effects can make the optical literature perplexing. An effect or instrument described in terms of ray diagrams in one article may be discussed in terms of wavefronts in another. An original "understanding" may then appear to be wrong or at least irrelevant, whereas in fact it is merely incomplete. For this reason a knowledge of the vocabulary, scope, and relationships of the ray, wave, and quantum approaches is essential.

Many optical phenomena of interest to biologists are appropriately explained in terms of the wave theory of light. The limited resolving power of lenses, the operation of the phase contrast microscope, and the formation of diffraction patterns are a few of many examples. Difficulties in understanding these effects usually derive from lack of familiarity with the nature of wave motions (summarized in this chapter) and of the interactions of wave motions (considered in Chapter 3).

THE NATURE OF LIGHT

An instrumentalist need not be concerned with the question: "Which treatment of light is the true one?" Nevertheless, to the scientist who wishes to understand, as well as to exploit the properties of light, this question persistently presents itself. Rays possess physical reality only as directions of energy transfer; they imply nothing concerning the nature of the radiation itself. The question then remains: "Is light a wave motion, or is it a stream of particles?" Common sense suggests that it surely could not be both of these things and that, therefore, either the wave or quantum theory must be "wrong". In fact, as explained later, common sense turns out to be wrong!

The experimental evidence for both wave and particle properties of light is irrefutable. Furthermore, there are not "two kinds of light," since one and the same light beam can be made to display, successively, both typically wavelike and typically particulate properties. There thus exists unequivocally, an established *duality* of the nature of light.

Even in considering only the wave theory, mechanical analogies break down as a means of "explaining" light. That light as an *electromagnetic* rather than a mechanical fluctuation is difficult to visualize provides a clue that we, as macroscopic beings, are limited in perceptual ability to the visualization of macroscopic events. Observation of "extremely small" objects with the strongest possible microscopes reveals some properties which seem decidedly odd. In the electron microscope, for example it is possible to observe objects (macromolecules, virus particles, etc.) for which gravitation is negligible in comparison to forces of surface tension and weak chemical interactions. Even then the objects observed are extremely large in comparison to individual atoms, subatomic particles, or, in the limit, the massless photon. These smaller particles can be expected to deviate much further in behavior from what is "reasonable" in terms of the macroscopic.

The truly "unreasonable" element in the behavior of photons (or of any other type of "fundamental" particle) is that the behavior of a single photon cannot be specified precisely. To put it another way, the behavior of a single photon cannot be used to predict the behavior of a group of photons. For the behavior of a single photon, only a *probability distribution* can be specified; that is, a description of the behavior that is in fact observed for some very large number of photons as a group.

Consider a hypothetical experiment in which a very few photons travel toward a narrow slit. The arrival of the photons at a screen placed some distance beyond, as shown in Fig. 1-17a, is counted. The number of photons arriving is plotted as a function of distance along the screen. The expected distribution is one in which *all* photons arrive at positions corresponding to the geometrical image of the slit, as indicated in Fig. 1-17b. The result actually obtained, however, is plotted in Fig. 1-17c. Most of the photons have gone where they "ought" to have gone, but a few have landed in some rather curious positions. With a total flow of only a few photons, not enough of these oddities are recorded for a definite pattern to be detected; a larger number of photons must pass through in order to show what is really going on. Then the distribution obtained is that shown in Fig. 1-17d. This, as is shown in Chapter 3, is just the sort of pattern to be expected of a light *wave*.

What conclusion can be drawn from this result? Simply that *a large number of photons behave, collectively, as a wave*. The wave pattern can be said to represent the probability distribution for behavior of photons. Light "really is" a stream of photons but, except in the case of

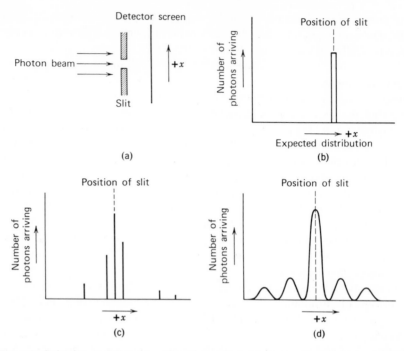

Fig. 1-17 Behavior of photons at a slit: (*a*) experimental arrangement; (*b*) expected distribution of photon arrivals; (*c*) observed distribution of arrivals of a small number of photons; (*d*) observed distribution of arrivals of many photons.

effects involving only very small numbers of photons, the wave theory is an accurate and convenient means of describing the behavior of light. This view of the interrelationships of photons and waves may not provide a really concrete physical picture of the nature of light, but at least it satisfies the even more fundamental demand for a unity of truth. Furthermore, it establishes a rational basis for treating light alternately as waves or photons, according to convenience.

REFERENCES

A short account of many of the properties of light discussed in this chapter may be found in almost any textbook of general physics.

An invaluable reference for many of the subjects discussed in this book is: F. A. Jenkins and H. E. White, *Fundamentals of Optics* (3rd ed.), McGraw-Hill, New York, 1957. Both geometrical and wave optics are presented from an essentially *experimental* point of view and a wide variety of optical instruments are briefly described. In particular, Chapter 1(Light Rays), Chapter 11 (Light Waves), Chapter 20 (The Electromagnetic Character of Light), and Chapter 30 (Photons) are of interest.

K. Ford, *The World of Elementary Particles*, Blaisdell, New York, 1963, especially pp. 113–130. The book provides a very readable account of the nature of the photon and of the dual character of light.

E. Ruechardt, *Light, Visible and Invisible*, the University of Michigan Press, Ann Arbor, 1958, especially Chapter 1 (Introduction) and Chapter 2 (Light Rays, a Useful Fiction). The book is a nonmathematical account, written by a physicist, of many aspects of the science of optics.

G. Feinberg, *Sci. Am.*, **219**, 50 (1968). This article, "Light", summarizes modern concepts of the nature of light and is the introductory article in an issue devoted entirely to the subject of light.

Although standard textbooks of optics tend to present this subject from a somewhat rigorous, mathematical approach, the reader of the present volume may find it helpful to consult such texts on specific subjects. Three excellent texts are:

J. Strong, *Concepts of Classical Optics*, Freeman, San Francisco, 1958. A general presentation which includes a series of appendices on topics of particular current interest. As the author states, "Some familiarity with electricity and magnetism, . . . and some knowledge of calculus, vectors and complex numbers are prerequisite" to the perusal of the text.

J. M. Stone, *Radiation and Optics*, McGraw-Hill, New York, 1963. A general development of wave optics from the point of view of electromagnetic theory, with emphasis on the use of Fourier transforms.

M. Born and E. Wolf, *Principles of Optics*, Macmillan, New York, 1964. This very authoritative text covers a great variety of topics in wave optics. A high level of mathematical sophistication is assumed.

CHAPTER 2

The Electromagnetic Spectrum

Electromagnetic radiation is not confined to the range of wavelengths visible to the eye; it exists as a continuum of wavelengths longer and shorter than those of the visible range. Correspondingly, in terms of the quantum nature of light, it may be said that there exists a continuum of quantum energies lesser and greater than those of visible light. Instruments of various types make use of a wide selection of wavelengths.

The division of the electromagnetic continuum into spectral regions is illustrated in Fig. 2-1. In this chapter a brief characterization of each of these regions of the electromagnetic spectrum is presented.

It should be understood, however, that the spectrum is not discontinuous in the manner which this treatment might suggest. The several regions are operationally defined by the methods that are used to produce

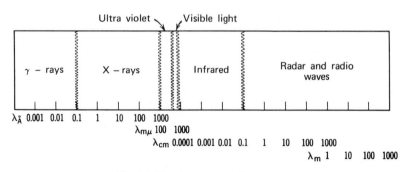

Fig. 2-1 The electromagnetic spectrum.

or to detect each type of radiation. Boundaries between the spectral regions are not precise, and certain wavelengths can be assigned to either of two adjacent regions.

GAMMA RAYS

The short wavelength, high quantum energy end of the electromagnetic spectrum consists of the γ rays, which are radiations of wavelength less than about 0.1 Å (10^{-9} cm). This type of radiation is produced in the course of *nuclear* transitions which may either be artificially induced (as in cyclotrons and similar devices) or spontaneous (radioactive decays of atomic nuclei). In such processes the emitted γ rays are usually accompanied by massive charged particles, the α and β rays. The latter, of course, are *not* part of the electromagnetic spectrum. Particles such as α rays (helium nuclei) and β rays (high energy electrons) have both charge and mass and are readily absorbed or decelerated by matter. γ rays may be separated from the other products of nuclear disintegration by taking advantage of this fact.

Interactions of γ rays with matter similarly are interactions with atomic nuclei. Ionizations result from such processes; hence γ rays, like other types of ionizing radiation, may be detected by various types of counting devices, such as the scintillation counter. They may also be detected by photographic emulsions.

X-RAYS

X-rays constitute the wavelength range from about 0.1 Å to about 100 Å. They are produced by bombarding matter with accelerated electron beams. X-ray generation and the nature of x-ray emission spectra are not discussed here, since these topics are considered in Chapter 28. Briefly, x-ray emission is associated with electronic transitions which involve the inner shells of atoms.

X-rays, like γ rays, are an ionizing radiation and may thus be detected by nuclear counters. Photographic emulsions are also widely used for the detection and quantitation of x-rays.

Quantum energies of radiations in the x-ray range correspond to the removal or rearrangement of inner shell electrons. Thus chemical reactions occur only as a *secondary* effect of the interaction of x-rays with matter. The ability of matter to absorb x-rays depends upon the *density* of electrons in the material; absorption is thus primarily a function of the average *atomic number* of the absorber. Specific chemical composition affects x-ray absorption only to an immeasurably slight extent at the wavelengths employed in studies of biological materials.

The refractive indices of materials with respect to x-rays are close to, but very slightly *less* than, unity. X-rays incident upon matter from air or vacuum thus enter an optically less dense medium, so that the phenomenon otherwise described as "total internal reflection" occurs at appropriate angles of incidence.

ULTRAVIOLET RADIATION

The ultraviolet range of radiations (UV) is loosely divided into three subregions which are: the vacuum ultraviolet, extending from about 100 to 2000 Å (1 to 2×10^{-5} cm), the far ultraviolet, extending from 2000 to 3400 Å (2 to 3.4×10^{-5} cm; 200 to 340 mμ), and the near ultraviolet, extending from about 3400 Å to the threshold of the visible at 4000 Å.

As the name implies, the vacuum ultraviolet includes a range of wavelengths that is absorbed by air (and, generally, by all matter), and which thus can be studied only in evacuated equipment. Since there has been little or no application of vacuum ultraviolet wavelengths to biological instrumentation, this region of the spectrum is not discussed further here.

The distinction between the near and far ultraviolet is a hazy one, but 3400 Å may be selected as the boundary wavelength, because it represents the approximate short wavelength limit for the use of optical parts made of *glass*. At shorter wavelengths, glass is essentially opaque, and optical parts must be made of *quartz*.

Ultraviolet wavelengths are produced in arc and spark discharges and are present at high intensities in the radiation which reaches the upper atmosphere from the sun. Wavelengths in the near ultraviolet are also present to some extent in radiations from certain heated filaments such as tungsten lamps.

The effect of absorption of ultraviolet quanta by atoms and molecules is to produce electronic transitions between energy levels. The absorption of ultraviolet light is thus dependent upon the chemical bonding of the absorber. Correspondingly, the majority of chemical compounds absorb specific, characteristic wavelengths somewhere in the ultraviolet range. The electronic transitions associated with absorptions and emissions of quanta are discussed in Chapter 7.

Detection of ultraviolet radiation is accomplished by photosensitive surfaces (i.e., by photocells of various types), by the excitation of visible emission from fluorescent screens, or by exposure of photographic emulsions.

VISIBLE LIGHT

Visible light consists of wavelengths in the surprisingly narrow range from 4000 Å to 7000 Å (400–700 mμ; 4 to 7×10^{-5} cm). These wavelengths are emitted by matter that is heated to high temperatures or which is energized by electric discharge. Occasionally, visible light is a by-product of chemical reactions; an example is the luminescence produced by the firefly.

The color divisions of the visible spectrum are (approximately): 400–450 mμ, violet; 450–500 mμ, blue; 500–560 mμ, green; 560–600 mμ, yellow, orange; and 600–700 mμ, red. The limits of the visible range vary somewhat among individuals.

The absorption of visible light, as of ultraviolet radiation, by atoms and molecules produces transitions of valence electron energy levels. The transitions induced by visible light correspond to lower quantum energies than those induced by ultraviolet frequencies. As described in Chapter 7, the energy so absorbed is dissipated by processes of reemission or photochemical reaction or is simply lost in the form of heat.

Visible light can be detected quantitatively by photocells (cf. Chapter 24) or by photographic emulsions (Chapter 21). Emulsions used to measure light intensities at the red (long wavelength) end of the spectrum require appropriate sensitization by combination with dyes.

INFRARED RADIATION

The infrared wavelengths include an extensive range from 7000 Å (0.7 μ; 7×10^{-5} cm) to about 1 mm ($10^3 \mu$; 10^{-1} cm). Wavelengths are commonly expressed in microns (1 $\mu = 10^{-4}$ cm) or in waves per centimeter. Thus a wavelength of 10,000 Å = 1 μ corresponds to $1/10^{-4} = 10^4$ waves/cm (10^4 cm^{-1}).

Infrared rays are radiated by heated objects, which may or may not be heated to a level of incandescence, so that visible light is produced at the same time.

The absorption of infrared quanta by matter induces molecular *vibrations*, and *rotations* and *translations* of atoms and molecules. Vibrations and rotations may involve portions of a molecule or the molecule as a whole, while translations, by definition, are movements of the entire molecule as a unit. Translations are associated with lower energies than are vibrations and rotations and thus are produced by the longer infrared wavelengths and by radio waves. The production of movements of molecules as a whole by the absorption of infrared accounts for the common indentification of this region of the spectrum as "heat rays." In fact, heat is a form of energy, and, since energy is conveyed by all types of electromagnetic radiation, heating occurs to some extent upon irradiation of matter by *any* wavelength range of the electromagnetic spectrum.

Absorption of specific infrared wavelengths is characteristic of specific chemical bonds (e.g., —C=O, —C—H, —C=C—, etc.) within molecules, the vibration or rotation of these groups being excited thereby. The infrared absorption spectra of compounds thus may be highly structured, their interpretation leading to deductions concerning chemical structure and conformation.

Infrared radiations are detected by rather nonspecific devices which measure total energy output, such as thermocouples, bolometers, and Golay cells. The shorter infrared wavelengths can also be recorded on appropriately sensitized photographic emulsions.

RADIO WAVES

Radio waves, and similar types of radiation, which extend from about 1 mm to many meters in wavelength, constitute the long wavelength, low energy end of the electromagnetic spectrum. For practical reasons, these radiations pertain to a quite different area of instrumentation than do other types of electromagnetic wave; only their fundamental similarity to the remainder of the electromagnetic spectrum is noted here.

WAVE AND PHOTON PROPERTIES OF ELECTROMAGNETIC RADIATIONS

At any wavelength the energy per quantum is given by the expression

$$(2\text{-a}) \qquad\qquad E = h\nu = \frac{hc}{\lambda},$$

where λ is the wavelength of the radiation, h is Planck's constant, and c is the velocity of light. As the inverse relation between wavelength and energy per photon shows, short wavelength, high frequency radiations (γ rays, x-rays) are of high quantum energy, while long wavelength, relatively low frequency radiations (radio waves, infrared) are of low quantum energy. This means that the number of photons corresponding to a given quantity of energy varies enormously over the electromagnetic spectrum. As Table 2-1 shows, one erg in the radio wave range represents about 10^{10} as many photons as does one erg of γ radiation. Thus relatively *few* photons are required to produce detectable effects during irradiation by γ or x-rays, whereas enormous numbers of photons are involved in any measurable interaction of infrared radiation. Since, as explained in Chapter 1, wave properties are in fact the description of the behaviour of very large numbers of photons, it is clear that infrared and radio wave radiations must exhibit the most evident wavelike properties, whereas quantum effects must be the most readily observed at short wavelengths. The ultraviolet and visible ranges constitute a medium region in which quantum effects are quite readily detectable, even though the wavelike properties of these radiations are generally more obvious.

Table 2-1   Numbers of Photons Associated with 1 Erg of Energy (= $1/h\nu$)

λ,cm	ν, sec^{-1}	Number of Photons
10^{-10} (γ-ray)	3×10^{20}	5×10^{5}
10^{-8} (x-ray)	3×10^{18}	5×10^{7}
2×10^{-5} (UV)	1.5×10^{15}	1×10^{11}
5×10^{-5} (visible, green)	6×10^{14}	4×10^{11}
10^{-4} (near IR)	3×10^{14}	5×10^{11}
1 cm (radio)	3×10^{10}	5×10^{15}

REFERENCES

E. Ruechardt, *Light, Visible and Invisible*, the University of Michigan Press, Ann Arbor, 1958, Chapter 10 (Two Kinds of Invisible Light) and Chapter 14 (X-rays, an Invisible Light).

R. A. Smith, F. E. Jones, and R. P. Chasman, *The Detection and Measurement of Infra-Red Radiation*, Clarendon, Oxford, 1956, pp. 56–169.

CHAPTER 3

Interference and Diffraction

The ability to interfere—that is, to *interact*—is the diagnostic property of wave motions. While the observed effects of wave interactions are often quite different from what might be expected intuitively, rational interpretations become clear when the nature of the interacting motions is considered carefully.

The science of microscopy and diffraction, so central in biological instrumentation, is largely based upon the understanding of light as a wave motion. The features of microscope images, it is seen later, may be interpreted as originating from the interference of light waves which have interacted with the specimen. The theoretical aspects of microscopy and diffraction often seem difficult to the biologist. Frequently, however, the basic difficulty is mere unfamiliarity with the nature and interactions of wave motions. Appreciation of the behavior of light waves must be based upon an understanding of the nature of wave interactions in general.

This chapter describes the nature of wave interactions (with the exception of phenomena associated with polarization, which are considered in Chapter 4) both in general and as specifically applied to electromagnetic radiations. The topics considered here can be regarded as fundamental to many of the discussions to be presented in the remainder of this text. Appreciation of the properties of individual wave motions and of the use of the wave equation (1-1) for the description of wave motions, as presented in Chapter 1, is assumed.

(1-1) $$y \text{ (or } A) = A_0 \sin (\omega t - \alpha),$$

where y and A are the displacement by the wave and the instantaneous amplitude, respectively, A_0 is the amplitude, ω the frequency, and α the phase of the wave.

PRINCIPLE OF SUPERPOSITION

The effect of interacting wave motions is governed by the principle of superposition, which states that, *if two or more wave motions act at a point, the net disturbance at that point is the sum of the disturbances produced by each wave motion individually.* In other words, wave motions act *independently* of each other. The resultant disturbance, however, may be of so complex a form that the independence of the component wave motions is revealed only by a careful analysis.

The statement of the principle of superposition, though absolutely fundamental as a description of wave interactions, is deceptive in its simplicity. It is the (not directly observed) amplitudes rather than the observable intensities of component light waves which add when light beams interact. Thus, since intensity is proportional to the square of amplitude, the observed summation is not merely additive.

When two or more wave motions interact, the net disturbance produced by their superposition is termed the resultant wave motion (or simply the *resultant*).

INTERFERENCE, DIFFRACTION, AND RELATED TERMS

The distinction between the two terms, "interference" and "diffraction" which are applied to the interactions of light waves, is sometimes unclear. *Interference* refers to the interaction of light waves (or of any wave motions) *per se*; experimentally, the term applies to situations in which waves interact over a wide front. "Interference" is used at times to denote any overlapping of wave motions, and at other times applies more specifically to those interactions of waves which are observable as differences in frequency distribution or intensity or both. If the reader is aware of this ambiguity, it should cause no difficulty. For example, see the discussion of interference of light of different frequencies on pp. 69–70.

Diffraction is a term which applies specifically to the bending of light which is observable when narrow portions of a wavefront are examined. Both "refraction" and "diffraction" are terms which signify a bending of light, but, whereas the term refraction is applied to bending at points of discontinuity of the refractive index, "diffraction" is applied specifically to the spreading out of light beams which is a consequence of the wavelike properties of light.

Interference between portions of a diffracted beam becomes visible as differences in intensity which constitute a *diffraction pattern*.

INTERFERENCE OF WAVE MOTIONS

Some simple examples of the interactions of overlapping wave trains are considered in this section. Polarization effects are ignored in the discussion; that is, wave motions are tacitly considered to be in either of two states which are equivalent from the point of view of diffraction effects: (a) polarized in the same plane, or (b) unpolarized.

The interaction of two overlapping wave trains which are exactly *in* phase is particularly simple to consider. For such waves, the phase difference is zero; that is, $\alpha_1 - \alpha_2 = 0$. The corresponding wave equations thus are:

(3-a)
$$A_1 = A_{0_1} \sin (\omega t - \alpha),$$
$$A_2 = A_{0_2} \sin (\omega t - \alpha).$$

In this case, where the quantity $(\omega t - \alpha)$ (called the *argument* of the sine) is the same for both waves,

(3-b)
$$A = A_1 + A_2 = (A_{0_1} + A_{0_2}) \sin (\omega t - \alpha).$$

That is, the amplitude of the resultant of *in*-phase beams is simply the arithmetic sum of the individual amplitudes. However, the intensity of the resultant is then greater than the sum of the component intensities. For example, if $A_{0_1} = A_{0_2}$, $A = 2A_0$, so that the resultant intensity is proportional to $(2A)^2$, or twice the sum of component intensities. This conclusion appears to be a violation of the law of conservation of energy; however, a corresponding loss of energy density at other points in an experimental system can always be demonstrated.

The effect of overlapping of in-phase wave motions, termed *constructive interference*, could equally well be deduced directly from the principle of superposition without recourse to the use of wave equations. In terms of mechanical waves, taking for simplicity $A_1 = A_2$, it is evident that a displacement of the medium $+y$ produced by one wave motion is always accompanied by an equal displacement $+y$ by the other. The total displacement must thus be $+2y$ at all stages in the cycle of vibration. The constructive interference is illustrated schematically in Fig. 3-1.

The interference of two waves which are exactly *out* of phase, that is, with phase angle $180°$ or π, is almost equally simple to consider. Equations

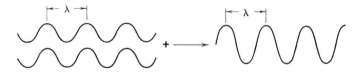

Fig. 3-1 Constructive interference.

for such a pair of waves are

(3-c)
$$A_1 = A_{0_1} \sin (\omega t - \alpha),$$
$$A_2 = A_{0_2} \sin (\omega t - \alpha - \pi).$$

Making use of the trigonometric identity

(3-d)
$$\sin (\theta - \pi) \equiv -\sin \theta,$$

the second equation of (3-c) can be written

(3-e)
$$A_2 = -A_{0_2} \sin (\omega t - \alpha).$$

The equation which describes the resultant vibration is then

(3-f)
$$A = A_1 + A_2 = (A_{0_1} - A_{0_2}) \sin (\omega t - \alpha).$$

The resultant in this case is a wave of diminished amplitude, and it vanishes in the particular case for which $A_{0_1} = A_{0_2}$. Note that, if $A_{0_1} > A_{0_2}$, the resultant wave is in phase with wave 1; if $A_{0_1} < A_{0_2}$, the resultant is in phase with wave 2.

Again, these results could be deduced directly from the principle of superposition. For $A_1 = A_2$ in equations (3-c), a displacement $+y$ of the medium by one wave motion is exactly balanced by a displacement of $-y$ induced by the other, resulting in zero net displacement, as indicated in Fig. 3-2. The wave motions are said to undergo complete *destructive interference*.

It may be noted that, whereas the popular use of the word "interference" corresponds approximately in meaning to "destructive interference", the former term applies, in optical contexts, to any overlapping of wave trains, regardless of whether the effect is an increase or decrease of intensity.

More generally, phase angles other than 0° or 180° must be considered, as must the interaction of more than two components of vibration. It is possible, though tedious, to obtain the resultant intensity and phase by taking the algebraic sum of the displacements due to each wave motion separately, and sketching the form of the resultant vibration. Figure 3-3 is an example of such a procedure; it illustrates the summation of three vibrations of equal amplitude which differ in phase, successively, by

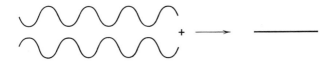

Fig. 3-2 Destructive interference.

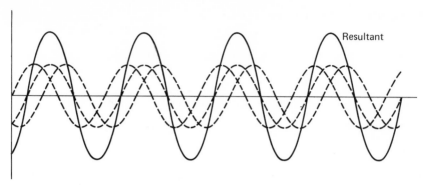

Fig. 3-3 Resultant of three wave motions.

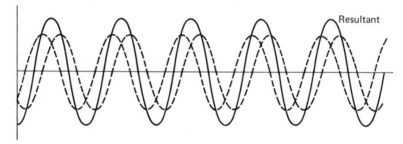

Fig. 3-4 Resultant of two wave motions.

60° ($\pi/3$). The resultant motion in this case is found to have twice the amplitude of any one of the three individual wave motions, and to be in phase with one of these components.

Figure. 3-4, illustrating the interference of two vibrations which are out of phase by 90° ($\pi/2$), provides an example of a resultant motion which is *not* in phase with either of its components.

Construction of a diagram such as Fig. 3-4 is a laborious affair; clearly, a systematic means of obtaining amplitudes and phases of a resultant vibration in the general case is required. These quantities may be obtained either by algebraic or graphical methods. Both are described here.

ALGEBRAIC COMPUTATION OF RESULTANT WAVEFORMS

Any two wave motions of identical frequency but arbitrary amplitude and phase may be described by the equations

(3-g)
$$A_1 = A_{0_1} \sin (\omega t - \alpha_1),$$
$$A_2 = A_{0_2} \sin (\omega t - \alpha_2).$$

According to the principle of superposition, the net displacement (instantaneous amplitudes) of these wave motions must be

(3-h) $\qquad A = A_1 + A_2 = A_{0_1} \sin(\omega t - \alpha_1) + A_{0_2} \sin(\omega t - \alpha_2).$

Making use of the trigonometric identity

(3-i) $\qquad \sin(A - B) = \sin A \cos B - \cos A \sin B,$

(3-h) becomes

(3-j) $\qquad A = A_{0_1}(\sin \omega t \cos \alpha_1 - \cos \omega t \sin \alpha_1)$

$\qquad\qquad + A_{0_2}(\sin \omega t \cos \alpha_2 - \cos \omega t \sin \alpha_2)$

$\qquad\qquad = (A_{0_1} \cos \alpha_1 + A_{0_2} \cos \alpha_2)\sin \omega t$

$\qquad\qquad - (A_{0_1} \sin \alpha_1 + A_{0_2} \sin \alpha_2) \cos \omega t.$

Assuming that consistent values of A' and θ can be found, (3-j) can be written in an alternative form as

(3-k) $\qquad A = A' \cos \theta \sin \omega t - A' \sin \theta \cos \omega t,$

in which

(3-l) $\qquad A_{0_1} \cos \alpha_1 + A_{0_2} \cos \alpha_2 = A' \cos \theta,$

(3-m) $\qquad A_{0_1} \sin \alpha_1 + A_{0_2} \sin \alpha_2 = A' \sin \theta.$

Equation 3-k may be arranged, with the aid of the trigonometric identity (3-i), to give

(3-n) $\qquad\qquad y = A' \sin(\omega t - \theta)$

Equation 3-n is a wave equation of the same form as (1-l) but represents the resultant wave. The frequency ω remains that of both components, while new values, A' and θ, of amplitude and phase are characteristic of the resultant motion and require evaluation.

Thus far, it has been *assumed* that it is possible to find values of A' and θ such that (3-n) can be made equivalent to (3-h). It is now shown that this is indeed the case. The amplitude A' of the resultant is obtained by, first, squaring and adding (3-l) and (3-m), which gives

(3-o) $(A')^2 \cos^2 \theta + (A')^2 \sin^2 \theta = A_{0_1}^2 \cos^2 \alpha_1 + 2 A_{0_1} A_{0_2} \cos \alpha_1 \cos \alpha_2$

$\qquad\qquad + A_{0_2}^2 \cos^2 \alpha_2 + A_{0_1}^2 \sin^2 \alpha_1$

$\qquad\qquad + 2 A_{0_1} A_{0_2} \sin \alpha_1 \sin \alpha_2 + A_{0_2}^2 \sin^2 \alpha_2.$

Making use of the trigonometric identity

(3-p) $\qquad\qquad \sin^2 \theta + \cos^2 \theta \equiv 1,$

equation 3-o reduces to

(3-q) $\qquad (A')^2 = A_{0_1}^2 + A_{0_2}^2 + 2A_{0_1}A_{0_2}(\cos\alpha_1 \cos\alpha_2 + \sin\alpha_1 \sin\alpha_2).$

Making use of the identity

(3-r) $\qquad\qquad \cos A \cos B + \sin A \sin B \equiv \cos(A-B),$

(3-q) in turn reduces to

(3-s) $\qquad\qquad \boxed{(A')^2 = A_{0_1}^2 + A_{0_2}^2 + 2A_{0_1}A_{0_2}\cos(\alpha_1 - \alpha_2)},$

from which the resultant amplitude may be computed. Note that (3-s) gives the *square* of the resultant amplitude; thus A' is obtained as the square root of that equation.

For the evaluation of the resultant (relative) phase θ, (3-m) is divided by (3-l):

(3-t) $\qquad\qquad \boxed{\tan\theta = \dfrac{A_{0_1}\sin\alpha_1 + A_{0_2}\sin\alpha_2}{A_{0_1}\cos\alpha_1 + A_{0_2}\cos\alpha_2}}.$

Equations 3-s and 3-t contain, in addition to A and θ, respectively, *only* parameters which characterize the original waves, specifically A_{0_1}, A_{0_2}, and the sines and cosines of the phase angles α_1 and α_2. These known quantities are simply substituted in (3-s) and (3-t) in order to obtain the characteristic parameters of the resultant wave. Note that, while the absolute values of the phase angles α_1 and α_2 are unknown, the arbitrary choice of some value, such as $0°$, for one of these angles determines the value to be used for the other, since the phase *difference* between the component waves is known.

The superposition illustrated by Fig. 3-4 is now computed as an example of the use of this algebraic method for obtaining resultant amplitudes. Appropriate equations for the component wave motions in that case are

(3-u) $\qquad\qquad\qquad\qquad A_1 = A_0 \sin(\omega t - \alpha),$

$$A_2 = A_0 \sin\left(\omega t - \alpha - \frac{\pi}{2}\right).$$

That is,

$$A_{0_1} = A_{0_2} = A_0,$$

while $\alpha_1 =$ (say) $0°$ and thus $\alpha_2 = \pi/2$. That is to say, wave 2 leads wave 1 by $\pi/2$. Then, since $\sin 0° = 0$, $\cos 0° = 1$, and $\cos(\pi/2) = 0$, (3-s) and (3-t) become

(3-v) $$(A')^2 = A_0^2 + A_0^2 + 0 = 2A_0^2, \qquad A' = \sqrt{2}A_0,$$

and

(3-w) $$\tan \theta = \frac{0 \cdot A_0 + 1 \cdot A_0}{1 \cdot A_0 + 0 \cdot A_0} = \frac{A_0}{A_0} = 1, \qquad \theta = 45° \text{ or } \pi/4.$$

The appropriate wave equation for the resultant vibration depicted in Fig. 3-4 is thus

(3-x) $$A = \sqrt{2}A_0 \sin\left(\omega t - \frac{\pi}{4}\right).$$

The interference of *three* wave motions could be approached in two ways; either the treatment leading to (3-s) and (3-t) could be extended to yield two other equations containing terms in $A_{0_1}, A_{0_2},$ and A_{0_3} and in $\alpha_1, \alpha_2,$ and α_3, or the interference of the resultant of two components with the third could be described by the direct use of (3-s) and (3-t).

The interference of larger numbers of wave motions is most conveniently treated by the method of complex amplitudes, as described in Appendix C. That approach is essentially identical with the one described here, but it makes use of the *complex* form of the wave equation which is more readily handled algebraically than is the sine form.

GRAPHICAL METHOD FOR DETERMINING RESULTANT VIBRATIONS

In the graphical method of determining resultant amplitudes and phases, each component wave motion is represented by a *vector*, the *length* of which is proportional to the *amplitude* of the vibration, while the *orientation* represents the *phase* of the wave. The properties of vectors are reviewed in Appendix B.

Beginning with one vector which is oriented in an arbitrary direction successive vectors, representing successive components of vibration, are drawn head-to-tail, with positive phase angles represented, by convention, as counterclockwise rotations. A vector drawn from the tail of the first component to the tip of the last one represents the resultant vibration both in amplitude and phase.

The example illustrated by Fig. 3-4, and treated above by the algebraic method, may be considered. Each vector is of equal length proportional, on any scale selected, to the amplitude A_0. The orientation of the first is arbitrary; the second is drawn 90° counterclockwise thereto. The resultant drawn from the tail of vector 1 to the tip of vector 2 is, as may be verified by direct measurement or by trigonometry, of length $\sqrt{2}A_0$, and makes an angle of 45° to the direction of vector 1, as illustrated by Fig. 3-5.

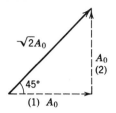

Fig. 3-5 Vector addition of two wave motions.

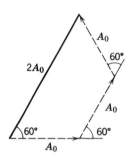

Fig. 3-6 Vector addition of three wave motions.

The superposition of three vectors, each of amplitude A_0 and differing successively in phase by 60°, as illustrated in Fig. 3-3, may be considered an application of the graphical method to the superposition of three vibrations. As shown by Fig. 3-6, the resultant is of amplitude $2A_0$ and is parallel to (i.e., in phase with) *one* of the three component vibrations.

For an example of the use of the graphical method in the addition of large numbers of components of vibration, see the discussion of interference filters in Chapter 22.

DETERMINATION OF PHASE BY SUPERPOSITION OF SINE AND COSINE FUNCTIONS

It is interesting to note that the phase of a wave may be represented by the relative amplitudes of superposed sine and cosine functions. As illustrated by Fig. 3-7, sine and cosine are functions of identical form which differ in phase by 90°. As drawn, these functions are of equal amplitude (i.e., $A = B$) as well as identical frequency. The curves of Fig. 3-7 could be represented by a pair of wave equations of sine form; that is,

(3-y) $y_1 = A \sin (\omega t)$ for the sine function,

            $y_2 = B \sin \left(\omega t + \dfrac{\pi}{2}\right)$ for the cosine function.

As shown in Fig. 3-4, the resultant of the two functions is readily seen to be a vibration of amplitude $\sqrt{2}A = \sqrt{2}B$ and relative phase $(-\pi/4)$.

In a more general case, sine and cosine functions are of amplitudes A and $B = qA$, where q is *any* numerical factor (i.e., not necessarily an integer). From (3-t), the phase θ of the resultant vibration is then

(3-z) $\tan \theta = A \left[\dfrac{\sin \alpha + q \sin (\alpha - \pi/2)}{\cos \alpha + q \cos (\alpha - \pi/2)} \right],$

or, since $\sin (\alpha - \pi/2) = -\cos \alpha$ and $\cos (\alpha - \pi/2) = \sin \alpha$.

(3-aa) $\tan \theta = A \left[\dfrac{\sin \alpha - q \cos \alpha}{\cos \alpha + q \sin \alpha} \right].$

For any position in the cycle at which $\alpha = 0$, and thus $\sin \alpha = 0$ and

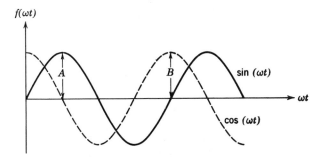

Fig. 3-7 Relation of sine and cosine functions.

$\cos \alpha = 1$, (3-aa) reduces to

(3-bb) $\tan \theta = -Aq.$

That is, the *phase* of the resultant wave is a function of q, the ratio of the amplitudes of sine and cosine functions contributing to the wave.

RESOLUTION OF COMPONENT VIBRATIONS

The foregoing has shown that the superposition of two or more vibrations of the same frequency produces a resultant wave which can be described as a *single* vibratory motion. That is, the resultant vibration can be specified by a wave equation containing particular single values of ω, A, and α. As long as the component wave trains continue to overlap, the amplitudes and phases associated with the individual component vibrations are unobservable. Conversely, any single wave motions may be resolved (analyzed) as the resultant of two or more components of the same frequency which differ in amplitude or phase, or both, from the original wave. Thus *a vibratory motion may always be analyzed as the sum of other sinusoidal motions which, collectively, are as "real" as the observed motion.* If a device acts in such a way as to physically remove one of the components of vibration from a light beam, the persisting wave properties of the light will be the resultant of the characteristics of the remaining components of the beam. A simple example is the phenomenon of dichroism, which is considered in Chapter 5. A dichroic material "mysteriously" converts unpolarized light into plane polarized light. In fact, it does so by removing (in this case, by a mechanism of absorption) all components of vibration in a given azimuth (i.e., in a given direction). The transmitted components of vibration constitute a plane polarized beam. Other examples of transformations of the properties of light waves are similarly found to be altogether rational when the properties of the final light beam are analyzed as those of the remaining components of vibration.

THE CONCEPT OF OPTICAL PATH

As noted in Chapter 1, refractive index is an *inverse* measure of the velocity of light in a medium:

$$(3\text{-cc}) \qquad \frac{n_1}{n_2} = \frac{v_2}{v_1} \quad \text{or} \quad n = \frac{c}{v} .$$

Thus the *velocity* of light in any medium is less than that in air. (X-ray *phase* velocities constitute an exception to this statement; see p. 610.) The *frequency* of vibration, an inherent characteristic of electromagnetic radiation as of vibratory motions in general, remains constant in all media, however. A consequence of the lowered velocity and invariant frequency of light in media is that more cycles of vibration are "squeezed" into a given path length in media than occur along the same path length in air. Thus, if two light beams travel along geometrically equivalent courses through air and through some other medium, the relative phase of the two beams shifts continuously as the distance of travel increases. This effect is illustrated in Fig. 3-8 for wave motions traveling identical distances AD. A beam traveling entirely in air ($n = 1.00$) is compared with two beams traveling through media for which the refractive indices are $n = 2.00$ and 1.83, respectively. All three beams are in phase up to position B. In air, the distance BC represents three com-

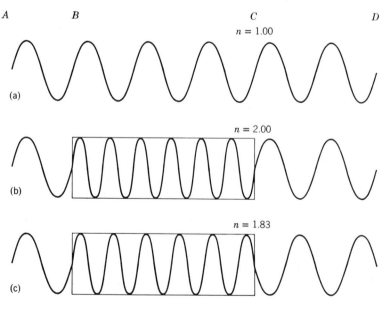

Fig. 3-8 Transmission of a wave motion by media of different refractive indices.

plete cycles of vibration, whereas, in the medium for which $n = 2.00$, BC represents six complete cycles of vibration. The two beams are again in phase along the distance CD; they have suffered a relative phase shift which is an integral number of wavelengths, that is, 3λ, 6π rad, or $3 \times 360°$.

In the block of index 1.83, as shown in Fig. 3-8c the distance BC represents five and one half cycles of vibration, so that the beam emerging from this block is exactly *out* of phase with the other two. Variations in the length or refractive index, or both, of the blocks would similarly produce other phase relationships between emerging beams, as would variation in the frequency (and thus the wavelengths) of the radiation.

The relative phases of the emergent beams shown in Fig. 3-8 are determined by the differences in *optical path* through which the beams travel. Optical path is defined as the product (*refractive index × distance*). Thus, for any system in which light travels physical path lengths l through media of refractive index n,

(3-dd) $$\text{Optical path} = n_1 d_1 + n_2 d_2 + \cdots = \sum nd.$$

Optical path differences between beams are designated Δnd or simply (P.D.).

The *phase difference* between beams of any wavelength λ for an optical path difference of (P.D.) is designated δ and is given by

(3-ee) $$\delta = \text{phase difference (rad)} = \left(\frac{2\pi}{\lambda}\right) \times \text{(P.D.)}.$$

A numerical example of path and phase differences may be considered. As shown in Fig. 3-9, the optical path of beam A is $8 \text{ cm} \times 1.0 = 8 \text{ cm}$, while that of beam B is $2 \text{ cm} \times 1.0 + 4 \text{ cm} \times 2.0 + 2 \text{ cm} \times 1.0 = 12 \text{ cm}$. Thus the optical path difference is $12 \text{ cm} - 8 \text{ cm} = 4 \text{ cm}$. For a vibration

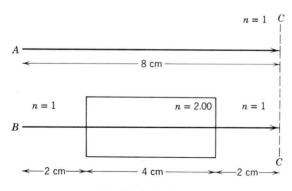

Fig. 3-9 Optical path.

of wavelength 1 cm the corresponding phase difference between beams A and B at position C would be

(3-ff) $$\delta = \frac{4\text{cm}}{1\text{cm}} = 4\lambda \quad \text{or} \quad 8\pi \text{ rad.}$$

The two beams would thus be *in* phase at position C. For another wavelength $\lambda = 6$ cm the phase difference δ' between beams A and B at position C would be

(3-gg) $$\delta' = \frac{4\text{cm}}{6\text{cm}} = \frac{2\lambda}{3} \quad \text{or} \quad \frac{4\pi}{3} \text{ rad.}$$

Thus, beams of 6 cm wavelength would be shifted in relative phase by 240° at position C; that is, beam B differs in phase by 240° with respect to beam A at that point. Note that, whereas optical path differs with wavelength only to the extent that the refractive index of the transmitting medium differs with wavelength, a given path difference represents different phase shifts for light of different wavelengths.

ORIGIN OF INTERFERENCE COLORS

Under suitable conditions the interaction of two monochromatic light beams can lead to variations in resultant intensity, as determined by δ, the phase difference between the beams. For example, if $\delta = 90°$, corresponding to a path difference of $\lambda/4$, the amplitude of the resultant of two beams of equal amplitude, A_0, is $\sqrt{2}A_0$, while the corresponding intensity is proportional to $(\sqrt{2}A_0)^2$, or $2I_0$ (cf. Fig. 3-4). For other values of δ the corresponding intensities range from zero to $4I_0$.

Since δ is a function of wavelength, the overlapping of *polychromatic* beams produces different degrees of constructive and destructive interference for each of the component wavelengths. If the resultant intensity is enhanced, in this way, over a narrow band of wavelengths, but is reduced or eliminated in other regions of the spectrum, *color* is observed. As discussed below, this condition is met when values of the optical path difference between the beams are *small*. The precise color obtained varies as a function of the path difference, of the initial distribution of wavelengths in the overlapping beams, and of the angle of observation.

A numerical example may be considered. Two coherent polychromatic beams, each of intensity $\frac{1}{2}I_0$, interact after traversing optical paths which differ by 4000 Å (400 mμ). This path difference represents one wavelength for violet light of $\lambda = 4000$ Å, which accordingly appears in the resultant beam with an intensity of $2I_{0_{4000}}$. Ultraviolet light of $\lambda = 2000$ Å, if present, is similarly reinforced, while infrared radiation at 8000

Å, for which the path difference is $\lambda/2$, is removed from the resultant by complete destructive interference. For green light of $\lambda = 5000$ Å the path difference is $5\lambda/4$, so that $\delta = 360° + 90°$; light of this wavelength accordingly appears in the resultant with an intensity of $\sqrt{2}/2I_{0_{5000}}$. Resultant intensities at other wavelengths can be computed from the corresponding values of the phase shift according to the methods which have been described on p. 36. In general, intensities at wavelengths between 4000 and 5000 Å are greater in the resultant beam than in the components; those at wavelengths longer than 5000 Å progressively diminish, falling to zero at $\lambda = 8000$ Å, and increasing again at still longer wavelengths. The net effect of interference is a resultant beam which is enriched in the near ultraviolet, violet, blue, and green portions of the spectrum. The beam is blue-violet in color, its precise hue being determined by the relative intensities of each wavelength which are present in the interfering beams. Similarly, in a system for which the optical path difference between overlapping beams is about 6000 Å, a reddish orange color is produced.

Interference colors can be observed when light beams which differ in optical path interact over a broad wavefront; for example, in the light which is reflected from thin films, as discussed below. These colors are observed *only* for optical path differences which are small. Larger path differences represent integral multiples of many different wavelengths, so that light in different portions of the spectrum is reinforced. For example, when the path difference is 4000 Å, light of $\lambda = 4000$ Å is the *only* visible wavelength for which the path difference is $m\lambda$. (Since $m = 1$, the violet light is said to reinforce *in the first order*). For a path difference of 12,000 Å, however, *two* visible wavelengths reinforce completely: 4000 Å in the *third* order, and 6000 Å in the *second*. Larger values of (P.D.) produce a series of reinforcements of wavelengths which span the visible spectrum; for example, at path differences of 45 μ (450,000 Å) there is tenth order reinforcement for light of $\lambda = 4500$ Å, ninth order for 5000 Å, eighth order for 5625 Å, and so forth. The combination of these wavelengths is then perceived as a "higher order white." Mixtures of wavelengths easily cause the observed interference color to appear "muddy", hence pure and striking colors are not produced in orders higher than the second or third; that is, as a consequence of optical path differences greater than 1.5 or 2 μ.

THE FORMATION OF INTERFERENCE COLORS BY THIN FILMS

The beautiful colors of thin films, such as oil slicks floating on water, result from interference between light reflected from the upper and lower surfaces of the film, respectively. Figure 3-10 illustrates a pair of beams

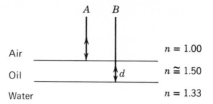

Fig. 3-10 Reflection at the surfaces of a thin film; normal incidence.

which are reflected at normal incidence from a thin film. (Although the rays reflected from each surface are shown separately, interference between coincident rays is, of course, the case actually considered.) Beam *B* of Fig. 3-10 travels farther than beam *A*, the difference in their optical paths so introduced being $(2n$ $)$, where *d* is the thickness of the film and *n* is the refractive index of the film material. This quantity might appear to be the complete difference in optical path between the beams which leave the film, but in fact it is no^t. As proved on p. 485, a shift in phase of 180° $(=\pi)$ must occur upon reflection at *one* side of any interface. It is found experimentally that this phase shift occurs when light is reflected from a surface of a medium of higher refractive index (i.e., at "rare-to-dense" rather than at "dense-to-rare" interfaces). Beam *A* of Fig. 3-10 is reflected in a medium of refractive index 1.00 from a medium of index 1.50, and thus suffers such a change, while beam *B*, reflected in a medium of index 1.50 from one of index 1.33, does not. Thus an *additional* phase shift of $\lambda/2$ is introduced, giving total phase and path differences between the two beams of

$$(3\text{-hh}) \quad \delta = \left(\frac{2\pi}{\lambda}\right) 2nd + \pi \quad \text{and} \quad (\text{P.D.}) = 2nd + \left(\frac{\lambda}{2\pi}\right)\pi = 2nd + \frac{\lambda}{2}.$$

If the quantity $(2nd + \lambda/2)$ is equal to $m\lambda$, where *m* is any integral number, beams *A* and *B* are reflected from the thin film *in* phase, and they interfere constructively. For $2nd + \frac{1}{2}\lambda = (m - \frac{1}{2})\lambda$, however, *A* and *B* leave the film *out* of phase and interfere destructively. The conditions for observation of minima and maxima of intensity are thus

(3-ii)

$$\text{minima at } d = \frac{m\lambda}{2n} \quad \text{and}$$

$$\text{maxima at } d = \frac{(m + \frac{1}{2})\lambda}{2n}$$

.

Note that destructive interference at the minima is complete only if the (total) intensities reflected at the two surfaces are equal. The relative intensity reflected or refracted is a function of the refractive indices of the two media. For normal incidence, *reflectance* [the fraction of intensity reflected (I_{refl}/I_0)] is given by the expression

(3-jj)
$$\frac{I_{refl}}{I_0} = \left(\frac{n_2 - n_1}{n_2 + n_1}\right)^2.$$

(For other angles of incidence the reflectance is a function of the state of polarization of the light also.) (For further discussion of relationships between reflected and refracted intensities see pp. 484 ff.)

Extremely thin films, for which d is less than $\lambda/4$, are virtually invisible, since the 180° phase shift between the light reflected from the upper and lower surfaces results in destructive interference at *all* wavelengths. Thicker films reinforce successively longer wavelengths, so that a succession of colors, ranging from blue through green and yellow and red, is observed as thickness increases. This sequence tends to repeat in a second order and again in a third order as film thickness increases further; however, films which are more than a few wavelengths thick produce only a higher order white as explained above on p. 45.

Observation of interference colors is helpful in the preparation of ultrathin sections for electron microscopy. These sections, which are embedded in plastics of various types, must be less than about 1000 Å $(0.1\,\mu)$ thick, while sections only 200 or 300 Å thick are considered desirable for most purposes. Sections which are easily seen are excessively thick, since their visibility depends upon higher order white interference effects. Red and green colors likewise signify excessive thickness, while those which appear yellow may be satisfactory. The thinnest specimens, as required for high resolution studies, produce only a slight "gray" reflection.

The observation of interference colors is not confined to viewing at normal incidence, the latter being merely the angle at which *maximum* phase shift between reflections from the upper and lower surfaces of a given film is obtained. Figure 3-11 illustrates two beams, Ⓐ and Ⓑ which

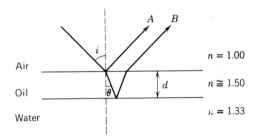

Fig. 3-11 Reflection at the surfaces of a thin film; oblique incidence.

are reflected from the upper and lower surfaces, respectively, of a film upon which light is incident at any angle *i*. (Note that, although rays Ⓐ and Ⓑ are not superposed as at normal incidence, it is nonetheless the difference in optical path between these rays upon which the interference effects depend. Some other incident ray, parallel to that shown, is reflected from the upper surface in coincidence with ray Ⓑ, while a third incident ray is reflected from the lower surface in coincidence with ray Ⓐ. Interference between the coincident rays is observable only to the extent that the two originate from a very small, effectively coherent region of the light source (cf. pp. 70–71). For extremely thin films, this is in fact the case.)

Consideration of the geometry of Fig. 3-11 gives conditions for the formation of maxima and minima which are (again, for a film of refractive index *higher* than that of the surrounding media)

(3-kk)

$$d = \frac{m\lambda}{2n \cos \theta} \qquad \text{for minima,}$$

$$d = \frac{(m+\frac{1}{2})\lambda}{2n \cos \theta} \qquad \text{for maxima,}$$

where θ is the angle of *refraction*, as obtained from Snell's law (1-b).

The "retardation colors" observed in images of birefringent specimens viewed in the polarizing microscope (cf. p. 326) are another example of the formation of interference colors. A list of the colors which result from specific values of the optical path difference is given in Table 15-2.

LENS COATING

A practical application of thin film interference effects is the coating of lenses for the purpose of reducing light losses. As (3-jj) shows, the intensity of stray light originating from reflection at interfaces may reach a troublesome level. Reflective losses of intensity may be minimized by causing reflected amplitudes to undergo destructive interference.

For instruments to be used with visible light, the thickness and refractive index of the coating film are so chosen that reflected light in the middle of the visible spectrum—a wavelength in the yellow or green—destructively interferes. (Many photosensitive devices, as well as the eye, respond maximally to wavelengths in this range.) Ideally, the refractive index of the coating is so chosen that the amplitudes of light reflected from the two surfaces of the film are exactly *equal*, thus producing *complete* destructive interference of the chosen wavelength. This is the case if

(3-ll) $(n_{\text{coating}})^2 = n_{\text{glass}},$

as may be established by evaluation of (3-jj) at each surface of the coating. Thus, if the refractive index of the coating is n and that of the glass is n^2, then

(3-mm)   reflectance at air to coating surface $= \dfrac{(n-1)^2}{(n+1)^2}$.

(3-nn)   reflectance at coating to glass interface $= \dfrac{(n-n^2)^2}{(n+n^2)^2}$.

The quantities given in (3-nn) do in fact reduce to those given in (3-mm). In practice, however, the square-square root relation of indices cannot be met exactly, since the indices of all available coating materials are somewhat higher than the values specified by (3-ll).

Since n is greater than one, the refractive index of the coating is intermediate between that of air and glass. Hence there is a rare-to-dense reflection at both coating surfaces; consequently *both* beams suffer a phase change of 180° upon reflection. Accordingly, conditions for formation of maxima and minima of reflected intensity are the *reverse* of those which apply to reflection from an oil film on water; that is,

(3-oo) $d = \dfrac{m\lambda}{2n}$ for maximum reflectance,   and

(3-pp) $d = \dfrac{(m+\frac{1}{2})\lambda}{2n}$ for minimum reflectance (at normal incidence).

In summary, lenses are coated by material of index $\sqrt{n_{\text{glass}}}$ to a thickness given by (3-pp). For example, if light of wavelength 5600 Å is to be removed by destructive interference at the surface of a lens constructed from glass of refractive index 1.56, the appropriate coating material would be of index $\sqrt{1.56} = 1.25$, and of thickness

(3-qq) $d = \dfrac{\frac{1}{2} \times 5600 \text{ Å}}{2 \times 1.25} = 1120 \text{ Å}.$

Although approximately complete elimination of reflected intensity by coatings occurs only at a single wavelength, an appreciable fraction of reflected intensities of other wavelengths is also removed. Significant reflection occurs only at the extremes of the visible spectrum; thus a bluish cast is commonly observed on the surface of coated lenses.

HUYGENS' PRINCIPLE

An object of finite size, for example, a rectangular aperture one cm in diameter, may cast a *sharp* shadow. As the size of the object (diameter of the slit) is reduced, however, the shadow tends to become hazier. If a slit less than about 0.2 mm in diameter is illuminated by a narrow beam of light, the edges of the geometrical shadow of the slit may be found to be surrounded by bright and dark fringes (i.e., by maxima and minima of intensity). Since these fringes may be seen to extend *inside* the geometrical shadow of the slit, it must be concluded that *illumination has bent around the edges of the slit.* This effect is an example of the *diffraction* (or "bending") of light.

The origins of diffraction phenomena may be visualized in terms of Huygens' principle, which states that *every point on a wave front acts as a source of secondary wavelets* and *the position of the wavefront at any time may be found by drawing the envelope of secondary wavelets at that time.* The direction in which the wave travels is given by the vector drawn from the origin of the secondary wavelet to its point of tangency with the wavefront. Ordinarily this direction is that of the normal to the wavefront. (An exception exists, however, in the case of the extraordinary wavefronts formed within birefringent media, as discussed in Chapter 6 and illustrated by Fig. 6-11). Huygens' principle governs the behavior of both mechanical waves and light waves.

An infinite number of sources of secondary disturbance may be considered to exist along any plane wavefront; six are illustrated in Fig. 3-12a. The condition of the wave after an interval of time equal to one period of vibration is illustrated in Fig. 3-12b. Each source of secondary wavelets has generated a trough (dotted line) and a crest (solid line); the envelope of the latter is shown. Across the greater part of the wavefront the envelope of secondary wavelets is *parallel* to the old wavefront, but at the edges the secondary wavelets spread outward without interference from adjacent wavelets. The result is that some portion of the intensity of the beam spreads outward at the edges; the narrower the beam, the greater is the proportion of total energy so lost from the forward direction. However, unless special arrangements are made for isolating beams which are narrow with respect to the wavelength of the

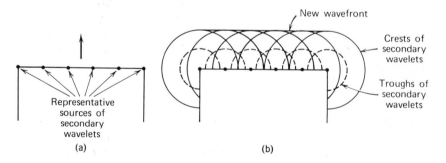

Fig. 3-12 Huygens' principle.

light, the fractional intensity so lost is negligible. Consequently, the light appears simply to travel in straight lines, as postulated in geometrical optics.

In Fig. 3-12 the light is represented as traveling through a medium of uniform refractive index; more generally, the secondary wavelets travel with velocities appropriate to the medium. As before, the envelope of these wavelets describes the positions of the wavefront as a whole, but it is more complex in form than in a uniform medium. For example, Fig. 3-13 illustrates the troughs and crests generated by a point source of disturbance in a region which contains media of two different refractive indices. That the waves are refracted according to Snell's law (1-b) can be verified by drawing in *rays*, normal to the wave surfaces, as is done in Fig. 3-13.

PROPERTIES OF SECONDARY WAVELETS

The properties of secondary wavelets must be more precisely specified in order to explain in detail the behavior of light in terms of Huygens' principle.

In Figs. 3-12 and 3-13, only those portions of the wavelets traveling in the *forward* direction are shown. However, positions of the secondary wavelets could be drawn in through a full 360°, making it appear that wave motions should travel *backwards*, as well as in the forward direction! Yet the energy of light beams, as of other wave motions, advances in a given direction in an orderly fashion. The solution to this dilemma is that the amplitude of the wavelets is described by

(3-rr) $$A = A_0(1 + \cos \theta),$$

where A_0 is the amplitude of the wavelet in the forward direction and θ is the angle between any other direction and the forward direction.

Note that, since $\cos 180° = -1$, the amplitude of secondary wavelets is *zero* in the reverse direction. (In general, it is important to distinguish

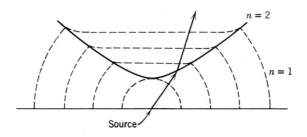

Fig. 3-13 Distortion of wavefront by passage through media of different refractive indices.

Framingham St...
Framingham, Massachusetts

diagrams such as Figs. 3-12 and 3-13, which indicate the positions of wavefronts, and diagrams in which either the amplitude or the intensity of light is plotted as a function of direction.)

A further specification of the properties of the secondary wavelets is that they are advanced in phase by 90° with respect to the main beam.

The reader may ask: "Do the secondary wavelets really exist?" In one sense the answer to this question is that the secondary wavelets are simply a convenient visual "crutch" for understanding the properties of waves, just as the wave theory itself is merely a convenient way of describing the behavior of large numbers of photons. Nevertheless, secondary wavelets are "real" in precisely the same sense that the individual components of any observed resultant motion are real (cf. discussion on p. 41). The isolation of a very limited portion of a mechanical wavefront even allows the direct observation of what is, essentially, a secondary wavelet.

DIFFRACTION PATTERNS

The preceding discussion may seem to imply that light intensity merely spreads out from the edges of beams, producing uniformly graded intensities out of the forward direction. In fact, however, the phase relationships between portions of the diffracted beam are such that interference causes the intensity of the beam to vary with direction in a regular manner. The resultant distribution of intensities is known as a *diffraction pattern*. In the following sections some simple types of diffraction pattern are described and the general features of diffraction patterns are discussed.

DIFFRACTION PATTERN FORMED BY A PAIR OF SLITS

The origin of the double slit diffraction pattern is particularly simple to consider. Experimentally, as indicated in Fig. 3-14a, light from a limited region of a source is first isolated by a slit or series of slits. In this way *coherent* illumination of the slits is provided. (This aspect of the formation of diffraction patterns is discussed on p. 65.) The light beam is then rendered parallel by a lens, forming a plane wavefront which illuminates the diffracting slits. The nature of the action of the lens need not be considered at this point, except to note that its use is geometrically equivalent to placing the light source at an infinite distance from the slits. Note that the slits in Fig. 3-14 extend in a direction perpendicular to the plane of the page.

In terms of Huygens' concepts each of the slits may be considered the source of a single secondary wavelet. Thus maximum intensity (positive interference) is observed on a screen placed to the right of these slits

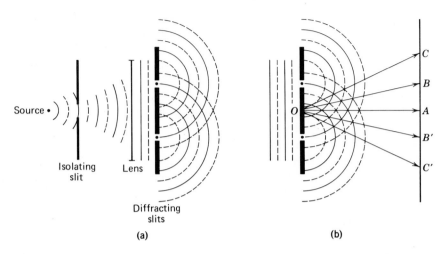

Fig. 3-14 Diffraction at a pair of slits: (*a*) illumination of the slits; (*b*) interference to form the diffraction pattern.

wherever crests overlap with crests or troughs with troughs. Zero intensity (destructive interference) is observed at positions where crests meet troughs. As shown in Fig. 3-14b, lines (rays) can be drawn in directions along which crests and troughs reinforce or cancel. For example, line OA, which is perpendicular to the plane of the slits, is the direction in which crests overlap crests and troughs overlap troughs after experiencing an *equal* number of cycles of vibration. A line of *maximum* intensity is thus formed on the screen through point A and parallel to the direction of the slits. This bright line is called a *zero order fringe*, since it results from the interference between beams which have *no* difference in optical path. Along line OB, however, light from the lower slit has traveled half a wavelength farther than that from the upper slit, while the reverse is true along OB'. Fringes of zero intensity are thus observed along lines passing through positions B and B'. Similarly, at C and C' the light arriving from one slit has traveled a full wavelength farther than that from the other, so that positive reinforcement occurs at these points. The fringes which pass through C and C' (perpendicular to the plane of the page) are termed *first order fringes*, because they result from interference between beams which differ in optical path by *one* full wavelength. At other positions, farther from A, a series of minima result from interference between rays which differ in optical path by $\frac{3}{2}, \frac{5}{2}, \ldots, (2n + 1)/2$ wavelengths, while a series of subsidiary maxima, or *higher order fringes*, results from interference between beams which differ in path by $2, 3, \ldots, n$ wavelengths.

The resultant intensity of the bright fringes formed by interference between the two beams decreases with distance from the central point A. A typical tracing of resultant intensity, as along line CAC' of Fig. 3-14, and as recorded by a densitometer (cf. p. 471) is shown in Fig. 3-15. In this type of diffraction pattern a graduation of intensity between the positions of minima and maxima can be observed; that is, the maxima are not "sharp". This feature may be contrasted with the case of multiple slit diffraction patterns, as discussed on p. 59.

The spacing of maxima and minima in the double slit pattern can be deduced from the geometry of the ray diagram, Fig. 3-16. (Note that this figure is equivalent to Fig. 3-14, except that consideration of the secondary wavelets is omitted.) As shown, the diffracting slits are located at N and P, M is the midpoint between the slits, and $MA = D$ is the slit-to-screen distance. At point A on the screen, rays arriving from N and P have traveled identical optical paths and are *in* phase (as is the case at point A in Fig. 3-14b). At any other point X on the screen, separated from A by a distance x, rays from N and P differ in optical path. The value of this path difference is

$$(3\text{-ss}) \qquad\qquad \Delta = PC = d \sin \theta,$$

where d is the separation of the slits, and θ is the angle between MX and the normal to the plane of the slits. Note that, since the slit-screen distance D is always very much larger than either d or x, rays from *both* slits may be considered to leave the plane of the slits at the same angle θ, while angle CNP is also equivalent to θ.

If the quantity Δ, given by (3-ss), represents an *integral* number of wavelengths, the rays NX and PX must arrive at X *in* phase, thus interfering to produce a maximum of intensity. Accordingly, maxima of the diffraction pattern are observed whenever

$$(3\text{-tt}) \qquad\qquad \boxed{d \sin \theta = m\lambda} \,.$$

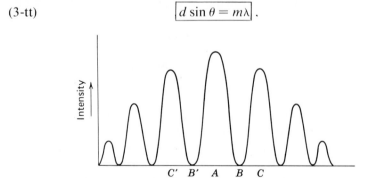

Fig. 3-15 Densitometer tracing of the double slit diffraction pattern.

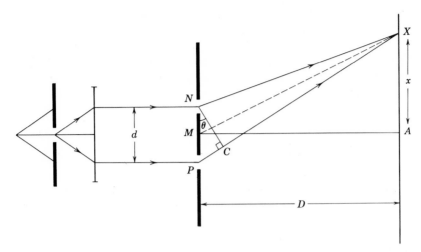

Fig. 3-16 Geometric representation of the formation of the double slit diffraction pattern.

where m is any integer. Similarly, if rays NX and PX are exactly *out* of phase at x, they interfere to produce zero intensity. The condition for minima (zeros) of the diffraction pattern is thus

(3-uu) $$\boxed{d \sin \theta = (m + \tfrac{1}{2})\lambda} \,.$$

Note that, since at small values of θ, the approximations

(3-vv) $$\sin \theta \cong \tan \theta \cong \frac{x}{D}$$

may be made, the conditions 3-tt and 3-uu may be written

(3-ww) $$x = m\lambda \frac{D}{d} \qquad \text{for maxima, and}$$

$$x = (m - \tfrac{1}{2})\lambda \frac{D}{d} \qquad \text{for minima,}$$

where D is the slit-screen distance, d the slit spacing, and x the distance on the screen from the central maximum of the diffraction pattern to the point considered.

In the case considered in Figs. 3-14 and 3-16 the zero order fringe is centered with respect to the slits because the optical paths from N and P to the central point A are identical. However, if these paths are caused to differ, as by interposing a medium of increased refractive index in the path of the beam from one of the slits, the zero order fringe, together with the accompanying system of higher order fringes, is *displaced in the direction of the path of higher refractive index.* Thus, for example, Fig. 3-17 shows a system similar to that illustrated in Fig. 3-16, but the open slit at *N* is replaced by a strip of transparent material the refractive index of which is such that the optical path of transmitted rays is increased by one half wavelength with respect to passage through air. Slit *P*, however,

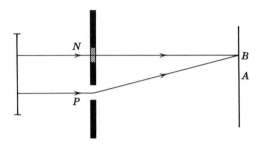

Fig. 3-17 Effect of an increase in optical path on the double slit diffraction pattern.

remains open as before. A minimum of intensity must now be observed at *A*, since rays reaching that point from slit *N* have traveled one half wavelength farther than those from *P*. The zero order maximum of the pattern is now displaced to a position at which the physical path length from slit *P* is half a wavelength longer than that from slit *N*. This condition is satisfied at point *B*, formerly the locus of a minimum, as shown in Fig. 3-14.

SINGLE SLIT DIFFRACTION PATTERNS

A single coherently illuminated slit produces a diffraction pattern of the form shown below in Fig. 3-19. Like the double slit pattern, it consists of a central fringe of maximum intensity surrounded by subsidiary maxima which are, in this case, of generally rather rapidly diminishing intensity. The similarity of the two types of patterns may seem puzzling in terms of the treatment of double slit diffraction patterns given above, in which each slit was regarded as the origin of a *single* secondary wavelet. A single slit, however, can be regarded as the source of a large or infinite number of secondary wavelets; interference among these wavelets is then considered.

An experimental arrangement for observation of single slit diffraction patterns is diagramed in Fig. 3-18. A slit *NP* of width *d* is illuminated by a parallel beam proceeding from a limited region of an intense light source (e.g., either a conventional high intensity source or a laser beam). A lens placed immediately beyond the slit focuses the rays diffracted from the slit in any given direction θ at a corresponding position X on the screen beyond. (Again, the nature of the action of the lens is not considered at this point.) The resultant amplitude and thus the intensity at X are, in general, functions of the path differences (such as MC of Fig. 3-18) between the rays reaching X. The interaction of an infinite number of rays (i.e., of rays from *all* portions of the slit) must be considered; thus derivation of an expression for the location of maxima and minima of the single slit pattern is more complex than the corresponding derivation for the two-slit experiment considered above.

An understanding of the formation of the single slit diffraction pattern may be obtained by considering the case in which the path difference MC, shown in Fig. 3-18, is exactly half a wavelength. Rays NX and MX then destructively interfere at X. Similarly, successive pairs of rays originating at successive positions below N and M, respectively, also differ in path by exactly $\lambda/2$ and thus also interfere destructively. Since *all* rays from the slit can be accounted for in this manner, the resultant intensity at X is zero. A minimum of intensity is likewise obtained at X if the distance MC is exactly one complete wavelength. In that case a ray originating halfway between N and M is exactly out of phase with NX, while ray MX is exactly out of phase with a ray originating halfway between M

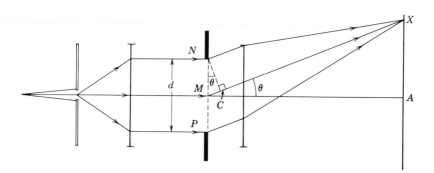

Fig. 3-18 Diffraction at a single slit.

and P. Again, *all* rays from the slit can be accounted for as members of pairs which interfere destructively. In general, then, minima of the diffraction pattern occur whenever the distance MC is any integral number of half wavelengths. Since, from Fig. 3-18,

(3-xx) $$MC = \tfrac{1}{2}d \sin \theta,$$

the condition for minima of the single slit pattern is

(3-yy) $$d \sin \theta = m\lambda \quad \text{for} \quad m = 1, 2, 3, \ldots$$

The phase shift corresponding to path difference MC, a quantity conveniently designated γ is

(3-zz) $$\gamma = \frac{2\pi \, (\text{path difference})}{\lambda} = \frac{\pi d \sin \theta}{\lambda}.$$

Expressions for amplitude and intensity at X as a function of θ may be obtained by integration of *all* contributions to amplitude at X. The integration, which is given in Appendix D, yields the expressions

(3-a*)

(3-b*)

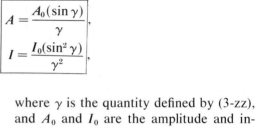

$$A = \frac{A_0 (\sin \gamma)}{\gamma},$$

$$I = \frac{I_0 (\sin^2 \gamma)}{\gamma^2},$$

where γ is the quantity defined by (3-zz), and A_0 and I_0 are the amplitude and intensity, respectively, at the center of the diffraction pattern. Equation 3-a* is also derived by another, quite different, approach on p. 228.

As obtained in Appendix D by differentiation of (3-a*), maxima of the single slit diffraction pattern occur when

(3-c*) $$\tan \gamma = \gamma.$$

These maxima are approximately but not precisely equidistant from the surrounding minima.

Figure 3-19 is a plot of the intensity of the diffraction pattern as a function of the quantity γ. As in all diffraction patterns, the central maximum of intensity is called the *zero order* image, since it results from the interference of beams with path differences of zero. Similarly, the successive subsidiary maxima are termed

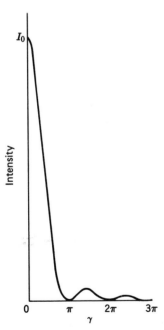

Fig. 3-19 Intensity of the single slit diffraction pattern.

first, second, . . . , nth order maxima, since they result from the interference of beams which differ in path by one, two, . . . , n complete wavelengths. As Fig. 3-19 shows, the intensity of the maxima is greatest in the zero order and falls off quite rapidly in the subsidiary maxima.

RELATION OF SINGLE AND MULTIPLE SLIT DIFFRACTION PATTERNS

A question which may arise is: "What happened to the diffraction patterns of the *individual* slits in the two-slit diffraction experiment? (as discussed on pp. 56–57). Is the treatment which views each slit as the source of a single wavelet valid?"

In fact, the form of the double slit diffraction pattern, as of any other diffraction pattern, is determined ultimately by the principle of superposition. The two-slit diffraction pattern is the sum of the two overlapping single slit patterns, the intensity at any point on the screen being determined by the phases and amplitudes of all rays arriving at that point. At certain positions the amplitudes contributed by each slit reinforce each other, while at other points they partially or completely cancel. The treatment of the double slit diffraction pattern given above simply located those positions of reinforcement or cancellation. Derivation of an expression for the intensities in the diffraction pattern would have required consideration of the distribution of amplitudes proceeding from each slit.

Equations 3-tt, 3-ww, and 3-c* predict only the *positions* of maxima of diffraction patterns; these equations make no specifications concerning the sharpness of the maxima. In fact, single slit patterns contain obvious gradations of intensity; maxima are not sharply defined. When contributions of additional slits are added to the intensities observed on a screen, the maxima tend to become sharper as the number of diffracting slits is increased. The limiting case is that of the *diffraction grating*, considered below, in which diffraction from many thousands of parallel slits produces extremely well-defined maxima, while intensities at other positions are essentially zero.

THE DIFFRACTION GRATING

A diffraction grating is a set of many equally spaced, parallel, identical slits or, equivalently, of parallel rulings on a reflecting surface. Each slit or ruling may be considered the source of a single secondary wavelet, so that rays travel in all possible directions from each opening of the grating, as indicated by Fig. 3-20a. Interference between the members of parallel sets of rays from all the openings then determines the directions in which resultant diffracted intensity may be observed.

Light from the grating spacings is collected by the lens (which may be the eye) and focused at a plane beyond the lens. Figure 3-20b, for example, shows those rays which leave the grating perpendicular to its surface. There is *no* difference in optical path between the rays, which thus arrive *in phase* at O, regardless of wavelength, and reinforce to produce a bright zero order spot at that central position. The *proportion* of intensity from the grating which appears in the zero order spot, or in any other maximum of the grating diffraction pattern, is determined by the *proportion of amplitude leaving each slit in the corresponding direction* and is thus a function of the *structure* (i.e., the dimensions and contours) of the individual openings.

Rays leaving the grating in other directions, as illustrated in Fig. 3-20c, are tilted with respect to the plane of the focusing lens, so that rays originating from successive slits undergo successive differences in optical path before reaching the lens. As shown, $ABDF$ represents a portion of the wavefront leaving the grating at some angle θ. (Note that θ may be described either as the angle between wavefront and grating or as the angle between individual rays and the normal to the grating.) The corresponding path difference between rays contributing to the wavefront which reaches the lens is some value Δ such that $BC = \Delta$, $DE = 2\Delta$, $FG = 3\Delta$, and so forth. (Note that Δ is the path difference between ad-

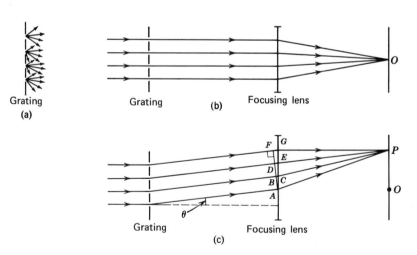

Fig. 3-20   The diffraction grating.

jacent rays in all cases.) If $\Delta = m\lambda$, as it does for *each* wavelength at a set of specific values of θ, rays from each opening of the grating are *in phase* upon arrival at P and reinforce to form a bright spot at that position. (Note that, as is brought out in Chapter 8, a lens serves to focus sets of parallel rays at a point *without introducing differences in optical path*.) The different integral values of m correspond to different orders of diffracted maxima; for example, $\Delta = 2\lambda$ for a second order maximum. Corresponding positions of complete destructive interference are those for which $\Delta = (m + \frac{1}{2})\lambda$.

As shown in Fig. 3-20, the path difference between rays leaving the grating from adjacent slits is $AC \sin \theta = d \sin \theta$, where d is the grating spacing. (Or, for rays traveling in media other than air, $nd \sin \theta$.) The *grating equation*, or condition for the formation of diffracted maxima, is thus

(3-d*) $$ \boxed{m\lambda = d \sin \theta} \,, $$

where m is any integer, d is the grating spacing, and θ is the angle between rays leaving the grating and the normal to its surface or, equivalently, between the grating and the wave surface leaving the grating.

Figure 3-21 illustrates oblique incidence of light upon a diffraction grating. The path difference between rays which are incident upon adjacent slits is obtained by dropping the perpendicular EH from one slit to the ray incident upon the next. $HF = d \sin i$ (where i is the angle between incident rays and the normal to the grating or, equivalently, between the incident wavefront and the grating surface) is then the difference in optical path between successive incident rays. This path difference must be added to that arising from diffraction. The general form of the grating equation is thus

(3-e*) $$ m\lambda = d(\sin i + \sin \theta) . $$

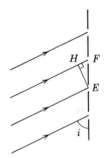

Fig. 3-21 Oblique illumination of a diffraction grating.

Equation 3-d* is a special case of (3-e*) which applies when (as is very often the case) gratings are illuminated *normally*.

Equations (3-d*) and (3-e*) specify the *positions* of diffracted maxima, but it should be understood that they fail to describe the distribution of intensities produced by gratings. As noted above, the diffraction pattern formed by a large number of slits differs from that formed by one or a few slits in that it is very much sharper. That this is so may be made evident by considering a specific example of diffraction by slits so spaced that the amplitudes of some specified wavelength diffracted by adjacent slits are only slightly out of phase with each other. If only a few slits diffract, these amplitudes produce a resultant intensity of appreciable magnitude. When many slits diffract, however, the light from pairs of slits separated by some distance is exactly out of phase. Net zero intensity results from the cancellation of amplitudes diffracted by successive pairs of slits, so that only light from those slits which are "unpaired" can contribute to resultant intensity. A simple example is illustrated in Fig. 3-22, which shows diffracted beams leaving a series of thirteen slits at an angle such that they differ in phase successively by ($\lambda/8$). Beams A and E, B and F, C and G, D and H, and I and M each cancel, since they are all of equal amplitude, and the members of each pair differ in phase by exactly ($\lambda/2$). Thus, only beams J, K, and L are "left over" to contribute (according to their relative phases) to resultant intensity. For larger numbers of diffracted beams the proportion of amplitudes left over is correspondingly smaller. The net effect is that detectable intensities are diffracted only at angles which very nearly satisfy (3-e*) or (3-d*). The distribution of intensities between the allowed maxima is, however, determined by the size and structure of the spacings themselves, as noted above (cf. also the discussion on p. 522).

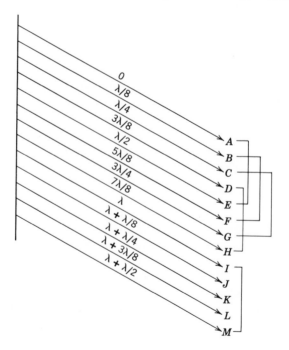

Fig. 3-22.

If a grating is illuminated by monochromatic light, a discrete series of well-spaced maxima result. If the grating is illuminated by polychromatic light, the angular position of maxima varies with the wavelength. A grating is thus a device which may be used to isolate monochromatic light; light leaving the grating at any angle θ, corresponding to the desired wavelength, is isolated by interposing an aperture beyond the grating. The use of diffraction gratings as monochromators is discussed further in Chapter 23.

EFFECT OF SLIT WIDTH UPON DIFFRACTION PATTERN SPACING

From (3-yy) the positions of minima in the diffraction pattern formed by a single slit have been specified as

$$(3\text{-}f^*) \qquad\qquad \sin\theta = \frac{m\lambda}{d}.$$

A most important point revealed by this equation is that a given minimum (or maximum), observed at position X on a screen after diffraction through an angle θ, moves, approximately, to position $2X$ (corresponding to diffraction through angle 2θ) if the width of the slit is reduced from d to $\frac{1}{2}d$. That is, *the narrower the diffracting aperture, the more widely spread is the diffraction pattern.* This conclusion is not limited to the patterns formed by diffraction from slits, but is applicable to apertures of any shape or to any obstacle encountered by light waves. Understanding of diffraction phenomena and, in particular, of the theoretical limitations upon the resolving power of microscopes, depends absolutely upon appreciation of the inverse relationship between the dimensions of object and diffraction pattern.

In gratings the narrower the individual slits, the larger is the proportion of intensity diffracted out of the forward direction. The intensity of higher diffracted orders is, in general, thereby enhanced relative to that of the zero order. Reflection from grating rulings may, however, favor diffraction of light into particular orders.

THE AIRY DISC

The form of patterns produced by diffraction at slits has been considered in some detail. While slit diffraction patterns are relatively simple to consider mathematically, diffraction at a small circular opening (aperture) is a topic of much greater interest in microscopy. A circular aperture produces a diffraction pattern which is (in two dimensions) a series of concentric rings of varying intensities. The plot of intensities along any line through the center of the pattern is of the same general form shown for the single slit pattern in Fig. 3-19, but it differs in details of the distribution of intensities. The pattern is called an Airy disc (named after Sir George Airy, who originally formulated the expressions given below for the distribution of intensities within the pattern). The Airy disc is a section through a three-dimensional diffraction figure which consists of concentric shells of intensity levels.

In considering diffraction at a circular opening, the net disturbance produced by rays from all regions of the aperture at any point X of the diffraction pattern must be computed. The contributions from each element of aperture area are then integrated, just as amplitudes contributed by all portions of a slit aperture are integrated in the treatment given in Appendix D. Details of the integration are of some mathematical complexity, and will not be given here (e.g., see [1]). However, some features of an alternative derivation which makes use of Fourier series are described in Chapter 10. The resulting expressions for amplitude and intensity of the diffraction pattern are

(3-g*)
$$A = \frac{A_0 2 J_1(\gamma)}{\gamma} \quad \text{and}$$

$$I = I_0 \left[\frac{2 J_1(\gamma)}{\gamma} \right]^2 \quad,$$

where A_0 and I_0 are the amplitude and intensity, respectively, at the center of the diffraction pattern; γ, as defined on p. 58, is the quantity $(\pi d \sin \theta)/\lambda$ in which d now represents aperture diameter and $J_1(\gamma)$ is the *first order Bessel function* of γ. The nature of Bessel functions is described in mathematical textbooks (e.g., see [2]). The appearance of Bessel functions in (3-g*) may alarm those who are unfamiliar with this class of function, but knowledge of the properties of Bessel functions is not required for an understanding of the form of diffraction patterns. It is only necessary to appreciate that minima are formed at any position X of the diffraction pattern at values of γ such that $J_1(\gamma) = 0$. Maxima of the diffraction pattern are correspondingly observed at values of γ such that maxima of $J_1(\gamma)$ occur. The relationship of the quantities d, θ, and X is illustrated in Fig. 3-23. Note that, since $\gamma = (\pi d \sin \theta)/\lambda$, larger values of X or θ correspond to minima or maxima of a given order when the aperture diameter d is reduced. That is, just as noted for the case of diffracting slits, the smaller the aperture, the more widely spread is its diffraction pattern. Values of $(d \sin \theta)/\lambda$ which correspond to minima and maxima of the diffraction pattern are listed in Table 3-1. (Note that all of these values are nonintegral.) A plot of the intensity of the diffraction pattern as a function of γ is shown in Fig. 3-24, and a photograph of the Airy disc is shown in Plate I. Highly coherent laser light was used to

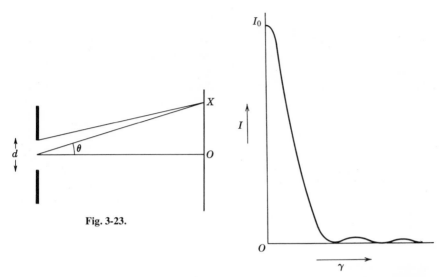

Fig. 3-23.

Fig. 3-24 Intensity distribution in the Airy disc.

record the image shown in the plate, thus forming a well-defined pattern in which many maxima of intensity appear. Conventional light sources form similar but less sharp patterns in which only a few maxima are discernible.

Table 3-1   Values of $(d \sin \theta)/\lambda$ which Correspond to Minima or Maxima of the Airy Disc

Order	For Minima	For Maxima
Zero (central spot)	—	0
1st	1.220	1.635
2nd	2.233	2.679
3rd	3.238	3.699
4th	4.241	4.710

FRESNEL DIFFRACTION

The diffraction patterns considered above have been ones which are formed when slit, grating, or aperture is illuminated by parallel light, either from a collimated (i.e., parallel) beam or from a source assumed to be infinitely distant. These cases, in which incident wavefronts are *planar*, are examples of so-called *Fraunhofer* diffraction. Alternatively, diffraction may occur at obstacles placed in a *divergent* beam, that is, at obstacles

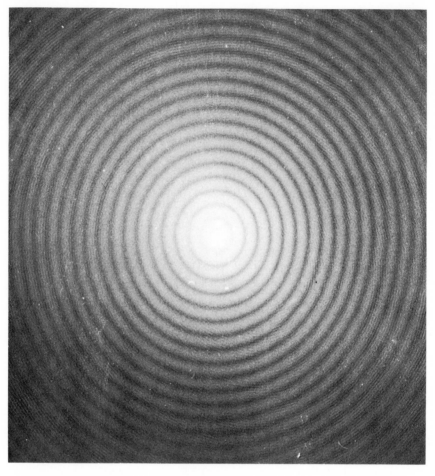

Plate I   The Airy disc. Image of a small circular hole as formed by laser light.

illuminated by a source placed at a finite distance. The resulting effects are then termed *Fresnel* diffraction. A short account of the nature of Fresnel diffraction effects is given here. Methods of accounting for the intensity distribution in the fringes formed by Fresnel diffraction at edges are described in many books on optics. An account of these approaches is unnecessary for an understanding of the material presented in this book and is omitted here.

Light which leaves a point source situated in a medium of uniform refractive index travels outward in all directions, forming a spherical wavefront with a radius which increases constantly with time. Thus, as shown in Fig. 3-25, the wavefront of an apertured beam originating at S is, at some instant of time, $ABCD$. According to Huygens' principle, all points of this wavefront act as secondary sources of wave disturbance. A point P located anywhere to the right of S thus receives light which originates at *all* portions of the wavefront $ABCD$. Contributions to the amplitude at P differ in optical path, however, and therefore in phase.

Figure 3-26 shows a section taken through the three-dimensional wavefront illustrated in Fig. 3-25, including points S, M, N, and P. Light which reaches P from N has traveled some optical path x, while that from M has traveled through $(x + \Delta)$. The summation of the resulting contributions to amplitude at P, which might appear to be extremely complex, is facilitated by considering successive *zones* of the wavefront $ABCD$. As illustrated in Fig. 3-27, the wavefront may be divided into annuli, such that the average path from successive annuli to point P differs successively by half a wavelength. Similarly, the cylindrical wavefront emerging from a slit may be divided into rectangular zones, each of which also differs by half a wavelength in average path to point P, as shown by Fig. 3-28. Contributions from each zone differ in phase by π (or any other phase angle, depending upon the size of zone selected) and possess amplitudes which are proportional to the area of each zone. The contributions of these amplitudes at P can then be summed graphically by the method described on p. 39. This method can be extended to

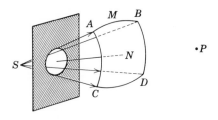

Fig. 3-25 Wavefront of an apertured beam, originating at a point source.

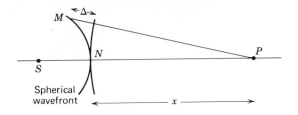

Fig. 3-26.

the subdivision of wavefronts into infinitesimally small zones, yielding a smooth curve from which amplitude at P can be obtained directly. In the case of cylindrical wavefronts, the smooth curve is known as *Cornu's spiral*.

Application of these methods to the problem of diffraction at a straight opaque edge shows that a series of bright and dark fringes is formed outside the geometrical shadow. The spacing of the fringes is approximately proportional to the square root of the wavelength of the illuminating light [cf. p. 410 especially (18-e)]. A slight intensity also appears just inside the geometrical shadow. A densitometer tracing of a typical set of *Fresnel fringes* is given in Fig. 3-29. The values of the spacings and intensities

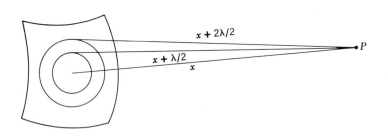

Fig. 3-27 Zones of a spherical wavefront.

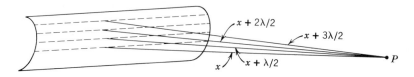

Fig. 3-28 Zones of a cylindrical wavefront.

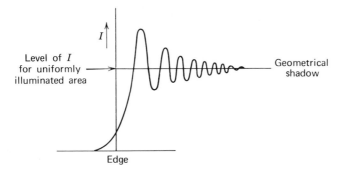

Level of I
for uniformly
illuminated area

Geometrical
shadow

Edge

Fig. 3-29 Densitometer tracing of Fresnel fringes.

of Fresnel fringes in images of slits, pinholes, etc., approach those obtained with parallel illumination (Fraunhofer fringes) as the source of illumination is moved away from the obstacle.

INTERFERENCE BETWEEN WAVE MOTIONS OF DIFFERENT FREQUENCIES

Only interference between wave motions of a single frequency (wavelength) has been considered thus far. What effects are observed when wave trains of different frequencies overlap? In the case of mechanical vibrations, complex waveforms can be observed. The resultant waveforms can be deduced from the principle of superposition, as in the overlapping of waves which differ only in amplitude and phase. In principle, analogous waveforms result from the superposition of electromagnetic radiations of different frequencies, but the resultant complex waveforms are not experimentally observable in this case. This is so because detectors, including the eye, "see" only the individual components of complex wave forms; resultant disturbances are analyzed by detectors into their individual sine wave components. For example, the energy of a light beam might be divided equally between radiation of, for example, 5500 Å and 8000 Å. A photocell would read, for the beam, the sum of the intensities of the separated wavelengths while the eye, which is insensitive to infrared radiation at 8000 Å, would perceive no difference between the beam and one containing *only* green light of 5500 Å at the same intensity. In this sense it may be said that *light beams of different frequencies do not interfere.* Nevertheless, the forms of the resultants of wave motions of different frequencies must be considered in discussing the phenomena of coherency and monochromaticity, and in accounting for the differences between the phase and group velocities of electromagnetic radiations.

COHERENCE AND MONOCHROMATICITY

As is discussed in Chapter 7, light beams originate as a series of finite pulses of radiation emitted by individual atoms. Such an understanding of the nature of the emission of light raises a fundamental question in considering diffraction effects: If light is indeed produced as a series of separate pulses by *independently* emitting atoms, how can a light beam, as a whole, be associated with a *regularly* varying phase? Clearly, if the phase of a beam cannot be defined, interference effects should be unobservable, since the properties of the macroscopic beam should be simply the resultant of the interference between pulses emitted with random phases by individual atoms. It is in fact so that, were atoms to emit altogether independently, the wavelike properties of light would be unobservable. Emission from individual atoms could be considered to possess wavelike character, but the individual wave trains would be of such brief duration and low intensity as to escape detection. That is, as stated in Chapter 1, wave properties of light are observed *only* in effects in which many photons participate.

In fact, interactions between adjacent atoms ensure that emission from atoms within limited regions of any source tends to be coupled. Wavelike properties (i.e., diffraction effects) can be observed when illumination is provided from any area of a light source sufficiently limited in size that emission occurs essentially in phase from all parts of the source. Light beams in which phase relationships are sufficiently constant for the demonstration of diffraction effects are said to be *coherent*, while beams which are incapable of revealing diffraction effects are *incoherent*.

When two incoherent beams overlap, a definite phase relation persists between the beams over an interval of the order of time occupied by the process of emission; that is, for about 10^{-8} sec. Subsequently a different (although again a definite) phase relation is substituted. There thus results a very rapidly changing succession of resultant intensities ranging from a minimum value (corresponding to a 180° phase angle between the beams) to a maximum (corresponding to zero phase angle). The *average* value of these intensities is perceived over any time interval long enough for direct observation of the resultant beam. Mathematically, it can be shown that the observed average intensity is simply the *sum* of the intensities of the component beams.

In the descriptions of diffraction experiments given earlier in this chapter it was stated that illumination was to be provided by using a slit to isolate light from a limited portion of a source. It should now be clear why this stipulation is made: illumination of diffracting slits directly from an extensive region of a conventional source (i.e., heated filament, discharge lamp, etc.) fails to produce a detectable diffraction pattern, since all portions of the diffracting slit(s) are then illuminated simultaneously by light from portions of the source so far separated as to emit with no fixed relationship of phases. The laser (described in Chapter 22) is a source which emits with constant phase over an extended area, and it may be used to demonstrate diffraction effects with considerable ease. For example, a sharp double slit diffraction pattern may be obtained simply by attaching a pair of slits directly to the emitting surface of a laser.

The terms lateral coherence and longitudinal coherence may be defined. *Lateral coherence* implies that a condition of equal phase persists across a wavefront and is thus implied in formulating the concept of a wavefront. *Longitudinal coherence* implies the persistence of definite phase relationships within a wave train over a period of time; that is, successive crests and troughs arrive uninterruptedly at a given point at intervals of constant length. The *coherence length* of a beam is defined as the greatest length, measured along the direction of propagation of a beam, over which such a definite relation of phases exists, while the *coherence time* is the time required for light to travel the coherence length.

The subject of monochromaticity has been introduced briefly on p. 18, where it was stated that a pulse of electromagnetic radiation which is of *finite* duration is equivalent to the superposition of sine waves of a *range* of frequencies and of *infinite* duration. Thus, insofar as a pulse of radiation is of limited duration, it is never truly monochromatic. The form of the experimentally realizable finite pulse may be deduced by applying the principle of superposition to the equivalent series of idealized sine waves.

At first sight it might seem that the coherency (ability to form diffraction patterns) and monochromaticity (degree to which only a single wavelength is present) of a light beam would be quite unrelated properties. Actually, these terms describe the same property from different points of view. Pulses emitted by a coherent source are, effectively, pulses of prolonged duration, since definite phase relationships must persist for a finite time interval, and are thus highly monochromatic.

THE INTERFEROMETER

The interferometer is an instrument in which the phase difference between two light beams is deduced by the observation of their interference pattern. These instruments are used for very accurate determinations of either the thickness or the refractive index of specimens or of the wavelength of light. Interferometric refractometry is described on p. 581; the general features of these instruments are discussed here.

The two light beams which form the interference pattern observed in the interferometer must be coherent (as in any diffraction experiment); that is, they must originate from a common source. Consequently, a "beam splitter" is an essential element of the instrument. The beam splitter may act by physically selecting two portions of an incident wavefront (i.e., by aperture splitting, as in the Rayleigh interferometer, shown in Fig. 26-7); by division of amplitudes at a partially reflecting ("half-silvered") plate (i.e., by amplitude splitting, as in the Michelson or Jamin interferometers, shown in Figs. 3-30 and 26-6, respectively); or by the diffraction of a portion of incident amplitude at an object (i.e., by diffraction splitting, as in the phase contrast microscope discussed in Chapter 13). In all cases the split beams travel along physically separate paths in the course of which a specimen of unknown optical path may be interposed, and they are reunited later by means of either lenses or mirrors. The fringe system formed by interference between the superposed beams is then observed, either directly or with the use of a lens system.

Many types of interferometer have been devised. One of them, the Michelson interferometer, is described here. Two other types, the Rayleigh and Jamin interferometers, are described in Chapter 26; the interference filter, discussed in Chapter 22, is an adaptation of the Fabry-Perot interferometer. Other interferometric systems are described in Chapter 14, which discusses the interference microscope.

A diagram of the Michelson interferometer is shown in Fig. 3-30. In

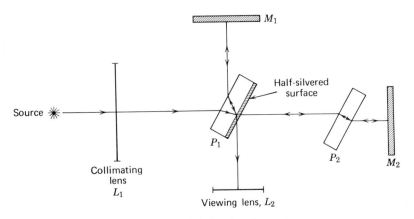

Fig. 3-30 The Michelson interferometer.

this instrument, light from a source is rendered parallel by a collimating lens, and is incident upon the half-silvered glass plate P_1 (i.e., the plate P_1 is lightly coated with reflecting material; hence the amplitudes reflected and transmitted at the second surface of the plate are equal). The reflected beam travels to the (fully silvered) mirror M_1, from which it is reflected back to P_1. The transmitted beam passes through the unsilvered glass plate P_2 to the (fully silvered) mirror M_2, from which it, also, is reflected back to P_1 through P_2. Half of the intensity from M_1 is transmitted from P_1 and recombines with the half of the intensity from M_2 which is reflected from P_1. The resulting interference pattern is observed by means of the lens L_2. If the positions of M_1 and M_2 are adjusted, and if the thicknesses of P_1 and P_2 are identical, the two superposed beams are *in* phase and thus form an interference pattern in which the zero order fringe is centrally located.

The interference pattern which is formed consists of a system of higher order fringes surrounding a central zero order fringe (as in the single and double slit diffraction patterns). While the zero order fringe formed by monochromatic light is indistinguishable from the higher order fringes, this central fringe may easily be located in a system which is illuminated by white light. Since the images of the source formed by *all* wavelengths superimpose *only* at the center of the interference pattern, the white light image consists of a central white fringe surrounded by overlapping series of (colored) higher order fringes which are formed by light of different wavelengths. Thus the zero order fringe of the monochromatic pattern can be identified by locating the position of the white fringe formed when the same instrument is illuminated by white light.

PHASE AND GROUP VELOCITIES

The distinction between phase and group velocities, which has been noted on p. 19, may be discussed at greater length here.

The individual sine wave components of any pulse of electromagnetic radiation travel with a phase velocity (v_ϕ) characteristic of each wavelength, while a pulse, as a whole, travels with a group velocity, (v_g), which, except for transmission through a vacuum, is different from v_ϕ. The origin of the group velocity may be understood as that of the resultant of the superposition of sine waves of different velocities.

Figure 3-31 shows portions of two wave trains, A and B, at some instant of time, $t = 0$. Wave A, *in the medium considered*, is of wavelength λ, with crests located at positions $1, \ldots, 6$. Wave B, in the same medium, is of wavelength λ' such that, when its crest $1'$ is coincident with crest

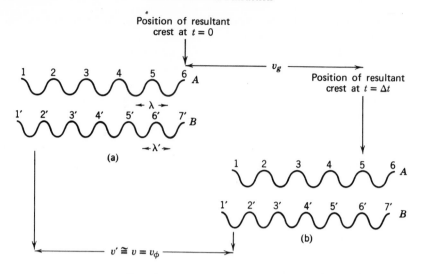

Fig. 3-31 Phase and group velocities.

1 of wave A, crest $7'$ of wave B is coincident with crest 6 of wave A. Therefore superposition of these two wave trains produces a resultant disturbance with maxima at positions 1-1' and 6-7'. Note that the resultant motion is not of simple sinusoidal form, since subsidiary maxima occur at intermediate positions.

If the two wave trains travel with identical velocities, the *relative* positions of crests and troughs in each do not change; hence the resultant waveform necessarily travels with the same velocity as the components; that is, the group and phase velocities are identical. This can be the case only for waves traveling through a vacuum; in any medium there is some dependence, however slight, of velocity upon wavelength.

Figure 3-31b shows the same wave trains after an interval of time Δt, during which train A travels with velocity v and train B with velocity $v' < v$ (as, in fact, electromagnetic waves of longer wavelengths generally travel through media with higher velocities). v and v' are considered to differ only slightly, so that both may be represented as the phase velocity v_ϕ. The value of Δt in Fig. 3-31b has been so chosen that crest 6 has "got ahead of" crest 7', while crest 5 has "caught up with" crest 6', so that a maximum of the resultant wave motion is now formed at the position where 5 and 6' coincide. The quantity t may be evaluated by noting that the distance which must be gained by wave A, relative to wave B, as crest 5 catches up with crest 6' is $(\lambda' - \lambda = \Delta\lambda)$. The corresponding rate of gain by wave A is given by the difference in velocities of the two waves; that is, $v' - v = \Delta v_\phi$. Thus

$$(3\text{-h*}) \qquad \Delta t = \frac{\Delta \lambda_\phi}{\Delta v_\phi}.$$

During the interval Δt, crest 6 (or any other crest of wave A) travels a distance $v\,\Delta t$, while each crest of wave B travels the very nearly identical distance, $v'\,\Delta t$. At the same time, the crest of the resultant wave travels that distance *minus* the separation of crests, that is, a distance $(v\,\Delta t - \lambda)$. Its velocity v_g is thus

$$(3\text{-i*}) \qquad v_g = \frac{v_\phi \Delta t - \lambda}{\Delta t} = v_\phi - \frac{\lambda}{\Delta t},$$

or, substituting from (3-h*),

$$(3\text{-j*}) \qquad v_g = v_\phi - \frac{\lambda\,\Delta v_\phi}{\Delta \lambda}.$$

For wave trains which differ only infinitesimally in wavelength, and thus in velocity, (3-j*) may be expressed as

$$(3\text{-k*}) \qquad v_g = v_\phi - \lambda \frac{dv_\phi}{d\lambda}.$$

It is important to remember that the quantity λ in (3-k*) and (3-j*) is the wavelength *in the medium of transmission* rather than the in vacuum wavelength, $\lambda_0 = c/v$.

Note that, whereas phase velocities may be deduced from the refractive indices corresponding to each wavelength, *direct* measurements of the speed of transmission of electromagnetic radiations yield the corresponding *group* velocities.

REFERENCES

General References

F. A. Jenkins and H. E. White, *Fundamentals of Optics* (3rd ed.), McGraw-Hill, New York, 1957. Various topics considered in the present chapter are discussed in the following chapters or pages: Chapter 11 (Light Waves), Chapter 12 (The Superposition of Waves), Chapter 13 (Interference of Two Beams of Light), pp. 232–239, 244–246, Chapter 14 (Interference Involving Multiple Reflections), pp. 261–264, 270–271, Chapter 15 (Fraunhofer Diffraction by a Single Opening), pp. 288–297, 302–303, Chapter 16 (The Double Slit), pp. 311–322, Chapter 17 (The Diffraction Grating), pp. 328–335, and Chapter 18 (Fresnel Diffraction).

The works by J. Strong, *Concepts of Classical Optics*, Freeman, San Francisco, 1958, by J. M. Stone, *Radiation and Optics*, McGraw-Hill, New York, 1963, and by M. Born and E. Wolf, *Principles of Optics*, Macmillan, New York, 1964, are largely devoted to the discussion of interference and diffraction phenomena.

M. Francon, *Diffraction-Coherence in Optics*, Pergamon Press, Oxford, 1966. Some specific applications to optical instrumentation are considered, and an up-to-date bibliography of work in wave optics is given.

P. Baumeister, *Military Standardization Handbook*, Chapter 20 (Optical Design), 5 Oct. 1962.

Specific References

[1] M. Born and E. Wolf, *op. cit.*, pp. 395–398. Derivation of the mathematical description of the Airy disc.
[2] I. S. Sokolnikoff and R. M. Redheffer, *Mathematics of Physics and Modern Engineering* McGraw Hill, New York, 1958. pp. 159–165.

CHAPTER 4

Polarized Light Beams

Light beams are characterized by their frequency, amplitude, phase, and state of polarization, characteristics described in Chapter 1. Variations in the properties of light as a function of frequency (wavelength) were described in Chapter 2, and Chapter 3 dealt with phenomena which are observed when two or more light beams differ in phase. In considering those aspects of the behavior of light, the radiation in many cases may be assumed to be *unpolarized*, so that considerations related to the state of polarization of the beams could be ignored.

When light impinges on matter, the electric vector of the radiation interacts with the electrons of the medium; the relative orientations of the two, and, thus, the state of polarization of the light, become considerations of importance. Note that since light waves are transverse vibrations, they can vibrate in various planes at right angles to the direction of propagation. In this respect they differ notably from longitudinal waves, which must invariably vibrate in the single plane which is parallel to the direction of propagation. The interactions of polarized light with matter are described in Chapter 6; in this chapter the properties of polarized light beams themselves are considered. An introductory definition of states of polarization has already been given on pp. 15–17.

POLARIZATION

Although individual light quanta may be considered to be associated with an equivalent electric moment, polarization phenomena are most simply considered in terms of the wave picture of the nature of light. As discussed in Chapter 1, a plane-polarized beam is one for which the electric vector associated with the light wave vibrates in (any) single direction in the plane perpendicular to the direction of propagation of the light. The *plane of polarization* is defined as that which contains the azimuth of vibration of the electric vector of the light wave and its direction of propagation. It is important to be aware that, when polarization phenomena are considered, the phrase "the direction of a light beam" may be ambiguous. Directions should be specified either as those of *propagation* (transmission) or of *vibration* (polarization). Conventionally, light beams are considered to propagate in the $+z$ direction and thus to vibrate in azimuths in the xy plane. Plane-polarized light is represented in Figs. 4-1a and 4-1b, in which the direction of propagation is that perpendicular to the plane of the page.

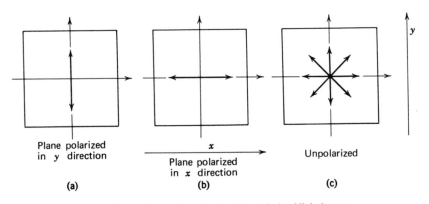

Plane polarized in y direction (a)

Plane polarized in x direction (b)

Unpolarized (c)

Fig. 4-1 Projections of polarized and unpolarized light beams.

An *unpolarized* light beam is associated with an electric vector which vibrates in all possible directions in the plane perpendicular to the direction of propagation, as represented schematically in Fig. 4-1c. The nature of unpolarized light is discussed more fully on p. 89. An additional possibility is that of *circular* or *elliptical* polarization; the nature of circularly and elliptically polarized beams is explained on pp. 81–87.

As shown in Chapter 1 (Fig. 1-13), a light beam which is plane polarized in any arbitrary direction may be regarded as the resultant of two wave motions which are plane polarized along any two mutually perpendicular reference axes, x and y. If θ is the angle between the x axis and the plane of polarization of a wave of amplitude A, then the amplitudes A_x and A_y of the components of vibration in the x and y directions, respectively, are

(4-a)
$$A_x = A \cos \theta,$$
$$A_y = A \sin \theta.$$

TRANSMISSION OF PLANE-POLARIZED LIGHT

The significance of the statement that a device "transmits light which is plane polarized in a given azimuth" should be clear if the nature of vibratory motions is understood. The statement does *not* mean that the light will not be transmitted if it is not initially plane polarized in the specified azimuth. It means, rather, that the device transmits all components of vibration which can be resolved into the given azimuth. Consider, for example, the incidence of four vibratory motions polarized in the azimuths shown in Figs. 4-2a through 4-2d on a device which completely transmits light which is plane polarized in the azimuth of Fig. 4-2a. As shown in Figs. 4-2b′ and 4-2c′, respectively, the motions represented by Figs. 4-2b and 4-2c can be partially resolved as vibrations

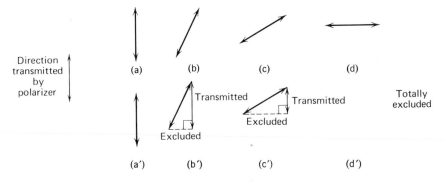

Fig. 4-2 Transmission by an analyzer.

in the plane of Fig. 4-2a; thus *part* of the amplitude of each is transmitted. Only the vibration represented by Fig. 4-2d, which has *no* component of motion in the direction of Fig. 4-2a, is completely excluded by the device.

APPLICABILITY OF THE TERM "INTERFERENCE OF POLARIZED LIGHT"

Although discussions of the interference of polarized light may be found in some books, statements also occur in the literature to the effect that "light rays cannot interfere unless they are polarized in the same plane." The latter statement is quite correct in an operational sense; the observed intensity of two superposed light beams which are polarized in different planes is simply the sum of the individual intensities. That is, *no* variations in total intensity occur as a result of the interaction of the polarized beams.

Polarized Light Beams

It is possible, however, to diagram the *form* of the resultant wave motion which is produced when the waves overlap, and it is to considerations of this sort that the phrase "interference of polarized light beams" can be applied. While the resultant modes of vibration cannot be observed directly, they provide a convenient pictorial representation of the associated light beam. For example, (as discussed further below), the phrase "a circularly polarized beam of light" signifies "a beam in which the component overlapping wave trains are of such amplitudes, phases, and azimuths of polarization as to produce a resultant circular motion when the combined waveform is projected in the plane perpendicular to the direction of propagation."

In order to observe, experimentally, interference between polarized light beams (i.e., in the sense of observing variations in resultant intensity) it is necessary to interpose, physically, some device which brings part of all component beams into the same azimuth of polarization.

EQUATION FOR THE FORM OF A RESULTANT BEAM AS PROJECTED IN THE *xy* PLANE

The resultant of two superposed plane polarized wave motions may be characterized by an equation which describes its projection in the *xy* plane, that is, in the plane perpendicular to the direction of propagation. Wave equations for the component motions, polarized in the *yz* and *xz* planes, respectively, are

(4-b)
$$y = A_y \sin(\omega t - \alpha_y),$$
$$x = A_x \sin(\omega t - \alpha_x).$$

These expressions may be expanded with the use of the trigonometric identity: $\sin(A - B) \equiv \sin A \cos B - \cos A \sin B$, giving

(4-c)
$$\frac{y}{A_y} = \sin \omega t \cos \alpha_y - \cos \omega t \sin \alpha_y,$$

(4-d)
$$\frac{x}{A_x} = \sin \omega t \cos \alpha_x - \cos \omega t \sin \alpha_x.$$

Multiplication of (4-c) by $(\sin \alpha_x)$ and of (4-d) by $(\sin \alpha_y)$, with subsequent subtraction of (4-c) from (4-d) yields, after application of the identity just cited,

(4-e)
$$\sin \omega t = \frac{(x \sin \alpha_y / A_x) - (y \sin \alpha_x / A_y)}{\sin(\alpha_y - \alpha_x)}.$$

Similarly, multiplication of (4-c) by $(\cos \alpha_x)$ and of (4-d) by $(\cos \alpha_y)$ followed by subtraction of (4-d) from (4-c) and rearrangement, yields

(4-f)
$$\cos \omega t = \frac{(y \cos \alpha_x / A_y) - (x \cos \alpha_y / A_x)}{\sin (\alpha_x - \alpha_y)}.$$

By making use of the trigonometric identity $\sin (A - B) \equiv -\sin (B - A)$, (4-f) can be rewritten

(4-g)
$$\cos \omega t = \frac{(x \cos \alpha_y / A_x) - (y \cos \alpha_x / A_y)}{\sin (\alpha_y - \alpha_x)}.$$

Using the identity $\sin^2 + \cos^2 \equiv 1$, squaring, and adding (4-e) and (4-g) leads to

(4-h)
$$\sin^2 (\alpha_y - \alpha_x) = \frac{x^2}{A_x^2} + \frac{y^2}{A_y^2} - \frac{2xy}{A_x A_y} (\sin \alpha_y \sin \alpha_x)$$

$$+ (\cos \alpha_x \cos \alpha_y).$$

Finally, use the identity $\cos (A - B) \equiv \sin A \sin B + \cos A \cos B$ leads to the more compact expression

(4-i)
$$\boxed{\sin^2 (\alpha_y - \alpha_x) = \frac{x^2}{A_x^2} + \frac{y^2}{A_y^2} - \frac{2xy}{A_x A_y} \cos (\alpha_y - \alpha_x).}$$

For the overlapping of mutually perpendicularly polarized beams that are of amplitudes A_x and A_y, and which differ in phase angle by $(\alpha_y - \alpha_x)$, the *form* of the resultant vibratory motion can be deduced from (4-i). Certain examples are considered below.

PLANE-POLARIZED LIGHT

If the resultant of two overlapping wave trains is to be a plane-polarized motion, (4-i) must reduce to the equation of a straight line. Some examples follow.

Consider the interference of two wave motions such that $A_x = A_y = A$, and which are in phase with each other, that is, for which $(\alpha_y - \alpha_x) = 0$. Equation 4-i then becomes

(4-j)
$$0 = \frac{x^2}{A^2} + \frac{y^2}{A^2} - \frac{2xy}{A^2} = x^2 - 2xy + y^2,$$

which reduces to

(4-k)
$$x = y.$$

This solution is the equation of a straight line, inclined at 45° to the x and y axes, as illustrated by Fig. 4-3. That is, the interference between two in-phase, plane-polarized beams of equal amplitudes produces a resultant beam which is also plane polarized. The resultant beam is in phase with the component motions.

A second example is that of interference between two in-phase plane-polarized vibrations of *unequal* amplitudes. If, for example, $A_x = 2A_y$, then (4-i) becomes

(4-l)
$$0 = \frac{y^2}{A_y^2} + \frac{x^2}{4A_y^2} - \frac{xy}{A_y^2} = 4y^2 + x^2 - 4xy = (2y - x)^2,$$

which reduces to

(4-m)
$$y = \frac{x}{2}.$$

This solution is also the equation for a straight line, but one which is inclined at 26°34′ to the x axis, as illustrated by Fig. 4-4.

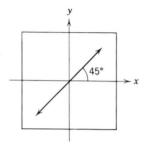

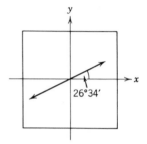

Fig. **4-3** Resultant of two in-phase plane polarized motions of equal amplitudes.

Fig. **4-4** Resultant of two in-phase plane polarized motions of unequal amplitudes.

A third example in which a plane-polarized vibration results is that in which the two component vibrations are not in phase, but are exactly *out* of phase; that is, $(\alpha_x - \alpha_y) = 180°$. If, as in the second example considered, $A_x = 2A_y$, then (4-i) becomes

(4-n)
$$0 = \frac{y^2}{4A_y^2} + \frac{x^2}{4A_y^2} + \frac{xy}{A_y^2} = x^2 + 4xy + 4y^2,$$

which reduces to

(4-o)
$$y = -\frac{x}{2}.$$

The resultant azimuth of polarization, shown in Fig. 4-5, again makes a 26°34′ angle with the x axis, but lies in the pair of quadrants opposite to those of the previous example.

CIRCULARLY POLARIZED LIGHT

In the preceding section, only those special cases of overlapping polarized motions have been considered for which the left side of (4-i) is zero More generally, *any* phase relation may exist between component vibrations, and a plane-polarized resultant motion is not obtained.

Consider, for example, the interference of two beams for which $A_x = A_y = A$, but $(\alpha_y - \alpha_x) = 90°$. Equation 4-i then becomes

(4-p) $\qquad 1 = \dfrac{y^2}{A^2} + \dfrac{x^2}{A^2}$ or $y^2 + x^2 = A^2$.

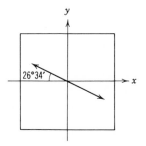

Fig. 4-5 Resultant of two plane polarized motions of opposite phase and unequal amplitudes.

Equation 4-p will be recognized as that of a circle of radius A, as sketched in Fig. 4-6. That is, the electric vector of the resultant wave motion, as viewed along the direction of propagation, traces out a *circular* path.

There exist two possibilities for the nature of the circular path illustrated by Fig. 4-6. The electric vector may trace out the circle in a clockwise or counterclockwise sense. Clockwise rotation (i.e., as seen in viewing the beam while looking toward its source) is produced when the component beam polarized in the y direction is 90° *ahead* of the component polarized in the x direction. This condition is designated right-hand circular polarization. When the y component of motion *lags* the x component by 90°, left-hand circular polarization results.

In order to understand how these rotations occur in the stated senses, the resultant vector can be considered at a number of stages during one-half cycle of a right-hand circularly polarized vibration. This is shown in Figs. 4-7 through 4-11. Each figure shows the motions traced out in the xy plane by a pair of vectors of equal amplitudes, A, and which vibrate in the x and y directions respectively. The tip of the resultant vector is designated thus: ⊛.

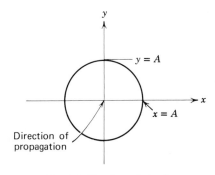

Fig. 4-6 Circular polarization.

In Fig. 4-7 the phase angle of the x vector is 0°, and that of the y vector +90°. The tip of the resultant vector is at $+A$ on the y axis. In Fig. 4-8 the phase angles of the x and y vectors are +45° and +135°, respectively; the tip of the resultant vector is at a position $+A$ along an axis tilted at 45° with respect to the x and y axes. Similarly, Fig. 4-9 shows the x and y vectors with phases of +180° and +90°, while phases of +225° and +135° appear in Fig. 4-10, and +270° and +180° in Fig. 4-11. The tip of the resultant vector is rotating in a *clockwise* direction during this sequence; the successive positions are shown in Fig. 4-12. A comparable sequence of vector diagrams, given in Fig. 4-13, shows the interference of two plane-polarized beams to produce a resultant *counterclockwise* motion.

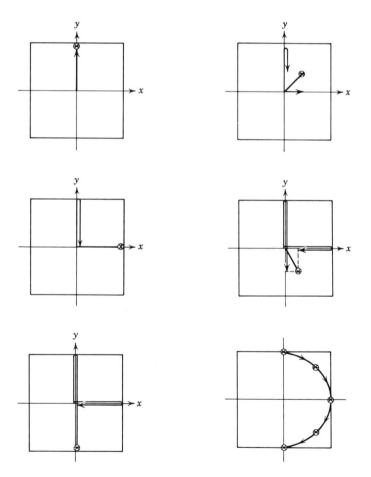

Fig. 4-7 to 4-12 Interference to form a right hand circularly polarized motion.

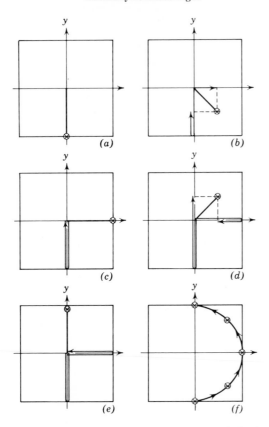

Fig. 4-13 Interference to form a left hand circularly polarized motion.

As viewed in three dimensions, the path of the tip of the electric vector of a circularly polarized motion is a *helix* of circular cross section. This is illustrated in Fig. 4-14.

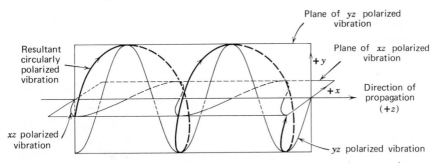

Fig. 4-14 Circularly polarized resultant and component motions, as seen in perspective.

ELLIPTICALLY POLARIZED LIGHT

If the phase angle between two plane-polarized beams is 90° but the beams differ in amplitude, or, in the most general case, if the polarized beams differ in *both* phase and amplitude, an *elliptically* polarized motion results. That is, the projection (in the plane perpendicular to the direction of propagation) of the path of the electric vector of the resultant motion is an ellipse. Any degree of ellipticity of the figure and all orientations of the ellipse may be obtained.

Consider, for example, the case in which $(\alpha_y - \alpha_x) = 90°$ and $A_x = 2A_y$. Equation 4-i then becomes

(4-q) $$1 = \frac{y^2}{A_y^2} + \frac{x^2}{4A_y^2} \quad \text{or} \quad 4A_y^2 = 4y^2 + x^2.$$

Equation 4-q describes an ellipse the major axis of which is coincident with the x axis, as shown by Fig. 4-15. In three-dimensional view the course of the tip of the electric vector would be similar to that shown for a circularly polarized motion in Fig. 4-14, but with compression of the helix in the y direction, as illustrated in Fig. 4-16. Elliptical motion, like circular motion, may occur either in the right- or left-handed sense.

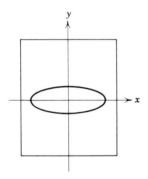

Fig. 4-15 Elliptical polarization.

In general, the "interference" of plane-polarized motions produces a resultant elliptical motion in the xy plane. Plane-polarized light, which is formed by the interference of beams which are exactly in or exactly out of phase, is a special case of the ellipse in which the length of the minor axis is zero. Likewise, circularly polarized light, which is formed by the interference of components of equal amplitude which are 90° out of phase, is another special case, in which the major and minor axes of the projected ellipse are of equal lengths.

It should be emphasized that circularly polarized light and elliptically polarized light are not phenomena which can be observed directly. As stated on p. 79, the individual plane-polarized components of such beams do not interfere in the operational sense of producing variations in intensity. Thus the intensity of an elliptically polarized light beam is simply the sum of the intensities of its components. The term "circularly polarized light" is simply a convenient way of specifying a light beam which is the combination of two beams of equal intensity and which are plane

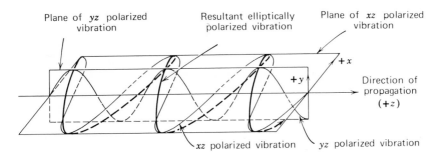

Plane of *yz* polarized vibration

Resultant elliptically polarized vibration

Plane of *xz* polarized vibration

+*x*

+*y*

Direction of propagation (+*z*)

xz polarized vibration

yz polarized vibration

Fig. 4-16 Elliptically polarized resultant and component motions, as seen in perspective.

polarized in mutually perpendicular directions and 90° out of phase with each other. Likewise, "elliptically polarized light" denotes a light beam which is the resultant of two beams plane polarized in mutually perpendicular directions, the amplitudes and phases of which are in an arbitrary but definite relation to each other.

The distinction between circularly polarized light and *unpolarized* light is sometimes found confusing; these two types of beam are in fact distinctly different. In terms of the wave picture of light, unpolarized motions are represented by Figs. 1-7 and 4-1c, and circularly polarized motions are comparably represented by Figs. 4-14 and 4-6. However, since the form of wave motions cannot be directly observed, the distinction between circularly polarized and unpolarized light must also be subject to experimental test. This can be done with the use of a quarter wave plate as described on p. 114.

Note, also, that the phenomenon of elliptically polarized light bears no special relation to the "index ellipse" and "index ellipsoid" which are used to characterize birefringent materials in the manner described in Chapters 6 and 15.

RELATION OF PLANE POLARIZED AND CIRCULARLY POLARIZED WAVE MOTIONS

A consideration of the interference of waveforms shows that, just as circularly or elliptically polarized beams are the resultant of a pair of overlapping plane-polarized light beams, so also *a plane-polarized motion may be regarded as the resultant of two circularly polarized motions of equal amplitude and of opposite senses.* The azimuth of the plane-polarized motion is determined by the position of equal phase of the circular motions.

This equivalence may be demonstrated by considering the instantaneous orientations of the electric vector associated with a pair of LH and RH circular motions. The circular vibrations are represented in Fig. 4-17 at an instant in which the two motions are in phase; one *quarter* cycle later, as shown in Fig. 4-18, the circular motions are 180° out of phase with each other. Figure 4-19 shows the positions traced out by the tips of the electric vectors after completion, successively, of $\frac{1}{8}$, $\frac{1}{4}$, $\frac{3}{8}$, and $\frac{1}{2}$ cycle of vibration from the initial in-phase position shown in Fig. 4-17. The resultant electric vector associated with the overlapping pair of circular vibrations can be found by addition. The resultants, plotted for 0 to $\frac{1}{2}$

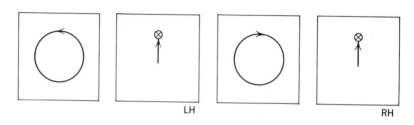

Fig. 4-17.

cycle, are shown in Fig. 4-19b. As illustrated, the *tip* of the resultant vector traces out a *linear* path in the *xy* plane and thus represents a plane-polarized motion.

Another example of overlapping pairs of circularly polarized motions is illustrated in Fig. 4-20, in which the orientation of the in-phase position differs from that chosen for Fig. 4-19. The azimuth of the resultant plane-polarized motion is correspondingly altered.

Two circularly polarized motions of *equal* amplitude and opposite sense interfere to form a plane-polarized motion, but two circularly polarized motions of *unequal* amplitudes and opposite sense interfere to form an elliptically polarized motion. The *ellipticity θ* of the resultant motion

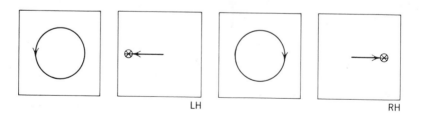

Fig. 4-18.

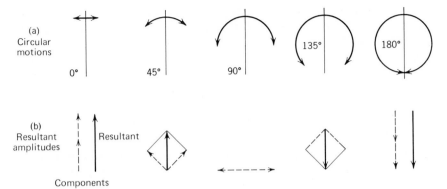

Fig. 4-19 Interference of circularly polarized motions to form a plane polarized motion.

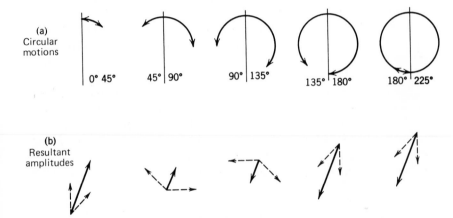

Fig. 4-20 Interference of circularly polarized motions to form a plane polarized motion.

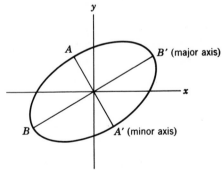

Fig. 4-21.

is defined as the angle whose tangent is the ratio of the minor to the major axis of the ellipse. For example, the ellipticity of the motion illustrated in Fig. 4-21 is given by

$$(4\text{-r}) \qquad\qquad \tan \theta = \frac{AA'}{BB'}.$$

THE NATURE OF UNPOLARIZED LIGHT

For some purposes unpolarized light may be considered the resultant of vibrations in any two (arbitrarily selected) mutually perpendicular directions. This is an oversimplified picture, however. Should any definite phase angle *persist* between the two components, a plane, circularly, or elliptically polarized beam, not an unpolarized beam, would result. In fact, unpolarized light must be regarded as the resultant of a pair of mutually perpendicularly polarized beams which *vary* continuously in *both* amplitude and relative phase. An equivalent way of describing unpolarized light is as an elliptically polarized motion of continuously varying ellipticity.

If the reader is aware of these correct descriptions of unpolarized light, confusion concerning the nature of polarized motions can be avoided. For the applications considered in this book, however, it is sufficient to think of unpolarized light as being resolvable as a pair of mutually perpendicularly polarized motions. Over any finite period of time the average amplitudes of the resolved motions are equal.

REFERENCES

F. A. Jenkins and H. E. White, *Fundamentals of Optics* (3rd ed.), McGraw-Hill, New York, 1957, pp. 226–230 (especially fig. 12-K), Chapter 27 (Interference of Polarized Light), and pp. 489–490.

H. S. Bennett, in R. M. Jones, Ed., *McClung's Handbook of Microscopical Technique*, 3rd ed., Hafner, New York, 1950, pp. 595–597, 615–621.

J. Strong, *Concepts of Classical Optics*, Freeman, San Francisco, 1958, pp. 27–28.

CHAPTER 5

The Production of Plane-Polarized Light

The production and detection of plane polarized beams of light are described in this chapter. Four types of interaction exist which can be exploited in the production of plane polarization. They are: scattering, reflection, dichroism, and birefringence. Of the four, only dichroism and birefringence are of importance in biological instrumentation.

A convenient notation for the representation of azimuths of polarization of light beams is summarized in Fig. 5-1.

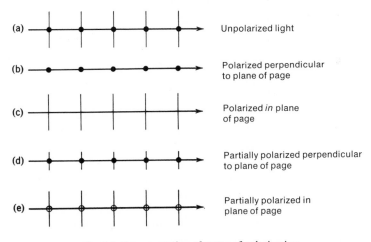

(a)	Unpolarized light
(b)	Polarized perpendicular to plane of page
(c)	Polarized *in* plane of page
(d)	Partially polarized perpendicular to plane of page
(e)	Partially polarized in plane of page

Fig. 5-1 Representation of states of polarization.

POLARIZATION BY SCATTERING

The light scattered from collections of small particles, such as macro-molecules in solution, is partially plane polarized. This effect is utilized as a means of studying the properties of the scatterers, rather than as a source of polarized beams.

POLARIZATION BY REFLECTION

Light reflected from the surface of a dielectric medium (i.e., a material which is electrically nonconducting, e.g., glass) is partially polarized to a degree dependent on the angle of incidence. At a specific angle of inci-dence, known as *Brewster's angle*, the components of the incident light which are reflected and refracted at the surface differ in direction by exactly 90°. Plane polarization of the *reflected* beam is then complete. The value of Brewster's angle is specified by

(5-a) $\tan \angle i = n$

where n is the refractive index of the reflecting medium.

In accounting for the production of a plane polarized reflected beam upon incidence at Brewster's angle, it may be noted that, since light is a transverse wave motion, it can have *no* longitudinal component of vibra-tion. Thus the direction of vibration *in* the plane of incidence for the trans-mitted beam, cannot occur in the reflected beam when the two are 90° apart. Accordingly, the beam reflected at Brewster's angle is completely polarized in a plane perpendicular to that of incidence. Reflection at Brewster's angle is diagramed (according to the convention defined in Fig. 5-1) in Fig. 5-2. Since the amplitude reflected is normally less than that transmitted, the refracted beam remains only *partly* polarized in the plane of incidence.

Reflection and refraction at an angle other than Brewster's angle are represented in Fig. 5-3. *Only* those components of vibration of the refrac-ted beam which are perpendicular to the direction of the reflected beam may contribute to the latter. Since both reflected and refracted beams lie in the plane of incidence (which is the plane of the page in Fig. 5-2), the component of vibration *perpendicular* to the plane of incidence can be included in the reflected beam without loss of amplitude. The component of vibration of the refracted beam *in* the plane of incidence, however, can only partially be resolved into a component of vibration perpendicular to the direction of propagation of the reflected beam. This resolution is

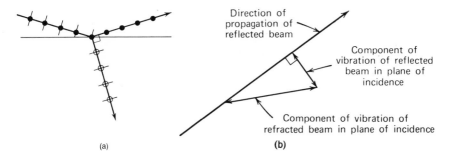

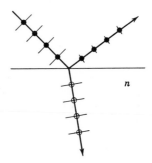

Fig. 5-2 Reflection and refraction of unpolarized light.

shown by Fig. 5-2b. Thus, in general, the reflected beam is partially polarized in an azimuth which is *perpendicular* to the plane of incidence.

Correspondingly, to the extent that amplitude polarized perpendicular to the plane of incidence is included in the reflected beam, the remaining amplitude of the refracted beam is *partially* polarized *in* the plane of incidence.

Because of the generally low intensity of the reflected beam, the reflection of light at Brewster's angle is rarely an efficient method for producing plane polarized light. However, it is a convenient way of obtaining or detecting polarized light when specialized equipment is unavailable. In particular, a pile of plane glass plates illuminated at Brewster's angle produces a series of

Fig. 5-3 Reflection at Brewsters angle.

reflected beams, all of which are polarized in the same azimuth, and which collectively may produce a moderate level of intensity.

POLARIZERS AND ANALYZERS

When light which is already plane polarized is incident upon a surface at Brewster's angle, the reflected intensity varies between zero and some maximum value, according to the azimuth of polarization of the incident beam. As illustrated by Figs. 5-4a, 5-4b, and 5-4c, respectively, incident light which is plane polarized *in* the plane of incidence produces zero reflected intensity, light polarized perpendicular to the plane of incidence produces maximum reflected intensity, and light polarized in intermediate azimuths produces intermediate levels of intensity.

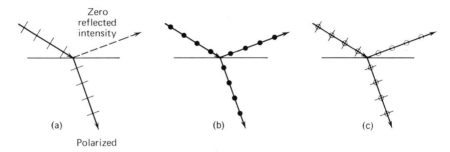

Zero reflected intensity

Polarized

(a)   (b)   (c)

Fig. 5-4 Reflection of plane polarized light at Brewster's angle.

A reflecting interface which is used in this way serves as an *analyzer*; that is, it is a device which reveals the state of polarization of an incident beam by producing a variation in intensity. Had unpolarized light been incident upon the same surface, also at Brewster's angle, the surface would have acted as a polarizer. "Polarizer" and "analyzer" are thus terms which refer to the *use* of polarizing devices to produce or to reveal polarization, respectively. Any pair of polarizing devices, including not only reflecting plates but also the "polaroids" and prisms to be described below, can thus be used as polarizer and analyzer.

As is shown in Chapter 15, the intensity of polarized light passed by an analyzer is given by the *law of Malus*, which states that

$$(5\text{-b}) \qquad\qquad I = I_0 \cos^2\theta,$$

where I_0 is the incident intensity, I the transmitted intensity, and θ the angle between the azimuth of polarization of the incident light and that completely transmitted by the polarizer. (Or, since all polarizers are in fact partially absorbing, I_0 may represent the intensity transmitted at $\theta = 0°$, and I may represent that transmitted at any other value of θ.) For "parallel polarizers," $\theta = 0°$, $\cos^2\theta = 1$, and $I = I_0$; that is, light passed by the polarizer is completely transmitted by the analyzer. For "crossed polarizers", $\theta = 90°$, $\cos^2\theta = 0$, and $I = 0$; that is, light passed by the polarizer is completely excluded by the analyzer.

VECTORIAL ANALYSIS OF POLARIZERS AND ANALYZERS

The roles of polarizer and analyzer in the transmission of intensity can be described by vector diagrams such as Fig. 5-5. Light leaving a polarizer vibrates in an azimuth such as that indicated by Fig. 5-5a. According to the treatment presented on pp. 81–82, however, this polarized beam can also be regarded as the resultant of two in-phase polarized beams

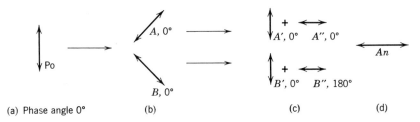

Fig. 5-5 Effect of crossed polarizer and analyzer.

which vibrate in mutually perpendicular azimuths, for example, beams A and B, as shown in Fig. 5-5b. A and B, in turn, can each likewise be resolved as a pair of mutually perpendicularly polarized motions, A', A'' and B', B'', respectively, as shown in Fig. 5-5c. Since A and B lie in opposite quadrants, the components of one of these beams are in phase with each other, while the components of the other differ in phase by 180° (cf. p. 82).

If the analyzer is *crossed* with respect to the polarizer, as shown by Fig. 5-5d, the two components, A' and B', are excluded. Components A'' and B'', vibrating in the transmission azimuth of the analyzer, are *equal in amplitude* and *differ in phase by 180°*. Thus they cancel by destructive interference; the net effect is that *no* intensity is passed by the analyzer.

A similar treatment may be applied when polarizer and analyzer are not exactly crossed. In Fig. 5-6, for example, the transmission azimuths of polarizer and analyzer, as shown in (a) and (d) of the figure, make an angle of 67°30′. Light from the polarizer may be resolved, as shown by Fig. 5-6b, into a component of motion, A, which makes a 45° angle with the transmission angle of the analyzer and a 22°30′ angle with that of the polarizer, and a second component, B, of *smaller amplitude*, which vibrates in the azimuth perpendicular to that of component A. As shown in Fig. 5-6c, A and B can be resolved, as in the previous example, into four components of motion, A', A'', B', and B''. A' and B' vibrate in a direction perpendicular to that transmitted by the analyzer, by which

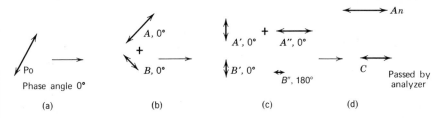

Fig. 5-6 Effect of polarizer and analyzer at arbitrary mutual orientation.

they are therefore excluded. A'' and B'', vibrating *in* the transmission azimuth of the analyzer, are again 180° out of phase with each other. Since they are of *unequal* amplitudes, however, there remains a resultant amplitude C which is transmitted by the analyzer, as shown in Fig. 5-6d.

For most purposes the foregoing treatment is an unnecessarily round-about way of describing the action of polarizers and analyzers. It serves, however, to establish a fact which is important in explaining certain effects obtained in polarization microscopy (cf. p. 324), namely, that *components of the light which is passed by the analyzer have undergone a mutual phase shift of 180°* $(\Pi, \frac{1}{2}\lambda)$.

DICHROISM; POLARIZATION BY SELECTIVE ABSORPTION

Certain crystals have the property of absorbing, preferentially, light which is polarized in a particular azimuth, while components of vibration which are polarized at right angles thereto are transmitted. Such crystals are said to be *dichroic*. (Relatively few crystals, however, exhibit dichroism.)

When unpolarized light enters a dichroic medium, the component of vibration polarized in the preferred direction is substantially transmitted, while the component polarized at right angles thereto is partially removed by absorption. After passage through an appreciable thickness of the dichroic material, the emergent light is substantially plane polarized. (Further discussion of dichroic absorption is given on p. 145.)

Dichroism is exploited in so-called *polaroids*. These are sheets of transparent plastic containing very many dichroic crystallites. The plastic sheets are formed under stress, which results in a parallel alignment of the crystallites. Consequently, the material as a whole is dichroic.

Polaroids can provide an inexpensive means of producing and analyzing plane polarized light and are satisfactory for use in many types of optical instruments. The limitations of dichroic polarizers, however, are such that birefringent polarizing prisms, to be described below, are usually required when the highest sensitivity is sought.

The shortcomings of dichroic polarizers—reduced intensity of the transmitted beam, limited spectral range, and, in certain wavelength regions, incomplete polarization—arise from the absorptive nature of their action. The azimuth of polarization removed by the polarizer is successively absorbed from the beam as it passes through the dichroic material; at least in principle, absorption is never absolutely complete. Furthermore, no substance is perfectly dichroic, but light polarized in the azimuth at right angles to that preferentially absorbed is itself absorbed to some degree, which may be appreciable. The intensity transmitted by the polarizer thus may be reduced to an undesirable extent (i.e., in excess of the 50% reduction in intensity which must inevitably accompany polarization of an initially unpolarized beam). Finally, the properties of dichroic polarizers are dependent upon the availability of suitable dichroic materials. Many substances display a high degree of dichroism for light in a limited wavelength range, but have little preferential absorption, or are nonabsorbing in other spectral regions. A particular advantage of dichroic polarizers is that they can be wide aperture systems. In contrast, birefringent prisms effect complete polarization with very little reduction in intensity of the transmitted component, and can be used over a wide range of wavelengths.

The term "circular dichroism" refers to the difference in the absorption of left-hand and right-hand circularly polarized light observed in certain substances. This phenomenon, which is exploited in studying the structure of the absorber, is discussed in Chapter 27.

BIREFRINGENT POLARIZING PRISMS

The property of birefringence may be exploited in the production of plane polarized beams of light. Birefringence is the condition of possessing *different refractive indices* with respect to light which is polarized in different azimuths. This phenomenon is described at length in Chapter 6.

Light entering a birefringent medium is resolved into two mutually perpendicular components which, because of the differences in refractive indices for the two, travel along *physically divergent paths*. (Certain specific directions of travel through the medium are exceptions; see pp. 106 ff.) Even at normal incidence, one of the beams, termed the extraordinary ray (*E* ray), diverges from its initial course, while the other, termed the ordinary ray (*O* ray), behaves according to the law of refraction at all angles of incidence. The physical separation of the mutually perpendicularly polarized *O* and *E* rays is exploited in several different types of polarizing prisms, which remove one of these components from the on-going beam.

The best known type of polarizing prism is the *Nichol prism*, which is diagramed in Fig. 5-7. This prism consists of a block of the highly birefringent crystalline mineral, calcite (a form of calcium carbonate), which is cut in two and rejoined by a layer of the transparent cement, Canada balsam. The index of refraction of calcite is approximately 1.49 for the *E* ray and 1.66 for the *O* ray; that of Canada balsam is 1.55 for both rays. As shown in Fig. 5-7, the *O* ray is incident upon the Canada balsam layer, which is an optically less dense medium with respect to that ray, at an angle greater than the critical angle, and is thus removed from the on-

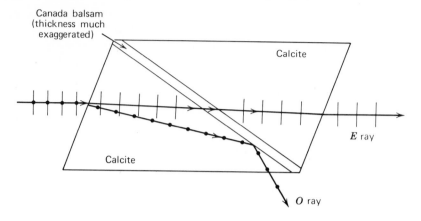

Fig. 5-7 The Nichol prism.

going beam by total internal reflection. (The phenomenon of total internal reflection has been described on p. 6.) The E ray is refracted as it enters the Canada balsam (which is an optically more dense medium with respect to this ray), and is incident upon the dense-to-rare Canada balsam-calcite interface at an angle which is less than the critical angle. The on-going beam from the prism thus consists of the plane polarized E ray. (As is made clear in Chapter 6, the "extraordinary" character of the E ray exists only with respect to its passage through calcite. In air, the separate O and E rays differ only in their planes of polarization.)

A properly constructed polarizing prism, in which stray reflections are avoided, produces a completely plane polarized beam. As noted above, reduction of intensity by absorption is minimal, while polarization can be effected at all wavelengths which are transmitted by the calcite. For these reasons, prisms are the most suitable type of polarizer and analyzer when very precise measurements of polarization effects are to be made. One disadvantage of polarizing prisms is their inability to produce beams of wide angular aperture (since prisms can be cut only at angles which result in total internal reflection of one ray from the cement layer); another is the difficulty in obtaining large crystals of calcite. Polaroids, which can readily be made of any size, avoid these disadvantages. Also, it is found experimentally that transmission by prisms is not the same at angles 180° apart, a difficulty which can also be avoided by the use of polaroids.

REFERENCES

F. A. Jenkins and H. E. White, *Fundamentals of Optics* (3rd ed.), McGraw-Hill, New York, 1957, Chapter 24 (The Polarization of Light).

J. Strong, *Concepts of Classical Optics*, Freeman, San Francisco, 1958, Chapter 6 (Polarized Light and Dielectric Boundaries). Also, Fig. 7-20 on p. 140 shows how a Nichol prism is cut from a block of calcite.

Birefringence and Optical Activity

When matter is illuminated by polarized light, certain effects are observed which may be related to the structure of the specimen at the atomic, molecular, or macromolecular level. Preferential interactions with polarized light include the phenomena of *dichroism, birefringence,* and *optical activity,* all of which indicate the presence of *asymmetric* structure in the specimen.

Dichroism is the preferential absorption of light according to its azimuth of polarization; it is discussed on pp. 96 and 97. The physical basis of dichroic absorption is considered in Chapter 7, circular dichroism in Chapter 27.

Birefringence is the property of possessing two extreme values of the index of refraction for light which is polarized in mutually perpendicular planes. *Optical activity* is the property of rotating the plane of polarization of plane-polarized light during its transmission. *Optical rotatory dispersion* (ORD), the variation of optical activity with incident wavelength, is discussed in Chapter 27.

Birefringence and optical activity are discussed in this chapter. A knowledge of the properties of polarized light beams, as presented in Chapter 4, is essential for understanding of the material to be considered.

WAVE SURFACES AND INDEX ELLIPSOIDS

If a point source of monochromatic light exists in air, a surface of equal phase may be drawn at a distance from the source which is equal to any definite number of wavelengths. The surface is spherical, since light travels through air with uniform velocity in all directions. A similar surface of equal phase could be drawn for light proceeding from a point source embedded in any other transparent medium of uniform refractive index. The surface would again be spherical, but then, for any given number of cycles of vibration, the radius of the sphere would be *smaller*, since the wavelength of light in any medium of refractive index greater than one is *less* than that in air. For example, the radius of the spherical wavefront in a medium for which $n = 2$ would be half as great as in air, as shown in Fig. 6-1.

Similarly, surfaces representing refractive index as a function of direction can be drawn for various media, as shown in Fig. 6-2. Again, both surfaces are spherical, but the radius of the sphere representing air is now only half as great as that representing the medium for which $n = 2$. In general, wave surface and refractive index diagrams may be drawn for any substance; the dimensions of the two are always inversely related inversely, in accordance with the definition

(6-a) $$n = \frac{c}{v},$$

where v is the velocity of light in the medium of refractive index n. The wave surfaces (velocity surfaces) shown in Fig. 6-1 may be thought of as *loci of points of equal phase* and are a form of physical representation of wave motions. The spheres of Fig. 6-2 are a special case of a type of figure known as an *index ellipsoid*. Index ellipsoids, and *index ellipses*, which are plane sections through the ellipsoid, do not represent any physical entity, but are simply a convenient means of representing variations in refractive index with direction. As noted in Chapter 4, index ellipses should not be confused in any way with the phenomemon of elliptically polarized light.

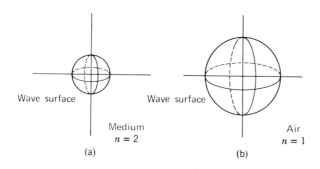

Fig. 6-1 Wave surfaces.

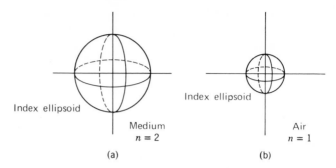

Fig. 6-2 Index ellipsoids.

OPTICAL ANISOTROPY; THE OPTIC AXIS

The media considered in Figs. 6-1 and 6-2 are optically isotropic; that is, refractive index is *invariant with direction*. Amorphous media, which are composed of randomly oriented molecules, and crystals with a very high degree of symmetry are optically isotropic. Other media, in which there is some preferential orientation of the component atoms and molecules, are optically *anisotropic*. Anisotropic media include the majority of crystals and various ordered biological structures, for example muscle fibers or chloroplasts.

Light transmitted in anisotropic media travels as *two* beams which are plane polarized in mutually perpendicular directions. In general, the two beams travel at different velocities and diverge physically. These media are characterized by two different indices of refraction and thus are birefringent.

In any birefringent medium there exists at least one direction in which the two components of transmitted light are indistinguishable, so that the medium, when viewed in this direction, appears to be isotropic. This direction is called the *optic axis* of the material. It is important to understand that the optic axis is *not* an axis in the usual sense of that term; that is, it is not a specific line around which an object is considered to rotate but is, in fact, a direction. (Certain axes of symmetry with respect to the atomic structure of birefringent materials are, however, parallel to the optic axis.)

Although the optic axis is usually a simple linear direction, radial, tangential or helical optic axes are occasionally encountered. For example, cell wall structures often possess a helically oriented optic axis.

Birefringent substances may be either *uniaxial* or *biaxial*. Only phenomena associated with uniaxial materials are considered in this chapter, since biaxial structures are almost unknown in biological systems.

ORDINARY AND EXTRAORDINARY RAYS

One of the two light rays transmitted by a birefringent medium travels with a velocity which does not vary with direction. Its wave surface is therefore a *sphere* such as those shown in Fig. 6-1. This ray is known as the *ordinary ray* or *O ray*. The ordinary ray is plane polarized, and it vibrates in a direction *perpendicular* to the plane defined by the ray and by the optic axis of the medium. Note, particularly, that the ordinary ray invariably vibrates in a direction perpendicular to the optic axis.

The second component of light transmitted in a birefringent medium is the *extraordinary ray* or *E ray*. The velocity of the extraordinary ray *varies* with direction in such a manner that its wave surface is an ellipsoid. The *E* wave always vibrates in a direction *parallel* to the plane defined by the direction of propagation of the ray and by the optic axis of the medium. Thus the orientation of the *E*-ray vibrations with respect to the optic axis *varies* with the direction of transmission.

An example of the relationship between the direction of the optic axis and the azimuths of vibration of the *O* and *E* rays is given in Fig. 6-3. (In Fig. 6-3 the notation given in Fig. 5-1 is used to indicate the states of polarization of the rays.) The optic axis of the medium shown in Fig. 6-3 is parallel to the plane of the page and is oriented in the direction shown. Unpolarized incident light is divided into an *O* ray, which vibrates in a

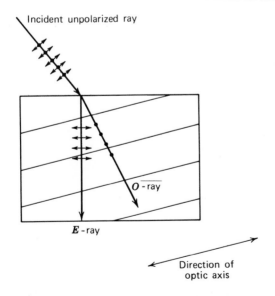

Fig. 6-3 Azimuths of vibration within a birefringent medium.

direction perpendicular to the plane of the paper (i.e., perpendicular to the optic axis), and an E ray vibrating *in* the plane of the paper. Both rays (like any transverse vibration) vibrate in azimuths perpendicular to their respective directions of propagation. Note that, if the optic axis of the specimen were tilted with respect to the plane of the page, the O- and E-vibration azimuths would be correspondingly tilted also.

Light which travels along (i.e., parallel to) the optic axis can vibrate *only* in directions perpendicular to the axis; in this special circumstance, O and E rays cannot be distinguished and, as stated above, the material appears to be optically isotropic. For light incident in any other direction, the O- and E-ray velocities differ. The E-ray velocity is smaller in positively birefringent materials and larger in negatively birefringent materials.

WAVE SURFACES AND INDEX ELLIPSOIDS OF OPTICALLY ANISOTROPIC MATERIALS

Figures 6-4a and 6-4b, respectively, show sections through the wave surfaces which would be formed by a point source of light embedded in positively or negatively birefringent material, respectively. These sections are in a plane which contains the assumed point source of light.

Index ellipsoids may also be constructed that correspond to the spheres illustrated in Fig. 6-2, and which characterize the birefringence of anisotropic media. In Figs. 6-5a and 6-5b, respectively, index ellipses (central section through the corresponding three-dimensional ellipsoids) are shown for positively and negatively birefringent media.

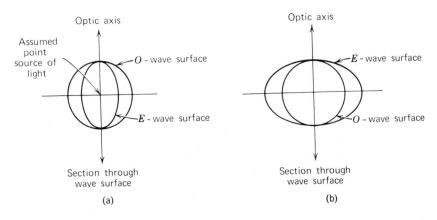

Fig. 6-4 Sections of wave surfaces: (*a*) in a positively birefringent medium; (*b*) in a negatively birefringent medium.

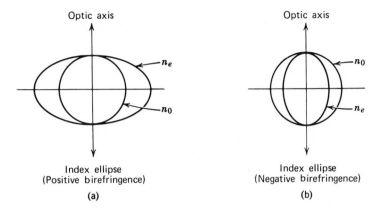

Fig. 6-5 Index ellipses.

A section through the *wave surface* formed in a *biaxial* anisotropic medium is shown in Fig. 6-6 for comparison.

The birefringent properties of a medium with respect to a given incident wavelength are fully determined by obtaining the size and shape of the index ellipsoid and its orientation with respect to some direction specified by the macroscopic features of the specimen (i.e., crystal face, fiber axis, etc.).

QUANTITATIVE EXPRESSION OF BIREFRINGENCE

The birefringence of a material is defined quantitatively by

(6-b) $$B \equiv n_e - n_0,$$

where n_0 is the (uniform) refractive index for the ordinary ray, and n_e

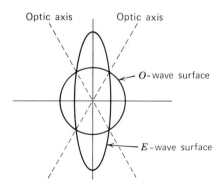

Fig. 6-6 Section of wave surfaces formed in a biaxial anisotropic medium.

is the *extreme* (maximum or minimum) value for the refractive index for the extraordinary ray. Note that the signs of birefringence specified in Figs. 6-4 and 6-5 are consistent with (6-b).

THE PHYSICAL BASIS OF BIREFRINGENCE

The behavior of light in birefringent media may be visualized by applying Huygens' concepts of the nature of propagation of waves. As explained on p. 50, each point of an incident wavefront may be considered as a source of secondary wavelets. New positions of the wavefront are then found by drawing the mutual tangent or "envelope" of all secondary wavelets. New ray directions are given by the vector drawn from the source of a secondary wavelet to its point of tangency with the envelope.

As discussed on p. 143, the transmission of light is associated with transitory perturbations of the electrons in the transmitting medium. The susceptibility of an electronic configuration to distortion is termed its *polarizability*. The more polarizable an electronic structure, the more extensively can light interact with it and, thus, the more slowly is light transmitted. (In the limit, the interaction is so extensive as to result in *absorption* of light quanta.)

Most chemical groups are *asymmetric* with respect to polarizability. For example, the electrons of carbon–carbon bonds are most easily displaced along the direction of the bond, and least easily displaced at right angles to the bond. Therefore, when carbon–carbon bonds or other asymmetric groupings are regularly oriented, the polarizability of the medium as a whole varies with direction. A structure containing parallel chains of carbon–carbon bonds tends (i.e., apart from the effects of substituent groups) to be most polarizable in a direction *parallel* to the carbon chains. Correspondingly, the velocity of light tends to be *lowest* and the refractive index *highest* in that direction.

Any line parallel to the optic axis of a birefringent medium is an axis of rotational symmetry with respect to electronic configuration. The ordinary ray, which invariably vibrates in a direction perpendicular to the optic axis, thus invariably encounters a medium of *uniform* polarizability, whatever its direction of transmission. That is, the ordinary ray encounters a symmetrical electronic environment and propagates with equal velocity in all directions. Consequently the wave surface associated with the propagation of the *O* ray is a *sphere*, just as are wave surfaces within optically isotropic media.

The extraordinary ray, however, encounters an asymmetric electronic environment. The angle between its azimuth of polarization and the optic axis of the medium *varies* with the direction of propagation. Accordingly the polarizability of the medium and thus the velocity of the E ray also vary with the direction of propagation. The resulting wave surface is not a sphere but an *ellipsoid*. Cross sections through typical extraordinary wave surfaces are shown in Figs. 6-7a and 6-7b for positively and negatively birefringent media, respectively. In the medium considered in Fig. 6-7a polarizability is *least* in the direction perpendicular to the optic axis, so that wave velocity is corresponding *greatest* along the optic axis. For the medium considered in Fig. 6-7b the reverse is true.

WAVE SURFACES WITHIN BIREFRINGENT MEDIA

Figures 6-8a and 6-8b show sections through combined O- and E-wave surfaces within positively and negatively birefringent media, respectively. The pairs of surfaces are tangent at the optic axis (except in the special case of optically active media) since the O and E waves are indistinguishable for this one direction of propagation. Correspondingly, when light is incident upon a medium parallel to its optic axis, no double refraction and no differences in O- and E-ray velocity result. As shown in Figs. 6 9a and 6-9b for negatively and positively birefringent materials, respectively, the new wave surfaces remain perpendicular to the direction of propagation.

When light is incident *perpendicular* to the optic axis, the O and E rays travel at different velocities but do not diverge. This is illustrated in Fig. 6-10 for two possible orientations of the optic axis of a *negatively* birefringent material. The difference in velocities is accounted for by the separation of the O- and E-wave surfaces. The two rays remain parallel

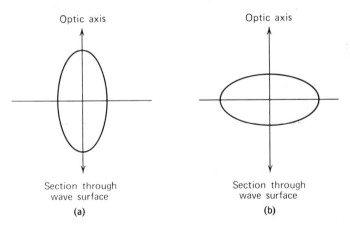

Optic axis

Optic axis

Section through
wave surface

Section through
wave surface

(a)

(b)

Fig. 6-7 Sections of E wave surfaces: (*a*) in a positively birefringent medium; (*b*) in a negatively birefringent medium.

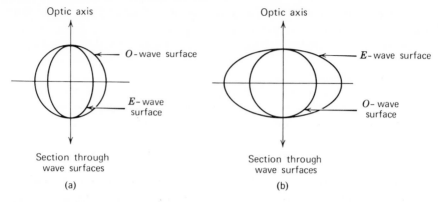

Fig. 6-8 Sections of O and E wave surfaces: (*a*) in a positively birefringent medium; (*b*) in a negatively birefringent medium.

because the vectors drawn from the origins of the secondary wavelets to the corresponding points on the *O*- and *E*-wave envelopes are coincident; specifically, they are both perpendicular to the wave envelopes.

The incidence of light at an arbitrary angle with respect to the optic axis is illustrated in Fig. 6-11, again for a *negatively* birefringent medium. The optic axis of the medium shown in Fig. 6-11 lies in the plane of the page but is tilted with respect to the incident illumination. Again, the *O* and *E* rays travel at different velocities, with corresponding separations of the *O*- and *E*-wave surfaces. In this case. however, the rays also diverge. The vector *OA* from the point of origin of a secondary wavelet to

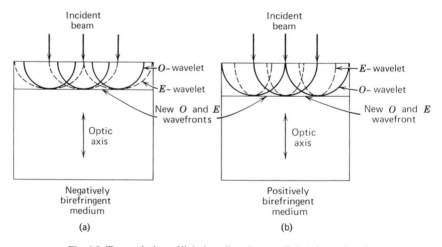

Fig. 6-9 Transmission of light in a direction parallel to the optic axis.

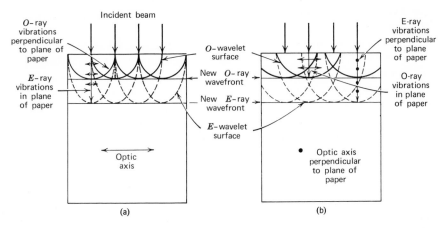

Fig. 6-10 Transmission of light in directions perpendicular to the optic axis of a negatively birefringent medium.

the point of tangency with the O-wave envelope is perpendicular to the O wavefront, but the corresponding vector, OB, drawn to the position of tangency with the E-wave envelope is no longer perpendicular to the wavefront. Thus the E ray diverges even, as shown, for normal incidence of the illumination.

EXPERIMENTALLY OBSERVED TYPES OF BIREFRINGENCE

Manifestations of birefringence may be grouped in four categories: intrinsic birefringence, form birefringence, birefringence of flow, and strain birefringence.

Intrinsic birefringence originates from the inherent asymmetry of polarizability of chemical bonds, as discussed above. It is a property of

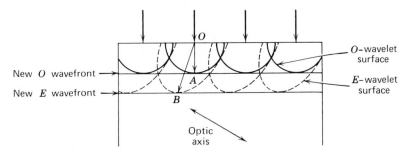

Fig. 6-11 Transmission of light incident at an arbitrary angle with respect to the optic axis of a negatively birefringent medium.

all but the most highly symmetrical crystals, of fibers, and of many biological structures.

Form birefringence results from regular arrangements of objects which are of dimensions comparable to the wavelength of light, but which need not themselves be . intrinsically birefringent. For form birefringence to be observed, the refractive index of the oriented objects must differ from that of the surrounding medium. *Birefringence of flow*, which arises when a preferential arrangement of structures is induced by a moving stream of liquid, is a special case of form birefringence. A familiar example of flow birefringence is that observed in suspensions of the rodlike tobacco mosaic virus particle.

Strain birefringence is produced by mechanical stress, which may cause a preferential alignment of particles, inducing form birefringence, or may induce an actual distortion of chemical bonds, imposing an intrinsic birefringence. Strain birefringence may be a problem in the production of glass lenses, since glass which hardens under strain may contain irregular birefringent areas.

Further discussion of types of birefringence and of the measurement of their effects is given in Chapter 15.

INTERFERENCE OF POLARIZED LIGHT IN BIREFRINGENT MEDIA

Observation of interference between light beams which have traveled through birefringent media *in a direction perpendicular to the optic axis* is an important means of characterizing these media. As shown in Fig. 6-10, light incident perpendicular to the optic axis forms ordinary and extraordinary rays which travel coincident paths *at different velocities*. Upon emerging from the birefringent medium into air (or any other optically isotropic medium) the O and E rays, still coincident, travel with identical velocity. The two rays differ in optical path, however, since the ray that traveled more slowly in the birefringent medium executed more cycles of vibration during its passage through the medium than did the faster ray. (Recollect that, while the wavelength of light changes during its passage through a medium, the *frequency* of vibration remains invariant.) The emergent light thus consists of rays that are polarized in mutually perpendicular planes and which differ in phase. The resultant motion is thus, in general, elliptically polarized; in specific cases, plane or circular polarization could result instead (cf. pp. 82–87).

The phase angle between the transmitted O and E rays, and thus the form of the resultant vibration, is determined by the extent of *retardation* between the rays during passage through the birefringent medium. Retardation, designated Γ, is the difference in path length between positions of equal phase of the polarized beams traveling through or emerging from a birefringent medium. Γ, defined by (6-c), is a *length* and is usually expressed in millimicrons.

(6-c) $$\Gamma \equiv (n_e - n_0)t,$$

where t is the distance the light has traveled through the specimen. Note that the retardation progressively increases as light travels through the birefringent specimen. This increasing relative retardation is illustrated for a specific case by Fig. 6-12 (in which the two rays are shown as separate for ease of illustration). (The wavy lines shown in Fig. 6-12 are *not* ray trajectories but are in fact plots of the electric field as a function of position at a given instant of time.)

Retardation is frequently expressed also in terms of fractions of a cycle of vibration, as represented by the phase angle δ. Then

(6-d) $$\delta = \frac{2\pi\Gamma}{\lambda} = \frac{2\pi(n_e - n_0)t}{\lambda} \quad \text{cycles}$$

for any given wavelength λ.

Since, as explained in Chapter 4, rays polarized in different planes do not interfere in the sense of producing variations in *intensity*, the phase difference between the O and E rays cannot be *directly* detected. However, when the beams pass through an appropriately oriented analyzer, a component of vibration of each may be brought into the same azimuth; these components can then interfere, giving a resultant intensity which is determined by their relative amplitudes and phases. One example of the use of an analyzer for this purpose is diagramed in Fig. 6-13.

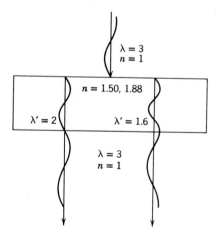

Fig. 6-12 Relative retardation.

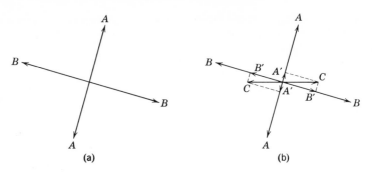

Fig. 6-13.

THE QUARTER WAVE PLATE

The effect of the *thickness* of a birefringent specimen which is illuminated perpendicular to its optic axis by initially plane-polarized light may be considered. The azimuths of vibration for light transmitted by the specimen might be AA and BB, as shown in Fig. 6-13a. Incident light polarized in either of these azimuths would be transmitted without alteration, whereas light polarized in any other azimuth, such as CC of Fig. 6-13b, is resolved into components of vibration $A'A'$ and $B'B'$ upon entering the specimen. For incident light which is plane polarized at 45° to AA and BB, $A'A' = B'B'$; that is, the mutually perpendicularly polarized beams traveling through the specimen are of equal amplitude (i.e., in the absence of dichroic absorption). Beams $A'A'$ and $B'B'$ travel at different velocities, experiencing a relative retardation. A succession of plane, circular, and elliptically polarized waveforms of the emerging beam result as the thickness of the specimen, and thus Γ, increases, in accordance with the effects described in Chapter 4. For example, when $\Gamma = 0$, no phase shift is introduced between the beams traveling through the specimen, and the emergent radiation is polarized in the same plane as the incident radiation. For a specimen thickness such that $\Gamma = \lambda/4$, the emergent beams differ in phase by a quarter cycle, and a circularly polarized motion results. For $\Gamma = \lambda/2$, the two plane-polarized motions leaving the specimen are exactly *out* of phase with each other, and interfere to produce a waveform which is plane polarized in a direction *perpendicular* to the azimuth of the incident light. The complete series of waveforms corresponding to certain values of Γ and δ is shown in Fig. 6-14. Note that the variation is one of *waveforms* only and *cannot be observed directly*. Also, if the plane of polarization of

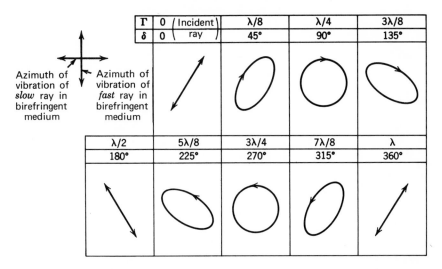

Fig. 6-14 Variation of resultant wave forms according to relative retardation of components.

the incident light is not oriented exactly at 45° to the transmission azimuths of the specimen, the O and E rays transmitted by the specimen are of unequal amplitudes; a succession of elliptical waveforms, approximating the series shown in Fig. 6-14, then results.

The quarter wave plate is a slab of birefringent material which is cut with parallel plane faces parallel to the optic axis. The thickness of the plate is such that, for a selected wavelength (usually either that of the mercury green line or the sodium D line), Γ is equal to $\lambda/4$. (Note that Fig. 6-12 shows a quarter wave plate.)

On direct examination, the light leaving a quarter wave plate is indistinguishable from the incident light. Introduction of an analyzer, so oriented as to transmit components of vibration polarized in an azimuth at 45° to the azimuths of vibration of the beams transmitted by the quarter wave plate, brings a *portion* of the amplitude of each of those beams into the *same* azimuth of vibration. The various azimuths of vibration are shown in Fig. 6-15. The components of amplitude transmitted by the analyzer (shown in Fig. 6-15e) interfere to give a resultant intensity the level of which is determined by their relative amplitudes (and thus by the azimuth of the incident plane-polarized beam) and by the phase angle (i.e., 90°).

The use of the quarter wave plate in distinguishing states of polarization of light beams is explained below; other applications of the device are described in Chapter 15.

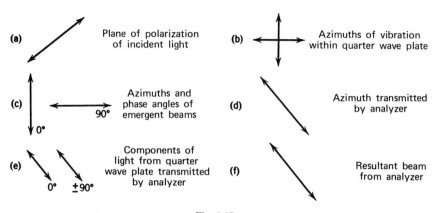

Fig. 6-15.

EXPERIMENTAL CHARACTERIZATION OF LIGHT AS UNPOLARIZED, CIRCULARLY POLARIZED, OR ELLIPTICALLY POLARIZED

A combination of a quarter wave plate and analyzer, arranged as shown in Fig. 6-16, may be used to distinguish unpolarized, circularly polarized, and elliptically polarized light beams.

If incident light is *unpolarized*, the intensity of the transmitted beam is unaffected by rotation of the analyzer.

For any fixed orientation of the quarter wave plate, transmission of *circularly* polarized light varies as the analyzer is rotated, zero intensity being observed once during each rotation through 180°. This may be explained by noting that, just as the quarter wave plate converts plane polarized into circularly polarized light, so, also, it converts circularly polarized into plane-polarized light. Equivalently, it may be said that the 90° phase shift between the mutually perpendicularly polarized components of the circularly polarized beam is either increased to 180° or reduced to 0° by the action of the quarter wave plate. In either case, a plane-polarized motion results. These relationships are illustrated by Fig. 6-17. Transmission of the resultant wave motion varies with the

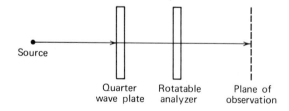

Fig. 6-16 Arrangement of optical components for determinations of state of polarization.

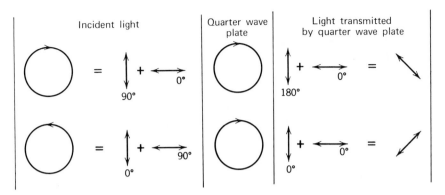

Fig. 6-17.

orientation of the analyzer in the same way as for any other plane-polarized motion. That is, successive maxima and extinction of intensity are observed at orientations separated by 90°.

The effect of *elliptically* polarized light on the quarter wave plate-analyzer combination is similar to that of circularly polarized light. However, in addition, variations in intensity (but not total extinction) may be observed with the analyzer alone.

These experimental relationships are summarized in Table 6-1.

Table 6-1 Experimental Discrimination between Unpolarized, Plane Polarized, Circularly Polarized, and Elliptically Polarized Light Beams

Unpolarized light	Neither analyzer alone nor analyzer-quarter wave plate combination varies intensity.
Plane polarized light	One position of the analyzer, used alone, produces extinction.
Circularly polarized light	Analyzer alone is without effect; one position of analyzer produces extinction when used in combination with quarter wave plate.
Elliptically polarized light	Transmitted intensity varies with position of analyzer, used alone, but no position of extinction is obtained. Insertion of quarter waveplate with optic axis parallel to analyzer position of maximum intensity produces extinction.
Mixture of unpolarized light with plane, circularly, or elliptically polarized light	Observations generally similar to the above, except that complete extinction cannot be obtained.

OPTICAL ACTIVITY

The phenomenon of optical activity consists of the *rotation* of the azimuth of vibration as it travels through a medium. Optical activity occurs in molecules which contain asymmetric carbon atoms, that is, carbon atoms to which four *different* substituents (atoms or radicals) are attached. Solutions of organic compounds are optically isotropic but are often optically active. In crystals, optical activity is invariably accompanied by birefringence, although many birefringent solids are not optically active.

Wave surfaces in negatively and positively birefringent optically active crystals are shown in Figs. 6-18a and 6-18b, respectively. The degree of separation of the ordinary and extraordinary wave surfaces at the optic axis is a measure of the extent of optical activity. The separation, which is slight, is exaggerated in Fig. 6-18.

The rotation of the plane of polarization by optically active media may be accounted for quite simply. It has been shown (p. 87) that a plane-polarized motion is the resultant of a *pair* of left- and right-hand circularly polarized motions of equal amplitude. In optically *in*active media these circularly polarized components of motion travel with equal velocities. The electronic structure of optically active materials, however, is such that vibrations traveling in left-hand or right-hand helical paths encounter different polarizabilities and thus different refractive indices of the medium. Correspondingly, the velocities of the two components differ. As the beams travel through the optically active medium with different velocities, their relative phase shifts progressively. As illustrated by Figs. 4-19 and 4-20, the azimuth of the resultant plane-polarized motion accordingly shifts continuously.

The measurement of optical activities and of the variation of optical activity as a function of the wavelength of the transmitted light are discussed in Chapter 27.

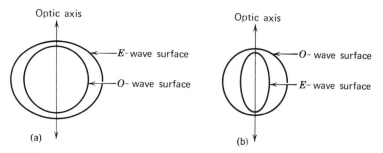

Fig. 6-18 Wave surfaces in optically active birefringent media: (*a*) negative birefringence; (*b*) positive birefringence.

EFFECTS OF APPLIED ELECTRIC AND MAGNETIC FIELDS

The application of strong electric or magnetic fields to matter may have the effect of inducing optical anisotropy or optical activity, or of altering these properties if they are already present. These effects are summarized briefly here.

Application of a strong electric field induces *electrical double refraction*, in which the optic axis is parallel to the applied field. The induction of electrical double refraction in liquids or isotropic solids is known as the *Kerr electro-optic effect*. The induction of variations in the birefringence of anisotropic crystals is known as the *Pockels effect*.

The Kerr effect may be exploited to produce an electro-optical shutter in the manner shown in Fig. 6-19. The arrangement consists of a crossed polarizer and analyzer between which is placed a cell (Kerr cell) containing a liquid (usually nitrobenzene) and two parallel plate electrodes. When no voltage is applied to the plates, as in Fig. 6-19a, the plane-polarized light from the polarizer is transmitted by the cell without alteration and is blocked by the analyzer. When voltage is applied to the electrodes, as shown in Fig. 6-19b, the liquid behaves like a birefringent medium, the optic axis of which is perpendicular to the direction of transmission. Consequently a component of vibration is introduced in an azimuth perpendicular to that of the polarizer. This component is transmitted by the analyzer.

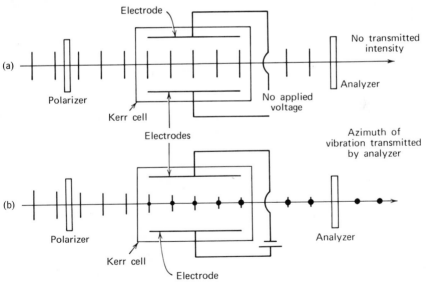

Fig. 6-19 Kerr electro-optical shutter: (*a*) no applied voltage; (*b*) applied voltage.

When very short interruptions of light beams are required, the delay in operation of mechanical shutters may be intolerable. The electro-optical shutter then provides a useful alternative. For example, electro-optical shutters may be used in measurement of phosphorescence from excited states of relatively short lifetimes, or in "Q-switched" lasers.

The magnetic analog of electric birefringence can be observed upon application of a strong magnetic field to a specimen. The *Faraday effect*, discussed on p. 598, is a rotation of the plane of polarization by a magnetic field applied in a direction parallel to that of transmission.

REFERENCES

F. A. Jenkins and H. E. White, *Fundamentals of Optics* (3rd ed.), McGraw-Hill, New York, 1957, Chapter 26 (Double Refraction), Chapter 28 (Optical Activity), and Chapter 29 (Magneto-optics and Electro-optics).

H. S. Bennett, in R. M. Jones, Ed., *McClung's Handbook of Microscopical Technique*, 3rd ed., Hafner, New York, 1950, pp. 601–611.

J. Strong, *Concepts of Classical Optics*, Freeman, San Francisco, 1958, Chapter 7 (Double Refraction-Calcite and Quartz).

J. Stone, *Radiation and Optics*, McGraw-Hill, New York, 1963, Chapter 17 (Double Refraction).

Interactions of Light with Matter

When light is incident upon matter, it may be reflected, transmitted or absorbed. Light energy which is absorbed may subsequently be re-radiated, transferred by a nonradiative mechanism, degraded to heat, or may initiate a photochemical reaction. The process of radiation may be one of fluorescence or phosphorescence. The nature of each of the processes mentioned is discussed in this chapter.

GENERAL PHENOMENA OF ABSORPTION AND EMISSION

The phenomena of absorption and re-emission are intimately related to the structures of atoms and molecules, and will be discussed in some detail below. Here it may be said that atoms and molecules consist of nuclei which are surrounded by a number of electrons equal to Z, the atomic number of the element. Molecules consist of two or more nuclei which *share* certain of their electrons. Basically, the phenomena of absorption and emission are explained, respectively, as the uptake and release of energy by atoms or molecules. The nature of observed absorptions and emissions is accounted for by the fact that *atoms and molecules can exist only at certain discrete energy levels, rather than at a continuum of energy levels.* Light is absorbed only when it is of a wavelength such that its quantum energy is exactly equal to that required for raising of the energy of the absorbing atom or molecule to an allowable level of higher energy. Similarly an emitted quantum of radiation corresponds exactly to the energy loss of the emitter.

In general, the energy of molecules is distributed between four categories, those of *electronic, vibrational, rotational,* and *translational* energies. The electronic energy level of a system is determined by the configuration of its electrons. In other terms the electronic energy is a function of the *orbitals* occupied by the electrons of the atom or molecule, as discussed below on p. 126. Vibrational energy implies a relative motion between the constituent atoms of a molecule, i.e., the periodic shortening and lengthening of chemical bonds. Rotational and translational energies are associated with movements of the molecule as a whole.

GROUND STATES AND EXCITED STATES

At low temperatures, atoms and molecules tend to exist in their *ground state,* or lowest energy level; an increasing proportion of particles is found in higher energy states as the temperature is raised. States of higher energy are referred to as *excited states* and may be specified as excited electronic states, excited vibrational states, etc. At ordinary temperatures, molecules exist in the ground electronic state but may occupy a number of excited vibrational levels within that state. (The specification of vibrational energy levels does not, of course, apply to single atoms.) Generally speaking, molecules in excited vibrational states are susceptible to chemical reaction, a fact which accounts for the acceleration of chemical reactions as the population of excited vibrational states is increased by increasing the temperature.

The distribution of particles between the energy levels of any system is given by the *Boltzmann distribution function,*

$$(7\text{-}a) \qquad \frac{n}{n_0} = C \exp\left[\frac{-(E - E_0)}{kT}\right],$$

where n is the number of atoms in any energy level, n_0 is the number of atoms in the ground state (or other reference state), E is the energy of the level, E_0 is the energy of the ground state, k is Boltzmann's constant $(1.38 \times 10^{-16}$ erg/degree), T is the absolute temperature, and C gives the ratio of the number of energy sublevels present in the two states.

ENERGY LEVEL DIAGRAMS

Diagrammatic representations of energy levels are extremely useful in describing the phenomena of absorption and emission. The nature of such diagrams is considered here.

The energy level diagram of a single electron within an atom, as shown in Fig. 7-1, is considered first. The ground state of the atom is indicated by G, and its successive excited electronic states by $E_1^*, E_2^*, \ldots$ Higher

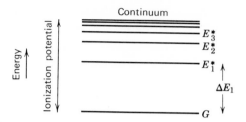

Fig. 7-1 Energy levels of an atomic electron.

excited states of the electron differ successively, as shown, by progressively smaller increments of energy; at some limiting value of the energy the electron escapes altogether from the atom and may then exist at a continuum of energy levels. The energy which must be supplied to the electron to just free it from the atom is the *ionization potential* (indicated in Fig. 7-1). The smaller amount of energy, ΔE_1 required to raise the electron from its ground state to the first excited state is also shown.

The energy level diagram of an electron is equivalent to that of a single electron atom, that is, that of the element hydrogen. Energy level diagrams for larger, many electron atoms contain additional levels which may represent the *interactions* of electrons. Details of atomic energy level diagrams need not be considered here.

As in the case of atoms, the ground state of the *molecule* is the electronic configuration of lowest energy, and a number of quantized excited states are conditions of higher energy. The upper limit of the energy levels of the molecule is determined by the energy required for ionization or for rupture of a chemical bond between nuclei. In the case of a regular array of molecules, that is, of crystalline or para-crystalline structures, certain additional types of energy level exist; electrons may acquire energies that are sufficient for their dissociation from specific molecules and yet remain in association with the array as a whole. There exists a narrow continuum of energies which constitute the *exciton band* of an array. (As discussed on p. 142, energies in this range may function in the radiationless transfer of energy within a crystal.) A somewhat more extensive continuum of energies constitutes the *conduction band*. Electrons with energies in the conduction band may conduct electric current within a crystal. An example is the possibility of obtaining a photocurrent through the grains of photographic emulsions when these absorb light. If electrons receive energy above a certain value, the "work function" (cf. p. 20), they become sufficiently energetic to escape completely from the material. These general features of the electronic energy levels are illustrated by Fig. 7-2.

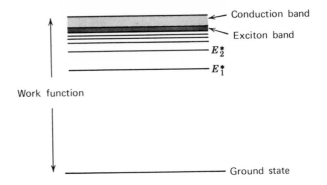

Fig. 7-2 Electronic energy levels.

Figure 7-3 shows a type of molecular energy level diagram in which both electronic and vibrational levels are represented. The ground electronic state G includes vibrational levels v_0, v_1, v_2,... As shown, the excited states E' and E'' are mutually translated, thus avoiding confusion from the overlap of their respective sets of vibrational sublevels $(v_0', v_1', v_2',...)$ and $(v_0'', v_1'', v_2'',...)$. Rotational sublevels occur within each vibrational level but are not shown in Fig. 7-3. Overlapping of vibrational energy levels of different excited electronic states is a common feature; it occurs whenever the ground vibrational levels of the respective electronic states are of comparable energy. Diagrams of the type shown in Fig. 7-3 are known as *Jablonski* diagrams.

The physical significance of the vibrational energy levels is emphasized by the *Franck-Condon* type of diagram, shown in Figs. 7-4 and 7-5. As the nature of this type of diagram is often misunderstood, it should be considered carefully. The Franck-Condon diagram is *a plot of the energy content of a diatomic molecule as a function of the separation of the two nuclei.* Thus, while the molecule is at any of the quantized vibrational levels, the energy content is *constant* and the interatomic separation r

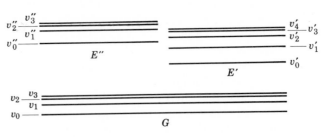

Fig. 7-3 Molecular energy level diagram (Jablonski diagram).

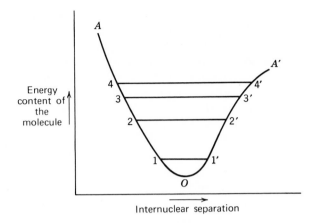

Fig. 7-4 Franck-Condon diagram.

oscillates between two extreme values. For example, 1 and 1′, as shown in the lowest vibrational level of Fig. 7-4, are extreme interatomic separations corresponding to the energy content E_1. Higher vibrational levels correspond to larger total energy content and to more extensive oscillations of the internuclear distance. The spacing of the levels is actually very close, although in Fig. 7-4 only a few levels, indicated by

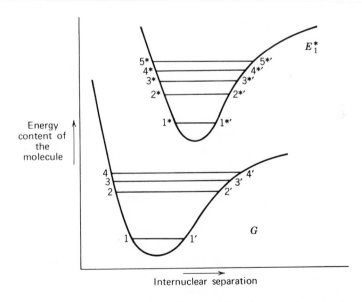

Fig. 7-5 Franck-Condon diagrams for two electronic states of a molecule.

the lines 4-4', 3-3' and 2-2', are illustrated. The envelope of all the quantized vibrational levels is the continuous curve AOA'. The closeness of approach of nuclei is limited by the repulsive force between these positively charged objects; the left side of the curve thus tends to become a vertical line at high energy levels. The maximum separation of nuclei becomes increasingly great at high energy levels; ultimately, the two nuclei become separated; that is, the chemical bond between the nuclei is broken. The right side of the curve thus becomes a horizontal line at high energy levels. The lower point O of the curves represents the limiting energy of the molecule in the absence of any vibration, a condition which, hypothetically, would occur at 0° absolute temperature. Vibrational level 1-1' is, however, the vibrational ground level of the molecule at the temperature to which the diagram applies. Note that the positions of vibrational sublevels thus vary slightly with the absolute temperature.

Diagrams like Figs. 7-4 and 7-5 apply strictly to *diatomic* molecules *only*. The vibrational levels of a triatomic molecule could be represented by an analogous three-dimensional diagram, but polyatomic molecules in which many internuclear distances must be considered cannot be thus graphically represented. In general, a molecule containing n atoms would require a hypothetical n-dimensional diagram. Diagrams such as Figs. 7-4 and 7-5 are, however, used symbolically to represent the energy states of polyatomic molecules since, in many cases, the vibrational variations of all possible internuclear distances need not be considered with respect to a given energy level transition.

ATOMIC ABSORPTION AND EMISSION SPECTRA

The classical example of the relation between absorption and emission by matter is that of the spectra of gaseous atoms; for example, of sodium vapor. The characteristic yellow color of many types of flame can be ascribed to the presence of sodium atoms at high temperatures. These very hot atoms are raised to excited electronic energy levels; subsequently their excess energy is dissipated by the emission of light quanta. The emitted quanta represent differences in energy levels between the excited states of the sodium atoms and the ground state. In the case of sodium, the transitions corresponding to wavelengths of 5890 and 5896 Å are of particularly high probability; other wavelengths characteristic of sodium are also emitted from the atoms, but with lower intensity.

The irradiation of sodium vapor by intense light results in the *absorption* of just those wavelengths which are emitted by the hot vapor.

In this case, excitation to the higher energy levels is produced by the absorption of photons, rather than thermally, but the energy levels attained by the atoms are precisely the same in both cases.

Similar effects are diagramed in Fig. 7-6 which illustrates the excitation of hydrogen atoms and their emission. A transition of high probability in this system is that between the ground state and the first excited state, as represented by line A. In a highly excited system, transitions between the first and second excited states, represented by line A', are also probable. Absorption and emission of the corresponding wavelengths, 1215 and 6562 Å, are thus observed. Emission of this type is called "resonance radiation" (cf. p. 138).

ATOMIC STRUCTURE AND ENERGY LEVELS

The existence of discrete, quantized energy levels within atoms is not intuitively obvious, but it may be accounted for by the methods of quantum and wave mechanics. Some of the results of these rigorous treatments are described here.

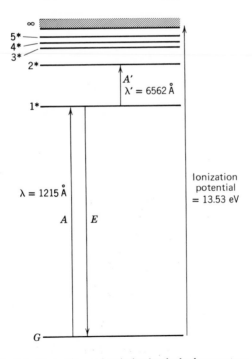

Fig. 7-6 Absorption and emission by the hydrogen atom.

According to the picture developed by classical physics, electrons should exist at definite locations at any given instant in time. In terms of the quantized atom first postulated by Bohr, the electrons travel through the space around the atomic nucleus in certain fixed orbits, as sketched for the hydrogen atom in Fig. 7-7a. This pictorial representation of the atom is not supported by more rigorous considerations; instead, the location of the electron must be described in terms of the *probability* of its occurrence in any given region. Thus the electron must be thought of as an "electron cloud," as indicated in Fig. 7-7b, rather than as a discretely located particle. The electrons of any atom may exist in one of a finite number of probability density distributions each of which is referred to as an electronic *orbital*. Corresponding to each orbital there is a specific value of the energy content of the system. Thus Fig. 7-7b is a representation of the *ground state* of the hydrogen atom, in which the orbital of the single electron constitutes a spherically symmetric and relatively compact shell. The probability of finding this electron at any position along an axis drawn through the nucleus is plotted in Fig. 7-7c. The total probability that the electron will occur at some distance is 1.0; that is, the electron must exist somewhere.

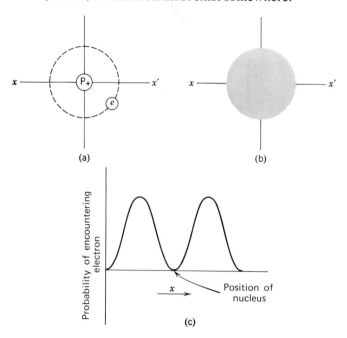

Fig. 7-7 Representations of electrons: (*a*) bohr hydrogen atom; (*b*) electron cloud; (*c*) plot of probability density.

The *electronic state* of a complete atom corresponds to the combined probability distribution for the location of all the electrons of the atom collectively. Since individual electronic orbitals are altered (perturbed) by the presence of other electrons, the overall energy level of the atom differs from the sum of the appropriate one-electron orbital energies.

The orbitals of electrons within any atom are characterized by four *quantum numbers*, which collectively designate the energy state of the electron. These quantum numbers are designated here as n, l, m, and s. The significance of these numbers is largely a formal, mathematical one, but it may be described in physical terms as follows.

The *principal quantum number, n,* describes the average distance of the electron from the nucleus. In the ground state of the hydrogen atom, n for the single electron has a value of one, and that electron lies as close as is possible to the nucleus. For successively higher values of n, that is, for the excited states of the atom, the electron lies at successively greater distances from the nucleus. The energy of the system correspondingly

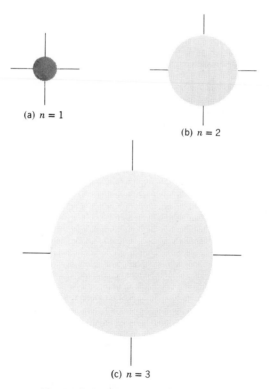

(a) $n = 1$

(b) $n = 2$

(c) $n = 3$

Fig. 7-8 Spherically symmetric orbitals.

increases. Considering only the spherically symmetric orbitals (those for which $l = 0$; see below) the form of the electron cloud for $n = 1, 2$, and 3 is as shown in Fig. 7-8. The orbitals of greater radius are correspondingly more diffuse (i.e., of lower electron density) as is evident from the corresponding probability density plots shown in Fig. 7-9.

The *azimuthal quantum number*, l, describes the asymmetry of the orbital. $l = 0$ for a spherically symmetric electron density distribution, while specific shapes of the electron cloud correspond to other values of l. For example, Fig. 7-10 shows the form of the electron density distribution that corresponds to $n = 2$ and $l = 1$. Note that, whereas the corresponding plot of electron density as a function of position along the XX' axis would be generally similar to that of the spherically symmetric orbitals (as shown in Fig. 7-7c or Fig. 7-9), the probability of finding this electron along the YY' or ZZ' directions is zero. The probability density distribution falls to zero (experiences a node) at the position of the nucleus. Such discontinuities in electron density distribution, while apparently fantastic in terms of a physical interpretation, are frequently encountered.

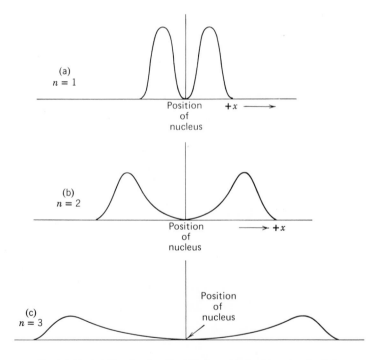

Fig. 7-9 Probability density plots for the orbitals shown in Fig. 7–8.

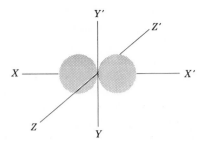

Fig. 7-10 Asymmetric orbital ($n = 2, l = 1$).

The azimuthal quantum number may assume values ranging from zero to $(n-1)$. Thus, for principal quantum number $n = 1$, only spherical orbitals exist. For $n = 2$, l may be zero or one, and orbitals may be of the forms sketched in Fig. 7-8 or 7-10, respectively. For $n = 3$, l may be 0, 1, or 2, and so forth. In general, states of higher l are states of higher energy, but differences in energy between the states of successive l values are smaller than those between states of successive n values. For simple atoms, all the l states associated with a particular n are of lower energy than any state of a higher n. Inversions occur, however, with atoms of complex structure, that is, for relatively heavy atoms.

Electrons for which $l = 0$, 1, 2, or 3 are called, respectively, s, p, d, and f electrons.† Thus s electrons are depicted in Figs. 7-7 and 7-8, and a p electron is shown in Fig. 7-10.

The *magnetic quantum number*, m, is a measure of the component of the azimuthal quantum number in the direction of an applied external magnetic field. For a given value of l, values of m may range between (-1) and $(+1)$. For example, an electron for which $n = 3$ and $l = 2$ (a d electron) may have $m = -2, -1, 0, +1$, or $+2$. States of different m but identical l and n are indistinguishable in the absence of an external magnetic field.

The *spin quantum number*, s, may be thought of as representing the clockwise or anticlockwise spin of the electron. s may assume values of $+\frac{1}{2}$ and $-\frac{1}{2}$.

While n, l, m, and s are the quantum numbers which are characteristic of individual electrons, or of the one-electron atom, hydrogen, the energy state of many-electron atoms as a whole is designated by certain other quantum numbers derived from the interactions of the individual elec-

---

†These designations originated historically from the description as "sharp," "principal," "diffuse," and "fine," respectively, of the lines of the hydrogen spectrum associated with transitions between levels to which the various l values are assigned.

tronic quantum numbers. In particular, the quantum number L for an atom represents a vector sum of electronic l values, while the quantum number S is the absolute value of the sum of individual electronic spins. (For a further consideration of electronic spins see below.)

The importance of the quantum numbers assigned to the electronic energy states becomes clear in view of the fact, stated by the *Pauli exclusion principle*, that *no two electrons present in an atom may share an energy level specified by identical values of all four quantum numbers.* Thus in a many-electron atom there exist "shells" of electrons, each of which corresponds to a value of n. For increasing values of n, there exists an increasing number of possible combinations of l, m, and s, leading to a correspondingly large number of possible energy states. In general, the electrons of an atom tend to assume configurations of minimum energy. In many-electron atoms, however, there may exist several states which do not differ greatly in energy. Therefore a number of different energy level transitions may occur with high probability for a given type of atom.

SINGLET AND TRIPLET STATES OF ATOMS AND MOLECULES

The spin quantum number S of an atom (or molecule) has been defined above as the absolute value of the sum of electronic spins within the system. The multiplicity of a system is defined as the quantity $(2S + 1)$. For example, a two-electron system in which the spins are antiparallel (paired) has $S = \frac{1}{2} + (-\frac{1}{2}) = 0$, and a multiplicity of one; such a system is designated as *singlet*. Similarly, all many-electron systems in which all the spins are paired also are singlet states. A two-electron system in which spins are parallel (unpaired) has $S = \frac{1}{2} + \frac{1}{2} = 1$, and a multiplicity of three; this is termed a *triplet* state. For equal values of the other quantum numbers, *states of higher multiplicity are states of lower total energy.*

Uncoupling of the paired spins of two electrons in a system constitutes a transition from a singlet to a triplet state. Figure 7-11, in which electronic spins are represented by arrows pointing up or down, represents, respectively, (a) the singlet ground state, (b) a singlet excited state, and (c) a triplet excited state of a system. It is found that *transitions in which*

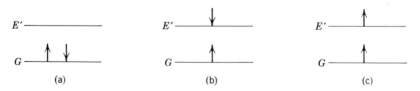

Fig. 7-11 Representation of singlet and triplet states: (*a*) singlet ground state; (*b*) singlet excited state; (*c*) triplet excited state.

electron spins must be uncoupled (or recoupled) *are less probable than transitions in which the multiplicity of the atom or molecule remains constant.* The decrease in probability for transitions between states of different multiplicities is found experimentally to be of the order of 10^5 or 10^6. Such transitions have been termed "forbidden transitions"; however, this term is a relative one. Forbidden transitions occur, but with low probability. Accordingly, the lifetime of excited states which can only decay by a forbidden transition is long.

As stated above, other things being equal, states of higher multiplicity are states of lower total energy. Nevertheless, the ground electronic states of molecules are normally singlet states as a consequence of the selection of orbitals described by the Pauli exclusion principle. Thus transitions between singlet states are the ones most commonly observed.

ABSORPTION SPECTRA OF MOLECULES

Absorption by atoms (at least in the gaseous state) tends to occur at a series of discrete wavelengths; that is, atomic absorption spectra may be classified as line spectra. Absorption by molecules, however, occurs over a continuum of wavelengths. A wavelength of maximum absorption is defined, but absorption of neighboring wavelengths is always to some extent appreciable. Typical atomic and molecular absorption spectra are diagramed in Figs. 7-12a and 7-12b, respectively.

The form of molecular absorption spectra is accounted for by the greater diversity of energy levels at which molecules may exist, as compared with single atoms. In general, the quantized vibrational sublevels of molecular energy are important.

That translational, rotational, and even vibrational sublevels are very closely spaced in comparison to electronic energy levels is made evident by a comparison of typical quantum energies for these transitions. As noted in Chapter 2, electronic transitions correspond in energy to quantum energies of wavelengths in the ultraviolet, visible, and near infrared. A typical electronic energy transition might thus be effected by the absorption of a quantum of light of wavelength 3500 Å. The corresponding quantum energy of this wavelength is

$$(7\text{-}b) \quad E = h\nu = \frac{hc}{\lambda} = \frac{6.6 \times 10^{-27} \times 3 \times 10^{10}}{3.5 \times 10^{-5}} = 6 \times 10^{-12} \text{ erg/quantum.}$$

Rotational transitions, on the other hand, are excited by infrared radiations of wavelengths longer than about 20 μ. Thus, for a typical transition at a wavelength of 50 μ,

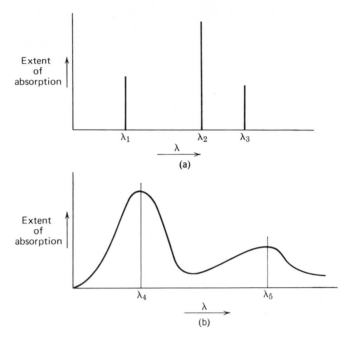

Fig. 7-12 Absorption spectra: (*a*) atomic; (*b*) molecular.

$$\text{(7-c)} \qquad E = \frac{6.6 \times 10^{-27} \times 3 \times 10^{10}}{5 \times 10^{-3}} = 4 \times 10^{-14} \quad \text{erg/quantum.}$$

Energy differences corresponding to rotational transitions are thus of the order of only one hundredth as great as those associated with electronic energy level transitions.

The means of representing molecular energy levels have been described on p. 122. The phenomena of absorption and emission may now be considered in some detail with respect to energy level diagrams.

The time which elapses during the absorption (or emission) of a photon is of the order of only 10^{-15} sec. This is a short time with respect to other electronic events, or to the time required for nuclear movements. (The generality that events of absorption and emission occur over periods which are short with respect to the time required for nuclear rearrangements is known as the *Franck-Condon principle*.) The time occupied by molecular vibrations and vibrational transitions is of the order of 10^{-12} sec. The lifetime of excited electronic states which can decay by an "allowable" transition (i.e., as explained on p. 131, for transitions to a state of identical multiplicity) is in the range from 10^{-9} to 10^{-7} sec. (If the decay of excited states is forbidden, their lifetimes may be much longer than this.) Two important consequences of these relative time values are: (a) internuclear distances do not change during the process of absorption or emission (since the time required for absorption is very much less than the period of vibration): and (b) since many cycles of vibration may occur during the lifetime of excited states, molecules undergo complete vibrational relaxation during their excited lifetimes. That is, the molecule, in general, decays to the lowest vibrational level of an excited electronic state before decaying to a state of lower electronic energy. (Note that this generality is not universal but applies to molecules in *solution*, as is the case in systems of biological interest.)

As shown in Fig. 7-13, the properties of molecular excitation are revealed by the Franck-Condon type of energy level diagram. Energy level transitions, designated by subscripts 1, 2, 3, and 4 in Fig. 7-13, occur between conditions of equal internuclear separation r and thus are represented by vertical lines. The wavelength of maximum absorption is that of quantum energy corresponding to a transition from the most probable internuclear separation in the ground state to an allowed vibrational level in the excited state. This is the transition designated λ_1 in Fig. 7-13. There exist, however, many other wavelengths which may be absorbed in effecting the electronic transition from G to E^*. They include transitions from other internuclear separations within the lowest vibrational

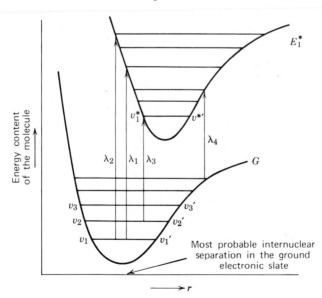

Fig. 7-13 Franck-Condon diagram representing light absorption by a molecule.

level (e.g., λ_2 of Fig. 7-13) and transitions from higher vibrational levels of the ground state to the various vibrational levels of the excited state (e.g., λ_3 and λ_4 of Fig. 7-13). Note that absorption of λ_3 or λ_4 is less probable than of λ_1 or λ_2, since relatively few molecules exist in the second or fifth vibrational levels at ordinary temperatures. Similarly, absorption of λ_3 is more probable than of λ_4.

As there are, in general, very many vibrational energy levels possible in both the ground and excited states, it will be clear that many different wavelengths correspond to possible transitions of the molecular energy level. The correspondingly large number of absorption lines in fact overlap to form a continuum. In spectra that show "fine structure" the increased probability of certain transitions is, however, revealed.

The transitions illustrated in Fig. 7-13 are also represented in a Jablonski diagram in Fig. 7-14.

The explanation for the complete *transmission* of certain wavelengths of light by any given substance becomes obvious from the foregoing discussion; if no possible transitions in the energy level of the molecule correspond to the quantum energy of incident light, absorption cannot occur.

Thus far, absorption has been described entirely as a quantum effect. It may be helpful to consider absorption phenomena from the point of view of the wave nature of light also. This is done on p. 143, after the extension of this discussion of quantum interactions to description of the *decay* of the excited state.

The quantitative expression and measurement of light absorption is discussed in Chapter 24.

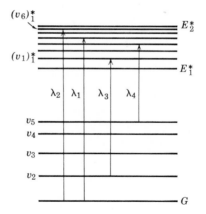

Fig. 7-14 Jablonski diagram representing light absorption by a molecule.

DECAY OF EXCITED STATES

A molecule which is raised to an excited electronic energy level by absorption of a photon must dissipate its excess energy by some means. A large fraction of the energy may be lost by reradiative or luminescent processes, including fluorescence, delayed fluorescence, or phosphorescence. Alternatively, nonradiative energy losses may occur, including energy transfers of various types, internal conversion processes, and photochemical reactions. The nature of each of these processes is now discussed.

FLUORESCENCE

Fluorescence may be described operationally as the "immediate" reemission of absorbed light energy. The wavelengths fluoresced extend to longer values than those absorbed, although there is generally some overlap between the shorter wavelengths fluoresced and the longer wavelengths absorbed by the same molecule. Absorption, fluorescence, and also phosphorescence spectra of a typical molecule are sketched in Fig. 7-15. When spectra are plotted on a frequency scale (rather than a wavelength scale, as shown) the absorption and fluorescence spectra are mirror images of each other.

Quantum mechanically, fluorescence is defined as the emission of light in association with "allowed" energy level transitions. Since the ground state of molecules is normally a singlet state (i.e., a state in which all electronic spins are paired), fluorescence is normally associated with singlet-singlet transitions. In Fig. 7-15 two absorption maxima at λ_1 and λ_2 are indicated. They correspond, respectively, to excitation to two different excited states, E_2^* and E_1^*, such that E_2^* is of higher energy than E_1^*. Each of the excited states is a singlet; otherwise, absorption of light of the corresponding wavelength would be highly improbable.

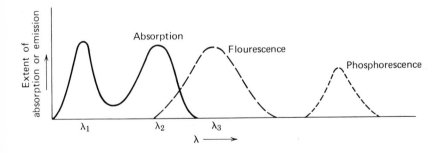

Fig. 7-15 Relationship between absorption, fluorescence and phosphorescence spectra.

Figure 7-16 is a diagram of the absorption of energy at the longer of the two wavelengths, λ_2, and of the subsequent fluorescent decay of the excited state of the molecule. As shown, photon absorption raises the molecule at the most probable internuclear radius to an excited vibrational level; in this case, to the level $4*4'*$ of the excited electronic state E_1^*. After absorption, the molecule undergoes many cycles of vibration during its excited lifetime. Vibrational relaxation by quantum losses of vibrational energy thus takes place. Fluorescent decay to the ground state then occurs after a total excited lifetime which is of the order of 10^{-8} sec. At the time of the decay the molecule has reached the lowest vibrational level of the excited state, in which, as shown, the internuclear separation oscillates between extreme values of r' and r''. The wavelength of *maximum* fluorescent intensity is λ_3, corresponding to decay from the most probable internuclear separation in the excited states. A continuum of other wavelengths, such as λ_4, are also fluoresced, each representing a transition from the lowest vibrational level of the excited state to a quantized vibrational level of the ground state. Since appreciable loss of vibrational energy thus precedes the emission of fluorescence, the wavelength of fluoresced radiation must in general, be *longer* than that of the light initially absorbed.

Figure 7-17 illustrates the absorption of light at λ_1, the *shorter* of the

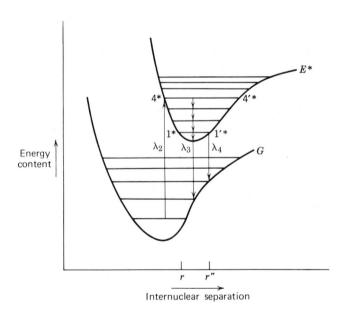

Fig. 7-16 Franck-Condon diagram representing absorption and fluorescence.

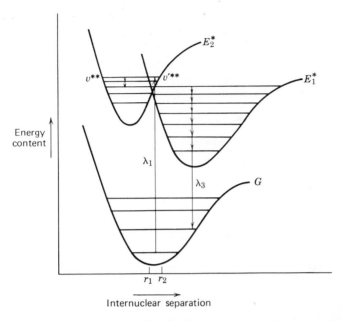

Fig. 7-17 Fluorescence by a molecule following excitation to a higher excited state.

two wavelengths of maximum absorption shown in Fig. 7-15. By this absorption the molecule is excited to electronic state E_2^* which, like E_1^*, has an excited lifetime that is 10^3 or 10^4 times as long as the period of vibration of the molecule. Decay of vibrational energy proceeds from the level $v^{**}v'^{**}$ to which the molecule is excited. A range of values of the internuclear separation (between r_1 and r_2, as indicated in Fig. 7-17) is common to both the excited states, E_2^* and E_1^*. It is found that a molecule may rather readily decay to an excited electronic state of lower energy during the process of vibrational relaxation. A change in electronic state at a condition of constant internuclear separation is thereby acheived without the emission of a photon. The process is termed "*internal conversion.*" The time occupied by the process of internal conversion is found to be of the order of 10^{-13} sec, that is, somewhat shorter than the period of molecular vibration. Once the molecule enters the lower excited state E_1^*, by the internal conversion route, it adopts the modes of vibration characteristic of E_1^*. The subsequent decay of vibrational energy and the loss of electronic energy by emission of λ_3 and other wavelengths are identical with those which occur when E_1^* is excited directly. The fluorescence spectrum excited by light of wavelength λ_1 is thus *identical* with that emitted upon irradiation by light of wavelength λ_2.

It should be noted that, whereas "internal conversion" has been defined here as the exchange of electronic states at constant internuclear separation, this term is used more generally to describe losses of energy by the combination of electronic energy level exchange with the associated processes of vibrational relaxation.

In individual atoms vibrational sublevels of energy comparable to those of molecules do not exist. Reemission thus normally occurs directly from the excited state produced by absorption; that is, the absorbed and emitted quanta are of identical frequencies. The process is then said to be one of *resonance radiation.*

PHOSPHORESCENCE

Phosphorescence may be described operationally as the "delayed" emission of light following excitation. The delay time for phosphorescence is characteristic of the system observed and may range from about 10^{-4} sec to many seconds. Phosphorescence occurs at somewhat longer wavelengths than fluorescence by the same molecule, as has been illustrated in Fig. 7-15.

Quantum mechanically, phosphorescence is defined as the radiation emitted in association with forbidden transitions between molecular energy levels, that is, with transitions between states of different multiplicity. Since the ground state of molecules is commonly a singlet state, phosphorescent transitions are normally those which occur between a triplet state (a state in which there are two unpaired electron spins) and a singlet state. Obviously, transition to a triplet state is a prerequisite for the emission of phosphorescence. Direct excitation to the triplet state by absorption is improbable. The energy levels of excited singlet and triplet states are often very close, however, so that singlet and triplet states may share certain values of the internuclear separation. Processes of internal conversion then occur with a reasonably high probability. The process of internal conversion between states of different multiplicity is termed *"intersystem crossing."*

Whereas internal conversion between states of identical multiplicity occurs during a time of about 10^{-13} sec, intersystem crossing, a forbidden process, is only 10^{-5} or 10^{-6} as probable an event, and thus occurs only after 10^{-7} or 10^{-8} sec. The frequency of intersystem crossing is thus of the same order of magnitude as that of fluorescent decay from the singlet state. Thus the proportions of fluorescent and phosphorescent emission depend upon the specific values of the lifetimes in any given system. Once a molecule enters the triplet state by intersystem crossing, it loses energy by vibrational decay, reaching the lowest vibrational level of the triplet state. From there it may, after a lifetime which is now not 10^{-8} sec but 10^{-4} sec or longer (since the decay to the singlet state is also a forbidden process), decay by emission to a vibrational level of the ground state. The processes of absorption, intersystem crossing, and phosphorescence are shown in a Franck-Condon diagram in Fig. 7-18 and by Jablonski diagrams in Figs. 7-19. Since, as noted on p. 130, the energy of triplet states is in general less than that of the corresponding singlets, phosphorescent emission represents a smaller energy loss than the corresponding fluorescent emission. Thus phosphorescent emission, as shown in Fig. 7-15, occurs at longer wavelengths than does fluorescence.

While decay from the lowest triplet state may occur by phosphorescence, this process is delayed, so that nonradiative processes of energy loss become much more probable than in the corresponding decays from singlet states. Collisional losses of excitation energy (cf. p. 142) are of particular importance. In solutions at room temperature intermolecular collisions are frequent, and normally account almost completely for energy losses from triplet states. Thus phosphorescence is rarely observed at room temperature, although it may be observed from many compounds at low temperatures. Phosphorescent emission is usually studied at liquid nitrogen temperature ($-196°C$).

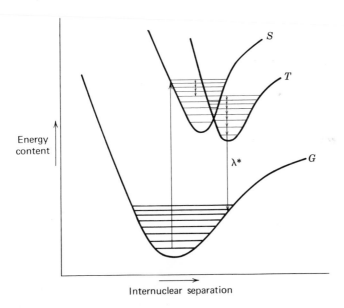

Fig. 7-18 Franck-Condon diagram, representing absorption, intersystem crossing, and phosphorescence.

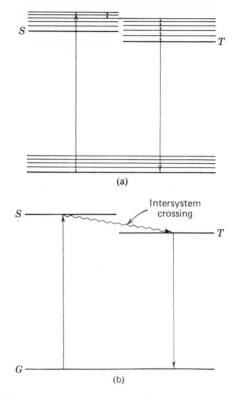

Fig. 7-19 Jablonski diagrams, representing absorption, intersystem crossing and phosphorescence.

DELAYED FLUORESCENCE

Delayed fluorescence is the emission of wavelengths characteristic of the fluorescence spectrum of a molecule after delays characteristic of phosphorescence. The effect is thought to occur by means of a two-photon mechanism, as summarized by

$$
\begin{array}{lll}
\text{(7-d)} & 2G + 2h\nu \rightarrow 2E_1^* & \text{(absorption)}, \\
& 2E_1^* \rightarrow 2T & \text{(intersystem crossing)}, \\
& T + T \rightarrow E_1^* + G & \text{(energy exchange)}, \\
& E_1^* \rightarrow G + h\nu' & \text{(fluorescence)}.
\end{array}
$$

Thus the characteristic mechanism of biphotonic delayed fluorescence is the exchange of energy between triplet states. That the extent to which delayed fluorescence occurs is often proportional to the square of incident intensity is evidence which supports this two-photon mechanism.

However, an alternative mechanism known as α-fluorescence is operative in some systems. Then molecules in the triplet state are returned by thermal excitation to the more energetic singlet state, from which they fluoresce.

EXCIMER FLUORESCENCE

At relatively high concentrations of fluorescing molecules the characteristic fluorescence may tend to disappear, and emission at some different wavelength is observed. This effect may be attributed to the formation of *excimers*. An excimer is a dimer which exists *only* in the excited state. The mechanism of excimer fluorescence is indicated by

(7-e)
$$A + h\nu \rightarrow A^*$$ (absorption),
$$A^* + A \rightarrow (AA)^*$$ (dimerization = excimer formation),
$$(AA)^* \rightarrow 2A + h\nu'$$ (excimer fluorescence).

Mixed excimers, that is, excited associations of two different molecules, may also occur.

ENERGY TRANSFER

Noncollisional transfer of energy between molecules or between portions of molecules has been observed in a number of cases. Such transfers may be definitively distinguished from the so-called *"trivial process"* in which energy fluoresced by one molecule in a solution is reabsorbed and then reradiated by a second molecule. The trivial process is summarized in

(7-f)
$$A + h\nu \rightarrow A^*$$
$$A^* \rightarrow A + h\nu',$$
$$h\nu' + B \rightarrow B^*,$$
$$B^* \rightarrow B + h\nu''.$$

(It may be noted that the trivial process may be, quantitatively, an important route for the dissipation of energy; its triviality exists only from the point of view of theoretical interest.)

Radiationless energy transfer can occur by a mechanism of *resonance* between donor and acceptor molecules which share a common vibrational frequency. In such a transfer the system (excited donor molecule + unexcited acceptor molecule) is quantum mechanically indistinguishable from the system (unexcited donor molecule + excited acceptor molecule). An equilibrium between the molecular states may thus be established:

(7-g)
$$A + h\nu \rightarrow A^*, \qquad A^* + B \rightleftharpoons A + B^*$$

Fluorescence characteristic of B^* rather than of A^* is subsequently observed if the lowest vibrational energy level of B^* is of significantly lower energy than that of A^*, that is, if the vibrational relaxation of B^* provides a *sink* for the energy of the system. The light then fluoresced by B^* is characteristic of that molecule, even though no wavelengths may have been present in the exciting radiation which would normally be absorbed by B. Although the same is true of the trivial process, resonance energy transfer can be distinguished experimentally from the reabsorption-reemission mechanism because transfer is independent of the geometry of the system and also tends to reduce the excited lifetime of the absorber.

Resonance energy transfer may occur with high efficiency but is effective only over very short distances. Theory, which is generally confirmed by experiment, predicts that separations r_0 for which resonance transfer occurs with 50% efficiency, should be of the order of 20 Å. Efficiency of transfer varies as a function of $(1/r^6)$ and thus rapidly becomes negligible at distances greater than r_0.

Resonance energy transfer may be observed not only between singlet states but also between triplet states of overlapping vibrational energy levels.

Energy exchange by *exciton transfer* can occur in systems that either are crystalline or, to some extent, regularly structured. For example, the peptide backbone of protein molecules may be sufficiently "crystalline" for this mechanism of transfer to apply. The energy is transported by electrons which have been raised to the exciton band (cf. Fig. 7-2) and may thus be transported over distances as large as several hundred angstrom units.

Collisional transfers of energy are quantitatively important processes, especially in the quenching of the long lifetime states from which phosphorescence may occur. Essentially the mechanism is simply one in which the energy of the excited state is transferred upon "contact" to the "quencher" molecule. The latter is thereby excited vibrationally, rotationally, or translationally, that is, by the reaction

(7-h) $$A^* + Q \rightarrow A + Q^*.$$

The action of the quencher molecule may, however, be more specific than is indicated by (7-h). For example, the quencher molecule may interact with the excited state A^* by an exchange of electrons, forming an ion pair. Dissociation of the ion pair may subsequently release the energy of the combination in the form of heat:

(7-i) $$A^* + Q \rightarrow (A^- Q^+) \rightarrow A + Q + \text{heat}.$$

PHOTOCHEMICAL REACTIONS

As Figs. 7-4 and 7-7 imply, the distribution of electron density in an atom (or molecule) varies markedly as different orbitals are assumed by the electrons of a system. For example, molecules that normally are planar in configuration may assume a three-dimensional structure in excited states. The chemical reactivity of atoms and molecules can be interpreted in terms of the localization of regions of deficient or of excessive electron density; that is, the chemical properties of any species are determined by its electron density distribution. From this point of view it is clear that the chemical properties of the excited electronic states of a molecule may be entirely different from those of the ground state. Correspondingly the photochemical reactions undergone by the excited molecule may differ completely from those catalyzed by heating, which normally excites only the higher vibrational levels of the electronic ground state. Photochemical processes may thus dissipate excitation energy before luminescent or other energy-loss processes can occur.

QUANTUM YIELDS

As discussed in the preceding sections, a number of different processes compete as routes for the decay of absorbed energy. The extent to which decay occurs by means of fluorescence, phosphorescence, transfer, etc., is expressed by the *quantum yield*, designated Φ, for the corresponding process. The fluorescence quantum yield, for example, is

(7-j) $$\Phi_{FL} = \frac{\text{number of quanta fluoresced}}{\text{number of quanta absorbed}}.$$

Quantum yields for phosphorescence or transfer are similarly specified. The quantum yield for photochemical reaction is, likewise,

(7-k)
$$\Phi_{Photochem} = \frac{\text{number of molecules reacted}}{\text{number of quanta absorbed}} = \frac{\text{number of moles reacted}}{\text{number of einsteins absorbed}}.$$

Means of determining quantum yield are described in Chapter 25.

ABSORPTION AND TRANSMISSION FROM THE POINT OF VIEW OF THE WAVE THEORY OF LIGHT

Absorption may be discussed in terms of the wave theory of light in the following manner. Atoms and molecules possess certain inherent modes of vibration, as determined by their electronic structure. If a light wave encounters a molecule for which an inherent mode of vibration is of the same frequency as the vibration of the light, then the energy of the light beam can be imparted to the molecule. The phenomenon of resonance produces a vibration of increased amplitude within the absorbing molecule. If, however, the frequencies of vibration of light and molecules are not matched, the light energy is immediately reradiated. The nonabsorbing molecule, like all points on an advancing wavefront, acts as a point source of light, which radiates in all directions.

According to this view, all media should diffusely scatter light which is not of a wavelength such that specific absorption occurs. The scattered light should, furthermore, be unaltered in wavelength from the incident light. This phenomenon is in fact observed and is known as the *Tyndall effect*. It frequently happens, however, that the scattered radiation is so interrelated in phase that, as a consequence of destructive interference, there is appreciable resultant amplitude and intensity only in the direction of the forward-going light beam. In the case of solutions or suspensions in which the dimensions of the solute particles are appreciable with respect to the wavelength of light, or of media in which atoms and molecules are regularly oriented, scattered amplitudes may reinforce in certain directions, so that appreciable scattered intensities are observed only in preferred directions. There exist also certain scattering processes (Compton effect, Raman effect) which are associated with changes in the wavelength of the scattered light; they are discussed on p. 149.

The wavelets scattered by the molecules of a medium are in general altered in *phase* with respect to the incident beam. Consequently, successive interference of the incident beam with the scattered wavelets produces a continuous shift in the phase of the on-going beam. As a result there is a closer spacing of troughs and crests in the on-going beam than there would be in the same wave train traveling through a vacuum. This statement is tantamount to saying that the *velocity* of the transmitted beam is reduced in comparison to that of the incident light, that is, the refractive index of the medium is greater than one. The extent of the phase changes that occur when light is reradiated by the atoms of the medium thus determines the velocity of light in that medium. Since media of high electron density offer the most numerous opportunities for reradiation of secondary wavelets, these media tend to have high refractive indices.

It is also possible to describe the absorption and transmission of light in a manner which combines a wave picture with present understanding of atomic and molecular structures. As explained on p. 126, the lowest energy state corresponding to any value of the principle quantum number n of an electron is an orbital of spherical symmetry, whereas orbitals of higher energy are in some way directional. The action of the electric vector (E vector) of an incident light wave can thus be viewed as a *distortion* of the electronic orbital in the direction of this vector. Alternatively, it may be said that the light induces an electric dipole within the molecule. If an incident photon is of precisely the energy required so to distort the electron cloud that it occupies a specific orbital of higher energy, absorption occurs. These effects are diagramed in Fig. 7-20. Figure 7-20a represents a light ray propagating in the $+z$ direction and which is plane polarized in the y direction. This wave motion is incident upon a spherically symmetric electron orbital such as that shown in Fig. 7-20b. (This might be the $1s$ electron of hydrogen, the outer $2s$ electron of helium, etc.). Absorption of the quantum of light results in the distortion of this orbital along the direction of polarization. As shown in Fig. 7-20c distortion of the electron cloud may result in the formation of a $2p$ orbital (for which $n = 2$ and $l = 1$). If the incident quantum energy is not exactly equal to the difference in energy between the s and the $2p$ orbitals, there is a transitory distortion of the orbital shown in Fig. 7-20b, but *no net work is done on the electron by the light*, and the electron cloud regains its original form as soon as the quantum has passed. If the "size" of the quantum is equated to the wavelength of the light, it can be estimated that

(7-1)
Time for passage of quantum $\cong$ time occupied by absorption or emission

$$\cong \frac{\text{"size" of quantum}}{\text{velocity of quantum}} \cong \frac{\lambda}{c} \cong 10^{-5} \text{ sec.}$$

Reduction in velocity of light transmitted by a medium is thus not a reduction of quantum velocity, which is inevitably c, but is a consequence of the delay of quanta in the production of highly transitory distorted states of electrons.

POLARIZATION EFFECTS ASSOCIATED WITH ABSORPTION AND FLUORESCENCE

Figure 7-20 depicted the absorption of light as a distortion of electronic orbitals in the direction of the E vector of the absorbed light. If incident light is unpolarized, distortions occur in all possible directions perpendicular to the direction of propagation of the beam; if incident light is plane polarized, all electronic orbitals are distorted in the single direction determined by the plane of polarization. For an isotropic system, such as an s electron, the extent of absorption is independent of the state of polarization of the incident light. However, if the initial distribution of electron density within the absorbing atom is asymmetric, distortion of electronic orbitals into configurations of higher energy may occur only

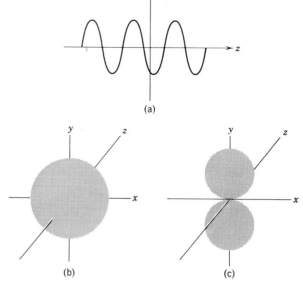

Fig. 7-20 Distortion of atomic orbitals by electromagnetic radiation: (*a*) polarized incident radiation; (*b*) initial spherically symmetric orbital (*c*) distorted orbital.

when the E vector of the incident light is appropriately oriented. Thus light polarized in an appropriate plane is preferentially absorbed by the molecule. If individual molecules are oriented at random, the population of molecules does not preferentially absorb according to the plane of polarization. If the electronically asymmetric molecules are uniformly oriented, however, absorption by the substance in bulk is determined by the plane of polarization. This is the phenomenon of dichroism described on p. 96.

In the absence of disturbances, light reemitted from an excited molecule tends to be polarized in the same plane as was the absorbed quantum. The polarization P of light leaving a specimen in a direction at right angles to the incident beam and to the direction of vibration of the incident polarized light, is

(7-m) $$\frac{I_{\|}-I_{\perp}}{I_{\|}+I_{\perp}} = P,$$

where $I_{\|}$ is the intensity polarized in the plane of the incident light, and $I_{\perp}$ is the intensity polarized in the plane perpendicular thereto. Intrinsically the value of P is determined by the nature of the electronic transitions associated with luminescence and may assume values which range between $+\frac{1}{2}$ and $-\frac{1}{3}$. (For a proof of this statement see [1].) However, extrinsic effects, including molecular rotations and interactions with other molecules, may reduce the value of P. The exploitation of measurements of polarization in the study of macromolecules is discussed on p. 571.

REFLECTION AND SCATTERING

According to Huygens' principle, every point on an advancing wavefront acts as a source of secondary wavelets. Thus molecules at the surface of an illuminated medium may be thought of as a series of point sources of light. Wavelets that leave individual molecules interfere with each other, producing resultant intensities in certain directions. When the resultant intensity travels in the direction of the incident beam or, more generally, in accordance with the law of refraction, the light is said to be *transmitted*. Intensity traveling in other directions is said to be *scattered* or *diffracted*. The terms "scattered" and "diffracted" are equivalent, except that the term "diffracted" tends to be applied when interference between individual amplitudes is considered. When sources of secondary disturbance are located along a plane, as within a crystal or at the surface of a medium, the reinforcement of scattered amplitudes results in the reflection of intensity; the direction of travel can be predicted by the law of reflection. Note that there is no fundamental physical distinction between the terms "scattered", "diffracted", and "reflected", but that each is appropriate to certain types of experimental arrangement.

Reflection is shown from the point of view of Huygens' principle in Fig. 7-21. In Fig. 7-21a a wavefront ABC travels toward the reflecting surface. The envelope $A'B'C'$ of secondary wavelets defines a subsequent position of the wavefront. (Note that, in Fig. 7-21, variations of *amplitudes* with direction, given quantitatively by the obliquity factor (3-rr), are indicated by the heaviness of the lines representing the positions of crests of secondary wavelets.) Figures 7-21b and 7-21c show instantaneous views as the wavefront is incident upon the reflecting surface. (Transmission, if any, is ignored.) Envelopes of the individual wavelets then form the bent wavefront $A''B''C''$. Figure 7-21d shows the reflected wavefront leaving the surface.

If no energy is imparted by the incident light to the molecules or atoms at the surface which serve as secondary sources (or if none of these particles, preexisting in an excited state, lose energy to the light beam), the wavelengths of the incident and scattered intensity must be identical. The term *"Rayleigh scattering"* applies to this type of light scattering, in which there is no change in wavelength. Since molecules are constantly in motion, there exist spontaneous density fluctuations within any medium.

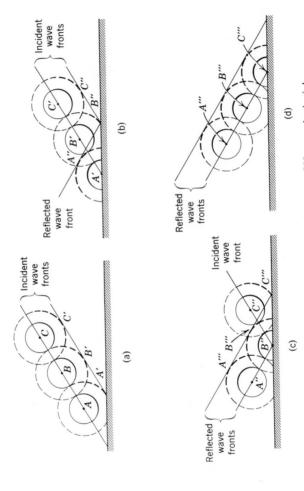

Fig. 7-21 Reflection, as accounted for in terms of Huygens' principle.

The scattering properties of media were related by Rayleigh to the mean square deviation of average density in the medium. He showed that for a "homogeneous" medium, or for any medium in which particle diameters are less than about $\lambda/20$, the intensity of scattered radiation varies as a function of $\lambda^{-1/4}$. Thus the relative proportions of wavelengths in a scattered polychromatic beam are altered by the process of Rayleigh scattering. Among other results, this treatment accounts for the blue color of the sky. (For a development of Rayleigh's theory see [2].)

Other types of scattering occur in which an exchange of energy with the medium results in a change in the frequency of the scattered light. The *Compton* and *Raman* effects are examples.

The *Compton* effect results from the "mechanical" interaction of incident x-ray quanta with the scatterer. Irradiation of matter by wavelengths in the wavelength range 0.1 to 1.0 Å produces a component of scattered intensity at the incident wavelength, but also a second component of slightly longer wavelength. The latter results from a slight loss of energy imparted to the electrons of the medium by the incident quantum. The "recoil electrons" so energized can be detected experimentally. The change in wavelength of the scattered x-rays is of the order of a fraction of an Angstrom unit. The magnitude of the Compton effect is negligible at longer wavelengths.

The *Raman* effect reveals rotational and, to some extent, vibrational transitions in molecular energy levels as shifts in the wavelength of scattered light. Figure 7-22a represents the spectral composition of an illuminating beam, and Fig. 7-22b shows the spectral composition of the scattered light. The intensity of light scattered at λ_0 is in general very much greater than that of the Raman lines. The differences between λ_0 and other scattered wavelengths, such as λ_1 and λ_2, correspond to possible rotational energy level transitions of the scattering molecules. Raman lines may result either from the gain of energy by the molecule from the beam or from loss of energy from the molecule to the beam. Thus Raman lines may be of shorter wavelength than the incident light. This is possible because appreciable numbers of molecules exist in excited rotational states at ordinary temperatures and thus are able to impart energy to the scattered beam. However, Raman lines which correspond to vibrational energy transitions usually are of appreciable intensity only on the longer wavelength, lower energy side of λ_0, as illustrated by λ_3 of Fig. 7-22b.

The value of λ_0 used in Raman spectroscopy is not critical, because it is the *change* of wavelength, not the absolute wavelength of scattered radiation, that is of interest. Observation of Raman spectra requires the use of monochromatic beams of very high intensity; thus the use of the laser (cf. Chapter 22) has aided the development of this type of spectroscopy.

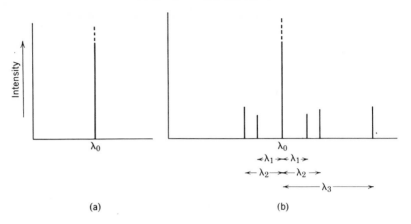

Fig. 7-22 The Raman effect: (a) incident wavelength; (b) spectral composition of scattered light.

PROPORTIONS OF REFLECTED OR REFRACTED AMPLITUDES AND INTENSITIES

Since intensity is proportional to the energy of a light beam and to the square of amplitude, it might appear that the sum of the fractional intensities reflected and transmitted at any interface would necessarily be *one* (in order to account for the conservation of energy). In fact, intensity is a measure of *energy density* rather than of total energy (cf. p. 14); hence the total energy of a light beam varies in proportion to its cross-sectional area. Except at normal incidence, the cross-sectional areas of incident and refracted beams differ. As shown in Fig. 7-23, the cross-sectional area of the incident beam is proportional to AB, that of the refracted beam is proportional to CD. Thus

(7-n) $$\cos i = \frac{AB}{d} \quad \text{and} \quad \cos r' = \frac{CD}{d},$$

where $d = AD$ and i and r' are the angles of incidence and refraction, respectively, so that

(7-o) $$\frac{AB}{CD} = \frac{\cos i}{\cos r'}.$$

That is, the cross-sectional areas of incident and refracted beams are related in the ratio of the cosines of the anges of incidence and refraction. (The cross-sectional area of a beam does not change upon reflection, since angles of incidence and reflection are equal.)

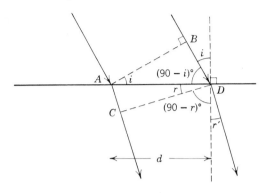

Fig. 7-23.

Furthermore, the intensity is proportional not only to A^2, but also to n, the refractive index of the medium, as given in Chapter 1:

$$\text{(1-q)} \qquad I \equiv \frac{cn}{8\pi} A^2.$$

By combining the considerations of (7-o) and (1-q) the total energy of the incident beam may be equated to the *sum* of the energies of the reflected and refracted beams. For incidence from a medium of refractive index n_1 upon a medium of index n_2, this gives

$$\text{(7-p)} \qquad \underset{\substack{\text{total energy of}\\\text{incident beam}}}{\frac{cn_1}{8\pi} A^2 d} = \underset{\substack{\text{energy of the}\\\text{reflected beam}}}{\frac{cn_1}{8\pi} (rA)^2 d} + \underset{\substack{\text{energy of the}\\\text{refracted beam}}}{\frac{cn_2}{8\pi} (tA)^2 d \frac{\cos r'}{\cos i}},$$

where r is the fractional amplitude reflected and t is the fractional amplitude transmitted (i.e., refracted). Equation 7-p reduces to

$$\text{(7-q)} \qquad 1 = r^2 + \frac{n_2 \cos r'}{n_1 \cos i} t^2.$$

The fractional amplitudes reflected and refracted are determined not only by the properties of the medium and the angle of incidence, but also by the plane of polarization of the incident light. Relations between these parameters are not derived here, but (7-r) is given as an example of expressions that apply to refraction and reflection at the surfaces of dielectric media:

$$\text{(7-r)} \qquad \begin{array}{l}\text{fraction of the amplitude of light}\\\text{polarized } \textit{perpendicular} \text{ to the}\\\text{plane of incidence which is } \textit{reflected}\end{array} = -\frac{\sin (i - r')}{\sin (i + r')} A_0.$$

Light which is incident along the normal cannot be distinguished as polarized perpendicular or parallel to the plane of incidence. Expressions for the proportion of reflected amplitude then reduce to that previously given on p. 47:

(3-jj)
$$\frac{I_{refl}}{I_0} = \frac{n_2 - n_1}{n_2 + n_1}.$$

REFERENCES

General References

D. M. Hercules, in D. M. Hercules, Ed. *Fluorescence and Phosphorescence Analysis*, Interscience, New York, 1966. Chapter 1 (Theory of Luminescence Processes) gives a particularly lucid account of fluorescence, phosphorescence, and related phenomena.

G. Oster, *Sci. Am.* **219**, 158 (1968). This article, "The Chemical Effects of Light," summarizes many of the concepts described in this chapter and discusses some applications of biological interest.

F. K. Richtmyer, E. H. Kennard, and T. Lauritsen, *Introduction to Modern Physics*, McGraw-Hill, New York, 1955. One of a number of modern physics texts which describes the development of knowledge of atomic structure from the study of spectra. See especially Chapter 5 (The Nuclear Atom and the Origin of Spectral Lines) and Chapter 7 (Atomic Structure and Optical Spectra).

V. Weisskopf, *Sci. Am.* **219**, 60 (1968). An article on "How Light Interacts with Matter."

Specific References

[1] G. Weber, in Hercules, *op. cit.*, pp. 220–222.
[2] M. Born, *Optik*, Springer, Berlin, 1933, Section 81.

CHAPTER 8

Geometrical Properties of Lenses

Many properties of lenses can be described in terms of the simple laws of geometrical optics, which were given in Chapter 1 (pp. 3–7). They are discussed in this chapter and the next one; features of lens action which must be accounted for in terms of the physical nature of light are described in Chapter 10. Topics considered here include: the nature of the focusing action of lenses, the location and magnification of images, and the origin of lens aberrations. A number of specific terms which are used to describe lenses and lens action are defined and discussed. Geometrical lens aberrations are described in Chapter 9.

CONVENTIONS FOR DIAGRAMS

In this book, lenses are represented digrammatically, as in Fig. 8-1. This convention is a simple way of representing lens action without introducing extraneous consideration of the nature of the specific lens or lens system used for focusing. The more common and more literal notation, which shows the actual bending of rays at curved surfaces, is used only when it is necessary to illustrate, specifically, the nature of the refraction occurring at a particular surface. (See, for example, Fig. 8-25b.)

Strictly speaking, Fig. 8-1 represents the *single principal plane of a thin lens.* (The

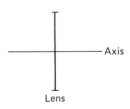

Fig. 8-1 Convention for representation of a lens.

153

concept of principal planes is developed on p. 170.) Nevertheless, the same symbol is used at times to represent lenses which are "thick" (p. 169), for systems of several lenses, and, in later chapters, for electron lenses.

Conventional representations of converging and diverging lens action (cf. p. 167) are shown in Figs. 8-2a and 8-2b, respectively. When neither of these symbols is included in a diagram, *converging* lens action is implied. Apertures (stops; cf. p. 177) are represented by thickened lines, as in Fig. 8-3, while filters are represented as shown in Fig. 8-4.

UNIT OF LENS ACTION

Elementary courses in general physics often seem to create the impression that the thin glass lens is the basic conceptual entity in the geometrical treatment of lenses. The thin lens is convenient to consider because it is the simplest illustration of the way in which refractive effects are exploited to produce focusing. The fundamental unit of optical lens action, however, is *not* the thin lens, but the *single curved surface* between media of different refractive indices. Furthermore, as the discussion of electron lenses in Chapter 17 should make clear, the use of glass or similar transparent refractive solids is not an essential feature of lens action.

OPERATIONAL DEFINITION OF A LENS

The action of an ideal lens can be defined in terms of three operational properties. In addition, it is always assumed that a lens possesses an axis of *rotational* symmetry.

When an object that has an axis of rotational symmetry is rotated about that axis through any arbitrary number of degrees, it is *indistinguishable* in its final orientation from its original orientation. Thus object *A*, shown in Fig. 8-5, has an axis of rotational symmetry which extends perpendicular to the plane of the page through point *O*. Objects *B* and *C* lack such an axis. Note that, if object *A* is a section of a three-dimensional

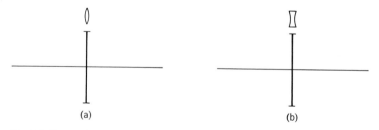

Fig. 8-2 Conventions for representation of lenses: (*a*) converging; (*b*) diverging.

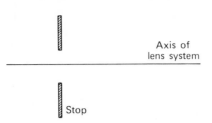

Fig. 8-3 Convention for representation of an aperture.

structure that has *different* curvatures above and below the plane of the page, as shown in Fig. 8-6, no rotational symmetry axis exists at right angles to that shown in Fig. 8-5. Very few objects possess axes of complete rotational symmetry; axes of *n*-fold symmetry ($n = 2, 3, 4, \ldots$) are much more common. For example, after rotation through 180° object *B* of Fig. 8-5 is indistinguishable from its original orientation, and thus is said to possess twofold symmetry. Similarly, object *C* of Fig. 8-5 has an

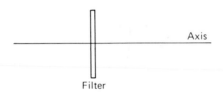

Fig. 8-4 Convention for representation of a filter.

axis of fivefold rotational symmetry. An ideal lens, however, must possess full rotational symmetry with respect to its refractive properties. Aspects of the symmetry properties of matter are described at greater length in Chapter 29.

In considering the threefold action by which the ideal lens is defined it is necessary to think of the specimen to be imaged by the lens as a point source of light or as a collection of point sources. The light emanating

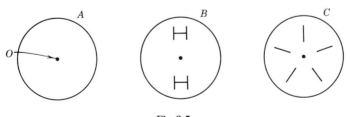

Fig. 8-5.

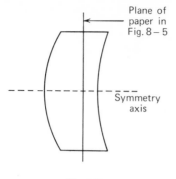

Plane of paper in Fig. 8 – 5

Symmetry axis

Fig. 8-6.

from each point of the object may be represented by any number of rays that diverge in all directions. Some of these rays encounter ("are collected by") the lens. (As shown in Chapter 10, *all* the rays would have to enter the lens in order to form a perfect image.) If the lens is ideal, then, for a point placed on the axis of symmetry, *those rays passing through the lens must be bent by it in such a way that they reunite at, or appear to diverge from, some other single point on the axis of the lens* (first condition). The two possibilities specified by this condition specify the formation of real or virtual images, respectively. Figure 8-7 illustrates the formation of a real image, and Fig. 8-8 the formation of a virtual image. Note that these figures follow the usual convention in which object points are represented as lying to the *left* of the lens. The distinction between real and virtual images is discussed at greater length on pp. 167–168. Note also that the action of the lens shown in both figures is *convergent*.

The action of a nonideal lens with respect to the first condition of ideal lens action is suggested (in an exaggerated fashion) by Fig. 8-9.

A second condition of lens action is that *rays originating from points which lie on a plane perpendicular to the axis, must be imaged in a plane which is also perpendicular to the axis.* The action of a lens which is ideal in this respect is diagramed in Fig. 8-10.

The action of a nonideal lens might be as shown in Fig. 8-11. Note that the lens shown in this figure satisfies the first condition for ideal lens action, since the rays from each object point unite at a single image point. However, the image points corresponding to a *linear* object lie along a

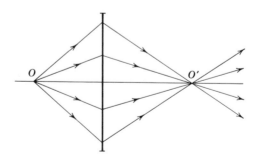

Fig. 8-7 Formation of a real image by an ideal lens.

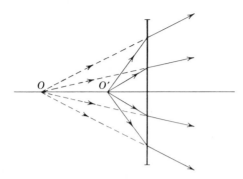

Fig. 8-8 Formation of a virtual image by an ideal lens.

curve. Finally, the third condition is that *the relative linear dimensions of the object must be preserved in the image.* Figure 8-12a, in which rays are omitted for clarity, indicates the imaging by an ideal lens from this point of view.

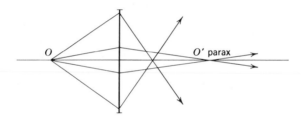

Fig. 8-9 A non-ideal lens.

Figure 8-12b provides an example of an image formed by a lens that is nonideal with respect to the third condition for lens action. The lens forms a planar image of a planar object in which single image points represent single object points. However, the magnification is *less* at the periphery

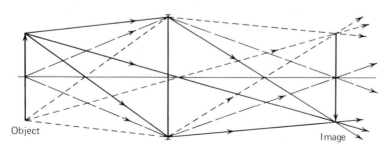

Fig. 8-10 Action of an ideal lens in forming a plane image.

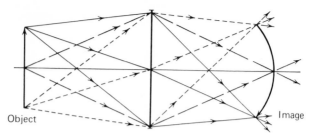

Fig. 8-11 Action of a non-ideal lens, forming a curved image.

of the image than close to the axis. This is indicated by the fact that the cross hatchings $A'A'$ and $B'B'$ of the image appear relatively closer to each other and farther from the axis than do the corresponding cross hatchings AA and BB of the object.

Note that the image formed by an ideal lens may be magnified, demagnified, erect, inverted, or rotated through some arbitrary angle; only *relative dimensions* must correspond faithfully between image and object.

It is possible to interpret an image that is *sharp*, even though curved or distorted, whereas, if no clear point-to-point correspondence exists between object and image, lens action has failed totally. Thus the first of the three requirements for lens action just considered is the most fundamental one.

Real lenses are nonideal, and they generally fail in all three of the conditions just described. For practical reasons most lenses, at least until recent years, have used surfaces of *spherical* curvature exclusively. Geometrical considerations to be developed below show that, that *in principle*, fulfillment of even the first condition for ideal lens action is impossible for these systems. However, even if all geometrical difficulties could be overcome, nonideal lens action would still be obtained as a consequence of the wavelike properties of light. The interaction of waves

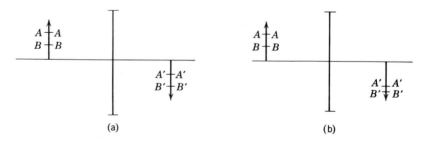

Fig. 8-12 Action of ideal and non-ideal lenses: (*a*) relative linear dimensions preserved by ideal lens; (*b*) relative linear dimensions distorted by non-ideal lens.

causes *point* objects always to be imaged as extended *discs* (the Airy disc; cf. p. 63 and Chapter 10).

IMAGING BY A SINGLE SPHERICAL SURFACE

The single refracting surface of spherical curvature is the fundamental "unit" of focusing action by glass lenses. Although images are rarely formed *inside* the glass, this is physically possible. For example, the curved end of a long, polished glass rod may refract rays to form an image within the rod. More often, rays refracted by one glass surface are refracted again at a second surface before forming an image. The image observed in air thus arises from the successive effects of two (or more) surfaces.

If the boundary between media of different refractive indices acts as a converging lens (i.e., with respect to the first condition for perfect lens action), all rays from an object point O reunite at the corresponding image point O', as shown in Fig. 8-13. In Fig. 8-13, the point A, at which the refracting surface crosses the axis (the lens axis; axis of symmetry) is known as the *vertex* of the surface. The distance from O to the vertex is the *object distance*, designated s; that from the vertex to O' is the *image distance*, designated s'. Any ray which contributes to the image is associated with a height h, the perpendicular distance from the axis of the refracting surface to the point where the ray intersects the surface. These conventions are illustrated in Fig. 8-14.

In order to demonstrate that a refracting surface acts as a lens (with respect to the first condition for ideal lens action) it is necessary to obtain an expression for s' which is not a function of h. If this can be done, it is established that the image point is the same for all rays which originate from a single object point and which are incident upon the surface, regardless of the angle between the rays and the axis of the system. Such an expression is now derived for refracting surfaces of *spherical* curvature.

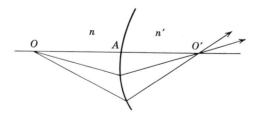

Fig. 8-13 Image formation by a single surface.

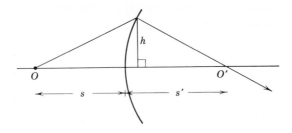

Fig. 8-14.

The law of refraction (1-b) applies at the refracting surface:

(1-b)
$$\frac{\sin i}{\sin r} = \frac{n'}{n},$$

where n is the refractive index of the medium of incidence and n' is the index of the medium of refraction. Since, however, sine terms do not readily cancel in subsequent rearrangements, it is convenient to make use of the expansion of the sine function as an infinite series, as given by

(8-a)
$$\sin i = i - \frac{i^3}{3!} + \frac{i^5}{5!} - \frac{i^7}{7!} + \cdots.$$

[Note that i must be given in *radians* in (8-a) and comparable expressions; for a discussion of angular measure see Appendix A.]

If consideration is limited to the imaging of *paraxial* rays (i.e., rays that travel "very near" the axis, and which thus make "very small" angles with the normal to the surface), then i^3 and higher terms of (8-a) are very small; then the approximation is valid:

(8-b)
$$\sin i \cong i.$$

Equation 8-b, in turn, leads to an approximate form of the law of refraction:

(8-c)
$$\frac{i}{r} \cong \frac{n'}{n}.$$

It is convenient to express the angles of incidence and refraction in terms of the angles between the incident and refracted rays and the lens axis, in the manner indicated by Fig. 8-15. In that figure, C is the center of curvature of the spherical surface: i, the angle of incidence, is then the external angle of a triangle with opposite angles α and β. Thus

(8-d)
$$i = \alpha + \beta,$$

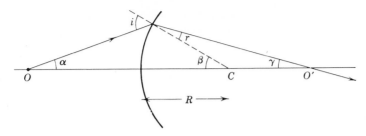

Fig. 8-15.

which is an exact equation. Another exact equation which also follows from the geometry of Fig. 8-15 is

(8-e) $$r = \beta - \gamma.$$

Substitution of (8-d) and (8-e) in the approximate expression (8-c) gives

(8-f) $$\frac{\alpha + \beta}{\beta - \gamma} = \frac{n'}{n} \quad \text{or} \quad n\alpha + n'\gamma = (n' - n)\beta.$$

If α is small, it is approximately true that

(8-g) $$\tan \alpha \cong \frac{h}{s} \cong \alpha.$$

[Note that (8-g) includes two assumptions: (a) that the horizontal distance from the vertex, that is, the point at which the refracting surface crosses its axis of symmetry, to the perpendicular h is negligible, and (b) that $\tan \alpha = \alpha$. The tangent approximation follows from the series expansion

(8-h) $$\tan \theta = \theta + \frac{\theta^3}{3} + \frac{2\theta^5}{15} + \frac{17\theta^7}{315} + \dots,$$

in which θ^3 and higher powers of θ are negligible for small values of θ.]
A similar approximation gives

(8-i) $$\tan \gamma \cong \frac{h}{s'} \cong \gamma.$$

The geometry of Fig. 8-15 gives the relationship

(8-j) $$\tan \beta = \frac{h}{R},$$

from which may be taken the approximation

(8-k) $$\frac{h}{R} \cong \beta.$$

Substitution of the approximate values for α, β, and γ given by (8-g), (8-k), and (8-i), respectively, in (8-f) gives

(8-l) $$\frac{nh}{s} + \frac{n'h}{s'} = \frac{(n'-n)h}{R}.$$

The term h appears as a common factor in (8-l) and cancels, giving the expression

(8-m) $$\boxed{\frac{n}{s} + \frac{n'}{s'} = \frac{n'-n}{R}}.$$

Equation 8-m was originally developed by Gauss; this and related expressions are sometimes referred to as *Gaussian lens equations*. In particular, the Gaussian image point is that specified by s' of (8-m), as opposed to the image of finite dimensions which is actually formed as a consequence of lens aberrations and of diffraction effects.

Equation 8-m provides, as sought, a means of determining the image distance s', which is independent of the height h at which rays are incident. It thus suggests that a single spherical surface acts as an ideal lens. This of course is true only to the degree that the approximations made in obtaining the equation are valid. All of these approximations are derived from the assumption that the image is formed by paraxial rays; that is, that angles of incidence and refraction are small. It is thus correct only to claim that *spherical refracting surfaces act as lenses for paraxial rays*. For rays which make large angles with the axis of symmetry, the first condition of ideal lens action is violated.

THIRD ORDER CORRECTION

It is possible to refine the approximations given in the preceding section to include third order terms; that is, to make use of the approximations

(8-n) $$\sin i \cong i - \frac{i^3}{6} \quad \text{and} \quad \tan i \cong i + \frac{i^3}{3}.$$

Use of these approximations leads to expressions in which h does *not* cancel out; an expression may be obtained that is analogous to (8-m) but contains a number of additional terms, all of which are functions of h. These additional terms can be grouped into five sums, known as the *Seidel sums*, which can be shown to represent, respectively, the five geometrical aberrations of spherical surfaces. (These aberrations are described in Chapter 9.) The sum corresponding to any one aberration can be made to disappear, but not all five sums can be made to vanish simultaneously.

Similarly, the description of lens action by the use of a fifth order approximation,

(8-o)
$$\sin i \cong i - \frac{i^3}{6} + \frac{i^5}{120},$$

results in the mathematical description of further aberrations. The latter become troublesome for rays which make larger angles with the axis than those for which the third order aberrations are of consequence.

The algebraic details of third and fifth order corrections are the consideration of lens designers and need not be described here. For most purposes it is sufficient simply to realize that (8-m) is an *approximation* which correctly applies only to paraxial rays.

The term "paraxial ray" has not been defined quantitatively here. In fact, a paraxial ray is one which travels sufficiently close to the axis to produce an image point smaller than some specified allowable dimension. That is, the quantitative definition of paraxial ray may be established individually for each case considered.

FOCAL POINTS AND FOCAL LENGTHS

The first focal point of a convergent surface is that point on the axis from which the divergent rays are rendered parallel after refraction by the surface, as diagramed in Fig. 8-16. Similarly, the first focal point of a divergent surface is that axial point toward which rays must converge in order to be refracted as a parallel beam. This is illustrated by Fig. 8-17.

The first focal points are located by substitution in (8-m). The image distance in this case is infinity; therefore the term n'/s' vanishes. f', the first focal length, is substituted for the object distance s. Note that the *first focal length is the distance from the first focal point to the vertex of the lens.* This quantity is taken as positive when F_1 is to the *left* of the vertex, and negative when F_1 is to the *right* of the vertex. (Very careful attention to these and similar conventions is required for the location

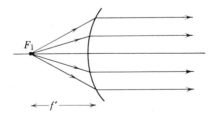

Fig. 8-16 First focal point of a converging surface.

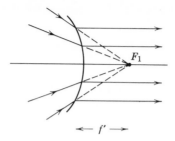

Fig. 8-17 First focal point of a diverging surface.

of images formed by *series* of refracting surfaces.) Substitution in (8-m) then yields

(8-p)
$$\frac{n}{f'} = \frac{n'-n}{R}.$$

Likewise, the second focal point F_2 of a convergent surface is that point to which initially parallel rays converge after refraction at the surface, as illustrated in Fig. 8-18. The second focal point of a diverging surface is that point from which initially parallel rays appear to diverge, as shown in Fig. 8-19. The *second focal length f''* is *the distance from the vertex to the second focal point*, and is taken as positive when F_2 is located to the *right* of the vertex, and negative when F_2 is to the *left* of the vertex. This convention, which applies to *image space*, is thus the reverse of that which applies to F_1 and f', which are located in *object space*. Note that "image space" and "object space" are not physically separate regions but are terms which refer to distances and sign conventions applied to image and object, respectively.

F_2 may be located by substitution in (8-m), now taking the object distance s as infinity, so that the term (n/s) becomes zero, while $s' = f''$, giving

(8-q)
$$\frac{n'}{f''} = \frac{n'-n}{R}.$$

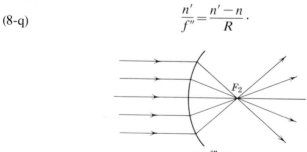

Fig. 8-18 Second focal point of a converging surface.

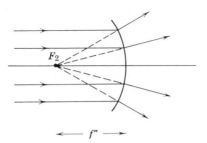

Fig. 8-19 Second focal point of a diverging surface.

Since $n \neq n'$, the two focal lengths f' and f'' are *not* equal for a single refracting surface as they are (*vide infra*) for lenses surrounded by air. If n and n' were equal, no refracting surface would exist.

Note that a short focal length f' or f'' implies that the refracting surface is highly convergent; that is, there is a *large* change in refractive index and/or a small radius of curvature. Thus a lens of "short focal length" is a "strong" lens.

THIN LENSES

The treatment of the single spherical surface can be extended to considerations of the thin lens. The thin lens may be considered to be a pair of superimposed spherical surfaces, that is, two surfaces of which the vertices are located at the same point on their mutual axis. The surfaces of a thin lens are separated by a medium of refractive index n' and may have different radii of curvature R_1 and R_2.

If parallel light were incident upon the first surface of a thin lens, the object distance, now designated s_1, would be infinity, and the single surface equation (8-m) would become

(8-r)
$$s_1' = \frac{n'R_1}{n' - n}.$$

The image which would be formed by the first surface if refraction by the second surface had not previously redirected the rays acts as the object for the second surface. It is important to understand that this *virtual object* may be treated, in terms of signs, distances, and magnifications, exactly as if it were a real object placed at the same position. A similar approach may be used in computing the positions of images formed by any series of refracting surfaces; the image, whether real or virtual, produced by one surface acts as the object for the next one.

Since the vertices of the two surfaces of a thin lens are assumed to coincide, the image distance for the first surface s_1', though numerically equal to the object distance s_2, is opposite in sign, on account of the reversal of sign conventions between object and image space, as noted on p. 164. For a lens illuminated by parallel light, the image distance s_2'' for the second surface is also the second focal length f_2 of the lens as a whole. Thus, by substitution in (8-m),

(8-s)
$$\frac{n'}{s_2}+\frac{n}{f_2}=\frac{n-n'}{R_2}.$$

[Note that n and n' have been interchanged in (8-s) to account for the fact that a glass-to-air refraction is now being considered.] The value $s_2 = s_1'$ from (8-r) may be substituted in (8-s) to give

(8-t)
$$\frac{n}{f_2}=\frac{n-n'}{R_2}+\frac{n'(n'-n)}{n'R_1}=(n'-n)\left(\frac{1}{R_1}-\frac{1}{R_2}\right)$$

as the equation which specifies the second focal length of a thin lens. A comparable equation for the first focal length may be derived in an analogous manner. If the lens is located in air or, more generally, if the refractive index on both sides of the lens is the same, the expressions for the first and second focal lengths become identical. Equation 8-t is then known as the *lens maker's equation*. In practical applications of this and other lens equations, adherence to a consistent system of sign conventions is important (see Appendix E).

Note that the focal lengths are measured from the lens to the corresponding focal points. If the position of the "lens," as opposed to that of either of its vertices, is ambiguous, the lens cannot be considered to be "thin," and (8-t) is inapplicable.

In terms of the concept of principal planes, developed on pp. 170–173, a thin lens is one for which the first and second principal planes coincide.

LOCATION OF IMAGES FORMED BY THIN LENSES

A simple formula can be developed for the location of images formed in air by a thin lens of any given focal length.

For the successive surfaces of the lens in air, (8-m) becomes, respectively,

(8-u)
$$\frac{1}{s_1}+\frac{n'}{s_1'}=\frac{n'-1}{R_1},$$

(8-v)
$$\frac{n'}{s_2'}+\frac{1}{s_2''}=-\frac{n'-1}{R_2},$$

in which s_1 is the object distance both for the first surface and for the lens as a whole, s_1' is the image distance for the first surface, s_2' is the object distance for the second surface, and s_2'' is the image distance for both the second surface and the lens as a whole. Since, as explained above,

(8-w)
$$s_1' = -s_2',$$

(8-u) becomes

(8-x)
$$\frac{1}{s_1} - \frac{n'}{s_1'} = (n' - 1)R_1.$$

Addition of (8-v) and (8-x) then gives

(8-y)
$$\frac{1}{s_2''} + \frac{1}{s_1'} = (n' - 1)\left(\frac{1}{R_1} - \frac{1}{R_2}\right).$$

Substitution of $n = 1$ in (8-t) shows that the right side of (8-y) is equivalent to $1/f = 1/f_1 = 1/f_2$ thus leading to the simple relation

(8-z)
$$\boxed{\frac{1}{s_2''} + \frac{1}{s_1} = \frac{1}{f}.}$$

From (8-z) the properties of images, as briefly summarized in the following section, may be deduced. A complete collection of definitions of terms, sign conventions, and equations is also given in Appendix E.

IMAGING PROPERTIES OF CONVERGING AND DIVERGING LENSES; REAL AND VIRTUAL IMAGES

A *real image* is one at which light rays physically reunite, so that a photographic plate placed at the position of a real image is exposed. A *virtual image* is one from which light rays appear to diverge; rays are not, in fact, concentrated at the position of a virtual image, so that a photographic plate placed at the position of the image is not exposed (i.e., by focused light rays). Note that both real and virtual images are ultimately perceived as a consequence of the formation of a real image on the retina of the eye. If a real image is formed in space by a lens, the rays that diverge from it must be reconverged (by the refractive action of the eye) in order to form the final real image on the retina.

A converging (positive) lens bends light rays toward the axis and is assigned a positive focal length (as stated on p. 163). A converging lens forms a real image of an object placed to the left of the first focal point and a virtual image of an object placed between the focal point and the lens. These statements may be verified by substitution of specific numerical values in (8-z), with due regard to the sign conventions that apply in object and image space, respectively.

The formation of a real image of an axial object point has been diagramed in Fig. 8-7. Figures 8-20 and 8-21 show the formation of real and virtual images, respectively, of extended objects by this type of lens. With reference to Fig. 8-21, it may be noted that rays from the object

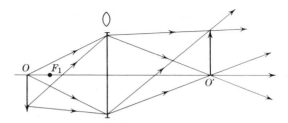

Fig. 8-20 Formation of an extended real image by a converging lens.

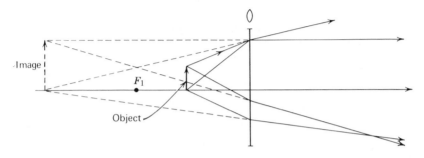

Fig. 8-21 Formation of an extended virtual image by a converging lens.

are so highly divergent upon reaching the lens, so that the converging action of the lens is insufficient to bend the rays from each object point far enough that they reunite to the right of the lens. Note that, with respect to the object, the image is erect in Fig. 8-21 but inverted in Fig. 8-20.

A diverging (negative) lens bends light rays *away* from the axis and is assigned a negative focal length. The formation of a virtual image of an axial object point by a negative lens has been diagramed in Fig. 8-8; Fig. 8-22 illustrates the formation of an image of an extended object by a negative lens. As suggested by Fig. 8-22, an object placed anywhere to

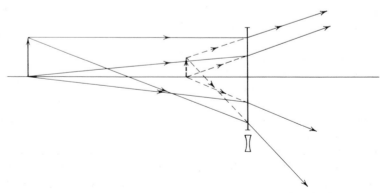

Fig. 8-22 Formation of an extended virtual image by a diverging lens.

the left of a diverging lens results in an erect virtual image. This statement may be verified by substitution of specific numerical values in (8-z) (with due regard to sign conventions).

CONJUGATE POINTS

When a real image is formed, the image and object positions are said to be *conjugate* points. As a consequence of the principle of the reversibility of rays (p. 3), an object placed in the position of a real image forms an image in the position of the original object. Thus object and image space are continuums of conjugate positions.

The positions of an object and its virtual image are not necessarily conjugate, however. This is evident from the fact that rays do not actually converge at a virtual image. The failure of an object placed at the position of a virtual object to form an image at the position of the original object may also be verified by calculation in specific cases.

LENSES IN SUCCESSION

When two or more lenses are arranged in succession along an axis (i.e. coaxially), the image formed by one lens constitutes the object for the subsequent lens, whether or not a real intermediate image is formed. Two well-separated converging lenses, and the formation of a real image between them, are shown in Fig. 8-23. Imaging by the same lens pair, placed closer together, is shown in Fig. 8-24. The image formed by the

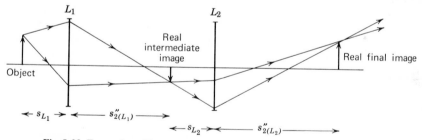

Fig. 8-23 Formation of intermediate and final images by successive lenses.

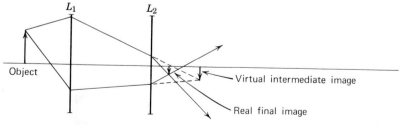

Fig. 8-24 Formation of intermediate virtual image and real final image by successive lenses.

first lens L_1 would then lie to the right of L_2. The convergent action of L_2, however, causes a final real image to be formed closer to the lens pair than the intermediate image would be.

THICK LENSES

A thick lens is one in which the vertices of the refracting surfaces are separated by an appreciable distance d. Introduction of this parameter leads to the development of equations describing the thick lens which are, in general, more complex in form than those that describe the thin lens. The two focal lengths of a thick lens in air, however, are identical, as for a thin lens, and the location of the images formed by a thick lens may be determined with the use of (8-z) *if the object distance, image distance, and focal length are appropriately defined.* For example, the definition of the object distance as the distance from the object to the lens is obviously inadequate when the position of the lens is ambiguous. In fact, distances with respect to thick lenses must be defined with respect to the positions of the *principal planes* of the lens. The concept of principal planes and their properties are described in the following section.

If the locations of the principal planes of a thick lens are known, image positions may be computed directly from

(8-z)
$$\boxed{\frac{1}{s_1} + \frac{1}{s_2''} = \frac{1}{f}},$$

where s_1 is the distance from the object to the first principal plane, s_2'' is the distance from the second prinicpal plane to the image, and the focal length f is the distance between either principal plane and the corresponding focal point. The sign convention applicable to these distances is the same as for single surfaces or thin lenses.

The positions of the principal planes may be computed if the index of refraction of the lens and the curvatures of its surfaces are known. The computation procedure is summarized at the end of the next section, and specific equations are given in Appendix E.

PRINCIPAL PLANES

The focal length of a thick lens is *not*, as might seem intuitively reasonable, the distance from the focal point to the lens surface or to the center of the lens, but to the corresponding prinicpal plane of the lens. The principal planes are the planes of effective redirection of the rays which pass through the lens. In general, a lens has *two* principal planes, although, in the special case of the thin lens, the two coincide at the position of the lens itself. The principal planes h_1 and h_2 are perpendicular to the lens axis (at least ideally) and cut the lens axis at the *principal points*, which are designated H_1 and H_2, respectively. These conventions are illustrated in Fig. 8-25a.

It should be understood that the principal planes have no physical existence and may, in fact, even be located outside the boundaries of the lens. Nevertheless, the refractive behavior of the lens is such that imaging occurs *as if* all bending of light rays occurs only at the principal planes. That two such planes should exist is not obvious but may be proved geometrically; the proof, however, is not given here. The utility of the concept is that it permits the definition of focal lengths and object and image distances for thick lenses and thus makes possible the direct calculation of image positions with the use of (8-z). The image positions so calculated, as for thin lenses, are of course the ideal or Gaussian image positions; in order to account for the effects of lens aberrations, individual rays must be traced through the lens in the manner described in the following section.

An additional useful property of the concept of principal planes is that, when the positions of these planes and of object and image planes is known, the rays passing through a lens may be sketched accurately without computing the angles of incidence and refraction at each surface. It is for this reason that the notation described at the beginning of this chapter is used throughout this book.

The action of the principal planes may be described in the following way. A ray which first approaches a lens traveling toward any given point of the *first* principal plane subsequently behaves as if it had in fact reached that point, had then traveled parallel to the axis between the first and second principal planes, and had then been redirected toward the image from the second principal plane. These hypothetical trajectories are illustrated for focusing by an equiconvex lens in Fig. 8-25a. As shown, the principal planes are symmetrically located with respect to the lens. The first principal plane h_1 crosses the axis at H_1, which lies to the left of H_2, where the second principal plane h_2 crosses the axis. A point on the axis to the left of H_1 is imaged. Physically, focusing occurs by refraction, in accordance with Snell's law, as illustrated in Fig. 8-25b. Note that rays

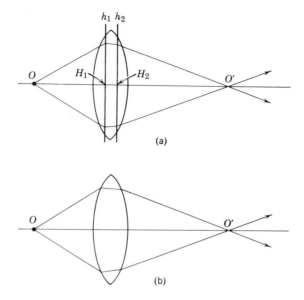

Fig. 8-25 Principal planes: (*a*) hypothetical action of principal planes; (*b*) actual refraction of rays by lens.

diverge from O and converge to O' at identical angles in both parts of the figure. Only in the region of the lens itself do the hypothetical and physical trajectories differ.

Figure 8-26 illustrates a lens for which the principal planes lie *outside* the lens limits. For the purpose of illustration, h_2 is shown to the *left* of h_1; this is typical of electron lenses, but it would not in fact occur with the type of glass lens shown here. (For the concavo-convex or "meniscus" lens shown, h_1 would lie to the left of the lens, but h_2 would fall within the lens boundaries.) Solid lines in Fig. 8-26 illustrate the hypothetical ray trajectories according to the principal plane concept; dotted lines give the physical paths of rays where they diverge from the former.

The focal points of lenses may be located experimentally by illuminating the lens with a parallel beam of light. The positions of the principal planes can be computed from the refractive index, thickness, and surface curvatures of the lens. In the procedure, first the two focal lengths corresponding to the individual surfaces are calculated by the use of (8-p). From these lengths, by substitution in an appropriate equation the focal length of the lens itself is calculated. Given f, the principal plane-to-vertex distances may be computed. The relevant equations, which are derived by geometrical manipulation, are all given in Appendix E. The manual

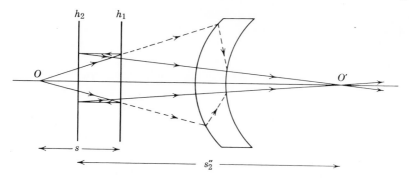

Fig. 8-26 Hypothetical action of principal planes.

calculation of thick lens parameters is straightforward but unquestionably tedious. Nevertheless, one computation of the properties of a typical thick lens is recommended as a useful educational experience!

The back focal length is a term not infrequently used in the literature. It refers to the *distance from the second vertex of the lens to the first focal point*, as shown in Fig. 8-27, and is useful as a directly measurable index of the converging power of the lens. Note that the distance from F_2 to the first vertex of the lens (which becomes the back focal length when the lens is illuminated from the opposite direction) may not be identical with this quantity.

RAY TRACING EQUATIONS

Equation 8-m and related expressions locate the ideal or Gaussian image plane and are useful in locating the approximate position of an image. The lens designer, however, wishes to determine the *quality*

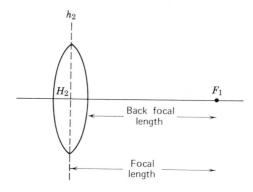

Fig. 8-27.

(i.e., the sharpness) of image formed by a lens of any given design. For this purpose the trajectories of individual rays must be traced through the lens. Whereas paraxial rays intersect in the ideal image plane, rays that make appreciable angles with the lens axis are found to intersect at image distances which vary as a function of the angular divergence.

A series of four equations may be obtained which describe the refraction of rays from an axial object point by a single refracting surface. These equations are developed here. As shown in Fig. 8-28, a point source of light at O is imaged by a single refracting surface at O'. Rays leaving O make an angle θ with the axis and later reach O' at some other angle θ'. The center of curvature of the refracting surface is at C, while the object and image distances s and s' and the radius of curvature, R, are as indicated. The quantities n, n', θ, s, and R are known; θ' and s' must be computed.

The law of sines states that for *any* triangle, in which sides A, B, and C are opposite angles α, β, and γ, respectively,

$$\text{(8-aa)} \qquad \frac{\sin \alpha}{A} = \frac{\sin \beta}{B} = \frac{\sin \gamma}{C}.$$

Application of (8-aa) to triangle OPC of Fig. 8-28 gives

$$\text{(8-bb)} \qquad \frac{\sin (180° - i)}{R + s} = \frac{\sin \theta}{R}$$

or, since $\sin (180° - i) = \sin i$,

$$\text{(8-cc)} \qquad \boxed{\sin i = \sin \theta \left(\frac{R + s}{R} \right)}.$$

A second equation is simply the *law of refraction*:

$$\text{(8-dd)} \qquad \boxed{\sin r = \frac{n \sin i}{n'}}.$$

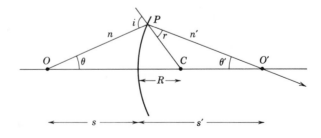

Fig. 8-28.

A third equation is obtained by equating the sum of the angles in triangle OPO' to 180°. In so doing, it is necessary to observe the convention that angles measured counterclockwise from the reference axis are positive, and those measured in a clockwise direction are negative. Hence the appropriate sum is

(8-ee) $$\theta + (180° - i) + r + (-\theta') = 180°.$$

Equation 8-ee reduces to

(8-ff) $$\boxed{\theta' = \theta + r - i}\,.$$

Finally, application of the law of sines (8-aa) to triangle POO' of Fig. 8-28 yields

(8-gg) $$\frac{\sin r}{s' - R} = -\frac{\sin \theta'}{R},$$

which rearranges to

(8-hh) $$\boxed{s' = R - \left(\frac{\sin r}{\sin \theta'}\right) R}\,.$$

Each of the four ray tracing equations (8-cc, 8-dd, 8-ff, 8-hh) is geometrically exact; no approximations have been invoked in their derivation. Application of these four equations in succession leads to an evaluation of the distance s', to the right of the vertex, at which a ray leaving the axis at angle θ recrosses the axis. A succession of values of s', corresponding to different values of θ, is thus found, in contrast to the single value of s' suggested by (8-m).

In tracing the courses of individual rays through the complete lens, the image formed by the first surface is taken as the object to be imaged by the second, as in the application of the idealized approximations. The four ray tracing equations are then applied, as before, at the second surface.

Equations 8-cc, 8-dd, 8-ff, and 8-hh apply only to the imaging of rays which originate from axial object points. In considering the imaging of extended fields it is necessary to trace rays which originate from off-axis points also. Two general categories of rays must then be considered: *meridional rays*, which lie in a plane containing the lens axis and thus intersect the axis at some points, and *skew rays*, which do not cut the axis at any point. (Rays originating from axial points are by definition meridional rays.) (For a description of the tracing of rays from off-axis points, see Conrady, Chapter 8.)

In evaluating lens designs a set of curvatures and refractive indices are chosen for a lens or a group of lenses, and individual rays are traced through the system, surface by surface. It is remarkable that the technology of lens design reached an advanced state at a time when these extremely tedious calculations could only be performed manually! More recently, computer techniques have been applied with great success to the problem of lens disign.

ORIGIN OF THE LENS ABERRATIONS

Application of the ray tracing equations shows that real lenses fall far short of satisfying the first condition for ideal lens action (i.e., that rays originating from a point object must reunite at a single image point). A specific example is now considered.

A lens has curvature $R_1 = 10.000$ cm and $R_2 = -5.000$ cm (i.e., the lens considered is biconvex and is more strongly curved on the right side). It is 0.6 cm thick and is constructed of a glass of refractive index 1.5180. For this lens, values of the image distance s_2'' may be calculated for an object placed 24 cm to the left of the first vertex. Results of the calculation for a series of values of θ, the angle of illumination, are given in Table 8-1. A decrease of the order of 10 percent in the image distance is found at $\theta = 3°$ as compared with the "Gaussian" image distance for paraxial rays. No well-defined image point thus exists; there is simply a concentration of light in the region of the image. In fact, image "points" of suitably restricted dimensions can be obtained only with combinations of more than one lens.

The defect exemplified by the data of Table 8-1 is an example of *spherical aberration*, one of the five third order geometrical errors which are revealed by ray tracing calculations. These aberrations and their correction are described in Chapter 9.

GRAPHICAL METHOD FOR IMAGE LOCATION

If the focal lengths and the positions of the principal planes of a lens are known, images may be located by a simple graphical procedure. [The image positions are those for paraxial rays, as given by (8-z).]

Rays which pass through the center of a lens essentially cross a pair of

Table 8-1   Image Distances Obtained by Ray Tracing

θ (Angle of Illumination)	s_2'' (Calculated Image Distance)
0° (paraxial rays)	8.793 cm (Gaussian image distance)
1°	8.709
2°	8.470
3°	8.054

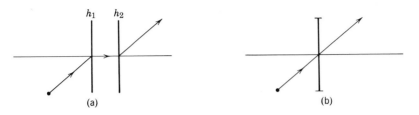

Fig. 8-29 Rays passing through the center of a lens: (a) thick lens; (b) thin lens.

parallel plane surfaces. Thus the ray trajectory after refraction by the lens is *parallel* to the original trajectory, as shown in Fig. 8-29a. Figure 8-29b illustrates the special case of a *thin* lens at which a central ray undergoes no lateral displacement.

As shown in Fig. 8-30, any ray traveling parallel to the axis as it leaves the object point O subsequently either crosses the axis of a converging lens at the second focal point F_2 (Fig. 8-30a) or is refracted by a diverging lens so as to proceed as if it had originated on the axis at F_2 (Fig. 8-30b). Similarly, rays leaving the *lens* parallel to the axis may be traced as shown in Fig. 8-31 if the position of the first focal point F_1 is known. The image point O' corresponding to the object point O is then located at the intersection of the rays drawn in Figs. 8-29 through 8-31, as shown in Fig. 8-32.

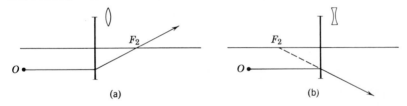

Fig. 8-30 Rays travelling parallel to the axis in object space: (a) converging lens; (b) diverging lens.

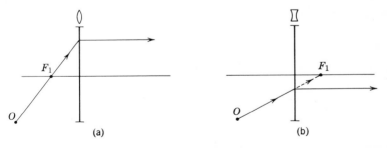

Fig. 8-31 Rays travelling parallel to the object in image space: (a) converging lens; (b) diverging lens.

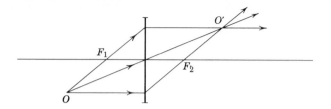

Fig. 8-32 Graphical location of image.

REFLECTION LENSES

Curved mirrors, like refracting surfaces, may act as lenses. Imaging by a concave mirror is illustrated in Fig. 8-33a, and that of a convex mirror in Fig. 8-33b. Combinations of refracting media with reflecting surfaces may be used also; they are sometimes referred to as "thick mirrors."

In certain spectral regions—the vacuum ultraviolet and part of the far infrared—no refractive materials are available for the construction of lenses, and mirror lenses must be used exclusively.

APERTURE

"Aperture" is a term used in optical instrumentation to signify a number of different but related entities. The term may mean the *diameter of a lens* (or of some other optical device such as a prism) *which is filled by light* and thus is useful to the system. It may refer to the aperture *angle*, which is usually specified as *the half angle between the extreme rays* which reach the optical system from a point on the object, as illustrated by Fig. 8-34a. (The angle of illumination specified in Table 8-1 is in fact an aperture angle.) Finally, the term "aperture" may be used to designate a stop; that is, a device which physically limits the lens aperture. This usage of the term is particularly common in electron microscopy. An aperture, in this sense, is a small *hole*, as illustrated by Fig. 8-34b. The term "stop" is more often used in light optics.

f NUMBER

The f number of a lens is defined as the quantity (f/A), the ratio of focal length to lens aperture (i.e., aperture in the sense of illuminated diameter). The square of the f number is then a measure of the light-gathering power of the lens. This quantity is useful in describing the lenses of cameras and enlargers, for which it gives a quantitative measure of the relative amount of light focused on a film; f numbers are discussed at greater length with respect to the photographic enlarger on p. 476.

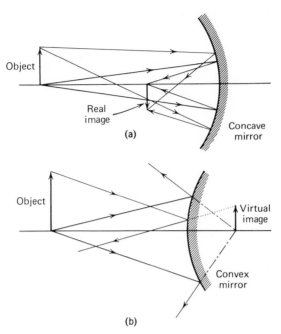

Fig. 8-33 Mirror lenses: (a) concave mirror; (b) convex mirror.

ZONES

A *zone* of a lens is the annulus at some specified radius (r to $r + \Delta r$) from the lens axis. Since all regions of a zone are equivalent (for an axially symmetric lens), the calculation summarized in Table 8-1 describes refraction of cones of light from the object by the complete 1°, 2°, and 3° lens zones. A similar use of this term has already been made on p. 67, where the division of a spherical wavefront into zones was described.

APERTURE STOPS, ENTRANCE AND EXIT PUPILS, AND CHIEF RAYS

In every lens system there is some physical limitation on the angular distribution of rays from the specimen which can contribute to the image.

Fig. 8-34 (a) Aperture angle; (b) stop.

In the simplest case the limitation is the boundary of the lens itself, as shown in Fig. 8-35. The limit could also be the edge of one lens in a system of several lenses, or it could be a stop. Whatever the physical limit is, it is called the *aperture stop* of the system. For example, the "limiting objective aperture" used in the electron microscope (p. 398) is the aperture stop of that system.

The larger the diameter of the aperture stop, the brighter the image can be, since more of the rays which proceed from the object are included in the image. Furthermore, the quality of the image is dependent upon the size of the angular aperture; this fact cannot be accounted for in geometrical terms, but it is discussed in Chapter 10.

Once the aperture stop of a lens system is located, the angular widths of the light beams that enter and leave the system are defined. In this connection the *entrance* and *exit pupils* of optical systems may be defined. The entrance pupil is the image of the aperture stop formed by all preceding lenses. If the aperture stop is the boundary of a single lens, or of the first lens of a system, or if it is a stop placed in front of the first lens of a system, then the aperture stop and the entrance pupil are one and the same. However, if the aperture stop is the boundary of any lens but the first one of a system, or if it is a stop placed to the right of any lens, then its image is formed (if it is illuminated from the right) by the lenses which lie to the left of it. The location and magnification of this image, as of any other image, can be computed. The entrance pupil, unlike the aperture stop, need not be physically present; an example is shown in Fig. 8-36.

Similarly, the *exit pupil* is the image of the aperture stop formed by all subsequent lenses. In Fig. 8-36 aperture stop and exit pupil are identical, while Fig. 8-37 shows a system in which aperture stop, entrance pupil, and exit pupil are all distinct.

The entrance pupil need not always be on the object side of a lens and the exit pupil on the image side. For example, in Fig. 8-38a an aperture stop, which is also the entrance pupil, is placed within the focal length of

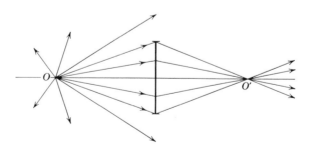

Fig. 8-35 Single lens, acting as aperture stop.

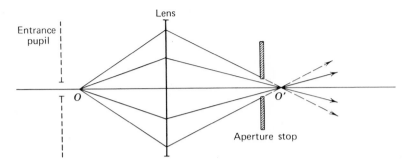

Fig. 8-36 Lens system in which aperture stop and exit pupil coincide.

the lens. The exit pupil, a virtual image of the stop, is then formed further to the left of the lens.

The *chief ray* is that ray which passes through the center of the aperture stop. This ray approaches the lens system in the direction of the center of the entrance pupil and, after passing through the center of the aperture stop, leaves the system as if proceeding from the center of the exit pupil. Tracing of chief rays thus provides a simple graphical way of locating parts of extended images. An example is shown in Fig. 8-38b (in which the Gaussian image plane has been located by calculation). Note that rays from two different object points have been traced through the lens system in that figure; it is important to distinguish the type of diagram shown in Fig. 8-38b from the more usual sort in which two or more rays are traced which originate from a single object point, and which thus must reunite at the image.

FIELD STOP

The *field stop* of an optical system is that element (either a lens boundary or an opaque stop) which limits the field of view in the image. The sizes and positions of the field and aperture stops are, in general, different,

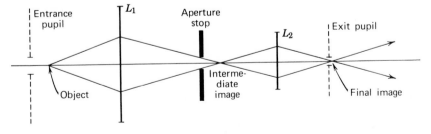

Fig. 8-37 Lens system, showing aperture stop, entrance pupil and exit pupil.

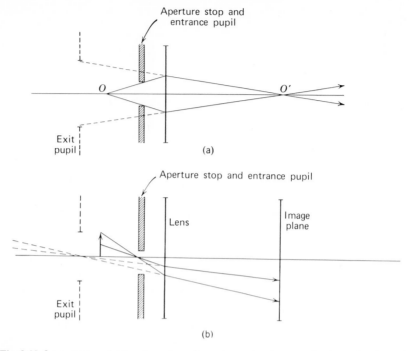

Fig. 8-38 Lens system: (*a*) location of aperture stop and entrance and exit pupils; (*b*) tracing of chief rays from two points on an object.

although the two may coincide. For example, the limiting objective aperture of the electron microscope normally serves as an aperture stop only, but may become a field stop when the microscope is far out of focus. That is, an *image* of this aperture can be obtained by extensive defocusing.

MAGNIFICATION

Enlargement by lens systems may be expressed in terms of either lateral or angular magnification.

Lateral magnification is defined as the ratio of image and object dimensions; that is,

$$(8\text{-ii}) \qquad\qquad m_{\text{lat}} \equiv -\frac{y'}{y},$$

where y' is a dimension of the image, and y is the corresponding object dimension.

Positive or negative values of m denote erect or inverted images, respectively.

A formula for the lateral magnification obtained upon imaging by a single spherical surface may be derived. In Fig. 8-39, POC and $OO'Q$ are similar triangles. Therefore

(8-jj)
$$-\frac{y'}{y} = \frac{O'Q}{PO} = \frac{OO'}{OC},$$

or, since $OC = s + R$ and $CO' = s' - R$,

(8-kk)
$$\boxed{\text{magnification} = -\frac{y'}{y} = \frac{s' - R}{s + R} \quad \text{for a single surface}},$$

where y' and y are image and object dimensions, respectively, s' and s are image and object distances, respectively, and R is the radius of curvature of the surface.

Similarly, an equation for the lateral magnification produced by a thin lens may be derived. In Fig. 8-40, POA and $AO'Q$ are similar triangles. Thus

(8-ll)
$$-\frac{y'}{y} = \frac{O'Q}{PO} = \frac{AO'}{OA},$$

or, since $OA = s$ and $AO' = s_2''$,

(8-mm)
$$\boxed{\text{magnification} = -\frac{y'}{y} = \frac{s_2''}{s} \quad \text{for a lens}},$$

where y' and y are image and object dimensions, respectively, and s_2'' and s are the image and object distances, respectively. If s and s_2'' are measured from the respective principal planes, (8-mm) applies to thick as well as thin lenses.

An equivalent formula for lateral magnification may be expressed in terms of the Newtonian convention for expressing object and image distances. As shown in Fig. 8-41, object and image distances, designated x and x', respectively, are then measured from the respective focal planes, rather than from the corresponding principal planes. In Fig. 8-41, ray PA travels parallel to the lens axis and is refracted through F_2 to P', while ray

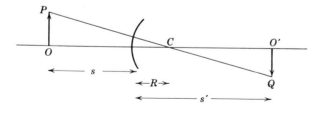

Fig. 8-39.

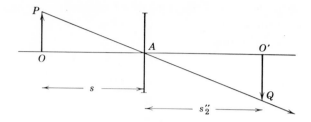

Fig. 8-40.

PB passes through F_1 and travels parallel to the axis after refraction. POF_1 and F_1CB are similar triangles, and $CB = y'$, so that

(8-nn)
$$\frac{CE}{PO} = \frac{f_1}{x} = -\frac{y'}{y}.$$

Likewise, triangles ACF_2 and $P'O'F_2$ are similar, and $AC = y$, so that

(8-oo)
$$\frac{O'P'}{AC} = \frac{x'}{f_2} = -\frac{y'}{y}.$$

Combination of (8-nn) and (8-oo) then gives

(8-pp)
$$M_{\text{lat}} = -\frac{y'}{y} = \frac{f_1}{x} = \frac{x'}{f_2},$$

where f_1 and f_2 are the first and second focal lengths, respectively, of the lens (and thus are equal for the usual case of a lens in air), x is the object distance measured from the first focal plane, and x' is the image distance measured from the second focal plane. Note that Fig. 8-41 illustrates a case in which $n_1 \neq n_2$, so that the two focal lengths of the lens are different.

Except when images are recorded photographically, the combination of lenses used in a microscope forms a virtual image which is focused by the eye of the observer, forming a real image on the retina (cf. p. 167). The

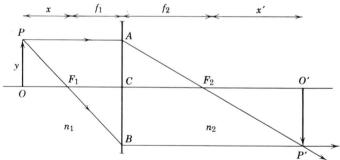

Fig. 8-41.

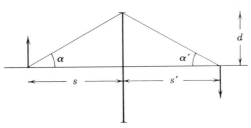

Fig. 8-42.

size of the object perceived by the eye depends upon the *angle* which the image subtends at the eye rather than its absolute size. For this reason, *angular* rather than lateral magnification is useful for specifying the enlargements produced by optical instruments, such as light microscopes or telescopes, in which the image is viewed directly. However, when photographic records normally are made, as in electron microscopy, lateral magnifications are specified.

The angular magnification is defined as the ratio of the angle subtended by the image at the eye to the angle which the object would subtend at the unaided eye. Thus,

$$(8\text{-qq}) \qquad M_{ang} = \frac{\alpha'}{\alpha},$$

where α is the aperture angle of the lens, and α' is the corresponding angle in image space, as shown in Fig. 8-42. The angular magnification of the compound light microscope is discussed further on p. 279.

For paraxial rays, $\tan \alpha' \cong \alpha'$. From the geometry of Fig. 8-42, $\tan \alpha = d/s$ and $\tan \alpha' = d/s'$, and

$$(8\text{-rr}) \qquad M_{ang} = \frac{ds}{ds'},$$

or, substituting $s'/s = M_{lat}$ from (8-mm),

$$(8\text{-ss}) \qquad M_{ang} = \frac{1}{M_{lat}}.$$

That is, *for paraxial rays only*, the angular magnification is the inverse of the lateral magnification. In fact, the quantity $(\tan \alpha'/\tan \alpha)$ is sometimes called the "paraxial magnification."

REFERENCES

F. A. Jenkins and H. E. White, *Fundamentals of Optics* (3rd ed.), McGraw-Hill, New York, 1957, Chapter 2 (Plane Surfaces), Chapter 3 (Spherical Surfaces), Chapter 4 (Thin Lenses), Chapter 5 (Thick Lenses), Chapter 6 (Spherical Mirrors), Chapter 7 (The Effects of Stops), Chapter 8 (Ray Tracing), and pp. 130–132.

L. C. Martin, *The Theory of the Microscope*, Blackie, Glasgow, and American Elsevier, New York, 1966, pp. 3–12.

A. E. Conrady, *Applied Optics and Optical Design*, Part I, Dover, New York, 1957. Chapter 1 (Fundamental Equations), Chapter 8 (Trigonometric Tracing of Oblique Pencils), and Chapter 9 (General Theory of Perfect Systems). The book is a classical treatise on practical lens design. While mathematical rather than descriptive in nature, being primarily devoted to the derivation of lens formulas, the book uses algebraic and trigonometric methods almost exclusively. The notations for image and object distances, angles, etc., used by Conrady are different from those in the present text; however, a list of equivalent notations can easily be compiled by careful study of the first few pages.

M. Herzberger, in J. Strong, *Concepts of Classical Optics*, McGraw-Hill, New York, 1963, Appendix L, pp. 537–542. This article, "Modern Trends in Methods of Lens Design" is an informative short account of that subject. Two interesting illustrations show forms of computed image points.

E. W. Silvertooth, in Strong, op. cit., Appendix M, pp. 544–552. This well-illustrated article, "Graphical Ray Tracing," is an excellent summary of practical approaches to lens design.

H. E. Bennett and J. M. Bennett, *Applied Optics*, Academic Press, New York (in preparation, 1969). Chapter 2 (Thin Lens Theory) contains detailed examples of both numerical and graphical lens calculations.

CHAPTER 9

Geometrical and Chromatic Lens Aberrations

Certain failures in lens performance, which may be accounted for in terms of the concepts of geometrical optics, are discussed in this chapter. The geometrical aberrations result from the failure of surfaces of spherical curvature to refract light rays in a manner which satisfies the conditions of perfect lens action given on pp. 154–159. In particular, rays from a point object are not reunited at a *single* image point, as demonstrated for a single spherical surface on pp. 159–162. The chromatic aberrations arise from the variation of the refractive index of lens materials as a function of the wavelength of incident light, so that the size and position of images are also functions of wavelength.

The technology of glass lenses has long ago developed to a high degree. Thus the geometric and chromatic aberrations are largely a matter of academic interest to the routine user of light microscopes and similar instruments. Understanding of the nature of these effects is important, however, when limiting resolution is sought, or when lenses are to be selected for special purposes (e.g., for the imaging of extended fields).

Electron microscope imaging systems are crude by comparison with their light optical counterparts. Aberrations of electron lenses cannot be reduced to a negligible level, but must be tolerated to some degree. The resulting effects are most easily understood, at least in a general way, by analogy with the corresponding aberrations of glass lenses. Thus the study of the geometrical aberrations of glass lenses is a subject of particular interest to the electron microscopist.

The material presented here requires an understanding of the laws of geometrical optics, as summarized in Chapter 1, and familiarity with the properties of lenses as described in Chapter 8.

Those aberrations are discussed here which can be described by making a third order correction in the derivation of lens equations (cf. p. 162). These defects are variously referred to in the literature as primary or third order errors; secondary or fifth order errors, as described by further refinement of the lens equations, are not considered here. Imaging as described by the theory of third order errors is in fact a reasonably close approximation to imaging as described by the *exact* methods of ray tracing. The origin of each type of defect is shown diagramatically, its effect on the image described, and the nature of correction indicated. Expressions which numerically relate lens parameters to the magnitude of each defect are not derived, however.

It is important at the outset to understand the general nature of the geometrical aberrations. Texts in elementary general physics often fail to emphasize the fact that the well known thin lens equations are in fact *approximations*. Subsequently, the student may be puzzled to learn that these lenses suffer from aberrations. It should be clear that surfaces of spherical curvature are not "naturally" perfect refractors (as the derivation on pp. 159–162 shows.) Lens aberrations are thus not a consequence of a failure of the law of refraction, but result from the operation of that law. Furthermore, although aberrations of course can arise from imperfections in the *construction* of lenses, inherent aberrations exist in technically perfect systems.

SPHERICAL ABERRATION

Spherical aberration results from the fact that the outer (i.e., marginal or peripheral) portions of spherical surfaces refract relatively *more strongly* than the paraxial portions. This is shown in Fig. 9-1 (in an exaggerated fashion) for the imaging of an axial point by a single converging

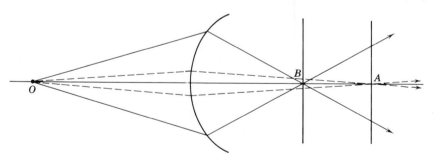

Fig. 9-1 Spherical aberration.

surface. In Fig. 9-1, the greater the distance from the axis, the more highly converging is the surface; thus the image moves to the *left* as zones of the surface farther from the axis are illuminated. A numerical example of the same effect has been given on p. 176.

In Fig. 9-1 *any* plane perpendicular to the axis and located between A and B could be called the "image plane." A lies in the plane in which paraxial rays from O reunite (the *paraxial* or Gaussian image plane, located as predicted by (8-m) for a single surface, or by (8-z) for a lens); B lies in the plane in which rays from O to the lens peripheri cross the axis (*marginal* or *peripheral* image plane). The appearance of these two extreme image planes is diagramed in Figs. 9-2a and 9-2b, respectively. (Note that Figs. 9-1, 9-2, and many other figures in this chapter are considerably exaggerated for purposes of illustration.) In each case the point image is surrounded by a circular haze contributed by out-of-focus rays that have passed through other zones of the surface. Figure 9-2 shows sections through the center of a three-dimensional zone of confusion which surrounds the ideal image point. The radius of the disc of confusion is minimal at an intermediate plane which is said to contain the "circle of least confusion."

When spherical aberration occurs at a diverging surface, the marginal rays are also refracted too strongly relative to the paraxial rays; again, the marginal ray image lies closer to the lens than the paraxial image. In this case, however, the image is formed to the left of the surface, so that the spherical aberration of peripheral zones moves the image to the *right*. This is illustrated (again in an exaggerated fashion) by Fig. 9-3.

Spherical aberration is expressed quantitatively in two ways. The longitudinal spherical aberration is the distance AB in Figs. 9-1 and 9-3; that is, the distance along the axis from the paraxial ray image [located by (8-m) or (8-z)] to the marginal ray image (located by ray tracing). Thus,

(9-a) $\boxed{\text{longitudinal spherical aberration} = s'_{\text{parax}} - s'_{\text{marginal}}}$.

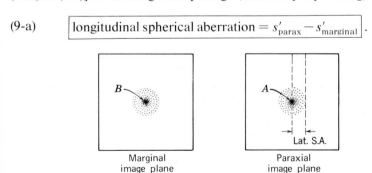

Marginal   Paraxial
image plane   image plane
(a)   (b)

Fig. 9-2 Spherically aberrant images of point objects.

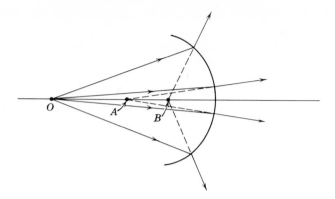

Fig. 9-3 Spherical aberration at a diverging surface.

The longitudinal spherical aberration may thus be given for the imaging of a specific object point, or it may be specified for *parallel* incident rays. In the latter case it is the difference between paraxial and marginal focal lengths; that is,

(9-b)   longitudinal spherical aberration $= f_{\text{parax}} - f_{\text{marginal}}$.

The lateral spherical aberration is the direct measure of what is seen in the paraxial image plane. It is the radius of the "zone of confusion" at the paraxial image plane, that is, the distance from the paraxial ray image point to rays originating from the margin of the surface or lens, as shown in Fig. 9-2a. If θ' is the angle made by the marginal rays with the axis after refraction by the lens or surface,

(9-c) $\boxed{\text{lateral spherical aberration} = (s_p' - s_m') \tan \theta'}$,

in which s_p' is the image distance for paraxial rays and s_m' is the image distance for marginal rays.

Third order theory shows that spherical aberration is a function of the square of the radius of the lens aperture (the square of the "semiaxis of illumination"). That is, spherical aberration increases as the first power of the area of the aperture of illumination. Specifically, when modified to describe spherical aberration, (8-m) becomes

(9-d) $\dfrac{n}{s} + \dfrac{n'}{s_h'} = \dfrac{(n'-n)}{R} + \left[\left(\dfrac{h^2 n^2 R}{2f''n'} \right) \left(\dfrac{1}{s} + \dfrac{1}{R} \right)^2 \left(\dfrac{1}{R} + \dfrac{(n'-n)}{ns} \right) \right]$,

where f'' is the second focal length of the surface, and s_h' is the image distance for rays incident upon the surface at any distance h from its axis.

For rays incident upon the surface parallel to its axis, (9-d) reduces to

(9-e)
$$\frac{n'}{s_h'} = \frac{n'}{f''} + \frac{h^2 n^2}{2f'' R^2 n'}.$$

Spherical aberration is generally most severe in the imaging of off-axis points. This may be demonstrated geometrically in the following manner. An auxiliary axis is defined, with respect to any given off-axis object point, as the line joining image and object points, and passing through the center of the lens. This is shown as line PAP' in Fig. 9-4. Rays from P which make small angles with the auxiliary axis behave like paraxial rays and are focused according to (8-z), but with the distances s and s_2'' measured along the auxiliary axis. In general, these lengths, PA and AP', are somewhat greater than the corresponding values OA and OO' for axial points.

Spherical aberration in the imaging of an off-axis point is illustrated (to an exaggerated degree) in Fig. 9-5. The extent of the defect varies with the distance of the object point from the axis according to the angle subtended by the object point at the lens.

In summary, spherical aberration is a defect which occurs throughout the image plane, including the axial region. The defect is of major importance in the images formed both by uncorrected glass lenses and by electron lenses. Figure 9-6a shows an object and Fig. 9-6b the corresponding image as formed by a (hypothetical) lens system which suffers severely from spherical aberration but is free from all other types of image defect.

The other geometrical errors of lenses, described below, differ from spherical aberration in that all disappear at the axis of the image.

CORRECTION OF SPHERICAL ABERRATION

Spherical aberration may be *limited* by *reducing the aperture* of a given lens, or by selecting a favorable *shape* of lens. For any given object

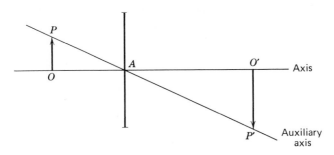

Fig. 9-4 The auxiliary axis.

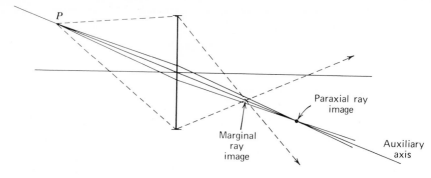

Fig. 9-5 Spherical aberration in the imaging of an off-axis point.

distance and lens zone, spherical aberration may be *eliminated* by the use of an appropriate *"doublet"* of positive and negative (converging and diverging) lenses.

A numerical example of a spherically aberrant lens has been considered on p. 176. This was a system for which $R_1 = 10$ cm, $R_2 = -5$ cm, $d = 0.6$ cm, and $n = 1.518$, used with an object distance of 24 cm. From these data the longitudinal spherical aberration (9-a), lateral spherical aberration (9-c), and diameter of the disc of confusion may be computed. These qualities are listed in Table 9-1 for a series of aperture angles. Table 9-1 makes plain that the lateral aberration and thus the diameter of the disc of confusion corresponding to each image point increase rapidly as the angular aperture of the lens is enlarged [in accordance with the square-function dependence of aberration upon h indicated by (9-d) and (9-e)]. At some small aperture angle, however, spherical aberration is reduced to a level which may be tolerated. The tolerable level is determined by the requirements of the specific lens system.

For example, if the angular aperture of the lens considered in Table 9-1 were restricted to 1°, images of point objects would appear in a plane as discs less than 0.1 mm in diameter in the plane of least confusion. Since

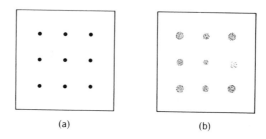

(a)   (b)

Fig. 9-6 (a) Object; (b) corresponding spherically aberrant image.

Table 9-1 Spherical Aberration of a Lens

θ	θ'	$\tan\theta'$	s_2''	Longitudinal Aberration	Latitudinal Aberration
0°	0°	0.000	8.79 cm	0.00 cm	0.000 cm
1°	2°44′	0.048	8.71	0.08	0.0038
2°	5°35′	0.098	8.47	0.32	0.0314
3°	8°42′	0.153	8.05	0.74	0.1132

this is less than the resolution limit of the eye (0.1–0.2 mm), haziness due to spherical aberration would be apparent in the image only if further magnification were produced by additional lenses. Thus, in principle, spherical aberration could be controlled to any desired level simply by reducing the angle of illumination of the lens system. Unfortunately, correction of spherical aberration in this manner is accompanied by loss of resolving power through the formation of an extended diffraction pattern corresponding to each image point. Limitations imposed by diffraction effects are discussed in Chapter 10. When small lens apertures are used, the adequacy of illumination may also be a problem.

A partial correction of spherical aberration can be achieved by selection of lens shape, a method referred to as *lens bending*. Each of the surfaces of a lens or lens system makes a contribution to the net spherical aberration; these contributions may be quite different for the successive surfaces. An example is given in Table 9-2, which tabulates the aberrations produced at the first surface, only, of the lens which is considered in Table 9-1. Note that, whereas the absolute magnitudes of lateral spherical aberration produced by the first surface alone and by the complete lens are similar (e.g., 0.103 cm and 0.113 cm, respectively, at 3°)

Table 9-2 Spherical Aberration at the First Surface of a Lens

$$(R = +10 \text{ cm}; n = 1.518)$$

θ	θ'	$\tan\theta'$	s_2''	Longitudinal Abberation	Latitudinal Aberration
0°	0°	0.000	150.0 cm	0.0 cm	0.000 cm
1°	0°9′	0.003	148.6	1.4	0.004
2°	0°19′	0.006	145.2	4.8	0.028
3°	0°31′	0.009	139.6	11.4	0.103

the magnification (s_2''/s) is about seventeen times as great for the first surface as for the complete lens. Thus the effective net aberration is contributed primarily by the action of the second surface.

In general, it is found that spherical aberration is minimal (but not absent) for a lens shape which is biconvex, but more highly curved on the left side (i.e., this is true in parallel illumination or in the imaging of objects located far to the left of the lens. The degree of spherical aberration obtained varies with the object position, however.) Concavo-convex lenses are highly aberrant, while convexo-concave surfaces are less so. The lens described by Tables 8-1, 9-1, and 9-2 would be of nearly optimal shape if illuminated in the reverse direction to that for which the calculations were made (i.e., if $R_1 = 5$ cm and $R_2 = -10$ cm). The size of the disc of confusion is then reduced by about 50 percent. Although this is a substantial improvement, corrections which reduce lens aberrations by orders of magnitude are required in lenses of high quality.

It may seem that the possibility of changing the spherical aberration produced by a lens simply by turning the lens around is a violation of the principle of reversibility of rays. (cf. p. 3.) In fact, the latter applies to individual rays, not to images as a whole. An object producing the distribution of light rays found in spherically aberrant images of points would in fact (i.e., in the absence of other aberrations) produce a perfect point image.

It should be noted that the data in Tables 9-1 and 9-2 describe only spherical aberration, since the imaging of an axial object point is considered. If, however, the imaging of an off-axis point had been computed instead, the data obtained would have described the superimposed effects of all the aberrations to which extra-axial points are subject. For angles other than 0°, the data of Tables 9-1 and 9-2 are obtained by the exact method of ray tracing, and thus include the effects of fifth and higher order spherical aberration in addition to primary or third order spherical aberration.

Spherical aberration can be corrected precisely for rays reaching the lens at any chosen value of h by substituting appropriate doublets of converging and diverging lenses for single lenses. As shown in Fig. 9-1, a converging single lens forms a marginal ray image which is located to the left of the paraxial ray image, while, as shown in Fig. 9-3 a diverging lens forms a marginal ray image which is located to the right of the paraxial image. When two such displacements of the image are superimposed, they tend to cancel, so that the final images formed by paraxial and marginal rays may coincide. Spherical aberration is thus eliminated entirely with respect to the chosen zone of the lens; that introduced by surrounding zones is reduced. An intermediate lens zone is always chosen for exact compensation such that spherical aberration remains minimal even for the least corrected zones of the lens.

It might appear that the net refractive effect of the doublet, as well as its aberration, would be eliminated by combining converging and diverging elements. The focusing power of a lens is not directly proportional to the extent of spherical aberration produced, however. Thus it is possible to design pairs of positive and negative lenses in which aberration is corrected, but which are of net converging or diverging action. Usually the two lenses are cemented together; three curvatures and two refractive indices may then be specified independently for the system.

The action of a corrected doublet is most conveniently illustrated by considering a pair of lenses which are separated, so that a real intermediate image is formed between them. A system of this sort is diagramed in Fig. 9-7. The converging lens L_1 focuses marginal rays from an axial object point O at position A, while paraxial rays from O are focused at B. The diverging lens L_2 forms a virtual image of B at position C. If L_2 were free of spherical aberration, rays from point A would appear to diverge from a point to the left of C. However, the rays which leave A are incident upon the margins of L_2, where they are refracted relatively strongly. The virtual image of point A is thus moved to the right, and it coincides with the paraxial image at C.

Figure 9-8 compares the spherical aberration present in a single lens with that of a doublet corrected at $h = 2.0$ cm (i.e., for matched curvatures, refractive indices, and object positions). As shown, spherical aberration increases rapidly in both lenses for large values of h. Even at the most unfavorable zone of the doublet, however, the aberration is less than 10% as great as in the uncorrected lens at $h = 2.0$ cm.

The spherical aberration of lens combinations may also be controlled by the use of aplanatic surfaces, as described in the following section.

The spherical aberration of electron lenses is discussed on p. 364.

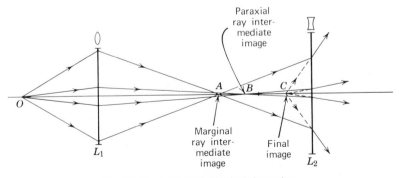

Fig. 9-7 Correction of spherical aberration.

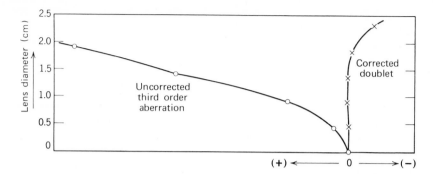

Fig. 9-8 Degree of spherical aberration produced by different zones of an uncorrected lens and of a comparable corrected doublet.

APLANATIC POINTS AND SURFACES

Light is refracted *without spherical aberration* from certain points and surfaces with respect to a single refracting surface of spherical curvature. These positions are known as *aplanatic points* and *aplanatic surfaces.*

One aplanatic point is the center of curvature of the surface. Rays which originate at this point are not refracted at all, since they must travel along radii and thus are incident normally on the surface.

Figure 9-9 shows a medium of index n' which is of spherical curvature, centered at C, and suspended in a medium of refractive index $n(<n')$. Points P and P' are so located that

(9-f) $$PC = \frac{n}{n'}R \quad \text{and} \quad P'C = \frac{n'}{n}R,$$

where R is the radius of curvature of the surface. It is shown here that P is an aplanatic point.

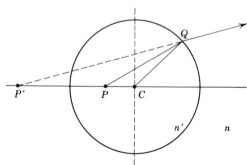

Fig. 9-9 Aplanatic points.

Refraction of a ray from P at some arbitrary point O on the surface may be considered. $CQ = R$, the radius of curvature. From the construction, as specified by (9-f).

$$(9\text{-g}) \qquad \frac{PC}{CQ} = \frac{(n/n')R}{R} = \frac{n}{n'} \quad \text{and} \quad \frac{CQ}{P'C} = \frac{R}{(n'/n)R} = \frac{n}{n'}.$$

Thus the triangles $P'QC$ and QPC are similar, since they also share the angle PCQ.

In triangle QPC, angle $PQC = i$, the angle of incidence of ray PQ at the surface. From the law of sines,

$$(9\text{-h}) \qquad \frac{\sin i}{PC} = \frac{\sin QPC}{CQ}.$$

From (9-g), $PC/CQ = n/n'$, so that

$$(9\text{-i}) \qquad \frac{\sin i}{\sin QPC} = \frac{n}{n'}.$$

But (9-i) is identical in form with the law of refraction (i.e., for refraction from a more dense to a less dense medium). Thus angle QPC must be equal to the angle of refraction. Since triangles QPC and $P'QC$ have been shown to be similar, this means that angle $QP'C$ also must be equal to the angle of refraction. Thus the ray refracted at Q travels along a path which is a projection of the line $P'Q$. In other words, a virtual image of P is formed at P'.

The choice of Q as the position of refraction was arbitrary (i.e., within certain broad limits; in fact, Q must be so located that the angle of refraction is less than 90°). Therefore *all* rays from P must appear to originate from P'; that is, there is *no* spherical aberration in the image at P'.

Similar considerations show that *all* positions lying on an arc of radius PC, and centered about C, are also aplanatic points, that is, rays proceeding from these points appear to originate from conjugate positions on the arc through P' of radius $P'C$.

Aplanatic surfaces are exploited in the design of oil immersion objectives (cf. p. 270) and in lenses used for ultraviolet microscopy (cf. p. 284). Unfortunately, however, chromatic aberration is not avoided in the system shown in Fig. 9-19. Since correction of chromatic aberration inevitably introduces some degree of spherical aberration, the general usefulness of the aplanatic principle in lens design is limited.

COMA

Coma is a defect of lenses which results in the imaging of a point as a "cometlike" blur. The head of the comet is formed by the chief ray from the object point (cf. p. 181), while rays entering the optical system away from the center of the entrance pupil form circular regions in the image, those from the marginal lens zones being of larger radius. Hence the comatic image diminishes in intensity toward its "tail."

The origin of coma is most conveniently illustrated by considering the imaging of an off-axis point, located at infinity, as shown in Fig. 9-10. This is an example of *negative* coma in which the magnification of marginal ray images is less than that of the chief ray. Note that the diagram represents a single lens with no external stops, so that the chief ray passes through the center of the lens. In Fig. 9-10, only those rays are shown which pass through the diameter of the lens *in* the plane of the paper. The refraction of these rays produces a line image which is in fact tilted through 30° with respect to the plane of the paper; only the chief ray lies in the plane of the paper of Fig. 9-10 (i.e., Fig. 9-10 shows a projection of the line image in the plane of the page).

Rays passing through the diameter of the lens which is at right angles to the plane of the page in Fig. 9-10 also form a line image, but one which is inclined at an angle of 60° to the first image. This is illustrated by Fig. 9-11, which is a view of the image plane showing only the chief ray, marginal rays, and intermediate rays from one lens zone for two mutually perpendicular lens diameters. Contributions to the image from rays which pass through other diameters of the lens are focused above and below the plane of Fig. 9-11 (i.e., to the right and left of the image plane represented in Fig. 9-11). The complete image is a cone of 60° angle, as shown in Fig. 9-12.

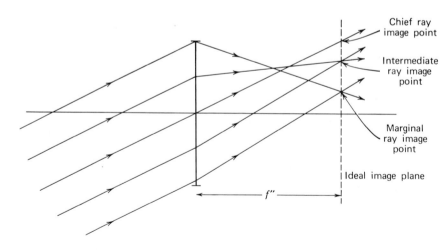

Fig. 9-10 Comatic imaging of an off-axis point.

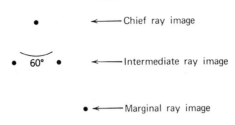

Fig. 9-11 Contributions to the comatic image.

The contributions to the comatic image from all parts of a given lens zone form a circular image the radius of which is determined by the radius of the zone. The contributions from different portions of the zone to the circular image occur in the manner illustrated by Fig. 9-13.

The 60° angle of the comatic figure is an inherent feature of this aberration, while the dimensions of the figure are determined by the square of the angular aperture of illumination of the lens and by the first power of y', the distance of the image point from the lens axis. Thus coma, like spherical aberration, increases with distance of the object point from the axis; unlike spherical aberration, the magnitude of coma is zero for axial points. Because of the first power dependence of coma upon y', this defect is of consequence even for points located quite close to the lens axis. Thus correction of coma must be effected even in instruments with relatively limited fields of view.

Coma is regarded by lens designers as a particularly objectionable effect, since the comatic image is asymmetric (unlike images which suffer from any of the other lens aberrations). Thus particular difficulty is encountered in specifying the location of comatic image points. Fortunately, the correction of coma is straightforward. As lens shape is shifted from concavo-convex to convexo-concave, coma changes from a negative to a positive effect. That is, the image of a point changes from a cometlike figure with its tail pointing toward the lens axis (in which the maximum magnification is that of the chief ray) to one with its tail pointing away from the axis (in which the chief ray is of minimum magnification).

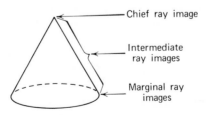

Fig. 9-12 The complete comatic image.

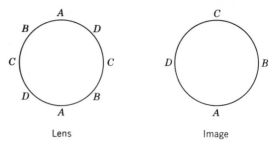

Lens   Image

Fig. 9-13 Contributions from different portions of a lens zone to the comatic image.

The magnitude of the aberration falls to zero for lenses of an intermediate shape, which is very nearly convexo-plane. Spherical aberration is also very nearly minimal for this shape of lens.

Acomatic systems are found to satisfy the Abbe sine condition, as derived in the following section.

THE SINE THEOREM, LAGRANGE'S THEOREM, AND THE ABBE SINE CONDITION

In Chapter 8, (8-kk) was derived; it gives the magnification produced by a single refracting surface:

(8-kk)
$$M = -\frac{y'}{y} = \frac{s'-R}{s+R},$$

where y is the object dimension, y' the corresponding image dimension, R the radius of curvature of the surface, s the object distance, and s' the image distance.

Rearrangement of the exact ray tracing equations (8-cc and 8-hh) gives

(9-j)
$$s+R = \frac{R\sin i}{\sin\theta} \quad \text{and} \quad s'-R = -\frac{R\sin r}{\sin\theta'},$$

where s, s', and R are as defined above, i and r are the angles of incidence and refraction, respectively, at the surface, and θ and θ' are the angles made by incident and refracted rays, respectively, with the lens axis (or, for extra-axial points, with the auxiliary axis; cf. p. 191).

Substitution of (9-j) in (8-kk) gives

(9-k)
$$\frac{y'}{y} = \frac{\sin r \sin\theta}{\sin i \sin\theta'},$$

and combination of (9-k) with the law of refraction yields

(9-l)
$$\boxed{y'n'\sin\theta' = yn\sin\theta}.$$

Equation 9-1 is known as the *sine theorem*. Note that all parameters in the sine theorem which refer to object space and image space, respectively, appear on opposite sides of the equation. The sine theorem applies to all rays passing through the optical system. For paraxial rays, however, the approximation $\sin \theta \cong \theta$ is appropriate, yielding, from (9-1),

(9-m)
$$\boxed{y'n'\,\theta'_p = yn\,\theta_p},$$

where θ_p and θ'_p are values of θ for paraxial rays. Equation 9-m is known as *Lagrange's theorem*.

In acomatic lens systems the magnification of rays from all zones of the lens must be equal to that for paraxial rays. That is, the condition

(9-n)
$$\frac{y'_m}{y} = \frac{y'_p}{y}$$

(where y'_m and y'_p are image heights for marginal and paraxial rays, respectively) must hold. Combination of (9-k), (9-l), and (9-m), with cancelation of the common factor $(y'n'/yn)$ gives

(9-o)
$$\boxed{\frac{y'}{y} = \frac{\sin \theta}{\sin \theta'} = \frac{\theta_p}{\theta'_p} = \text{constant.}}$$

Equation 9-o is known as *Abbe's sine condition*. In effect, the sine condition states that all rays must leave an (ideal) lens (or optical system) at distances from the axis which are proportional to the angles of incidence of the corresponding rays. Figure 9-14a illustrates the imaging, at F_2, of a ray which enters a lens parallel to the axis at some height h. For parallel rays, h, rather than $\sin \theta$ is proportional to $\sin \theta'$, and the appropriate special form of the sine condition then becomes

(9-p)
$$\frac{h}{\sin \theta'} = \text{constant} = f^*,$$

where f^* is the length of the ray from the lens to F_2. f^* is invariant with h *only if* the second principal "plane" of the lens is not a plane at all, but

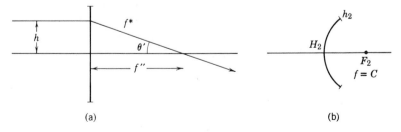

(a)   (b)

Fig. 9-14.

an arc of radius f^*, with a center of curvature at F_2. This condition, which is illustrated in Fig. 9-14b, can be shown to occur in coma-free systems.

LENS ASYMMETRY; ACTION OF CYLINDRICAL LENSES

Astigmatism is a third type of geometrical aberration. It should be understood, however, that the term "astigmatism" is used to refer to either of two quite different optical effects. The common property of the two types of astigmatism is that each produces *differences in focal length as a function of the angle at which rays are incident upon the lens*. In the eye, and also in electron lenses, what is commonly called "astigmatism" results from an asymmetry of the lens itself, that is, from a distortion from spherical to cylindrical or partially cylindrical form. (In this connection a lens of spherical curvature is sometimes referred to as a "round" lens.) This defect, which is accurately called *lens asymmetry*, is not an inherent property of spherical surfaces but a technical imperfection. Asymmetry is not a problem in glass lenses of good quality, but will be discussed here for the purpose of distinguishing fully between the two types of effect known as "astigmatism".

A spherical surface has equal curvatures in all directions, and, in the absence of third order errors, focuses all rays from a point object to a point image, as shown by Fig. 9-15, regardless of the angle at which rays are incident upon the lens. (Note that Fig. 9-15 and also Figs. 9-16, 9-18, and 9-21 to 9-24 are not sections of the lens system in a plane containing the axis, but projections of a three-dimensional system in which the object may be envisioned as lying above the plane of the page and the image as lying below the plane of the page.)

As shown in Fig. 9-16, rays incident upon a cylindrical refracting surface from a point lying on the axis AA' (perpendicular to the axis of the cylinder) form a line image CC' which is parallel to the axis of the cylinder. In Fig. 9-16 line BB' lies along the surface of the cylinder and is

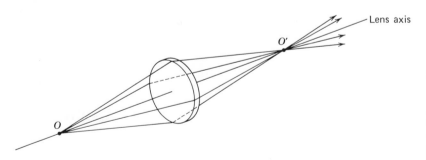

Fig. 9-15 Imaging of a point by a surface of spherical curvature.

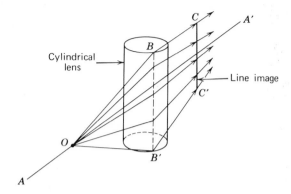

Fig. 9-16 Action of a cylindrical lens to form a line image of a point object.

parallel to the plane of the page. The group of rays shown leaving point O is tilted out of the plane of the page through a uniform angle such that the rays intersect the cylinder along BB'. Since the cylinder has no curvature in the direction of BB', rays are not refracted in that direction. However, the surface has spherical curvature in the horizontal direction, so that the rays tilted at angles other than that illustrated in Fig. 9-16 are refracted toward the line image CC', which lies in the plane containing AA'.

Figure 9-17 shows a section of the cylinder in a plane containing its axis. The rays that lie in the plane of this section travel parallel to their original course after leaving the cylinder. They are, however, deviated toward AA' to an extent which is dependent upon the thickness of the cylinder. The result of the deviation is a partial compression of the line image in the direction perpendicular to AA'. Rays that enter the cylinder in any section other than that shown in Fig. 9-17 are deflected above or below the plane of incidence, thus contributing also to the line image, as shown in Fig. 9-16. Rotation of the cylinder about AA' through 90° produces a 90° rotation of the line image. A refracting cylinder is often called a *cylindrical lens*. Note, however, that cylindrical lenses are *not* lenses in the sense defined in Chapter 8.

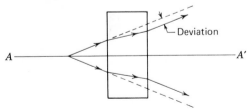

Fig. 9-17.

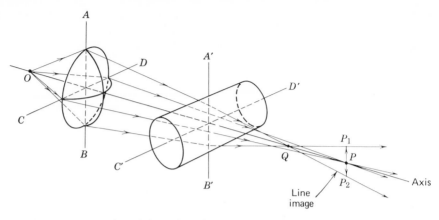

Fig. 9-18 Representation of the action of an asymmetric lens as that of a combination of spherical and cylindrical lenses.

An asymmetric lens is of partially cylindrical curvature and may be treated as a combination of a lens of spherical curvature with a cylindrical lens. The focusing action of the combination is illustrated in Fig. 9-18. As shown, light rays which are incident parallel to the axis are refracted in the direction of P by the spherical lens element. Rays that intersect diameter CD of the spherical element are incident upon the cylindrical element along its axis $C'D'$. These rays are not further refracted by the cylinder and thus are focused at P. Rays that are incident on the spherical surface along the diameter AB, at right angles to CD, enter the cylinder along $A'B'$ and experience an additional convergence, forming a focus at point Q. At the plane of P these rays have again diverged and form the line image, P_1P_2. Rays which enter the system along intermediate diameters of the spherical lens cross the axis at points between P and Q and contribute to the line images in the planes of P and Q. The nearest approach to a single image point is a "circle of least confusion" formed at a position intermediate between Q and P.

The effect of lens asymmetry on the appearance of the image is shown in Fig. 9-20d. The asymmetry of electron lenses, and its correction, are discussed on pp. 371 and 409.

THIRD ORDER ASTIGMATISM

A perfectly ground spherical lens, although free of the asymmetry just described, nevertheless forms a pair of line images of point objects which lie off the lens axis. Whereas the line images formed by an asymmetric lens are parallel and perpendicular, respectively, to the axis of the cylindrical component of the lens, the line images formed by a spherical lens are parallel and perpendicular, respectively, to the perpendicular from the object point to the lens axis. This is illustrated by Fig. 9-19. The line images shown in Figs. 9-19b and 9-19c are in focus in different planes. Their length increases as the square of the distance of the image point from the lens axis, so that the astigmatic image is much more blurred at the periphery than in paraxial zones. The object is distorted in the radial (sagittal) or tangential direction at the respective image planes. These effects, and the quite dissimilar imaging of the same object by a partially cylindrical lens, are illustrated (to an exaggerated degree) in Fig. 9-20. Note that the effect of lens asymmetry, unlike that of third order astigmatism, is constant across the image plane.

How do the effects of third order astigmatism arise? Again, as in the case of lens asymmetry, the refraction of rays which are incident upon different lens diameters must be compared. The tangential image (Fig. 9-20b) is formed by rays incident upon that diameter of the lens which is parallel to the perpendicular from object point to lens axis. The imaging of these rays is shown in projection by Fig. 9-21. (The rays included in Fig. 9-21 cross diameter AA' which is parallel to OO'.) In the absence of coma and spherical aberration, all the rays which enter the lens along this "tangential" diameter converge to form a point focus (the "tangential" focus) at P.

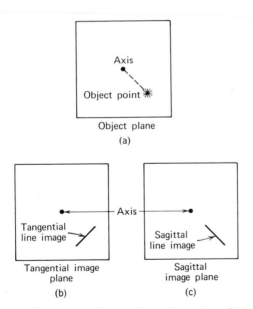

Fig. 9-19 An object and the corresponding line images formed by a lens which suffers from third order astigmatism.

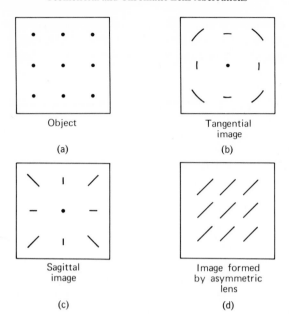

Fig. 9-20.

Similarly, rays which enter the lens along a diameter which is perpendicular to the perpendicular from object point to lens axis converge (in the absence of other aberrations) to a point focus at Q (the "sagittal" focus), as shown in Fig. 9-22. (The rays included in Fig. 9-22 cross diameter BB' which is perpendicular to OO'.)

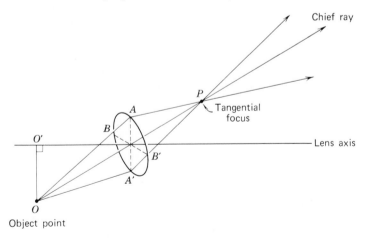

Fig. 9-21 Formation of the tangential image of an off-axis point.

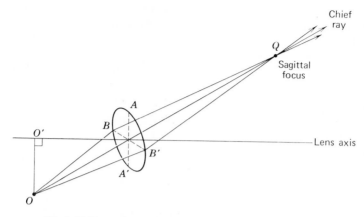

Fig. 9-22 Formation of the sagittal image of an off-axis point.

It is of interest to note that rays incident upon the tangential lens diameter AA' are meridional rays, whereas rays incident upon the sagittal lens diameter BB' are, in general, skew rays (cf. p. 175).

The two point images illustrated in Figs. 9-21 and 9-22 are shown superimposed in Fig. 9-23. At the plane in which the tangential rays focus, the sagittal rays are spread out into a line CC', which is the *tangential line image*. The tangential rays diverge again from P to form the *sagittal line image* DD' in the plane of focus of the sagittal rays. As shown, CC' is parallel to BB', while DD' is parallel to AA'.

How do rays that pass through lens diameters other than AA' or BB'

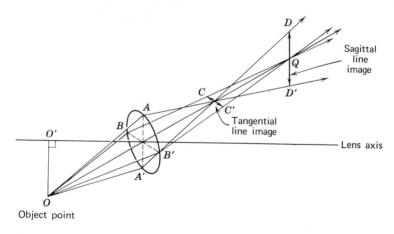

Fig. 9-23 Formation of the tangential and sagittal line images of an off-axis point.

contribute to the astigmatic images? Figure 9-24 illustrates a pair of rays which are incident upon an intermediate diameter EE'. These rays pass through some part of *each* of the line images, contributing to these images in the manner illustrated by Fig. 9-25. The rays from EE' are themselves focused (except for effects of spherical aberration and coma) at some intermediate position, designated E^* in Fig. 9-24. In general, the images formed by rays from all lens diameters at positions between CC' and DD' are *ellipses*; at one intermediate position a circular image is formed which is the circle of least confusion. The plane of circles of least confusion is the closest approach, in an astigmatic system, to a well-defined image plane.

In summary, rays from the tangential diameter of a lens (as defined with respect to the position of the off-axis object point) reunite at a point image in the tangential image plane, in which rays from the sagittal diameter are spread out to form the tangential line image. Similarly, the tangential rays diverge to form the line image in the sagittal image plane, while rays from intermediate lens diameters contribute to each of these images. Astigmatism is termed *positive* when the tangential focus lies to the *left* of the sagittal focus, and *negative* when the T focus lies to the *right* of the S focus. Figure 9-23 depicts a positive astigmatism, which is commonly found for single lenses.

In a qualitative sense, third order astigmatism can be accounted for as arising from the unequal curvatures which various parts of the spherical lens surface present to rays arriving from extra-axial points. Third order astigmatism, like the other geometrical errors, is thus a direct consequence of the law of refraction.

Quantitatively, the *separation* of tangential and sagittal image planes can be shown to be independent of lens aperture, but to vary as the *square*

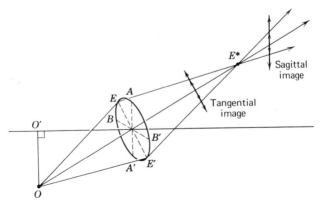

Fig. 9-24 Formation of the complete tangential and sagittal line images of an off-axis point.

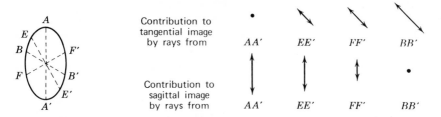

Fig. 9-25 Contributions of various lens zones to the tangential and sagittal line images.

of the distance from the (ideal) image point to the axis of the lens system. Thus astigmatism is a defect of relatively slight importance in optical systems which are used to view limited areas (such as high power microscope lenses) but is a problem of consequence in the design of wide field instruments (such as cameras). The *diameter* of the circle of least confusion is a function not only of the image dimension $(y')^2$, but also of the *first power* of the aperture of the lens.

The correction of astigmatism is achieved in lens systems by appropriate spacing of individual lenses and apertures. Correction must be considered in relation to the aberration of curvature of field, as described in the following section.

CURVATURE OF FIELD

A system free of spherical aberration, coma, and astigmatism satisfies the first condition for perfect lens action (i.e., it forms a point image of a point object) but may not form the image of a planar object in a plane perpendicular to the axis. Figure 9-26 illustrates the formation of an image

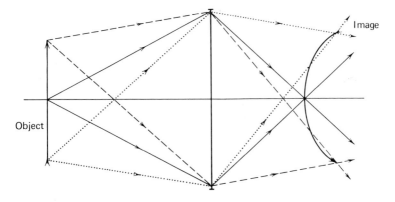

Fig. 9-26 Curvature of field.

that suffers from curvature of field; an image curved in the opposite sense has been illustrated in Fig. 8-11. Any desired zone of the image, but not the complete image, can be brought into sharp focus in a single plane.

In images that suffer from curvature of field, paraxial zones of the object are focused in the Gaussian plane located as specified by (8-z). It may be shown that (in the absence of other aberrations) other regions of the image lie out of the paraxial image plane by an amount which is proportional to the square of the image height y'. The positions of focus lie on a *parabola*; in three dimensions the locus of positions of sharp focus is a paraboloid of rotation. The focal surface is commonly known as the *Petzval surface*. Curvature of field is assigned positive sign when the Petzval surface is concave with respect to the object (as in Fig. 8-11) and negative when the Petzval surface is convex with respect to the object (as in Fig. 9-26).

If astigmatism is present in the image, the tangential and sagittal image surfaces are curved modifications of the Petzval surface. The three surfaces are coincident at the axis. For other zones of the image it can be shown that the *tangential* surface is invariably located three times as far from the Petzval surface as is the sagittal surface. (See Fig. 9-27.)

The degree of curvature of the Petzval surface is found to be invariant with respect to object and image positions, lens thickness, extent of lens bending, or the separation of individual lenses. It is, however, a function of the refractive index and focal length of each lens and is affected to some degree by appropriate location of apertures. The effect of curvature can be minimized by suitable correction of third order astigmatism in the manner explained below.

Figure 9-27 illustrates three possible types of correction for astigmatism in the presence of a given amount of curvature of field. In Fig. 9-27a both the astigmatism and the curvature of field are of positive sign. Both T and S surfaces are thus more highly curved than the Petzval (P) surface. The locus of sharpest images (i.e., of the circles of least confusion), as indicated by a dotted line in Fig. 9-27, lies between the T and S surface; when this surface is focused upon, in the system represented by Fig. 9-27a, the extra-axial defocus of the Petzval surface is exaggerated. It is clear, therefore, that curvature of field and astigmatism should be of opposite sign in a corrected system.

Figure 9-27b represents a system for which curvature of field is positive, as before, while astigmatism is negative but relatively slight. The intermediate surface of sharpest focus is curved, but less so than the Petzval surface. Because of the slight astigmatism it is also relatively unblurred.

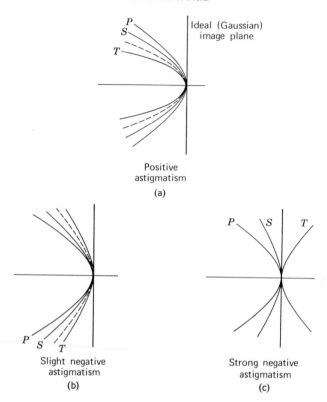

Fig. 9-27 Combined corrections of astigmatism and curvature of field.

Figure 9-27c represents a system in which third order astigmatism is negative and relatively strong. The $1:3$ ratio of separations of the T and S surfaces from the P surface is maintained, but the T and S curvatures are now of opposite sign. The resultant surface of sharpest focus can thus (for appropriate numerical values of the aberrations) be entirely flat. The sharpness of focus at the margins of the image is, however, inferior to that produced by the system shown in Fig. 9-27b, as a consequence of the greater astigmatism in this system. In short, the practical correction of astigmatism and curvature of field must involve a compromise between the flatness and the sharpness of the image. The nature of the correction selected varies with the purpose for which the lens is to be used.

It should be evident that the observed curvature of the image field is the combined result of the mathematically defined aberration of curvature of field with that of third order astigmatism.

DISTORTION

The third condition for perfect lens action states that the relative linear dimensions of the object must be preserved in the image (cf. p. 157). The aberration of *distortion*, which is a failure of this condition, occurs when the magnification of the image changes with distance from the axis of a lens system. In positive distortion (more commonly called *barrel distortion*), the magnification of the image is greatest at the axis; in negative or *pincushion distortion*, magnification is least at the axis. These effects are illustrated in Fig. 9-28.

The lateral magnification of images is given in Chapter 8 as

$$(8\text{-mm}) \qquad M = \frac{y'}{y} = -\frac{s_2''}{s},$$

where s is the object distance and s_2'' the image distance. Distortion arises when the value of the ratio s_2''/s, as measured along the chief ray (cf. p. 181) for extra-axial points, differs from that measured along the lens axis for axial points.

Figure 9-29 illustrates the imaging of an off-axis point Q by an unstopped lens. The chief ray from Q passes through the center of the lens, so that the ratio (s_2''/s) for that point is the same as for the paraxial point O. The image OQ' is thus undistorted.

In the system shown in Fig. 9-30 an aperture is placed to the left of the lens. The chief ray, which by definition is the ray that passes through the center of this aperture, is incident upon the lens at N, an off-axis position. In object space, QN makes an appreciable angle (γ) with the axis; in image space, NQ' makes a smaller angle (γ') with respect to the axis. Thus, from the geometry of Fig. 9-30,

$$(9\text{-q}) \qquad \frac{QN}{s_0} > \frac{Q'N}{s_2''} \quad \text{and thus} \quad \frac{s_2''}{s_0} > \frac{Q'N}{QN}.$$

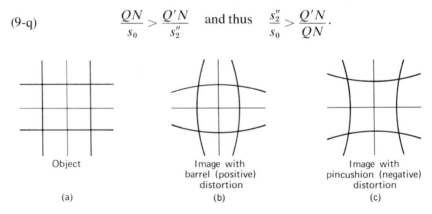

Object	Image with barrel (positive) distortion	Image with pincushion (negative) distortion
(a)	(b)	(c)

Fig. 9-28.

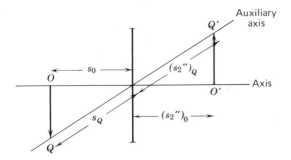

Fig. 9-29 Distortion-free imaging by an unstopped lens.

That is, since $s_2''/s_0 = M_0$ and $Q'N/QN = M_Q$,

(9-r) $$M_0 > M_Q$$

and barrel (positive) distortion is observed in the image.

Similarly, Fig. 9-31 illustrates a system in which an aperture is placed to the right of the lens. The ratio (s_2''/s_0) is then smaller than the quantity (QN/NQ^{**}), so that the lateral magnification of OQ is increased.

It can be shown that the distances $Q'Q^*$ and $Q'Q^{**}$ (in Figs. 9-30 and 9-31, respectively) are a function of the cube of the ideal image height y' $(= O'Q')$. This third power dependence indicates that distortion is negligible in paraxial regions of the image but increases to very large values in peripheral image zones. Correction is thus important in lens systems which are used to view extended fields, especially when quantitative measurements or composite images are to be made (cf. p. 396).

As Figs. 9-30 and 9-31 show, distortions of either sense can be produced by the introduction of apertures. Similarly, distortion can be cor-

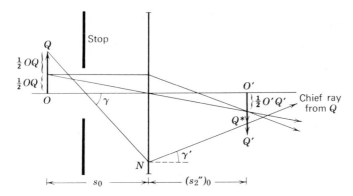

Fig. 9-30 Effect of a stop, placed between object and lens, in producing barrel distortion of the image.

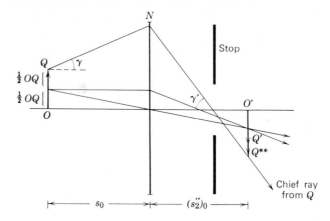

Fig. 9-31 Effect of a stop, placed between lens and image, in producing pincushion distortion of the image.

rected by positioning the apertures of a system in such a way that chief rays from extra-axial points pass close to the center of the lens system. Note that, whereas all the other lens aberrations may be minimized by reducing the lens aperture, distortion is produced or increased by small apertures.

THE CHROMATIC ABERRATIONS

The refractive index of all materials varies with the wavelength (frequency) of incident electromagnetic radiation. (This variation, known as *dispersion*, is also discussed in Chapter 23. Within the visible range the refractive index is normally *greater* for the shorter (blue, violet) than for longer (red) wavelengths. For a given single lens, therefore, the *focal length* with respect to *blue* light is shorter than that with respect to red. This effect is illustrated for the imaging of an axial object point by Fig. 9-32.

Figure 9-32 illustrates the *longitudinal chromatic aberration* of the lens. This aberration is defined as the *difference in focal length* with respect to the C and F lines of the solar spectrum (wavelengths 6563 Å and 4861 Å, respectively). The term "longitudinal chromatic aberration" is also used to refer to the axial distance between images of a point placed at any specified finite distance from the lens.

The images formed at positions *A* and *B* in Fig. 9-32 are of the form suggested by Figs. 9-33a and 9-33b, respectively. The image in plane *A* is a blue point surrounded by a series of colored halos the outermost of which is red, while the image at plane *B* is a red point surrounded by a

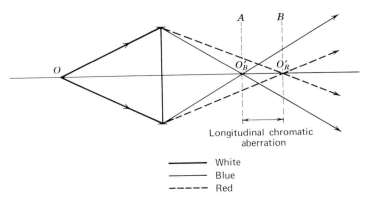

Fig. 9-32 Longitudinal chromatic aberration.

series of colored halos the outermost of which is blue. Such colored fringes are readily observed around the images formed by crude lenses.

If the object imaged by a chromatically aberrant lens is an extra-axial point, an additional, related defect can be recognized. This is the *lateral chromatic aberration*, which results in differences in magnification between the images formed by light of different wavelengths. Lateral chromatic aberration is illustrated in Fig. 9-34.

Imaging of white light by a *diverging* lens results also in the formation of a blue (shorter wavelength) image which is closer to the lens than the corresponding red (longer wavelength) image. In this case, however, the blue image lies to the right of the red image, as shown for the imaging of an axial point in Fig. 9-35.

Correction of the chromatic aberrations is effected by the use of positive-negative lens doublets in a manner analogous to that used for the correction of spherical aberration (cf. pp. 194–195). For any chosen pair of blue and red wavelengths, the positive element of the doublet produces a red image some distance to the right of the corresponding blue image. The succeeding negative lens, which would form a red image to the left of the corresponding blue image, causes the final images of the two colors to coincide. (Alternatively, the negative element may precede

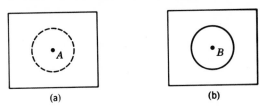

Fig. 9-33 Chromatically aberrant point images at two levels of focus.

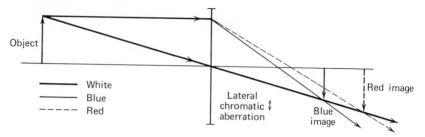

Fig. 9-34 Lateral chromatic aberration.

the positive element.) The C and F wavelengths are commonly those chosen for exact correction, but any other pair of wavelengths may be selected.

If the corrected doublet is to have net *converging* action, the *negative* element is made of a glass for which the difference in refractive index for light of different wavelengths is greater than that of the positive element (i.e., the glass of the negative element is of higher *dispersive power*; cf. p. 509). In this way, equal but opposite chromatic aberrations are produced by the two lenses of the doublet, but the radius of curvature of the negative element is larger (and its refracting power correspondingly less).

Among optical materials, the quantity $(n_F - n_C)$ varies from 0·008 for ordinary crown glass to as much as 0·047 for the densest flint glass.

Figure 9-36 illustrates the correction of chromatic aberration by a pair of lenses. (A pair of separate lenses, rather than the cemented doublet actually used, is shown for clarity in illustration. Similarly, a separated doublet for the correction of spherical aberration was shown in Fig. 9-8.) "Red" and "blue" light form intermediate images at O'_R and O'_B, respectively, but a single final image at O''. Other wavelengths are not focused exactly at O'', but form a final image which could lie either slightly to the right or slightly to the left of O'', depending upon the values of n and the spacings of the lens combination.

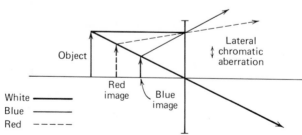

Fig. 9-35 Lateral chromatic aberration of a diverging lens.

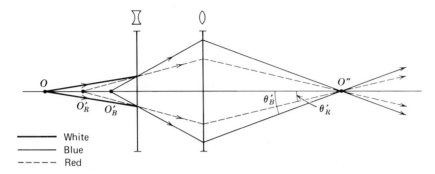

Fig. 9-36 Correction of longitudinal chromatic aberration.

Correction of chromatic aberration for three (or, in principle, m) different wavelengths can be achieved by the use of three (or, in principle, m) component lenses made from optical glasses of different dispersive powers.

In Fig. 9-36, $(s_2'')_R = (s_2'')_B$, but $\theta_R' \neq \theta_B'$. Thus $(\theta'/\theta)_R \neq (\theta'/\theta)_B$; that is, the angular magnifications of the red and blue images are unequal. The system shown is in fact corrected for longitudinal chromatic aberration, but not for lateral chromatic aberration. A system in which *lateral* chromatic aberration is corrected is shown in Fig. 9-37. Here angular magnification is the same for both colors, but the images are formed in different planes, so that halo effects occur in any one plane, that is, the longitudinal chromatic aberration is uncorrected. In practice, however, when one of the chromatic aberrations is corrected, the other is found to be very nearly so.

Chromatic aberration may, of course, be avoided absolutely by the use of monochromatic illumination.

The chromatic aberration of electron lenses is described on pp. 367.

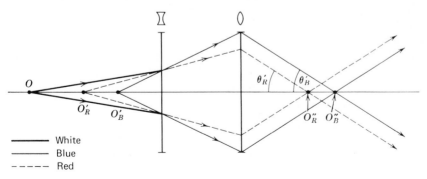

Fig. 9-37 Correction of lateral chromatic aberration.

ABERRATION IN REFLECTION LENSES

Reflection lenses (i.e., mirror lenses; cf. p. 177) are extremely advantageous in many applications because they are entirely free of chromatic aberration. (Note that this follows from the fact that angles of reflection are determined *solely* by the corresponding angles of incidence.) Reflection lenses thus are particularly useful in instruments such as spectrophotometers which employ a range of wavelengths.

These lenses suffer, however, from geometrical errors, of which third order astigmatism and spherical aberration are particularly troublesome. In general, corrections tend to be more difficult to effect than in refraction systems. However, the spherical aberration remains small if aperture angles are restricted to 10° or less. Aspheric mirrors have also been used to produce images of high quality.

APPLICATION OF CORRECTIONS

Six types of lens aberration (in addition to the purely technical defect of lens asymmetry) have been described in this chapter. They include the five monochromatic or "Seidel" aberrations which occur in the imaging of monochromatic light beams, and chromatic aberrations which occur only when (as is usually the case) the imaging beam contains a mixture of frequencies. In general, any or several, but not all, of the defects may be corrected simultaneously by the methods described. The choice of the corrections to be made depends upon the purpose for which the lens system is to be used. Spherical and longitudinal chromatic aberration are effects of considerable and roughly comparable magnitude which occur on the axis as well as in all other regions of the image plane. These errors must be corrected to some degree in virtually every optical system; a particularly high degree of correction is required when a limited area is to be viewed at high magnification, as in microscope objectives. Coma, while zero on the axis, becomes appreciable in zones of the image which are relatively close to the axis; its correction is also effected in almost every lens system. Third order astigmatism, curvature of field, distortion, and lateral chromatic aberration are defects of the image margins. These aberrations must be corrected in lenses which are used to view extended areas (e.g., camera lenses) but their correction is relatively less important in high power objectives.

A lens which is free of spherical aberration and coma is called an *aplanat*, and one in which astigmatism and curvature of field are controlled is called an *anastigmat*. Chromatic aberration is corrected for light of two wavelengths in an *achromat* and for three wavelengths in an *apochromat*.

REFERENCES

J. Strong, *Concepts of Classical Optics*, McGraw-Hill, New York, 1963, pp. 332–336, and Chapter 16 (Image Defects).

F. A. Jenkins and H. E. White, *Fundamentals of Optics* (3rd ed.), McGraw-Hill, New York, 1957, Chapter 9 (Lens Aberrations).

M. Herzberger, in R. K. Luneburg, *Mathematical Theory of Optics*, University of California Press, Berkeley, 1964, Supplementary Note II, pp. 411–431. This article, "Optical Qualities of Glass," discusses, especially, prospects for the correction of chromatic aberration. Data are given for a number of glasses, and a short account of the history of the development of optical glasses is included. Recent progress in the development of optical glasses is summarized in Fig. 14-15 (p.324) of the book by Strong, *op. cit.*

A. E. Conrady, *Applied Optics and Optical Design*, Part I, Dover, New York, 1957, Chapter 2 (Spherical Aberration), Chapter 4 (Chromatic Aberration), Chapter 5 (Design of Achromatic Object Glasses), and Chapter 7 (The Optical Sine Theorem).

L. C. Martin, *The Theory of the Microscope*, Blackie, Glasgow, and American Elsevier, New York, 1966, pp. 113–150. Aberrations are discussed with particular reference to microscope systems.

F. D. Smith, *Sci. Am.* **219**, 97 (1968). This article, "How Images are Formed," describes contemporary approaches to the design of aberration-free lenses.

CHAPTER 10

Interactions of Light Waves in the Imaging Process; Resolution

A geometrical description of lenses implies that, in the absence of all the aberrations described in Chapter 9, a point source of light (point object) is imaged as a discrete point. In fact, however, the geometrical treatment of lenses is adequate only when the imaging of spacings which are substantially larger than the wavelength of light is considered. Interactions between the portions of the light beam forming an image lead to variations in intensity which are *not* directly related to the structure of the object. In order to account for details of the distributions of intensities found in images of very small objects, and thus to assess the ultimate limitations of the imaging process, it is necessary to take the physical nature of light beams into account.

In this chapter, lenses are described from the point of view of their effects upon light waves. It is assumed throughout that lenses are free of geometrical aberrations, chromatic aberrations, and technical imperfections. The discussion specifically describes the action of convergent lenses only, but could readily be modified to apply to divergent lenses also. The resolution achieved by imaging systems is defined, and theoretical values of limiting resolving power are derived.

Understanding of the material presented in this chapter requires familiarity with the geometrical properties of lenses, as described in Chapters 8 and 9, and appreciation of the nature and interactions of light waves, as discussed in Chapters 1 and 3. Although a rigorous description of the interactions of light waves requires mathematical treatments of some complexity, the discussions in this chapter insofar as possible are descriptive in character. However, the usefulness of Fourier transform functions in the description of the imaging process, as of certain other optical effects, is so great that this approach is introduced here.

THE NATURE OF AN IDEAL LENS

A geometrical condition for perfect lens action is the requirement that rays originating from a point object reconverge at a single image point (cf. p. 156). From the point of view of wave optics, an ideal lens is a device which *converges the wavefront originating from an object point to a single image point without introducing differences in optical path between portions of the wavefront*. Thus the physical definition of a perfect lens includes an additional stipulation: rays from an object point must arrive at the image point *in* phase after traversing identical optical paths. This definition of an ideal lens is illustrated by Fig. 10-1a. The curved lines in Fig. 10-1a represent loci of positions of equal phase of the wavefront originating from O. These loci are concave with respect to O in the region to the left of the lens. Portions of the wavefronts are compressed within the lens (i.e., since the refractive index of the lens is greater than that of air, paraxial portions of the wavefront travel a greater

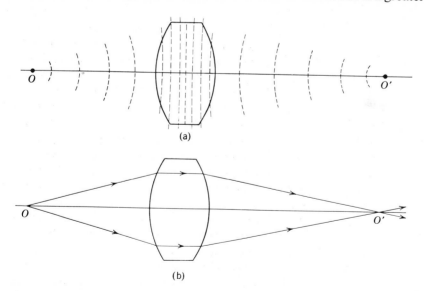

(a)

(b)

Fig. 10-1 Imaging by an ideal lens: (a) convergence of wavefronts; (b) refraction of rays.

optical path length per unit physical path in passing through the regions of the lens) with the result that wavefronts emerge from the lens convex with respect to the object point O, but with a center of curvature at the image point O'. Figure 10-1b represents the same system shown in Fig. 10-1a but from a purely geometric point of view.

The effect of lens aberrations may be described as a distortion of the wavefronts leaving the lens from the circular (spherical) form shown in Fig. 10-1a.

FORMATION OF A DIFFRACTION PATTERN IN THE REAR FOCAL PLANE OF THE LENS

When any specimen is illuminated, each point of the specimen can be considered to act as a source of light waves which may travel in all possible directions. (Note that the object points diagramed in the figures of Chapters 8 and 9 are so treated; rays originating from each object point represent specific possible directions of energy transfer.) The light which travels in any one direction from a series of object points then constitutes a parallel (collimated) beam; light from the specimen as a whole correspondingly may be described as a series of collimated beams traveling in all directions. By definition, each such beam must converge to a focus in the rear focal plane of the lens. The beam which is incident upon the lens parallel to its axis is focused on the axis at the second focal point, while collimated beams which are incident in other directions converge at other positions on the second (rear) focal plane. The intensity which can be observed at each of these positions is determined by the relative amplitudes and phases of the rays which converge at that position. It is clear that, in general, the focused diffraction pattern formed in the rear focal plane of the lens is related to the structure of the object.

It is particularly simple to consider the imaging of a specimen of *periodic* structure, as shown in Fig. 10-2, that is, of an object which is, essentially, a diffraction grating. As shown in Fig. 10-2 points $A_0, A_1, \ldots$, on the object are separated by a constant distance d. Light leaving each object point in two arbitrarily selected directions, AB and AC, is shown. [Note that *any* other direction(s) could equally well be chosen for illustration.] Light leaving the object in direction AB converges to position Ⓑ on the rear focal plane, while light leaving the object in direction AC converges to position Ⓒ. Subsequently, light diverges from these positions on the focal plane to the positions $A_0', A_1', \ldots$, on the image plane. The distinction between the origins of the intensities observed in the focal and image planes should be clear: light which leaves the specimen in any given direction is brought to a focus at the focal plane, while light which

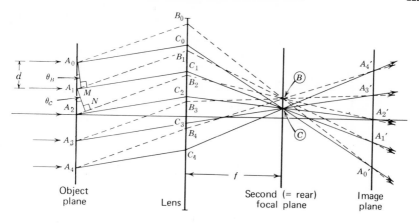

Fig. 10-2 Imaging of a specimen of periodic structure.

originates from any given object point is brought to a focus in the image plane.

The intensity which is observed at Ⓑ, Ⓒ and other positions on the focal plane is determined by: the *amplitude* leaving each object point in the various directions, AB, AC, etc.; and the relative phases of the collimated beams $A_0B_0, A_1B_1, \ldots$, or $A_0C_0, A_1C_1, \ldots$. Positions of maxima and minima in the diffraction pattern are determined by the relative phases of the successive beams which are radiated ("diffracted") in various directions. For example, the successive beams which contribute to the focused spot at point Ⓑ differ in path by A_1M $(= d \sin \theta_B)$, while those which contribute to the spot at C differ in path by A_1n $(= d \sin \theta_C)$. If these quantities are equal to an integral number of wavelengths, constructive interference occurs at the corresponding focal spot, and a maximum of intensity is observed. As described on pp. 59–63, the complete diffraction pattern produced by the diffraction grating-like specimen consists of sharp maxima, located as described in Chapter 3,

(3-d) $$m\lambda = d \sin \theta,$$

(3-e) $$m\lambda = d(\sin i + \sin \theta),$$

while intensity is essentially zero at all intermediate positions.

More commonly, the objects imaged by lenses are irregular structures. Nevertheless, a diffraction pattern is again formed in the rear focal plane; for example, as illustrated in Fig. 10-3. Points of the randomly structured object are not separated by any constant spacing and may vary both in the total amplitude scattered and in the angular distribution of scattered amplitudes. Nonetheless, as shown, light leaving object points A_1, A_2,

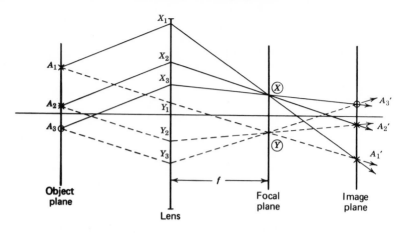

Fig. 10-3 Imaging of a specimen of irregular structure.

$A_3, \ldots$, in the directions AX and AY converges to the corresponding points, $\circled{X}$ and $\circled{Y}$, respectively, on the focal plane. Computation of the positions and intensities of maxima becomes a much more complex problem than for the gratinglike specimen, since the relative phase and amplitude contributed by each object point to each position on the focal plane must now be considered separately. Generally, the diffraction pattern formed from a randomly structured object is more diffuse than that formed from a regularly structured object.

ABBE'S DESCRIPTION OF OBJECT AND IMAGE

An object imaged by a lens forms a diffraction pattern in the rear focal plane of the lens. The distribution of amplitudes in the diffraction pattern is related in a highly specific, although indirect, way to the structure of the object, and thus also to the form of the image. Abbe developed a mathematical description of the relationships that exist between object, diffraction pattern, and image; it shows that the complete diffraction pattern is described by the *Fourier transform* of the mathematical function which specifies the distribution of amplitudes at the object. In turn, the image is the Fourier transform of the function which specifies the distribution of amplitudes in the diffraction pattern. The transform ("inverse transform") of a Fourier transform has the property of regenerating the original function. Thus the distribution of amplitude in an object and that in an image are identical—if the *complete* diffraction pattern is included in the focal plane. In fact, this condition can never be completely satisfied, since some of the light leaves the object in such directions that it does not enter the

aperture of the lens. The image is correspondingly a less than faithful "copy" of the object to the extent that it is formed from a less than
complete diffraction pattern.

Extensive familiarity with the properties of Fourier integrals is not
required for an elementary appreciation of Abbe's theory. It is simply
necessary to understand that the operations specified by these functions
have the property of regenerating the original when applied twice in
succession. That is,

(10-a) $\qquad f(x) \xrightarrow[\text{operation}]{\text{Fourier transform}} f'(x') \xrightarrow[\text{operation}]{\text{Fourier transform}} f(x).$

[Note that the designations $f(x)$ and $f'(x')$ are used in this chapter to represent entirely different functions of the respective variables x and x';
the function f' does *not* specify the first derivative of the function f.]

A specific formulation of Abbe's theory in mathematical notation may
now be considered. Figure 10-4 illustrates the imaging of a one-dimensional object, specifically, the imaging of a section of a slit. Distances
along the object plane in Fig. 10-4 are specified by the variable x, which
extends across the width of the slit from $x = -\frac{1}{2}d$ to $x = +\frac{1}{2}d$. Similarly,
distances along the focal and image planes are specified by the variables x'
and x'', respectively. The effective aperture of the focal plane is d', and the
diameter of the image of the section of the slit is d''. Amplitude distributions in the object, focal, and image planes can be expressed by functions
$f(x), f'(x')$, and $f''(x'')$, respectively. For example, in Fig. 10-4, $f(x)$ is as
given by

(10-b) $\qquad \begin{aligned} f(x) &= 0, & x < -\tfrac{1}{2}d, x > +\tfrac{1}{2}d, \\ f(x) &= A^*, & -\tfrac{1}{2}d < x < +\tfrac{1}{2}d. \end{aligned}$

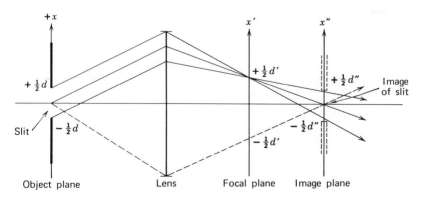

Fig. 10-4 Imaging of a section of a slit.

Abbe's theory then states that

(10-c)
$$f'(x') = C \int_{x=-\infty}^{x=+\infty} f(x) \exp\left(-\frac{2\pi ixx'}{\lambda f}\right) dx,$$

in which λ is the wavelength of light, f is the focal length of the lens, and C is an appropriate constant.

Inversion of the Fourier integral (10-c) gives

(10-d)
$$f(x) = C' \int_{x'=-\infty}^{x'=+\infty} f'(x') \exp\left(+\frac{2\pi ixx'}{\lambda f}\right) dx',$$

in which C' is an appropriate constant related to C.

Ideally, the amplitude distribution in the image plane, $f''(x'')$, should be that of the object plane, adjusted by a scale factor M, which gives the magnification. That is,

(10-e)
$$f''(x'') = f\left(\frac{x''}{M}\right).$$

Substitution of (10-d) in (10-e) then gives

(10-f)
$$f''(x'') = C'' \int_{x'=-\infty}^{x'=+\infty} f'(x') \exp\left(+\frac{2\pi ix'x''}{\lambda fM}\right) dx',$$

in which C'' is an appropriate constant related to C and C'.

Equations 10-c and 10-f constitute a mathematical statement of Abbe's description of object, diffraction pattern, and image for the simple case of a one-dimensional object. For the more usual cases of two- or three-dimensional objects, analogous expressions apply in which double or triple integrations must be performed over the x and y directions, or x, y, and z directions, respectively. A general proof that (10-c) and (10-f) do in fact describe the relations of the amplitude distributions is not given here. In the following section, however, the applicability of these equations to the imaging of slits is demonstrated.

Single slits and also series of slits (diffraction gratings) and circular apertures are particularly simple objects, for which $f(x)$ is easily deduced. It is, of course, true that for objects of complex structure (including the majority of microscope specimens) formulation of $f(x)$ is much more difficult.

IMAGING OF A SLIT

Whereas Fig. 10-4 depicted the imaging of a *section* of a slit, (10-c) and (10-f) can also apply to the imaging of a slit of infinite length when the constants C, C', and C'' are suitably adjusted. The following discussion demonstrates that these equations apply to the imaging of a slit by evaluating these functions for the two extreme cases of a very broad and a very narrow slit. This evaluation requires only an elementary knowledge of integral calculus; conclusions derived from the evaluation are in any case summarized at the end of the section.

For a single slit of width d, the function $f(x)$ has been given by (10-b). This function is plotted in Fig. 10-5a. The corresponding distribution of amplitudes in the focal plane, $f'(x')$, is given by (10-k), which is obtained by evaluation of (10-c) in the following manner. Substitution of (10-b) in (10-c) first gives

(10-g)
$$f'(x') = C \int_{x=-1/2d}^{x=+1/2d} A * \exp\left(-\frac{2\pi i x x'}{\lambda f}\right) dx.$$

(Note that the limits of integration are here reduced, that is, to $-\frac{1}{2}d$ to $+\frac{1}{2}d$, since $f(x)$ is zero outside this range.) Holding x' constant, and noting

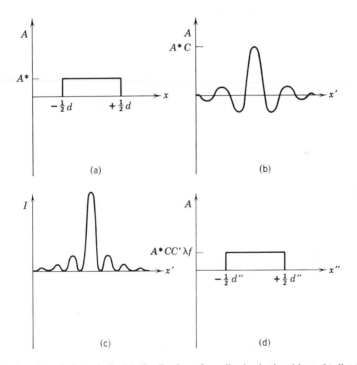

Fig. 10-5 Imaging of a broad slit: (a) distribution of amplitudes in the object; (b) distribution of amplitudes in the focal plane; (c) distribution of intensities in the focal plane; (d) distribution of amplitudes in the image.

that A^*, λ, and f are also constant, it becomes obvious that (10-g) may be integrated according to the form

(10-h)
$$\int e^{ax}\,dx = \frac{e^{ax}}{a}.$$

Equation 10-g thus becomes

(10-i)
$$f'(x') = \frac{CA^*\lambda f}{2\pi i x'}\exp\left[-\frac{2\pi i x' x}{\lambda f}\right]_{x=-1/2d}^{x=+1/2d}$$

$$= \frac{CA^*\lambda f}{2\pi i x'}\exp\left(-\frac{\pi i x' d}{\lambda f}\right) - \exp\left(+\frac{\pi i x' d}{\lambda f}\right),$$

or, since

(10-j)
$$\sin x = \tfrac{1}{2}i(e^{ix} - e^{-ix}),$$

It follows that

(10-k)
$$\boxed{f'(x') = dCA^* \sin\left(\frac{\pi x' d}{\lambda f}\right)\left(\frac{\pi x' d}{\lambda f}\right).}$$

As shown by Fig. 10-5b, the function given by (10-k) has a central maximum CA^* and passes through successive zeros, with intermediate maxima of rather rapidly diminishing magnitude. Zeros of (10-k) occur when $\sin(\pi x' d/\lambda f) = 0$, that is, when the quantity $x'd/\lambda f$ is an integer.

The distribution of amplitudes in the single slit diffraction pattern has also been described in Chapter 3:

(3-a*)
$$A = \frac{A_0(\sin\gamma)}{\gamma},$$

where $\gamma = \pi d \sin\theta/\lambda$.

Equation 3-a* is derived (in Appendix D) by a method quite different from that used above to obtain (10-k). Nevertheless, by substituting the notations $f'(x') = A$ and $CA^*d = A_0$, and by noting that $\tan\theta \cong \sin\theta = x'/f$, it becomes evident that (3-a*) and (10-k) are completely equivalent.

The observed distribution of intensities in the focal plane should be proportional to $[f'(x')]^2$, a function which is plotted in Fig. 10-5c. The diffraction pattern formed in the rear focal plane of the objective lens of a compound microscope may be observed directly, by removing the ocular lens, or by using a special ocular which focuses an image of the focal plane rather than the intermediate image plane. That the observed pattern is in fact that plotted in Fig. 10-5c may be verified in detail by measurement.

Equation 10-k describes the diffraction pattern originating from a slit of *any* size. The extent to which the pattern is included in the focal plane varies, however, according to the effective aperture of the diffraction plane and to the width of the slit. For a *wide* slit, d is relatively large, so that zeros of the quantity $\sin(\pi dx'/\lambda f)$ are closely spaced. Thus all appreciable intensity of the diffraction pattern is confined to a limited region near the axis of the lens. This means that (in the limit) *all* of the diffraction pattern is passed by the system and can contribute to the amplitude distribution in the image plane (i.e., for a wide slit, the aperture of the focal plane is considered to be *large* with respect to the spread of the diffraction pattern). The amplitude distribution in the image plane, $f''(x'')$, may then be obtained by evaluating (10-f) over the complete range from $x' = -\infty$ to $x' = +\infty$. In so doing, the value of $f'(x')$ given by (10-k) is substituted in (10-f), giving

$$(10\text{-}l) \quad f''(x'') = C'' \int_{x'=-\infty}^{x'=+\infty} \frac{CA^*\lambda f}{\pi x'} \sin\left(\frac{\pi x' d}{\lambda f}\right) \exp\left[\frac{2\pi i x' x''}{\lambda f M}\right] dx'.$$

Expressions of the form of (10-l) cannot be integrated directly to give a finite value of $f''(x'')$. It may be noted however, that

$$(10\text{-}m) \qquad \int_{-\infty}^{+\infty} \frac{\sin au \, e^{ibu}}{u} \, du = 2 \int_{0}^{\infty} \frac{\sin au \cos bu}{u} \, du.$$

Therefore, for $u = (\pi x'/\lambda f)$, $du = (\pi/\lambda f) \, dx'$, $a = d$, and $b = 2x''/M$, (10-l) may be expressed as

$$(10\text{-}n)$$
$$f''(x'') = \frac{2CC''A^*\lambda f}{\pi} \int_{x'=0}^{x'=\infty} \frac{\sin(\pi x' d/\lambda f) \, \cos(2\pi x' x''/\lambda f M) \, (\pi d/\lambda f)}{\pi dx'/\lambda f} \, dx'.$$

The value of a definite integral of the form of (10-n) is given by

$$\int_{0}^{\infty} \frac{\sin au \cos bu}{u} \, du = 0, \quad \frac{b}{a} < -1, \quad \frac{b}{a} > +1$$

$$(10\text{-}o) \qquad\qquad\qquad = \frac{\pi}{2}, \quad -1 < \frac{b}{a} < +1.$$

Thus

$$f''(x'') = \frac{2CC''A^*\lambda f}{\pi} \frac{\pi}{2} = CC''A^*\lambda f, \quad -1 < \frac{2x''}{Md} < +1$$

$$(10\text{-}p) \qquad\qquad\qquad = 0 \qquad\qquad \frac{2x''}{Md} < -1, \frac{2x''}{Md} > +1$$

That is, $f''(x'')$ is zero except within the limits $-Md/2 < x'' < +Md/2$, while within these limits the function has a constant value. Comparison of $f''(x'')$ with $f(x)$ as given by (10-b) shows that these two functions satisfy (10-e). That is, the two functions are *identical* except for the magnification factor M and the corresponding adjustment in the level of constant amplitude. The form of the image of a broad slit is shown in Fig. 10-5d.

For an extremely narrow slit, the quantity d in (10-k) is very small, so that zeros of the quantity $\sin(\pi x'd/\lambda f)$, and thus also the corresponding diffraction pattern, are widely spread. In the limit the effective size of the diffraction pattern is very much *larger* than the aperture d' of the diffraction plane (the focal plane), so that the amplitude of the pattern is approximately *constant* across that plane. The amplitude distribution function in the diffraction plane is then given by

(10-q) $\qquad \begin{aligned} f'(x') &= 0, & x < -\tfrac{1}{2}d', \;\; x > +\tfrac{1}{2}d', \\ f'(x') &= A^{*'}, & -\tfrac{1}{2}d' < x < +\tfrac{1}{2}d' \end{aligned}$

which is analogous to (10-b). The corresponding values of $f''(x'')$ may then be obtained by substituting (10-q) in (10-f). By a series of operations which are mathematically identical with those in (10-g) to (10-k) [except that the effective limits of integration in (10-f) are now $-\tfrac{1}{2}d'$ to $+\tfrac{1}{2}d'$ rather than $-\infty$ to $+\infty$], $f''(x'')$ is obtained as

(10-r) $\qquad\qquad f''(x'') = \dfrac{\lambda f M}{\pi d' x''} \sin \dfrac{\pi d' x''}{\lambda f M}.$

Equation (10-r) is identical in form with (10-k) and thus describes an amplitude distribution like that shown in Fig. 10-5b. This distribution, however, exists in the *image* plane rather than in the rear focal plane, while its spacings are determined by the size of the effective aperture d' *of the lens system* rather than by the spacing d of the object. Thus, *for a sufficiently small specimen, the distribution of amplitude and intensity in the image plane is unrelated to the structure of the specimen.* Note that minima of the image occur where $\sin(\pi x''d'/\lambda f M) = 0$. The first minimum of the image thus occurs when

(10-s) $\qquad\qquad \dfrac{\pi x''d'}{\lambda f M} = \pi,$

that is, when

(10-t) $\qquad\qquad x'' = \dfrac{\lambda f M}{d'}.$

The appearance of d' in the denominator of (10-t) indicates that, when the aperture of the lens system is large, maxima of the image of a narrow slit (or other small object) are *closely spaced.* As the lens aperture is reduced, the image becomes more diffuse.

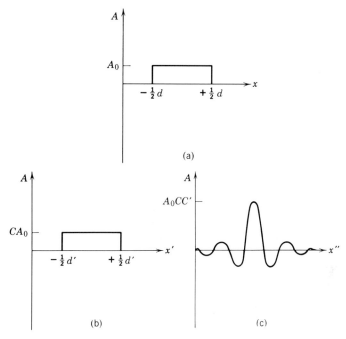

Fig. 10-6 Imaging of a very narrow slit: (a) distribution of amplitudes in the object; (b) distribution of amplitudes in the focal plane; (c) distribution of amplitudes in the image.

The amplitude distributions in the object, focal, and image planes for a narrow slit are shown in Fig. 10.6, which may be compared with Fig. 10-5.

In summary, it has been shown in this section that a *wide* slit forms a compact diffraction pattern which, in the limit, is completely included in the focal plane of the lens. The distributions of amplitude and intensity in the image plane resemble those of the object. A *narrow* slit forms a diffuse diffraction pattern which, in the limit, uniformly illuminates the focal plane of the lens. The corresponding distributions of amplitude and intensity in the *image* plane are then identical in form with the diffraction pattern produced in the *focal* plane by a wide slit. The dimensions of the pattern are, however, determined by the aperture of the lens system and not by the structure of the object.

IMAGE OF A CIRCULAR APERTURE

A circular aperture is a two-dimensional object and must be treated as such, whereas the slits discussed in the preceding section could be considered one-dimensional. Consequently, (10-c) and (10-f) must be modified to include integrations over both x and y dimensions in order to compute the Fourier transform and inverse transform of the amplitude distribution of a circular hole. Mathematical details of the procedure are not given here, but the overall features of the imaging of a circular aperture which are, in other respects, analogous to the imaging of a slit, are described.

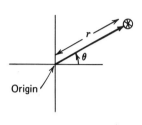

Fig. 10-7.

The amplitude distribution at a circular aperture is most simply described in terms of polar coordinates. It will be recollected that the location of points in two dimensions can be specified by the parameters r and θ in the manner illustrated by Fig. 10-7. Thus the amplitude distribution of the aperture is given by

(10-u)
$$f(r) = A_0, \quad r < \tfrac{1}{2}d,$$
$$f(r) = 0, \quad r > \tfrac{1}{2}d.$$

Note that (10-u) is comparable to (10-b) for a slit. Computation of the Fourier transform of (10-u) yields the expression given in Chapter 3:

(3-g*)
$$A = \frac{A_0 2 J_1(\gamma)}{\gamma} = 2A_0 \frac{J_1(\pi d \sin \theta / \lambda)}{\pi d \sin \theta / \lambda},$$

in which J_1 is a first order Bessel function. Equation 3-g*, describing the amplitude distribution in the rear focal plane of a lens which images an aperture, is analogous to (10-k), which describes the diffraction pattern of a slit. Zeros of the diffraction pattern occur at values of θ such that the first order Bessel function is zero. As given in Chapter 3, the first of these zeros occurs when

(10-v)
$$\sin \theta = \frac{1.220\lambda}{nd},$$

where θ is the angular width of the minimum (cf. Figs. 3-24 and 10-7). λ is the wavelength, n the refractive index of the medium in which the aperture and its diffraction pattern are located, and d is the width of the aperture. As noted in Chapter 3, the angular width of the diffraction pattern *increases* as the size of the aperture diminishes. Thus a large aperture forms a diffraction pattern which is essentially confined to the paraxial region of the rear focal plane of the lens. Essentially the complete Fourier transform thus contributes to the image, so that a faithful copy of the circular hole is produced in the image plane. The amplitude distribution in the image plane is given by

$$f''(r'') = KA_0, \qquad r'' < \frac{Md}{2},$$

(10-w)

$$f''(r'') = 0, \qquad r'' > \frac{Md}{2}.$$

Note that (10-w) is analogous to (10-p), which describes the image of a wide slit.

For a *small* circular hole the diffraction pattern in the focal plane is widely spread, and thus an incomplete Fourier transform determines the distribution of amplitudes in the image. In the limit the focal plane is uniformly illuminated by the central maximum of the diffraction pattern. Computation of the inverse transform then shows that a diffraction pattern (Airy disc) is formed in the image plane. As in the case of the narrow slit, the dimensions of the diffraction pattern are determined by the properties of the imaging system rather than by the structure of the object. In particular, the size of the pattern is such that the radius r'' of the first minimum is

(10-x)
$$\boxed{r'' = \frac{1.22\lambda s_2''}{n'A}},$$

in which s_2'' is the lens-to-image distance, n' is the refractive index of image space, and A is the diameter of the lens.

RESOLUTION

Resolution may be defined as the ability to distinguish closely spaced points as separate points. The *resolution limit* is the smallest separation of points which can be recognized as distinct. The resolution limit of optical lenses and thus of the optical microscope is of the order of $0.2\,\mu$, as is shown below. That of the eye is about 0.1 mm, and that of the electron microscope is of the order of 5 Å. The *resolving power* of an instrument specifies the resolution which *may* be achieved under optimum viewing conditions.

The distinction between resolution and resolving power is often misunderstood, and it should be noted carefully: resolving power is a property of the instrument and is a quantity that may be estimated on theoretical grounds. Resolution is equal to or poorer than the resolving power and is the quantity observed under any given set of experimental conditions. For the electron microscope, especially, it is true that the resolution achieved may be considerably inferior to the theoretical resolving power of the instrument. A further discussion of the distinction between resolution and resolving power is given on p. 452.

Various numerical quantities are used to express the resolving powers of different types of optical instruments. For lens systems and microscopes, resolving power is expressed in terms of the *minimum resolvable distance* d_{min}. Note that d_{min} is an inverse measure of resolving power; small values correspond to "high resolving powers." For telescopes it is convenient to express resolving power as the minimum angular separation of points which can just be distinguished. For diffraction gratings and prisms, which separate, physically, the component wavelengths of light beams (cf. p. 63 and Chapter 23) resolving power is defined as

$$(10\text{-}y) \qquad\qquad R.P. = \frac{\lambda}{\Delta\lambda},$$

where $\Delta\lambda$ is the minimum difference in wavelengths which can be distinguished at wavelengths close to λ. The expression given by (10-y), unlike d_{min}, specifies a quantity which is large at high resolving powers.

When an instrument is used to form images of extremely small objects, the question is often asked: "How much will it magnify?" While the level of direct magnification is of some importance from the point of view of technical convenience, a much more fundamental question is: "What is the resolving power of the instrument?" The image formed by any lens may *always* be increased in magnification by an additional lens or lenses; for example, by a photographic enlarger. Thus a succession of six glass lenses, each effecting magnification by 10×, could be used to produce a net magnification of one million. The later stages of magnification would be useless, however, because subsequent lenses cannot supply information lost from the diffraction pattern formed in the rear focal plane of the first lens. Magnification is said to be "useful" only when it serves to enlarge resolved detail to a size that can be comfortably observed by the naked eye. Beyond this point, magnification produces larger but increasingly blurred images. A familiar example of the futility of excessive magnification is provided by newspaper photographs which, even on relatively slight magnification, are found to consist of an array of black and white dots and to contain no more detail than is already evident upon casual inspection. (Here, however, as is sometimes true in microscopy also, the limitation originates from the properties of the object rather than from those of the imaging system.)

In certain experiments, for example, in dark field microscopy (cf. p. 282), it is possible to *detect* objects which are known from other types of measurement, or known by observation in systems of higher resolving power to be smaller than the resolution limit of the optical system. Such objects, however, are not in fact resolved, since it is impossible to determine whether an observed image point corresponds to one or a group of several objects.

The detectability of a point is determined by its *contrast*, that is, by the relative difference in intensity between the image point and its *surround* (background). While the term "contrast" is generally not used to imply any particular quantitative measurement, the *percent contrast* may be defined:

$$(10\text{-z}) \qquad \% \text{ contrast} = \frac{I_{\text{background}} - I_{\text{specimen}}}{I_{\text{background}}} \times 100.$$

The term "contrast" is also used to refer to the inherent properties of the *specimen* from which the contrast of the image is derived. The resolving power of an instrument is unrelated to the contrast of the specimen, but the contrast may critically determine the level of resolution achieved. Note that, although it may be true to say that "staining of the specimen improves the resolution of the image," it is always incorrect to assert that "resolving power is improved by staining the specimen." The contrast of specimens and images is discussed in Chapter 11.

For lens systems, resolution is defined in terms of *point* objects (as stated above), and calculations of the resolving power are thus based on consideration of closely spaced object points. In general, adjacent *lines* or other extended objects are more easily distinguished than are isolated points. Thus it may be possible to distinguish individual members of sets of lines separated by distances somewhat less than the theoretical value of d_{min}. Although this may be referred to loosely as "resolution," it is not in fact resolution in the strictest sense of that term.

ORIGINS OF LIMITATIONS OF RESOLUTION

Why should the resolution which can be achieved by a lens be limited? The computations given above provide an answer to this question. It was shown on pp. 226–231 that, whereas the image of a wide slit closely resembles the object, the image of a very narrow slit is unrelated to the structure of the object, but takes the form of a diffraction pattern. The diffraction pattern "image" is necessarily not confined to a line but spreads out on either side of the ideal (geometrical) line image. When two adjacent slits are imaged, the overlapping of diffraction patterns may confuse the images of the two. A similar overlapping of diffraction patterns (Airy discs) confuses the images of adjacent point objects. Equations 10-t and 10-x show that the spread of diffraction patterns *increases* as the lens aperture is *reduced*. Since the lens aperture, however large, must always be finite, some spreading of all images by diffraction is inevitable. The resolution limit is then determined by the difficulty of distinguishing separate images when the diffracted fringes overlap.

The reader may ask: "If (as defined above on p. 221) a geometrically ideal lens reunites rays from any object point *without* introducing differences in optical path, what is the source of the diffraction patterns associated with image points?" The answer is that the optical paths of all rays are in fact equal *only* at the ideal image point. In the surrounding space, amplitudes exist which originate from the object point, but which differ from each other in phase at any specific position. The mutual interferences of these amplitudes produce the observed diffraction pattern. This way of considering the nature of the image is illustrated in Fig. 10-8. Figure 10-8a shows a wavefront (locus of points of equal phase) z, z', z'' converging toward an image point O'. Each point on the wavefront may be considered to act as a source of secondary (Huygens') wavelets. The directions of travel of the secondary wavefronts are represented by rays which are perpendicular to the wavefront. This is shown in Fig. 10-8b, where rays originating at z, z', and z'' and traveling toward O' and

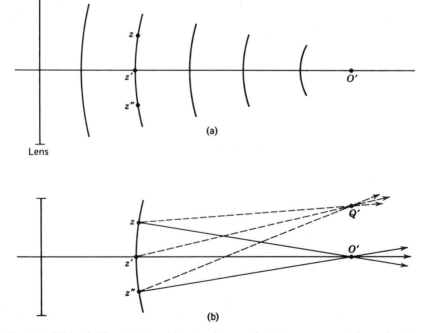

Fig. 10-8 Origin of diffracted intensities at an image point: (a) convergence of wavefronts to the image point; (b) illumination of a nearby point.

toward a nearby point Q' are shown. Rays zO', $z'O'$, and $z''O'$ are all exactly equal in length, since they are radii of the arc $z\ z'\ z''$, and thus in optical path. zQ', $z'Q'$, and $z''Q'$ are *not* radii of an arc and thus are of different lengths. Accordingly, they differ in optical path upon arrival at Q'. In this way there results an intensity at Q' that originates from the object point, and which constitutes a part of the diffraction image of that point. The intensity actually observed at Q' is determined by the values of the path differences for rays arriving at Q' from *all* points on the object.

APPROACHES TO THE ESTIMATION OF LIMITING RESOLUTION

In undertaking a quantitative estimation of the limiting resolving power of lenses (and thus of microscopes), assumptions must be made concerning the nature of the illumination of adjacent points at the limit of resolution. Two extreme cases are convenient to consider: those of coherently and incoherently illuminated points. Incoherently illuminated points are equivalent, in this context, to *self-luminous points* and are often referred to as such.

Coherently illuminated points receive light from portions of the same wavefront. The *amplitudes* contributed by each object point to any point in the image plane then interfere to produce a resultant intensity which is determined by the phase relationships, as well as the amplitudes, of the two. Alternatively, the *intensities* contributed by each incoherently illuminated object point to any point in the image plane can be added *directly*. In fact, light from closely spaced object points is always partially coherent to some degree; neither of these treatments of resolution is absolutely exact. Fortunately, however, both approaches predict quite similar resolving powers, so that either is adequate in estimating the general level of resolution to be expected in an optical system. Some familiarity with both approaches is desirable in gaining an appreciation of the physical nature of the imaging process.

ABBE'S THEORY OF THE RESOLUTION OF COHERENTLY ILLUMINATED POINTS

The concepts that are fundamental to an estimation of limiting resolution of coherently illuminated points have been introduced earlier in this chapter. As shown on pp. 222–224, and by Figs. 10-2 and 10-3, the illumination of *any* specimen, whether periodic or aperiodic in structure, results in the formation of a diffraction pattern in the rear focal plane of the imaging lens. (Note that Figs. 10-2 and 10-3 imply that the specimen is illuminated by *portions of a wavefront*; that is, coherently.) Abbe showed that the amplitude distributions of object, diffraction pattern, and image are related by Fourier transform functions, the amplitude distributions of object and image planes being, ideally, *identical* except for a scale factor representing magnification. Consideration of these ideas in terms of the physical nature of the imaging process (and without further reference to the Fourier transforms) leads directly to estimates of the resolution limit.

Figure 10-9 represents a structureless "specimen" illuminated by a parallel (collimated) beam of light. *No* intensity is diffracted (scattered) at the object plane. Consequently, light from the specimen enters the lens as a parallel beam and is brought to a focus on the axis at the second focal point, while other points on the second focal plane remain unilluminated. Subsequently, the beam diverges again, spreading out to illuminate the image plane uniformly. Thus light that does not interact with the specimen contributes to the central spot of the diffraction pattern and to the background intensity of the image.

Figure 10-9 may be compared with Figs. 10-2 and 10-3 in which light diffracted from the specimen contributes to off-axis maxima in the focal plane. That is, for collimated illumination *the central spot of the diffraction pattern in the rear focal plane of the lens represents light which has not interacted with the specimen, whereas the remainder of the diffraction pattern may provide information concerning the structure of the specimen.* The greater the amount of information concerning the structure of the specimen included in the focal plane (that is, the more complete the diffraction pattern), the more accurately does the amplitude distribution in the image plane represent the structure of the object. When objects are just resolved, their separation is evident in the image plane, but no other details of structure are revealed. Correspondingly, the diffraction pattern contains only a minimum of information specifying the spacing.

The condition for resolution is thus that *one diffracted order be included in the focal plane of the lens.* Inclusion of diffracted orders in the focal plane is determined, in turn, by the aperture angle of the lens.

These considerations are conveniently quantitated by consideration of the diffraction grating-like specimen shown in Fig. 10-2. An illuminated

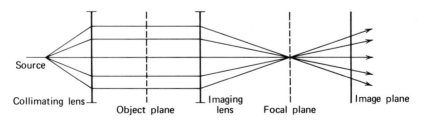

Fig. 10-9 Imaging of a structureless "specimen".

diffraction grating (or any other periodic object) forms a series of discrete diffraction maxima (cf. p. 61). For normal illumination the angular positions θ of the maxima of light of any given wavelength λ are as given by (3-c*). Thus, for a periodic specimen located in a medium of refractive index n,

(10-aa)
$$\sin \theta = \frac{m\lambda}{n\mathrm{d}},$$

where m is any integer, and d is the periodic spacing of the object. The larger the value of d, the smaller is the corresponding value of $\sin \theta$ and thus of θ. Correspondingly, a specimen of *large* periodicity can form *many* diffraction maxima within the aperture of the focal plane. Figure 10-10, for example, illustrates an object with a periodic spacing large enough for three diffracted orders to enter the lens and thus to contribute to the diffraction pattern. (For clarity, only rays which are scattered to the right by the specimen are shown in Fig. 10-10; the corresponding set of rays scattered toward the left are omitted.) The spacing d is easily

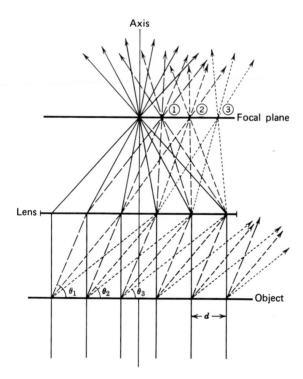

Fig. 10-10 Imaging of an easily-resolved periodic spacing.

resolved in the image. (Note that Fig. 10-10 and also Fig. 10-11 exaggerate the size of the spacing d with respect to the dimensions of the lens and to the object and focal distances. Relative dimensions are more accurately represented in Figs. 10-12 and 10-13.)

As d decreases, diffraction maxima become more widely spaced, so that fewer can be included in the aperture of the lens. In the limit, all diffracted intensity falls outside the lens aperture, and the periodicity fails to be resolved. The first diffracted order from a spacing which is just not resolved is shown in Fig. 10-11.

In considering Figs. 10-10 and 10-11 it is important to understand that amplitudes are scattered in all directions by both closely and widely spaced objects. However, when the number of slits (or other periodic diffracting objects) is very large, significant resultant intensity is produced only in those directions which satisfy (10-aa) (cf. p. 62 and Fig. 3-22).

For parallel illumination a structure is just resolved when the first order diffraction maxima just enter the lens, as illustrated by Fig. 10-12. The smallest resolved spacing thus is that for which the diffraction angle θ_1 equals the aperture angle α of the lens. Equation 10-aa then becomes

$$(10\text{-bb}) \qquad \sin \alpha = \frac{\lambda}{n d_{\min}} \quad \text{or} \quad d_{\min} = \frac{\lambda}{n \sin \alpha},$$

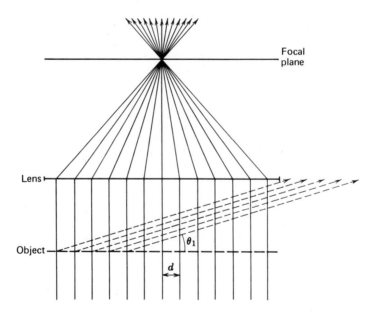

Fig. 10-11 Attempted imaging of a periodic spacing which is just not resolved.

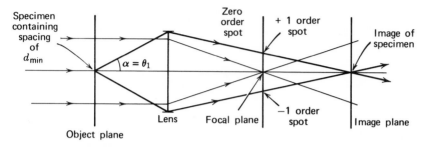

Fig. 10-12 Imaging of a spacing which is just resolved; parallel illumination.

in which n is the refractive index of the medium between object and lens, and d_{min} is the minimum resolvable spacing. The quantity ($n \sin \alpha$) in (10-bb) defines the *numerical aperture* (or N.A.) of the lens. Thus,

(10-cc) $$\boxed{d_{min} = \frac{\lambda}{\text{N.A.}} \quad \text{for parallel coherent illumination}}.$$

The value of α may, in the limit, approach 90°, so that $\sin \alpha$ approaches one. In air, the limiting value of numerical aperture is thus 1.0. However, if object and lens are immersed in a medium of higher refractive index, such as a transparent oil, the values of n and N.A. may be as large as about 1.6. For visible light, λ must be at least 4000 Å. Accordingly, from (10-cc), the optimum resolution (i.e., the smallest possible value of d_{min}) is given by

$$d_{min} = \frac{4000 \text{ Å}}{1.6 \times 1.0} = 2500 \text{ Å} \quad \text{or} \quad 0.25 \, \mu.$$

EXTENSION OF THE RESOLUTION LIMIT BY THE USE OF CONVERGENT COHERENT ILLUMINATION

Equations 10-bb and 10-cc apply for *parallel* illumination of the specimen. However, smaller values of d_{min} may be obtained (in suitable specimens) with *convergent* illumination. In this case the direct beam from an object point can enter the lens at one margin, while a single first order diffraction maximum enters at the opposite margin. The condition for resolution—that both scattered and unscattered light contribute to the image—is still fulfilled, while possible values of $\sin \theta$ are doubled. The minimum resolvable distance is correspondingly halved.

The case of limiting resolution in convergent illumination is illustrated by Fig. 10-13. For this case, (3-d*), for an obliquely illuminated diffraction grating, becomes, in a medium of refractive index n:

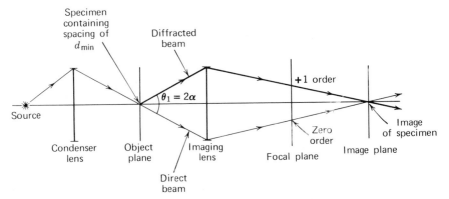

Fig. 10-13 Imaging of a spacing which is just resolved; tilted illumination.

(10-dd) $$m\lambda = nd(\sin i + \sin \theta),$$

in which i is the angle of incidence, θ is the angle of diffraction, and m is any integer. i may be at most the aperture angle α of the imaging lens, while, at limiting resolution, m is equal to 1, and θ, like i, is equal to α. Thus the expression for the minimum resolvable distance is

(10-ee) $$\boxed{d_{\min} = \frac{\lambda}{2n \sin \alpha} = \frac{\lambda}{2\text{N.A.}} \quad \text{for convergent coherent illumination}}.$$

For a microscope system, (10-ee) is sometimes expressed as

(10-ff) $$d_{\min} = \frac{\lambda}{\text{N.A.}_{\text{condenser lens}} + \text{N.A.}_{\text{objective lens}}}.$$

Note, however, that no additional resolving power can be gained by using a condenser of greater numerical aperture than the objective lens.

The ultimate resolving power under convergent illumination is obtained from (10-ee) as approximately $0.12 \, \mu$. Observation of spacings close to the limit of resolution shows, as predicted by (10-cc) and (10-ee), a marked variation in resolution according to the method of illumination of the specimen. However, lack of contrast may preclude the gain in resolution predicted by (10-ee) for specimens which scatter only an extremely small proportion of total amplitude into the first order.

THE RESOLUTION OF SELF-LUMINOUS POINTS (INCOHERENTLY ILLUMINATED POINTS)

Calculation of the minimum resolvable separation of incoherently illuminated points may be approached as follows. A point object is imaged by a lens as a diffraction pattern, the "Airy disc." The extended Airy disc images of adjacent points overlap, so that only the summed intensity can be observed. If the object points are sufficiently close, their images may be indistinguishable. A separation of the geometrical centers of the images is arbitrarily selected for which the overlapping diffraction patterns are considered to be just distinguishable. The corresponding spacing of object points is then related numerically to the properties of the imaging system.

The form of the Airy disc and the formation of an Airy disc as the image of a small hole (or of a point source of light) have been described on pp. 63 and 231, respectively. When two point sources of light are sufficiently far apart to be easily resolved, two well separated diffraction images are formed, as shown in Fig. 10-14. As the separation of object points is reduced, these extended images overlap. If the light from the two points is *incoherent*, the observed intensity at each position in the image plane is the *sum* of that associated with each object point individually. Figure 10-15a is a plot of the separate intensities associated with each point, and Fig. 10-15b is a plot of the summed intensity as actually observed in the image plane. If the slight dip in the center of the densitometer tracing shown in Fig. 10-15b can be detected, the two points are resolved. In this connection it must be remembered that the tracings in Fig. 10-15 are idealized. Actual densitometer tracings suffer from some level of background "noise," which has the effect of obscuring the small central dip (cf. p. 472 and, especially Fig. 21-6). In computing the limiting resolution, a realistic choice must be made concerning the separation of diffraction patterns which is just detectable. The criterion which is most commonly selected is the *Rayleigh criterion.* The Rayleigh criterion states that *two overlapping diffraction patterns are just distinguishable when the central maximum of one coincides with the first minimum of the other.* The criterion was first selected with reference to the overlapping

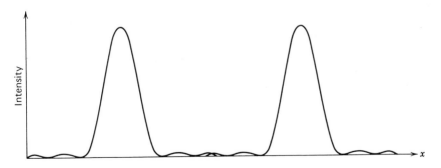

Fig. 10-14 Diffraction images of two easily-resolved points.

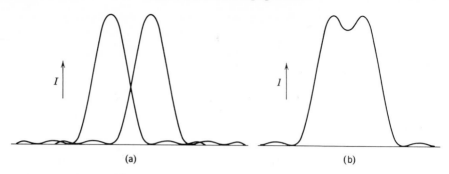

(a)   (b)

Fig. 10-15 Images of two incoherently illuminated points at the limit of resolution: (a) individual intensities; (b) summed intensities.

line images formed by spectroscopes, but it also constitutes a reasonable standard for the distinction of point images.

In evaluating d_{min} it is necessary to compute the separation of points, in object space, for which the Rayleigh criterion is just satisfied. Figure 10-16 represents the imaging of two closely spaced points P and Q which are separated by a distance y. The object plane, containing P and Q, is located close to the first focal point of the lens. As shown, rays from P illuminate the lens margins at the full angular aperture α and make angle α' with the axis at the corresponding image point P'. Q', the image of Q, is located a distance y' from P'. The refractive indices in object and image space are n and n', respectively, the diameter of the lens is A, and the image distance is s_2''. Application of the sine theorem (9-h) gives, for this system,

$$(10\text{-gg}) \qquad\qquad y = \frac{y'n' \sin \alpha'}{n \sin \alpha}.$$

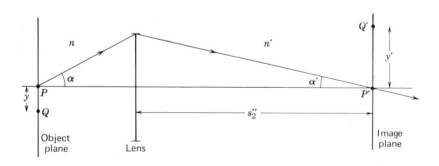

Fig. 10-16.

If the quantity y (the separation of P and Q) is such that the Rayleigh criterion is just satisfied (i.e., if $y = d_{min}$), then Q' must coincide with the first minimum of the diffraction pattern associated with P'. Also, correspondingly, P' must coincide with the first minimum of the diffraction pattern associated with Q'. The distance y' to the first minimum of the Airy disc has been given above as

(10-x)
$$r'' = y' = \frac{1.22\lambda s_2''}{n'A}.$$

Since P and Q are specified as being located close to the first focal plane, the image is formed far from the lens, and thus the angle α' is small. For small values of α', α', $\tan \alpha'$ and $\sin \alpha'$ are all approximately equal so that

(10-hh)
$$\frac{A}{2s_2''} \cong \sin \alpha'.$$

Substitution of (10-hh) in (10-x) then gives

(10-ii)
$$y' = \frac{0.61\lambda}{n' \sin \alpha'},$$

while substitution of (10-ii) in (10-gg) gives

(10-jj)
$$y = \frac{0.61\lambda n' \sin \alpha'}{n' \sin \alpha' n \sin \alpha} = \frac{0.61\lambda}{n \sin \alpha},$$

or

(10-kk)
$$\boxed{d_{min} = \frac{0.61\lambda}{\text{N.A.}} \quad \text{for incoherently illuminated points}}.$$

The difference in intensity which must be detected in order to achieve the resolution predicted by (10-kk) may be computed. For two points M and N separated by a distance which just satisfies the Rayleigh criterion, the total intensity at the image points M' and N' is compared with that at the position midway between the image points. At either geometrical image point:

intensity from point M at the geometrical image point $M' = I_0$
intensity from point N at the geometrical image point $M' = 0$

---

total intensity at M' (or N') $= I_0$

The intensity contributed by each diffraction pattern at the midpoint between M and N may be obtained by evaluating the quantity $J_1(\tfrac{1}{2}r_{1st\,min})/$

$(\tfrac{1}{2}r_{1st\ min})$; From Fig. 3-24, this value may be seen to be approximately $0.4I_0$. Thus,

intensity from M at midpoint $\cong 0.4I_0$
intensity from N at midpoint $\cong 0.4I_0$

total intensity at midpoint $\qquad \cong 0.8I_0$

The Rayleigh criterion thus demands that a difference in intensity of approximately 20% be just detectable.

EFFECT OF AN ALTERNATIVE CRITERION FOR RESOLUTION

Various alternatives to the Rayleigh criterion for resolution are sometimes stated in the literature. In general, the effects upon estimated resolving powers are quite slight. For example, it is sometimes considered that two points are just resolved if the intensity of the diffraction pattern from each is $0.5I_0$ at a position midway between the geometrical image points. Numerical summation of intensities then gives

intensity from point M at point $M' \cong 1.00I_0$
intensity from point N at point $M' \cong 0.02I_0$

total intensity at M' (or N') $\qquad \cong 1.02I_0$

and

intensity from point M at midpoint $= 0.50I_0$
intensity from point N at midpoint $= 0.50I_0$

total intensity at midpoint $\qquad = 1.00I_0$

The alternative criterion, despite its apparent similarity to the Rayleigh criterion, thus imposes the much more stringent condition that a variation in intensity of only about 2 percent be detectable. Unfortunately, the corresponding improvement in resolving power is much less profound. As shown in Fig. 10-17, the separation of points satisfying the alternative criterion is about nine tenths as great as that of points satisfying the Rayleigh criterion, and (10-kk) becomes

(10-ll) $$d_{min} = \frac{0.54\lambda}{N.A.}.$$

That is, approximately tenfold improvement in the ability to detect differences in intensity can be expected to reduce the minimum resolvable

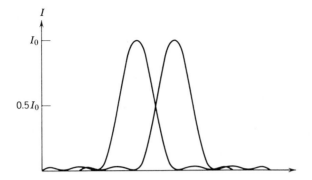

Fig. 10-17 Diffraction images of points satisfying the criterion that $I = 0.5I_0$ at midpoint.

separation of points only by about 10 percent. It may be concluded that the precise criterion selected for resolution is of little importance.

VALUES OF d_{min}; EXTENSION OF THE RESOLUTION LIMIT

Table 10-1 summarizes the expressions which have been derived above for d_{min}, the minimum resolvable separation of points.

A high power microscope objective has a numerical aperture of about 0·95, while the wavelength of white light can be taken to be 0·6 μ. Thus, under "typical" conditions, the resolution limit of the light microscope can be expected to be in the range

(10-mm)
$$d_{min} = \frac{(0.50 - 1.0) \times 0.60 \, \mu}{0.95} = 0.30 - 0.65 \, \mu.$$

In the limit, light at the short wavelength end of the visible spectrum might be used for imaging, and the numerical aperture of the lens system

Table 10-1   Expressions for d_{min}.

Equation	d_{min}	Applicable to
(10-cc)	$\lambda/\text{N.A.}$	Collimated coherent illumination
(10-ee)	$0.5\lambda/\text{N.A.}$	Convergent coherent illumination
(10-kk)	$0.61\lambda/\text{N.A.}$	Self-luminous or incoherently illuminated object; detectable $\Delta I = 20\%$
(10-ll)	$0.54\lambda/\text{N.A.}$	Self-luminous or incoherently illuminated object; detectable $\Delta I = 2\%$

increased to about 1.60 by means of oil immersion. The ultimate resolving power, according to the most optimistic of the estimates cited in Table 10-1, would then be

(10-nn) $d_{min} = \dfrac{0.5 \times 0.4\,\mu}{1.60} = 0.125\,\mu.$

These estimates are generally confirmed by experimental demonstrations of resolution in the optical microscope. The figure given by (10-nn) represents the effects of optimum values of the aperture angle (at best, slightly less than 90°) and of the refractive index (at most, about 1.6). Further extension of the resolution limit can thus be achieved only by reducing the wavelength λ.

While a continuum of wavelengths shorter than 4000 Å exists in the electromagnetic spectrum (cf. Chapter 2), two important types of difficulty tend to preclude their use for extension of the resolution limit. In order to be useful in microscopy, any radiation must (a) be capable of *interacting* with the specimen (i.e., must form an image in which *contrast* exists) and (b) must be *focusable*. γ-Ray microscopy is not feasible with respect to either of these requirements. X-rays interact with the electrons of specimens, while x-ray wavelengths, which are of the order of 1 Å, suggest corresponding values of d_{min} that would be extremely useful in the study of biological ultrastructure. Unfortunately, however, x-rays, like gamma rays, are not appreciably refracted by passage through matter. No really effective x-ray lenses exist, although various approaches to x-ray microscopy have been attempted. The latter are described in Chapter 28. Ultraviolet microscopy, which is discussed on p. 284, is a practical technique. Its principal merit, however, lies in the possibilities which it offers for observation of unstained biological structures that absorb at specific wavelengths in the ultraviolet range. The improvement in resolution that results from reduction of the wavelenth to 2000 Å is only twofold.

In summary, prospects for improvement of the resolution limit by the use of shorter wavelength electromagnetic radiations are not particularly bright. Appropriately accelerated electrons, however, are equivalent in wavelength to x-rays (cf. Chapter 16). Electron beams interact extensively with matter and, since they are *charged* (unlike electromagnetic radiation), can be focused. While resolutions achieved in the electron microscope fall far short of those predicted by the expressions listed in Table 10-1, the performance of the electron microscope in this respect nevertheless surpasses that of the light microscope by a factor of the order of one hundred.

HOLOGRAPHY

The description of object, diffraction pattern, and image as related by Fourier transform functions reveals that the diffraction pattern formed by illumination of any object contains all the information required to produce an image of the object. Unfortunately, much of this information is normally lost in the recording of the diffraction pattern, since it exists in the form of phases, which are not directly observable. Nevertheless, it is possible to preserve this information in a photographic record. The object must be illuminated by coherent light. The light scattered or reflected by the object is recorded on a photographic plate, which is simultaneously illuminated by a "coherent background" from the same source. The coherent background illumination does *not* interact with the object. The phase relationships of the imaging beam are thus recorded on the plate in the form of an interference pattern.

Subsequently, the photographic record, which is known as a *hologram*, is illuminated, also by coherent light. Illumination must take place in a system of appropriate geometry. When the coherency of both recording and imaging beams is adequate, the phase relationships of the original wavefront from the object are reestablished in the "reconstructed wave front." In this way a faithful image is reconstructed from the hologram in the absence of the object.

As initially described, holograms can be recorded, and the image reconstructed, in the manner diagramed in Fig. 10-18. In Fig. 10-18a the transparent object is illuminated by a point source of coherent light, and the hologram is recorded on a photographic plate placed some distance beyond the object. The coherent background illumination of the hologram

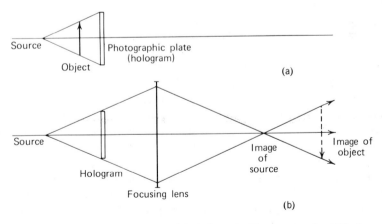

Fig. 10-18 Holography: (a) recording of the hologram; (b) reconstruction of the image.

is produced by light which passes through the object without interaction. In the reconstruction of the image, as shown in Fig. 10-18b, the object is removed, and the hologram is placed in the *same* position as before with respect to the light source. A lens placed beyond the hologram then forms an image in a position which is conjugate to that of the original object.

Alternatively, the image may be reconstructed without recourse to the use of lenses in the manner diagramed by Fig. 10-19. Figure 10-19a shows the recording of the hologram. Light from the coherent source is divided into two beams, one of which illuminates the hologram directly, forming the coherent background. The other illuminates the object, which may, in this case, be opaque, and is then reflected onto the hologram. Subsequently, as shown in Fig. 10-19b, the hologram is illuminated by the reference beam alone. As before, the hologram must be placed in the same position with respect to the source as during its recording. Looking through the hologram, the image may be seen (or photographed) in the position originally occupied by the object.

As may be demonstrated experimentally, and as is easily shown by mathematical analysis, the image so reconstructed (i.e., as shown in either Fig. 10-18b or Fig. 10-19b) is produced from only a part of the amplitude leaving the hologram. The remainder of the amplitude may be described as consisting of two components. One component, a wave *in* phase with the background illumination, may be reduced by appropriate photographic technique. The other, called the "conjugate wave," is proportional in amplitude but opposite in phase to the wave which forms the reconstructed image. The conjugate wave forms a second image of the object. Since this image is, in general, physically separated from the principal reconstructed image, interference between the two can be avoided in systems of suitable geometry.

An interesting feature of holograms derives from the fact that *all* positions on the hologram may receive light from *all* points of the object. Each point on the hologram is thus characteristic of the object as a whole. (The same is true of diffraction patterns in general). While the complete hologram must be illuminated in order to reconstruct an image with highest resolution, *any* portion of the hologram may be used to form an image of the *entire* object, although with impaired resolution. Holograms differ from focused diffraction patterns in that the intensities which arise from particular spacings in the object are not segregated in the hologram.

A most important potential application of holography derives from the fact that the wavelength of the light used to reconstruct the image may differ from that used to form the hologram. The distance between source and hologram must then be adjusted in proportion to the ratio of the two

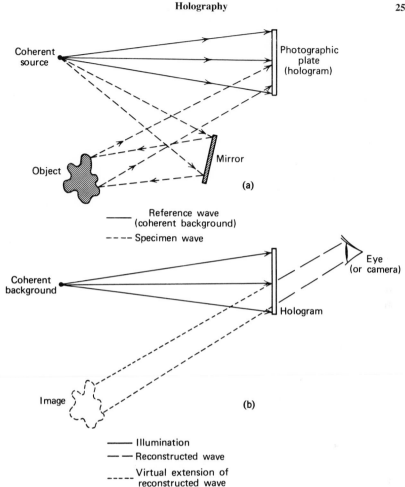

Fig. 10-19 Lensless Holography: (a) recording of the hologram; (b) reconstruction of the image.

wavelengths. In principle, the object could be illuminated, and the hologram recorded, by x-rays or by an electron beam. The hologram would
then be illuminated by a laser in a system of appropriately scaled-up
dimensions. The advantage of the procedure, in the case of x-rays, would
be the provision of an otherwise unavailable means of focusing. The
advantage in the case of electron imaging would be that the glass lenses
used in reconstructing the image could be corrected, as electron lenses
cannot be, for the effects of spherical aberration and other lens defects
(cf. Chapter 17).

Unfortunately, the possibilities for extending resolution by holographic methods are severely limited at least at present by the lack of coherent sources of x-rays or electrons. Without coherent illumination, satisfactory holograms cannot be formed. However, some possibilities exist for improving x-ray images by holography; they are discussed on p. 613.

REFERENCES

L. C. Martin, *The Theory of the Microscope*, Blackie, Glasgow, and American Elsevier, New York, 1966, Appendix I (Fourier Series). An excellent account of Fourier series and Fourier transforms, specifically as applicable to optics. Many examples are included.

J. M. Stone, *Radiation and Optics*, McGraw-Hill, New York, 1963, an account of the Abbe theory of resolution is given on pp. 224–232. Examples of functions and their Fourier transforms are given on pp. 213–216 and in Figs. 11-1 to 11-3.

D. Gabor, *Nature* **161**, 777 (1948). This classical paper, "A New Microscopic Principle," provides a very clear account of the basic idea of holography and its possible application to electron microscopy. Two other papers by the same author, "Microscopy by Reconstructed Wave Fronts," *Proc. Roy. Soc. (London)* **A197**, 454 (1949), and *Proc. Phys. Soc. (London)* **64**, 449, (1951), are more detailed accounts of the subject. All of these papers have been reprinted as Appendices in *An Introduction to Coherent Optics and Holography*, by G. W. Stroke, Academic Press, New York, 1966, pp. 181–242.

L. C. Martin, in G. Oster and A. W. Pollister, Eds., *Physical Techniques in Biological Research*, Vol. I, *Optical Techniques*, Academic Press, New York, 1955, pp. 341–351.

L. C. Martin, *The Theory of the Microscope*, Blackie, Glasgow, and American Elsevier, New York, 1966, pp. 160–189, 229–251, and Chapter 6 (Coherent Illumination). A detailed theoretical description of the images formed by various types of objects upon illumination by incoherent, partially coherent, or fully coherent light.

A. E. Conrady, *Applied Optics and Optical Design*, Part I, Dover, New York, 1957, Chapter 3 (Physical Aspects of Optical Images). Effects of the wave nature of light are considered with particular reference to the design of lenses.

B. M. Siegel, *Modern Developments in Electron Microscopy*, Academic Press, New York, 1964, pp. 36–45. A discussion of "Resolving Power of the Microscope" with particular reference to the electron microscope.

E. N. Leith and J. Upatnieks, *Sci. Am.* **212**, 24 (1965). "Photography by Laser." This article describes techniques for recording holograms.

K. S. Pennington, *Sci. Am.* **218**, 40 (1968), "Advances in Holography." Applications of holographic techniques are discussed.

D. R. Herriott, *Sci. Am.* **219**, 141 (1968). This article, "Applications of Laser Light," includes a section on the practice and applications of holography.

CHAPTER 11

Origins of Contrast

The imaging of object points by lenses has been considered from the geometrical point of view in Chapters 8 and 9, and from the point of view of the wavelike character of light in Chapter 10. However, neither of these approaches considers the question: How do image points become *visible* in an extended field? Geometrically perfect imaging of points separated by a distance greater than the resolution limit is to no avail if the intensity of the images of the points is identical with that of the background! Clearly, *contrast* must exist if instrumental resolving power is to produce resolution in the image.

"Contrast" specifies the differences in intensity between image points and their background or "surround." Various measures of this quantity may be specified; they include *percent contrast* and the *index of visibility*, which are defined by (10-z) and (11-a), respectively:

$$(10\text{-}z) \qquad \%\text{ contrast} = \frac{I_b - I_{sp}}{I_b} \times 100,$$

$$(11\text{-}a) \qquad \text{index of visibility} = \frac{I_b - I_{sp}}{I_b + I_{sp}},$$

where I_{sp} is the intensity in the image of the point considered, and I_b is that of the surround. Image contrast may be manifested either as a variation in the absolute intensity or in the relative intensities of light of different wavelengths (i.e., by the observation of color).

Contrast results from the interaction of light with the specimen and thus is determined by the properties of both specimen and imaging system. The characteristics of the specimen which give rise to interactions with the light beam are referred to as the *inherent contrast* of the specimen.

Several different types of interaction contribute to contrast. In this chapter these mechanisms are summarized, with particular emphasis on effects which produce contrast in the images formed by single lenses or in the "ordinary" compound light microscope. The subject of contrast is also discussed in parts of Chapter 12, in Chapters 13 through 15, which describe the enhancement of contrast by instrumental modifications, and in Chapter 19, where the contrast of electron images is discussed.

AMPLITUDE OBJECTS AND PHASE OBJECTS

Variation in image intensity may arise directly, by a loss (or gain) of amplitude at the specimen, or indirectly, as a consequence of alterations in the phases of beams which contribute to the image. Specimens which change the amplitude of the illuminating beam are called *amplitude objects*; those which produce changes in the relative phase of transmitted light are called *phase objects*. Although all real specimens are both amplitude and phase objects, to some extent, it is often convenient to consider the imaging of hypothetical specimens which are purely one or the other.

Loss of light intensity by *absorption* or *scattering* or both is the source of amplitude contrast, whereas inherent phase contrast is associated with differences in optical path (i.e., of refractive index, thickness or both) between adjacent regions in the specimen. In the ordinary light microscope, phase contrast is observed by defocusing or as a consequence of lens aberrations.

In ordinary and dark field light microscopes, specimens act almost exclusively as amplitude objects. The phase and interference microscopes, described in Chapters 13 and 14, respectively, are instruments in which phase objects are rendered directly visible. In the electron microscope, both amplitude and phase contrast effects may be of considerable importance.

ABSORPTION CONTRAST

Amplitude contrast may arise from either absorption or scattering of light but, in light microscopy, absorption is the "normal" source of contrast. Part of the amplitude of incident illumination is removed from the beam through absorption by the specimen. When rays from an absorbing point are reunited, the image appears correspondingly dark. The extent of absorption generally varies with the wavelength, so that a specimen may reduce the amplitude of only a limited range of wavelengths. The color of its image is then different from that of the surround.

Since the absorption of visible light is determined by the structure of the absorbing molecule, absorption contrast may provide information concerning the chemical structure of the specimen. Where inherent differences in absorption do not exist, they can often be introduced by *staining*. During the staining process, absorbing substances (chromophores) become specifically associated with certain regions in the specimen. Usually the appearance of a stained specimen represents the distribution, within the specimen, of chemical affinities for the stain or stains. Less often, particles of stain may physically absorb to surfaces of the specimen.

The nature of light absorption has been discussed in Chapter 7, and its quantitation is described in Chapter 24.

SCATTERING CONTRAST

Amplitude contrast by scattering occurs as a consequence of unequal angular distribution of the intensities scattered from different points in the specimen. An object which scatters part of the incident intensity beyond the aperture of the imaging system appears darker than a surround which scatters appreciably only *within* the lens aperture. Scattering contrast thus is increased by reducing the aperture of the lens or microscope (i.e., by reducing the size of the aperture stop). A short description of scattering effects has been given on p. 147.

A numerical example of scattering contrast is considered in Fig. 11-1. In this hypothetical example a background point B transmits 100% of incident intensity, of which 90% is scattered within an angle of not more than 30°, while 95% is included within an angle not greater than 60°. Specimen point S transmits 95% of incident intensity, scattering 60% within 30° and a total of 90% within 60°. The absorption contrast of the specimen is obtained by evaluation of (10-z) as

(11-b)
$$\frac{1.00I_0 - 0.95I_0}{1.00I_0} \times 100 = 5\%.$$

Figure 11-1a illustrates the imaging of S and B by a lens of 60° aperture. The intensities in the corresponding images are then

(11-c)
$$I_B = 1.00I_0 \times 0.95 = 0.95I_0,$$
$$I_S = 0.95I_0 \times 0.90 = 0.855I_0.$$

The percent contrast of point S is thus

(11-d)
$$\frac{0.95I_0 - 0.855I_0}{0.95I_0} \times 100 = 10\%.$$

Figure 11-1b shows the same system with the aperture angle of the lens

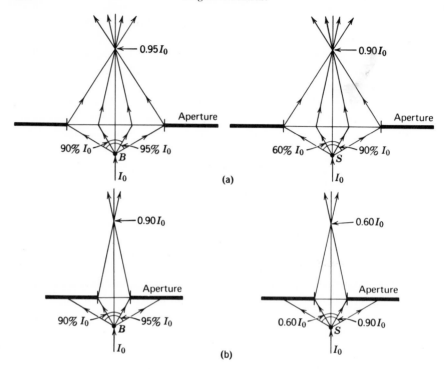

Fig. 11-1 Scattering contrast: (a) 60° aperture; (b) 30° aperture.

reduced to 30°. The intensities of the two image points are then

(11-e)
$$I_B = 1.00 \times 0.90 = 0.90I_0,$$
$$I_S = 0.95I_0 \times 0.60 = 0.57I_0.$$

Correspondingly, the percent contrast of point S is increased to

(11-f)
$$\frac{0.90I_0 - 0.57I_0}{0.90I_0} \times 100 = 36.67\%.$$

In fact, striking increases in contrast can often be observed by stopping down the lens aperture.

Scattering may also be interpreted as a type of phase contrast effect. As developed on p. 289, the light beam transmitted by a phase object may be regarded as the resultant of the "undeviated" or incident beam and a "deviated" or scattered component of amplitude. Loss of amplitude from the deviated component (i.e., by scattering outside the microscope aperture) results in a decrease in amplitude (and thus in intensity) as well as a change in phase of the resultant. This concept is illustrated by Fig. 13.6 on p. 292.

Scattering is of appreciable but secondary importance as a source of contrast in the light microscope. In electron microscopy, however, differential scattering is the *only* source of amplitude contrast and is of major importance. (cf. p. 426).

THE IMAGING OF PHASE OBJECTS

Figure 11-2 represents the imaging of a specimen which lacks amplitude contrast, that is, of a pure phase object. The refractive index of region A (the "object") is some value n; that of region B (the "surround") is some other value, $n_0 \neq n$. The object is imaged by beam AA' (solid lines) and a typical region of the surround by beam BB' (dotted lines). For a specimen of thickness d the two beams differ in optical path at the image plane by a quantity Δ such that

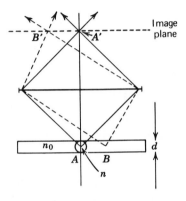

Fig. 11-2 Imaging of a pure phase object.

(11-g) $\Delta = (n - n_0)d.$

Thus there is a definite physical difference between the beams which are imaged at points A' and B', respectively. Nevertheless, *no* image contrast results (i.e., in the in-focus image formed by an ideal lens). This is the case because all the rays which contribute to A' or to B' are in phase with each other. Therefore no changes in intensity can result from interference at either position. It is evident that *pure phase objects are invisible when viewed in focus with a perfect lens.* Phase contrast is observed only when: (a) the object is viewed in a phase contrast or interference microscope (these modifications of the compound light microscope are described in Chapters 13 and 14; (b) the image is defocused (the effects which contribute to out-of-focus contrast are discussed in the following section); (c) the imaging lens suffers from aberrations.

The effect of lens aberrations is to cause the rays which pass through different zones of the lens to experience different optical paths. Interference between rays which reunite at the image plane produces variations in the resultant intensity. If the angular distribution of amplitudes leaving different object points were identical, no contrast would result, but, if the distributions for two points are different, the resultant intensity at their images is different also.

Thus it is evident that aberration phase contrast and scattering amplitude contrast are both dependent upon a single physical process: the scattering of light through a variable range of angles by different points in the specimen. Amplitude effects are observed in systems of narrow aperture, which exclude rays deflected through large angles, whereas phase effects are obtained in systems of wide aperture, which cause most of the scattered amplitude to be included in the image. This is an excellent example of the way in which instrumental factors may determine the nature of the observed image contrast.

The extent of the path differences introduced by the lens is greatest in uncorrected lenses (although small differences persist even in highly corrected lenses; for an example see Fig. 9-8). Aberration contrast in the images formed by high quality optical lenses is thus minimal. On the other hand, electron lenses invariably suffer from substantial aberrations; therefore aberration phase contrast may be an effect of consequence in electron images. A discussion of the effect is given on p. 437.

OUT-OF-FOCUS CONTRAST

By slight defocusing of the image, pure phase objects may become visible in well corrected ordinary microscope systems. The observed phase contrast results from interference between rays that have been transmitted by different points in the specimen, and which therefore may differ in optical path. Selective reflection or refraction at interfaces may enhance the path differences produced by differences in the refractive index.

Figure 11-3 illustrates the imaging of a pure phase object A and of two

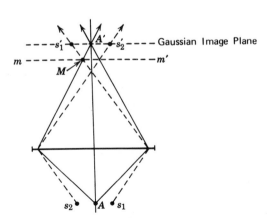

Fig. 11-3 Out-of-focus contrasting of a phase object.

nearby background points S_1 and S_2. At the image plane, rays from the corresponding object points reunite *in* phase, as discussed above, but at adjacent planes, such as *mm'*, rays from different object points intersect each other. For example, rays from S_1 in the surround (solid lines) and from the object A (dotted lines) intersect at point M. These rays differ in phase at M by an amount which is a function of the difference between the refractive indices at S_1 and A. Rays from other specimen points also contribute amplitude to M, producing a resultant intensity which is determined by the amplitudes and phases of all contributions.

In general, the effect of interference is that the image of any point, as observed in out-of-focus planes, is surrounded by a succession of minima and maxima of intensity. If any amplitude contrast exists, it tends to be highlighted by these focus "fringes." Comparable effects occur in the electron microscope, as described on pp. 402 and 433.

Reflection effects enhance out-of-focus contrast in specimens which are illuminated by convergent light. As shown in Fig. 11-4a, rays converge to a point O located on the boundary between media of different refractive indices n and $n' > n$. Certain rays (②,③) are incident upon the boundary from the medium of higher refractive index at an angle which is greater than the critical angle. Consequently, these rays are totally reflected from the interface (cf. p. 6). A concentration of light on the side of higher index results. When plane *mm'*, *above* the specimen, is focused upon, a bright line is observed on the high index side of the boundary, while a more diffuse line, located farther on the high index side, is seen in the more out-of-focus plane, *pp'*. This phenomenon, which is known as the *Becke line*, is exploited in measuring refractive indices of particulate specimens (cf. p. 584 and especially Fig. 26-8). The Becke line is also seen in images of planes which lie below the specimen, such as *qq'* in Fig. 11-4b. In this case the image is that of the virtual projection of rays such as ② and ③, which appear to originate from positions on the side of lower refractive index.

Lenslike effects can also contribute to out-of-focus contrast if the boundary between regions of different refractive index is curved. A boundary of spherical curvature, which is illuminated by a collimated beam, is shown in Fig. 11-5. Refraction of rays at the interface tends to produce a focused spot at the center of curvature. The image of this bright spot varies in size according to the level of focus. Spherical particles which are of higher refractive index than the surrounding medium thus tend to appear in out-of-focus images as a central bright spot surrounded by a dark boundary. Particles which are of lower refractive index than their surround appear dark at the center but are surrounded by a bright halo.

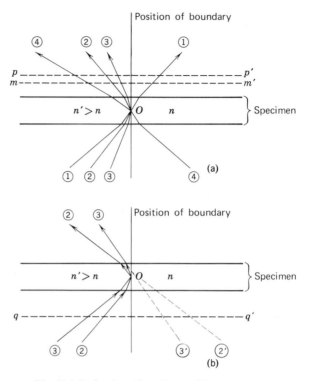

Fig. 11-4 Reflection effects in out-of-focus contrast.

It should be clear from this short discussion that out-of-focus effects, unlike amplitude contrast effects, are not directly related to the structure of the specimen. Because the same is true of phase contrast in general, phase images must always be interpreted with caution.

CONTRAST AND RESOLUTION

Whereas instrumental resolving power cannot effect resolution in the absence of adequate contrast, it is unfortunately true that *conditions which tend to enhance contrast are precisely those which tend to destroy resolving power.* Several examples are cited here.

Scattering contrast may be increased by reducing the aperture of the objective lens of a microscope. The minimum resolvable distance, however, is determined by the value of the quantity (λ/numerical aperture) (cf. Chapter 10), so that resolving power deteriorates as the aperture is reduced. If the nature of the specimen is such that the full resolving power of the microscope is not exploited, some extent of stopping down

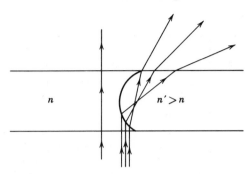

Fig. 11-5 Lens-like effect in out-of-focus contrast.

may be used to produce a net improvement in resolution. Ultimately, however stopping down the aperture reduces the resolving power to a level at which this quantity, rather than contrast, becomes the factor which limits resolution. A simple but striking experimental demonstration of this fact can be made by placing a pinhole aperture behind the objective lens of a compound microscope. (The experiment is described by Barer.)

Phase contrast is observed in ordinary light microscope images as a consequence of slight defocusing or of lens aberrations or both (cf. pp. 257 and 258). Both of these conditions are obviously detrimental to the resolving power of the system.

Theoretically, a twofold improvement in the resolution of periodic objects can be obtained by the use of tilted rather than collimated illumination (cf. p. 241). However, the amplitude scattered into the first diffracted order is often so small that the resulting variation in intensity at the image is undetectable. Lack of contrast then precludes exploitation of the full theoretical resolving power.

In the electron microscope, contrast, rather than instrumental resolving power, is usually the factor which limits resolution of detail in any given specimen. Thus reduction of aperture and slight defocusing can be used effectively to improve resolution at or above the level of about 15–20 Angstrom units. At very high resolution (5–10 Å), however, these conditions again become limiting. An additional problem, which has no counterpart in light microscopy, is that the molecules used as electron *stains* may be as large as (or even larger than!) the structures to be stained. Thus resolution may be lost simply by masking of the stained object.

NATURE OF THE IMAGE

Microscopes can be used over a wide range of magnifications, including 1.0. Thus the microscopist can observe a familiar object in a low power microscope and see a slightly enlarged image which is "the same as" the object. Increase of magnification may reveal details in the object which are invisible to the naked eye; it is natural to assume that they, also, are "the same as" the object. (At this stage it is necessary to establish that detail is not a consequence of damage to the specimen during preparation for microscopy.) But what is actually implied by the statement that "the image is the same as the object"?

Obviously, the image is a purely optical effect—even a real image lacks the mass, chemical composition, surface properties, taste, smell, or biological activity of the object. The "sameness" of object and image in fact implies that the physical interactions with the light beam that render the object visible to the eye (or which would render it visible, if large enough) are identical with those which lead to the formation of an image in the microscope. Except for minor differences between the eye and the microscope (i.e., in aperture, extent of absorption by lenses, etc.) it is quite correct, in the sense just given, to claim an identity of the object and the image formed in an ordinary light microscope.

Suppose, however, that the radiation used to form the image is a beam of ultraviolet light, x-rays, or electrons, or that the microscope employs some device which converts differences in phase to changes in intensity. The image then cannot possibly be "the same" as the object, even in the limited sense just defined! The eye is unable to perceive ultraviolet, x-ray, or electron radiation, or to detect shifts of phase between light beams. For example, there is a profound difference between the appearance of an "x-ray" and an ordinary camera portrait of the same subject!

This line of thinking reveals that the image must be *a map of interactions between the specimen and the imaging radiation*. The image reveals only those properties of the specimen which the microscope system is capable of revealing. Thus light micrographs, and direct visual observation reveal, primarily, the presence of objects which absorb visible light. Phase contrast and interference microscope images differentiate areas in which visible light travels with different velocities. Polarization microscopy reveals the presence of birefringent or dichroic objects. Ultraviolet microscopy reveals structures such as chromosomes that are transparent to visible light, but which specifically absorb light of shorter wavelengths. X-ray and electron micrographs are produced by areas of high mass density. *All* of these methods can produce "true" images which are, in some sense, "like" the specimen. However, each type of image, by itself, characterizes the specimen only incompletely.

The radioautogram is an extreme example of this concept of the image. In preparing a radioautogram, radioactive material is infiltrated into the specimen in such a manner that it becomes associated with specific sites. The corresponding positions of radioactivity are recorded on a photographic emulsion, which is subsequently developed. The resulting picture is an "image" of the specimen, but one obtained exclusively from the point of view of the location of radioactive atoms. This type of "image" is so specialized as to be, generally, uninterpretable without the aid of an additional image, the photomicrograph, upon which it is superposed.

The considerations just expounded may seem unduly academic from the point of view of ordinary light microscopy. The interpretation of the usual absorption contrast image seems straightforward—because this type of imaging is merely an extension, in scale, of the processes carried out by the unaided eye; the properties of absorption contrast are a matter of everyday experience. However, other types of microscope form images which are not directly related to unaided observations. These images extend our knowledge of the properties of specimens by revealing different types of interaction between specimen and beam. Unfortunately, such images can be misleading if they are interpreted without proper consideration of the nature of the interactions which produce them.

REFERENCES

R. Barer, *Lecture Notes on the Use of the Microscope*, 2nd ed., Blackwell & F. A. Davis, Philadelphia, 1959, pp. 16, 42–44. Also, the experiment referred to on p. 261 of the present text is described on pp. 64–65.

L. C. Martin, *The Theory of the Microscope*, Blackie, Glasgow, and American Elsevier, New York, 1966, pp. 369–375. A discussion of "The Interpretation of the Image" (in light microscopy).

C. Tanford, *Physical Chemistry of Macromolecules*, Wiley, New York, 1961, Chapter 5 (Light Scattering). Theories of light scattering by "small" and "larger" particles are discussed, with particular reference to biological macromolecules.

Unfortunately, very little discussion of *mechanisms* of contrast formation in the ordinary light microscope is to be found in the literature (i.e., as opposed to considerations of the effects of certain stains, and other preparative procedures, on the contrast in images of specific types of object). Barer, *op. cit.*, and Martin, *op. cit.*, briefly consider this aspect of the subject. The literature concerning phase contrast systems is much more abundant; the reader is referred to the bibliography for Chapter 13.

CHAPTER 12

Light Microscopy

The theory of light microscopy is for the most part that of lenses and of the imaging process, as presented in Chapters 8 through 11. Certain features of microscope systems as a whole are discussed in this chapter. Three special modifications of light microscopy are also described briefly; they are dark field, fluorescence, and ultraviolet microscopy. Neither the exposition of practical laboratory directions for the use of the light microscope nor the description of commercially available microscopes or accessories lies within the scope of this book.

THE COMPOUND MICROSCOPE

The compound microscope provides a means for convenient viewing of objects at high magnifications by using a second (ocular or eyepiece) lens system to remagnify the image formed by an initial (objective) lens system. Suitable illumination of the object generally requires not only a light source of high intensity but also the use of a lens (the condenser) for focusing light at or near the specimen. The optical components of the compound microscope are shown (in an extremely schematic fashion) by Fig. 12-1.

The *light source* is usually an incandescent tungsten filament. A rheostat may be used to vary the power supplied to this type of lamp and thus to control its brightness. When sources of high intensity are required, carbon or xenon arcs may be used, while mercury or sodium vapor lamps can serve as monochromatic light sources.

As shown in Fig. 12-1, light from the source may be converged by means of a *lamp lens*. The distance between lens and lamp is adjustable in some systems. The effective size of the source is controlled by a vari-

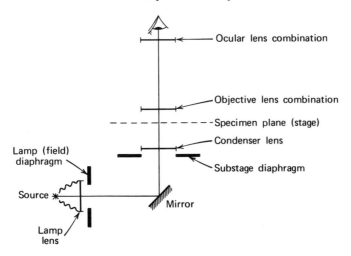

Fig. 12-1 Schematic of compound microscope.

able iris diaphragm (lamp or *field diaphragm*) which may also serve as the field stop of the microscope. In simple systems. however, the lamp lens and diaphragm are replaced by a diffusing ground glass plate which effectively acts as a light source of extended dimensions.

Light from the illumination system is collected by the *condenser lens*. The aperture of illumination is controlled by the setting of the variable iris diaphragm (*substage diaphragm*), which is located below the condenser lens. This diaphragm is placed, approximately, at the first focal plane of the condenser and serves as the aperture stop of the microscope. In microscopes used only at low magnifications, the condenser lens is sometimes omitted. Light from the source is then directed onto the specimen plane by means of a mirror, which may be curved in order to converge the illuminating beam. However, in the absence of a condenser lens, illumination of the specimen is insufficient for observations at high magnification. *Abbe* condensers make use of one or two uncorrected lenses to provide moderate or highly intense illumination of the specimen at medium or high magnifications, respectively. *Aplanatic* and *achromatic* condensers are also made; they are corrected for spherical aberration, and for both spherical and chromatic aberration, respectively. Methods of illumination are discussed in more detail on p. 266.

The object plane shown in Fig. 12-1 is the position of the microscope *stage*, upon which the specimen support (glass slide, etc.) rests. The stage may include provisions for calibrated translation of the object, or for its rotation about the axis of the lens system.

The *objective lens* forms a primary (intermediate) image, which is magnified further by the *ocular* (eyepiece lens). Neither objective nor ocular is a single lens; both are combinations of several lens elements. The objective lens is discussed on p. 269, and the ocular on p. 273.

The objective and ocular are connected by the tube of the microscope. The *optical tube length* (as distinguished from the mechanical tube length) is defined as the distance from the upper (second or "rear") focal plane of the objective to the intermediate image.

Focusing of the image is achieved by varying the distance between the specimen and the objective. Usually this is done by racking up and down the microscope tube, although fine adjustments of focus are sometimes produced by axial movements of the specimen stage.

When the microscope image is observed directly by eye, the ocular lens must form a virtual image. Depending upon the exact condition of focus selected, this image may appear to lie either at or beyond the plane of the specimen. Note that, if a real image were formed by the ocular, it would lie above the microscope tube. Rays originating from this image would be refocused by the eye on the retina. Thus the eye would have to be held some distance away from the real image which would then be viewed with considerable difficulty. A real image is required only if a photograph is to be recorded; then the ocular used for direct viewing must be replaced, either by an ocular designed for photomicrography or by the lens of a camera. (Real and virtual images have been discussed on p. 167.)

The ocular renders the light which forms the final image parallel or nearly so; thus the final image is formed at infinity or at some conveniently large distance, such as 25 cm (25 cm or, approximately, 10 inches is the "near point" for persons of standard vision; that is, the closest distance at which an object can be focused upon by the unaided eye). Figure 12-2a illustrates imaging by a compound microscope, and Fig. 12-2b illustrates the recording of a photomicrograph, with the use of a different ocular lens, in an otherwise identical system.

ILLUMINATION OF THE SPECIMEN IN THE COMPOUND MICROSCOPE

Since magnification is a linear measurement, the *area* of the specimen which is observed decreases as the *square* of magnification, m. Under uniform conditions of illumination the amount of light which is incident per unit area of the specimen is constant; thus the intensity observed in the image also decreases in proportion to m^2.

Image intensity increases with the first power of the area of the aperture of illumination, which is that of the substage condenser diaphragm

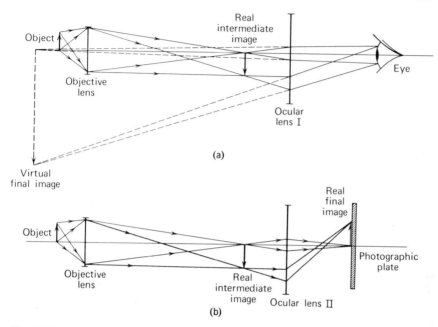

Fig. 12-2 Imaging in the compound microscope: (a) visual observation; (b) photographic recording.

and thus, in effect, that of the objective. (Since the angular aperture of the rays which form the intermediate image is relatively small, the ocular is normally not the limiting aperture or aperture stop of the microscope system.) Accordingly, the brightness of the image, for a given source and source-to-specimen distance, varies as described by

$$(\text{12-a}) \qquad \text{brightness} = K\left(\frac{\text{N.A. objective}}{M_{\text{total}}}\right)^2,$$

where K is a constant characteristic of the illumination system, M_{total} is the combined magnification achieved by the objective-ocular combination, and N.A. is the numerical aperture ($\equiv n \sin \alpha$) as defined on p. 241. Since the numerical aperture of the objective cannot be increased without limit, very intense illumination is required for observations at high magnifications.

Two systems of illumination are referred to frequently in the literature of microscopy. So-called *critical illumination* provides a focused image of the light source in the plane of the specimen, as shown in Fig. 12-3. The name of the method derives from the fact that this method of illumination was once believed to be optimal for high resolution microscopy. The

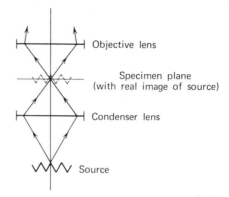

Objective lens

Specimen plane
(with real image of source)

Condenser lens

Source

Fig. 12-3 Critical illumination.

critical nature of critical illumination has long since been disproved, however. The illumination provided by this method is intense, but it is uneven if the output of the source itself varies from position to position. For this reason, critical illumination is most suitable at high magnifications, when very limited portions of the object plane are observed.

In the electron microscope, focusing of the condenser lens at the "cross-over" position constitutes a condition of critical illumination (cf. p. 388).

The method of *Kohler illumination* is advantageous in that it provides a uniformly illuminated field of view. This is particularly important when a nonuniform source, such as a ribbon filament, is required in order to produce very high intensity. The method is generally considered to be the optimal one for high resolution microscopy.

In Kohler illumination an image of the source is formed by the lamp lens at the first focal plane of the condenser, and an image of the field diaphragm is formed in the plane of the specimen by the condenser lens. The first focal plane of the condenser is the position of the substage diaphragm. It is sometimes simply stated that "an image of the source is formed at the iris diaphragm of the condenser lens." The significance of that condition is not evident, however, unless it is understood that imaging of the source in the focal plane is thereby implied! The effect of the condenser lens is then to collimate the light beam which originates from any point on the source. Thus light from each point forms a parallel beam which illuminates an extended area of the specimen. A second real image of the source is then formed in the rear focal plane of the objective.

Kohler illumination is illustrated in Fig. 12-4. In that diagram a *pair* of rays from each of two points on the source is traced through the microscope.

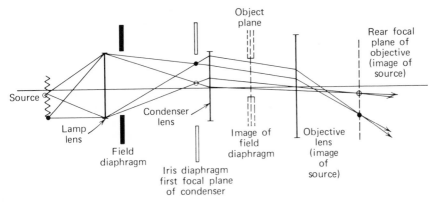

Fig. 12-4 Kohler illumination.

The formation of an image of the field diaphragm in the plane of the specimen is of practical use in locating the condition of Kohler illumination.

Critical illumination, in which distinct regions of the source are focused upon individual specimen points, tends to produce *incoherent* illumination of the specimen. That is, specimen points are effectively self-luminous to the extent that emission by adjacent regions in the source fails to be synchronized. Under Kohler illumination, however, each point of the object receives light rays from an extended region of the source. The phase (as well as the amplitude) of the resultant wave reaching the specimen is determined by the contributions from all points within that region of the source. Thus the phase of illumination received by adjacent points does not vary significantly, so that Kohler illumination tends to be *coherent*.

The terms *orthoscopic* and *conoscopic* are sometimes used in the literature to refer to illumination by collimated or convergent light beams, respectively.

THE MICROSCOPE OBJECTIVE; OIL IMMERSION

The objective lens, which forms the primary image, is the most critical optical element of the compound microscope. Practical microscope objectives are corrected for spherical aberration and coma, and for axial chromatic aberration either at two wavelengths (in *achromatic* objectives) or at three wavelengths (in *apochromatic* objectives). (The ocular lens corrects lateral chromatic aberration and, sometimes, astigmatism and curvature of field also.)

A single achromatic doublet (which is the simplest type of objective lens) corrected for spherical aberration also, performs adequately only at very low numerical apertures (about 0.1). Typically, several separate doublet or triplet lenses are used in the objective; a high power ("high dry") objective is shown in Fig. 12-5, and an oil immersion objective in Fig. 12-8. The planar front surfaces of these objectives are typical.

A *cover glass* (cover slip) is normally used in mounting the microscope specimen. As illustrated by Fig. 12-6, the cover glass has the effect of converging the cones of rays which originate from each point of the specimen. In this way the cover slip introduces aberrations, which must be corrected by the objective. The extent of the convergence is determined by the refractive index and thickness of the glass. Although the refractive index of different coverslips does not vary sufficiently to affect the quality of the image, the thickness, which may vary from 0.15 to 0.22 mm or more, is critical. The objective is usually corrected for use with a cover slip which is 0.18 mm thick; if different thicknesses are used, aberrations appear in the image. Compensation for thickness can be made by adjusting the mechanical tube length of the microscope or by use of a "correction collar" which changes the spacing between the elements of the objective lens. Either method produces satisfactory results, although the adjustment of the correction collar may be inconveniently critical. (An experimental study of cover slip effects is described by Barer.)

The numerical aperture may vary from 0.1 for very low power objectives to as much as 1.6 for high power immersion systems. As shown in Chapter 10, *resolution* is a function of the quantity (λ/N.A.); hence *large* numerical apertures correspond to *small* values of the minimum resolvable distance and thus to *high* resolutions. In general, the numerical aperture of an objective should match the magnification produced in the final microscope image. That is, detail which is just resolved should be sufficiently enlarged for comfortable viewing by the eye, but not so much

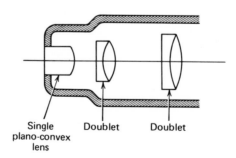

Single   Doublet   Doublet
plano-convex
lens

Fig. 12-5 "High-dry" objective.

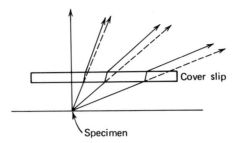

Fig. 12-6 Effect of coverslip.

that "empty" magnification results. (A discussion of magnification is given on p. 279.)

At the highest magnifications (ca. 1000×, *oil immersion* objectives are used. With this type of objective the value of the numerical aperture is increased by filling the space between specimen and lens with an oil of high refractive index. Preferably, the index of the oil should be identical with that of the lens itself.

Figure 12-7 illustrates the way in which the oil immersion method improves resolution. In Fig. 12-7a there is an air space between the cover slip and the lens. Ray ① from the specimen point *O* is refracted away from the normal to the cover slip surface as it enters the air space. Nevertheless, it enters the aperture of the objective. Ray ② is similarly refracted, but at a larger angle, so that it passes outside the objective lens. Ray ③ meets the dense-to-rare cover slip-to-air interface at an angle greater than the critical angle, and therefore is totally reflected within the cover slip (cf. p. 6). Figure 12-7b shows the course of the same rays when the space

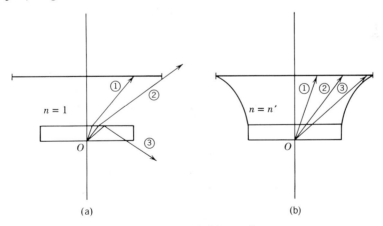

Fig. 12-7 Effect of oil immersion.

between cover slip and lens is filled with a medium of the same refractive index as the cover slip. The rays no longer undergo refraction before they enter the lens, since they travel through a medium of uniform refractive index. The angular size of the cone of light leaving point O which reaches the lens is thus substantially increased.

If the refractive index of the immersion oil is identical with that of the glass of the lens, rays are not refracted at all as they enter the lens, but only as they leave its upper surface. Therefore, if the specimen is placed at the aplanatic point of the first component of an objective, imaging by this part of the lens system is entirely free of spherical aberration. (The nature of aplantic refraction has been described on p. 196).

Note that no further improvement in resolving power would be achieved by using an immersion oil of index higher than that of the lens, since rays would then be excluded at the lens surface by total internal reflection within the oil.

The general design of a practical oil immersion lens is shown in Fig. 12-8a, and Fig. 12-8b shows the aplanatic refractions which occur at the first two elements of the system. The specimen is located at P, the aplanatic point of the hyperhemispherical lens L_1. Rays which are refracted by L_1 thus appear to proceed from P' (as was demonstrated on p. 197). P' is also the *center of curvature* of the first surface of the meniscus lens L_2. (A *meniscus* lens is any lens with surfaces of *opposite curvatures;* a positive meniscus lens is illustrated by Figs. 12-8a, 12-8b, and a negative meniscus lens is shown, for comparison, in Fig. 12-8c.) Rays enter L_2 along a radius of its first surface and accordingly experience no refraction at that surface. At the second surface of L_2 the rays are refracted aplanatically, so that they appear to diverge from P''. Refraction at the subsequent components of the oil immersion objective completes the convergence of rays from P, forming the intermediate image. These lenses also correct the chromatic defects which are introduced by L_1 and L_2, while introducing a minimum of spherical aberration. The fact that the beam is partially converged before entering these lenses aids in the control of spherical aberration.

The use of an oil immersion objective as a "dry" lens results in obviously defective images. The reason is that refraction then occurs at the *first* surface, also, of L_1, introducing spherical aberration for which the subsequent elements of the system provide no correction.

The objective lens is so located that the specimen lies just outside its first focal plane. Thus a real image is formed within the microscope tube; this image can be observed directly by removing the ocular lens.

High power objectives are lenses of short focal length (1.6–4.0 mm); low power objectives are lenses of moderate focal length (16–25 mm).

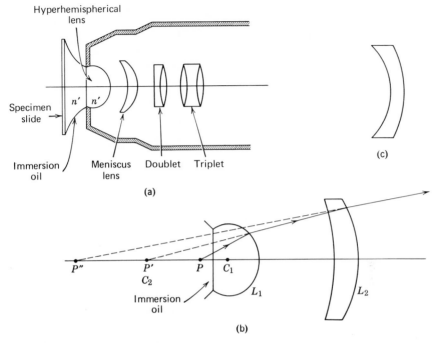

Fig. 12-8 (a) Oil immersion objective; (b) aplanatic refractions in the oil immersion objective; (c) negative meniscus lens.

Objectives to be used in a given microscope are usually so mounted as to be *parfocal*. That is, for a given object position, the images formed by different objectives of a parfocal set have an *identical* position. The action of parfocal high and low power lenses is compared in Fig. 12-9.

OCULAR LENSES

The ocular (eyepiece) lens serves to magnify the intermediate image formed by the objective. A second important function of these lenses is the correction of aberrations; in particular, of lateral chromatic aberration. A *compensating* ocular is one which is chromatically "overcorrected" (i.e., so corrected as to compensate for the residual chromatic aberration in the image produced by the objective). (An image formed by the ocular alone thus suffers from chromatic aberration opposite in sense to that produced by an uncorrected lens.) This type of ocular must be used in order to observe the full resolution provided by high quality (apochromatic) objectives. In principle, "matched" ocular and objectives should be used in order to obtain optimally aberration-free images; in practice,

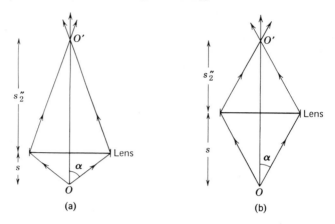

Fig. 12-9 Parfocal lenses: (a) high power; (b) low power.

however, many types of ocular provide very nearly equivalent corrections.

The simplest type of ocular, and one often used with low power objectives, is the *Huygenian* ocular. This system consists of two separate elements known as the *field lens* and the *eye lens,* respectively. Both lenses are of the approximately convexo-planar form for which spherical aberration is minimal. As shown in Fig. 12-10, rays from the objective are converged by the field lens, so that the intermediate image is formed at or very close to the focal plane of the eye lens. The eye lens thus renders the rays from the intermediate image parallel, forming a final virtual image at infinity. The same image location would result if the intermediate image plane (i.e., of the objective) were to coincide with the focal plane of a single ocular lens. However, the two-component system is a more favorable one from the point of view of controlling aberrations. Further corrections can be imposed by using either doublet or triplet elements in place of the single field and eye lenses.

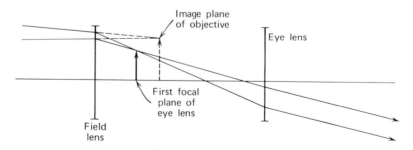

Fig. 12-10 Huygenian ocular.

Another common type of eyepiece is the *Ramsden* ocular, which is illustrated in Fig. 12-11. In this system the real intermediate image formed by the objective is located at or very near the first focal plane of the ocular. Again, the final virtual image is formed approximately at infinity. The field and eye lenses are separated by a distance equal to their (common) focal length, a condition which is optimal for the correction of lateral chromatic aberration.

These and other types of ocular used for direct viewing are all positive (i.e., magnifying) lenses. The *projection* oculars used for photomicrography may be negative (i.e., demagnifying) lenses. Correction of curvature of field is convenient in negative lens systems; the elimination of this defect is of particular importance in recorded micrographs, for which variations in focal level of different zones of the image cannot be compensated by continuous fine focusing during observation.

DEPTHS OF FIELD AND FOCUS

The formation of a planar image $A'B'C'$ of a planar object ABC is illustrated by Fig. 12-12. According to a geometric view, an ideal image would be *exactly* in focus only at the infinitely thin plane $A'B'C'$. Physically, however, the image points A', B', and C' are of finite extent. Even when free of aberrations, they form diffraction figures which extend in three dimensions. (The Airy disc is the section of the diffraction figure in the Gaussian image plane.) Thus there exists a region of finite width within which the image of ABC is observed to be in focus. The axial extent of this region (i.e., measured along the microscope axis) is the *depth of focus*, designated D'. Similarly, object points which lie slightly above or below plane ABC appear to be in focus at the ideal image plane $A'B'C'$. The maximum separation of points which simultaneously appear to be in focus, again measured along the lens axis, is the *depth of field*, designated D.

Note that the terms "depth of focus" and "depth of field" apply to image

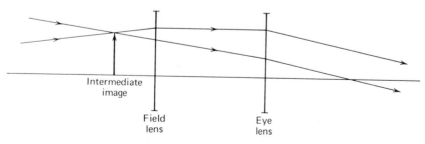

Intermediate
image

Field
lens

Eye
lens

Fig. 12-11 Ramsden ocular.

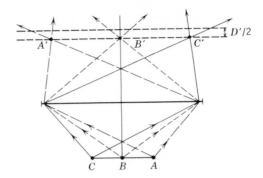

Fig. 12-12 Depth of focus.

space and object space, respectively. Unfortunately, confusion arises because depth of field is sometimes loosely referred to as "depth of focus"!

Figure 12-13 illustrates the effect of lens aperture upon the depth of focus. The imaging of a point in a given specimen plane by a lens of fixed focal length is shown in both parts of that figure. However, the lens aperture in Fig. 12-13b is only half as large as that in Fig. 12-13a. It is evident that, for any value of d_{min}, depth of focus is greater for a lens of reduced aperture.

The depth of field is also greatest for systems of narrow angular aperture, as shown by Fig. 12-14, which considers the imaging of three axial points A, B, and C. If the angular spread of rays contributing to the corresponding image points A', B', and C' is sufficiently small, all appear to

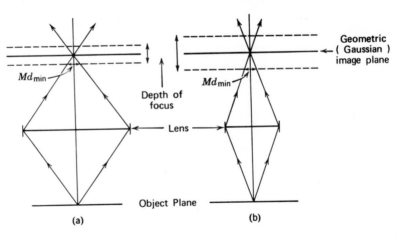

Fig. 12-13 Depth of focus: (a) for lens of wide angular aperture; (b) for lens of narrow angular aperture.

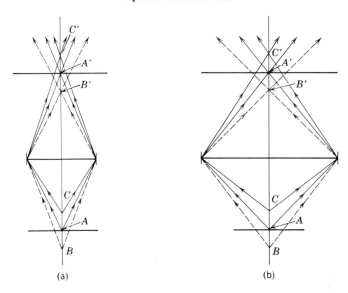

Fig. 12-14 Depth of field: (a) for lens of narrow angular aperture; (b) for lens of wide angular aperture.

be in focus at the plane of A'. If this condition is just met in a system of narrow angular aperture, as shown in Fig. 12-14a, the spread of rays which contribute to B' and C' is noticeable in the plane of A' when the aperture is increased, as shown in Fig. 12-14b.

A quantitative estimate of depth of field is considered in Fig. 12-15, which shows the rays which extend from two object points O and P to the lens margins. (The separation of O and P is much exaggerated; in fact, the marginal rays from each image point subtend very nearly the same angle α at the lens.) Rays from point O subtend a disc of diameter $y = Y_1 Y_2$ at point P. If the value of y is no greater than $d_{\min}$, the images of O and P both appear to be *in* focus at the geometrical image plane. That

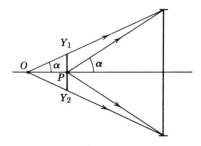

Fig. 12-15.

is, P appears *in* focus at the image plane of O, and vice versa, if

(12-b)
$$y = \frac{0.61\lambda}{n \sin \alpha}.$$

As shown in Fig. 12-15, the distance OP is one half the depth of field D, since a point placed a distance OP to the left of O would, like point P, appear to be in focus at the geometrical image plane of O. In triangle OPY,

(12-c)
$$\tan \alpha = \frac{\frac{1}{2}y}{OP} = \frac{y}{D}.$$

D is then evaluated by substituting (12-b) in (12-c):

(12-d)
$$D = \frac{y}{\tan \alpha} = \frac{0.61\lambda}{n \sin \alpha \tan \alpha} = \frac{0.61\lambda \cos \alpha}{n \sin^2 \alpha} \quad \text{or}$$

$$\boxed{D = \frac{0.61\lambda \cos \alpha}{n \sin^2 \alpha}}.$$

If α approaches $90°$, as for high resolution objectives, $\cos \alpha$ and thus also the depth of field approach zero. The lens then tends to form an in-focus image of a "single plane" through the specimen. This is sometimes referred to as an *optical section* of the object.

An expression which is comparable to (12-d) may be derived for the depth of focus, D'. As shown in Fig. 12-16, rays from an image point O' subtend a disc of diameter $z = Z_1 Z_2$ at the plane of a second image point P'. O' and P' appear in focus at either plane if the value of z is less than the diameter of the Airy disc. That is, if

(12-e)
$$z = \frac{0.61\lambda}{n' \sin \alpha'} = \frac{0.61\lambda M}{n \sin \alpha},$$

where M is the total magnification of the microscope.

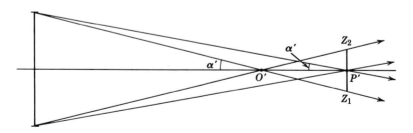

Fig. 12-16.

As shown in Fig. 12-16, the distance $O'P'$ is one half the depth of focus, D', since a point imaged a distance $O'P'$ to the right of P' would, like O', appear to be in focus at the plane of P'. In triangle $O'P'Z'$,

(12-f) $$\tan \alpha' = \frac{\frac{1}{2}z}{O'P'} = \frac{z}{D'}.$$

D' may then be evaluated by substituting (12-e) in (12-f), giving

(12-g) $$D' = \frac{0.61\lambda M}{\sin \alpha \tan \alpha'}.$$

Since, as given in Chapter 8,

(8-rr) $$M_{ang} = \frac{\tan \alpha'}{\tan \alpha},$$

(12-g) reduces to an expression which is independent of M, *if* the angular and total magnifications of the microscope system are approximately the same. As explained below, the total magnification is the product of the linear magnification of the objective and the angular magnification of the ocular. In light microscopy the aperture angle α is always appreciable; therefore it is reasonable to equate angular and total magnifications. The resulting expression for D' is identical in form with (12-d) for D. In the electron microscope, however, α is always extremely small; hence the angular and total magnifications cannot be equated. This case is considered on p. 413.

Values of the depths of field and focus in the light microscope, as estimated from (12-d), are of the order of $10\,\mu$ for low power objectives (N.A. = 0.25), $1\,\mu$ for high dry objectives, and $0.2\,\mu$ for oil immersion objectives. Approximate confirmation of these estimates can be obtained experimentally. For example, it is found (see [2]) that, for N.A. = 0.25, $D = 11.7\,\mu$ and $D' = 6.2\,\mu$.

MAGNIFICATION

In the compound microscope shown in Fig. 12-2a the intermediate image is located just inside the first focal plane of the eyepiece lens, so that the final virtual image is formed in a plane close to that of the object. The same microscope could be used to form an intermediate image *at* the focal plane of the eyepiece and, thus, a final image located at infinity. In such systems the linear ("lateral") magnification of the object (i.e., the quantity s_2''/s) is of little significance. In fact, it is the angular subtense of the image at the eye which is of importance. Thus the magnification of a compound microscope is expressed as the product of the *linear* magnification of the objective lens and the *angular* magnification of the eyepiece lens.

The linear magnification of the objective $(-s_2''/s)$ is most conveniently expressed in the "Newtonian" form, as given in Chapter 8:

(8-pp)
$$M_{obj} = \frac{-x'}{f_{obj}}.$$

The quantity x' is, precisely, the "optical tube length" (i.e., the distance from the rear focal plane of the objective lens to the intermediate image). The optical tube length is approximately equal to the more easily measured mechanical tube length, which is the distance from the top of the objective lens housing to the top of the tube into which the ocular is set. The mechanical tubes of compound microscopes are normally set at a standard value of 16.0 cm. Thus,

(12-h)
$$M_{obj} = \frac{-16\,cm}{f_{obj}} = \frac{\text{mechanical tube length}}{\text{focal length}}.$$

The angular magnification of the ocular is

(12-i)
$$\frac{\text{angular subtense of final image}}{\text{angular subtense of intermediate image if viewed directly}} = \frac{\tan \beta'}{\tan \beta},$$

where β and β' are the angles shown in Figs. 12-17a and 12-17b, respectively. Tan β is evaluated by considering that the intermediate image, of height y, is viewed at 25 cm distance (i.e., the near point for the normal eye). Thus,

(12-j)
$$\tan \beta = \frac{y}{25\,cm}.$$

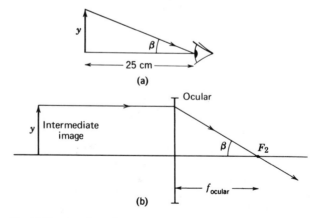

Fig. 12-17 Angular subtense: (a) by eye; (b) at microscope ocular.

A ray leaving the tip of the intermediate image parallel to the microscope axis intersects the axis at the second focal point of the ocular lens, as shown in Fig. 12-17b. The ray makes angle β' with the axis, so that

(12-k)
$$\tan \beta' = \frac{y}{f_{\text{ocular}}}.$$

Substitution of (12-j) and (12-k) in (12-i) then gives

(12-l)
$$M_{\text{ocular}} = \frac{25 \text{ cm}}{f_{\text{ocular}}}.$$

The total magnification of the microscope is then obtained from (12-h) and (12-l) as

(12-m)
$$\boxed{M_{\text{total}} = M_{\text{obj}} \times M_{\text{oc}} = \frac{-16 \times 25}{f_{\text{obj}} f_{\text{oc}}} = \frac{-400}{f_{\text{obj}} f_{\text{oc}}}.}$$

Unfortunately, confusion sometimes arises from an early convention used by instrument makers according to which the magnifications specified by (12-h) and (12-l) were "reversed." That is, it was considered that

(12-n)
$$M_{\text{obj}} = \frac{-25}{f_{\text{obj}}} \quad \text{and} \quad M_{\text{oc}} = \frac{16}{f_{\text{oc}}}.$$

Magnifications computed according to (12-n) were engraved on lens mountings. Note, however, that, if the magnifications of both lenses are specified consistently, the total magnification of the microscope is the same in both cases.

The total magnification achieved with an objective of given numerical aperture should usually be sufficient to render structure at the resolution limit comfortably visible, but not so great as to result in "softness" of the image. The minimum angular subtense resolved by the eye is about $0°1'$ of arc (approximately 3×10^{-4} radian), but a subtense of about $0°3'$ is required for comfortable viewing. The image begins to seem "soft" if the smallest detail subtends an angle of more than about $0°5'$. Combination of these values with the resolution limit of $\lambda/2$ N.A. ($\cong 3 \times 10^{-5}$ cm/N.A. for visible light) permits calculation of desirable levels of magnification.

When structure of size d_{min} is observed by the naked eye, a (very small) angle β is subtended such that

(12-o)
$$\beta \cong \tan \beta = \frac{d_{\text{min}}}{25 \text{ cm}} = \frac{3 \times 10^{-5}}{25 \text{ N.A.}}.$$

For these structures to be *resolved* by the eye, β must be increased to at least 3×10^{-4} radian by the microscope. Thus,

(12-p) $$M_{min} = \frac{3 \times 10^{-4} \times 25 \text{ N.A.}}{3 \times 10^{-5}} = 250 \text{ N.A.}$$.

Similarly, optimum magnification for comfortable observation of resolved detail lies in the range

(12-q) $$M_{opt} = (500 - 1000) \times \text{numerical aperture}$$.

Equation 12-q sets an upper limit of about 1500× to the useful magnification of the optical microscope; in fact, all detail present in images formed by lenses of maximum numerical aperture could be resolved, with careful observation by the normal eye, at a total magnification of not much more than 400×. Magnifications used in practice range up to about 1200×.

In most situations the light microscopist is unconcerned with the precise numerical level of magnification obtained, since the order of magnitude of the dimensions of objects is generally more or less obvious. When exact measurements are required, however, an ocular can be used in which a ruled scale (graticule) is incorporated. The graticule is placed at the first focal plane of the "eye lens" of the ocular (i.e., essentially in the plane of the "field lens" of the Ramsden type of ocular) so that an in-focus view of the rulings is superimposed upon the image. Rulings of the graticule can be calibrated with respect to observation of a slide ruled with calibrated spacings. If the magnification of the microscope is such that spacings on the graticule and calibration slide are very nearly coincident, the magnification of the microscope may be altered by varying its mechanical tube length so as to bring the two sets of spacings into exact coincidence. Alternatively, dimensions may be read from the graticule and corrected numerically by application of a factor determined by previous observation of the calibration slide. Appropriate calibration slides are obtainable commercially; usually they are ruled at 0.01 mm intervals.

DARK FIELD MICROSCOPY

Dark field microscopy is a method of obtaining high contrast images of specimens which scatter light strongly. In this method the direct beam from the light source is stopped, so that only scattered light can enter the microscope objective and contribute to the formation of the image. The advantage of the method is that relatively low intensities of scattered light can be focused with high contrast against the dark background, whereas removal of the same proportion of incident intensity from an ordinary "bright field" microscope image would produce an image point only slightly less bright than its surround, i.e., a much lower percent contrast.

Dark field conditions are generally produced by illuminating the specimen with a hollow cone of light of such an angle that the direct beam lies outside the aperture of the objective lens. Only light which is scattered by the specimen can then enter the objective aperture. A dark field system is diagramed in Fig. 12-18. Direct rays from the condenser, a lens of large numerical aperture, are shown by solid lines. These rays pass beyond the objective, which is a lens of smaller numerical aperture. Scattered rays, which are indicated by dashed lines, originate at the specimen, enter the objective, and form the image.

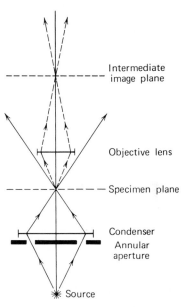

Fig. 12–18 Darkfield microscope.

As Fig. 12-18 suggests, a simple system for dark field observations consists of a compound microscope in which an annular aperture of the form shown in Fig. 12-19 is inserted just below the condenser lens. The numerical aperture of the objective must be reduced to a point at which the direct beam is excluded. For high resolution dark field microscopy, special condensers are available. They provide a wide cone of illumination, thus allowing the use of objectives of high numerical aperture. With the use of oil immersion, dark field microscopy can then be carried out at numerical apertures as high as 1.2. A typical dark field condenser is diagramed in Fig. 12-20. Variable dark field-bright field condensers are also made.

Fig. 12-19 Aperture for darkfield microscopy.

Because of the high contrast provided by dark field illuminations, it is sometimes possible to detect structures of dimensions less than d_{min} in dark field images. For example, intensely illuminated polystyrene spheres of about 1000 Å (0.1 μ) diameter (as known accurately both from light-scattering

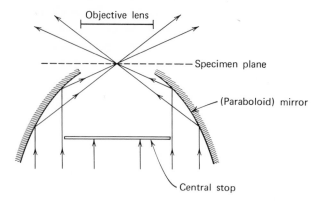

Fig. 12-20 Darkfield condenser.

measurements and from electron microscopy) can be observed in dark field images. However, as noted on p. 234, two or more adjacent particles of this size cannot be distinguished as separate (i.e., the particles are not resolved).

ULTRAVIOLET MICROSCOPY

Microscopic examination under ultraviolet light makes use of wavelengths in the range from 2500 to 3500 Å. Wavelengths in the vicinity of 2600 and 2800 Å are of particular interest on account of their selective absorption by nucleic acids and proteins, respectively. In fact, interest in biological ultraviolet microscopy depends more upon observation of specific absorption by biologically important substances than upon the roughly twofold improvement of ultimate resolving power which is possible at these shorter wavelengths.

Quartz optical parts are essential for ultraviolet microscopy. (Optical glasses are opaque to wavelengths shorter than about 3100 Å, whereas quartz transmits appreciably down to about 2200 Å.) For example, at 2200 Å, 1 mm of quartz transmits 94% of incident intensity, and is of refractive index 1.53. Dispersion (variation of refractive index with wavelength) is generally much greater in the ultraviolet than in the visible (cf. Fig. 23-1). For this reason adequate correction of the chromatic aberration of quartz lenses is difficult to achieve, so that ultraviolet microscope systems are usually designed for use with monochromatic illumination.

With monochromatic illumination it is possible to use *monochromat* lenses, which can be very highly corrected for spherical aberration. These objectives consist of a single hyperhemispherical lens followed by a series of meniscus lenses. The virtual image formed by each lens of the series is located at the center of curvature of the first surface of the succeeding lens (cf. Fig. 12-7b). Thus, imaging is entirely aplanatic, except for the very small contribution of a final focusing element.

The position of the focused ultraviolet image must be located by means of a fluorescent screen or photographically. Note that a microscope setting which produces an in-focus image with visible light forms an out-of-focus ultraviolet image, since the focal length of any refractive lens is always appreciably different for these shorter wavelengths. This difficulty is sometimes avoided by the use of reflection lenses, the focal lengths of which are independent of wavelength (cf. p. 218).

FLUORESCENCE MICROSCOPY

In the method of fluorescence microscopy, incident illumination is removed by selective absorption, while light which has been absorbed by the specimen and reemitted at an altered wavelength (i.e., phosphoresced

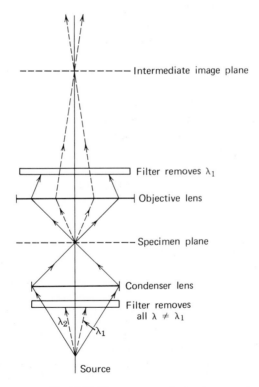

Intermediate image plane

Filter removes λ_1

Objective lens

Specimen plane

Condenser lens

Filter removes all $\lambda \neq \lambda_1$

λ_2 λ_1

Source

Fig. 12-21 Fluorescence microscope.

or fluoresced) is transmitted. Thus a colored image is formed against a dark surround. The particular interest of fluorescence microscopy derives from the specific nature of fluorescence emission. The method provides an extremely sensitive histochemical technique either when inherently fluorescent materials are present in the specimen (i.e., for the observation of autofluorescence) or when the object is stained with fluorescent dyes (fluorochromes).

A diagram of a system for fluorescence microscopy is shown in Fig. 12-21. The source must provide an intense ultraviolet beam. Either a (monochromatic) mercury arc or a xenon lamp-monochromator combination is used. A filter removes wavelengths which are not effective in exciting fluorescence. The incident beam then consists of some wavelength λ_1. Bright field illumination of the specimen is shown in Fig. 12-21, but dark field illumination is also suitable.

The light which is fluoresced by the specimen is transmitted by a second filter, which removes the incident wavelength λ_1, from the beam. Thus only light which has interacted with the specimen in such a way as to suffer a change in wavelength can contribute to intensity in the image plane. (Note that in Fig. 12-21 separated paths of incident and transmitted wavelengths are shown only for convenience; in fluorescence microscopy there is no difference in angular aperture with wavelength.)

A limitation of the method lies in its very sensitivity. Lenses, slides, coverslips, embedding media, and fixatives may all fluoresce to some degree, and certain materials used in microscopy (e.g., Canada balsam) fluoresce strongly. For this reason specially purified reagents and other supplies are required for fluorescence microscopy.

REFERENCES

General References

R. Barer, *Lecture Notes on the Use of the Microscope*, 2nd ed., Blackwell & F. A. Davis, Philadelphia, 1959. This excellent short text (59 pp.) describes both theoretical and practical aspects of light microscopy.

Photography through the Microscope, 4th ed., Kodak Scientific and Technical Data Book P2, Eastman Kodak Co., Rochester, N.Y., 1966. A very good short discussion of microscopy in general as well as of photomicrography in particular. Some especially interesting illustrations show the effect on the image of curvature of field (p. 8), of empty magnification (p. 12), and of spherical aberration due to use of a cover slip which is too thin (p. 14). Imaging by apochromatic and achromatic lenses is also compared on p. 19.

L. C. Martin, *The Theory of the Microscope*, Blackie, Glasgow, and American Elsevier, New York, 1966. Most of Chapter 3 (Further Properties of Optical Systems) is concerned with features of the compound microscope.

D. Birchon, *Optical Microscope Technique*, Newnes, London, 1966. Chapter 1 (The Optical Microscope), Chapter 2 (Microscope Equipment), Chapter 3 (The Use and Care of a Microscope), and Chapter 4 (Techniques of Illumination and Observation). Photographs of several types of microscope are included.

L. C. Martin, in G. Oster and A. W. Pollister, Eds. *Physical Techniques in Biological Research*, Vol. I, *Optical Techniques*, Academic Press, New York, 1955. An article on "The Light Microscope," pp. 325–375.

Specific References

[1] B. M. Spinell and R. P. Loveland, *J. Roy. Microscop. Soc.* **79**, 59–80 (1960), "Optics of the Object Space in Microscopy."

[2] M. Berek, *Marburg Sitzungs Ber.* **62**, 189 (1927). (Quoted on p. 203 of Martin, *op. cit.*)

CHAPTER 13

Phase Contrast Microscopy

The phase contrast microscope is a device which renders differences in refractive index between regions of a specimen visible as differences in intensity. More precisely, differences in optical path are converted into differences in intensity. The phase contrast microscope is thus a means of producing contrast in unstained specimens (i.e., where differences in absorption between regions of the specimen are negligible). In certain applications the instrument may be used to quantitate differences in optical path, but its principal function is qualitative; the interference microscope (Chapter 14) is the instrument of choice for quantitative applications.

A suitably adjusted phase contrast microscope can be used to observe specimen contrast by anyone familiar with the procedures employed in ordinary light microscopy. Observation does not demand an understanding of the optical principles upon which the instrument is based. However, a careful interpretation of the appearance of the specimen depends upon an appreciation of how contrast is effected in phase systems. Quantitative measurements require a knowledge of the characteristics of the particular microscope employed.

Understanding of the mechanism of action of the phase contrast microscope requires an appreciation of the mechanisms whereby contrast arises in microscope images generally, a grasp of the basic principles governing the interference of wave motions and of the analysis of wave motions into their component vibrations. The material presented in Chapters 3 and 11 is thus particularly relevant to the present chapter.

COMPARISON OF PHASE CONTRAST AND ORDINARY LIGHT MICROSCOPES

As discussed on p. 257, phase contrast, that is, variation in image intensity as a function of differences in optical path through the specimen, occurs to a limited extent in the ordinary light microscope as a result of defocusing, scattering outside the microscope aperture and/or of residual aberrations in the lenses. It must be emphasized that the phase microscope produces *in-focus* phase contrast and, furthermore, that this contrast is observed with lenses of full aperture which are free from geometrical aberrations.

With reference to the distinction between resolving power and resolution, first discussed on p. 233, it should be noted that the resolving power of the phase microscope can be no greater than the optimum achievable by the ordinary light microscope, but that the resolution achieved in the observation of a particular specimen may be much improved as a consequence of the enhancement of contrast.

VECTOR REPRESENTATION OF DEVIATED AND UNDEVIATED BEAMS

The mechanism of action of the phase contrast microscope can most readily be explained with reference to a vectorial representation of the light beams which pass through the microscope. As explained on p. 39 and in Appendix B, a wave motion may be represented by a vector, the length of which (with reference to some arbitrary scale) is proportional to the amplitude of the beam, and the orientation of which (with reference to some arbitrary azimuth) represents the phase of the beam.

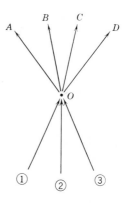

Fig. 13-1.

As shown in the ray diagram, Fig. 13-1, the incidence of a light beam upon any specimen point O results in the production of rays traveling in all possible directions. (Note that in Fig. 13-1, as in any ray diagram, no specification concerning the relative amplitudes of the rays is made.) The rays $A, \ldots D, \ldots$ leaving O collectively represent a wave motion of some definite amplitude and phase; this motion is conventionally designated the P *wave*, or *particle wave*, since it is the wave motion leaving the particle. Similarly, the incident beam, represented in Fig. 13-1 by rays ①, ②, and ③, is also a wave motion of definite amplitude and phase and is designated the S *wave* or *surround wave*. In the absence of specimen points such as O, the S wave propagates undisturbed through a region.

The S wave incident upon a specimen may be represented by a vector of unit length the phase of which is designated arbitrarily, in Fig. 13-2, by vertical orientation. A corresponding P-wave vector representing the wave motion beyond the specimen point may be drawn. The length of the P-wave vector is reduced from that of the S-wave vector in proportion to the amplitude transmittance by the specimen; that is, by the ratio

(13-a) $$\frac{\text{amplitude transmitted by specimen}}{\text{amplitude transmitted by surround}}.$$

(Note that this quantity is the *square root* of the transmittance ratio as measured experimentally, since the latter is a ratio of intensities.) If the specimen O is a pure phase object (cf. p. 254), the transmittance through O is equal to that of the surround, and the corresponding P-wave vector is equal in length to the S-wave vector. Light passing through a phase object travels through a greater or smaller optical path than that passing through the surround, and is thus shifted in phase relative to the S-wave. The P-wave vector is correspondingly rotated through some angle Δ. P- and S-wave vectors for a phase object are illustrated in Fig. 13-3. (The P-wave vector, of course, represents the wave leaving the particle as a whole; it is important not to confuse the direction of a vector, representing phase, with the directions of the rays represented in the beam of that phase.)

Any wave motion, such as the P-wave of Fig. 13-3, may be analyzed as the resultant of two (or more) other wave motions (cf. p. 41). As diagramed in Fig. 13-4, the P-wave can, in particular, be represented as the wave motion resulting from the interference of the incident S wave with a component of different amplitude and phase. The latter is termed the D *wave*. Figure 13-4 illustrates a case in which the phase angle Δ between the S and P waves is 45°. Geometrical computation utilizing the law of sines,* shows that the amplitude of the D wave is approximately 0.76 that of the U wave in this specific case. In Fig. 13-4 the S-wave vector represents the on-going beam of undeviated light and is accordingly redesignated the U *wave*. The S- and U-waves are identical, the former term being used to apply to the incident wave motion, while the latter

---

*The law of sines, which is applicable to any triangle, states that $A/\sin \alpha = B/\sin \beta = C/\sin \gamma$, where sides and angles are designated as shown in the accompanying figure.

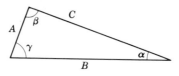

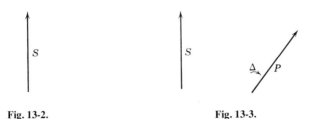

Fig. 13-2.   Fig. 13-3.

refers to a component of the transmitted wave motion. The *D* wave clearly represents the deviated beam which is produced by the interaction of incident light with the specimen.

In ordinary light microscopy, the *U* and *D* waves pass through the optical system without relative change of amplitude or phase and are united in the image plane to form a resultant, which in the case of a pure phase object is unaltered with respect to the resultant wave motion from the surround (i.e., in the absence of lens aberrations which may introduce slight relative changes of phase, in the manner discussed on p. 257). The component waves after transmission through the optical system are referred to as the *U'*, *P'*, and *D'* waves; in the ideal ordinary light

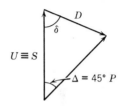

Fig. 13-4 D-wave as resultant of *U* and *P* waves.

microscope the amplitudes of *U'* and *D'*, and hence their intensities, are equal for phase objects. The *U*, *D*, *P* and *U'*, *D'*, *P'* wave motions in an ordinary light microscope can be diagramed as shown in Fig. 13-5. Note that Fig. 13-5b represents a slight loss of amplitude and variation in phase which affects each beam proportionately, but that the altered phase of *P'* relative to *P* has *no* effect upon the intensity observed at the image plane. The resolution of the *P*-wave motion (light leaving the specimen) into deviated and undeviated components may be compared with Abbe's concepts of image formation in microscopes, as expounded on p. 224.

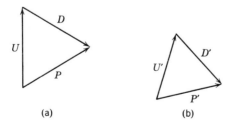

(a)   (b)

Fig. 13-5 Wave motions in the ordinary light microscope.

Clearly the U wave, or undeviated beam, is that which forms the zero order spot of the diffraction pattern formed in the rear focal plane of the objective. This wave contains no information concerning the structure of the specimen. The D wave consists of all diffracted orders, which contribute information to the image.

The interaction of light with a phase object may also be described as a *distortion* of the transmitted wavefront. (The rays shown in Fig. 13-1 are in fact, sets of normals to the incident and transmitted wavefronts.) This point of view is developed with respect to *electron* phase contrast on p. 431 and, especially, in Fig. 19-5, which see. Note that, to describe the transmission of light, Fig. 19-5 need be changed only by stating that regions of the specimen differ in refractive index, rather than "inner potential."

VECTORIAL REPRESENTATION OF PHASE CONTRAST EFFECTS IN THE ORDINARY LIGHT MICROSCOPE

If an object is imaged by a lens of low numerical aperture, some of the rays which contribute to the deviated wave (D wave) fall outside the aperture of the lens, while the U wave, which is of lower angular aperture, is transmitted essentially without loss of amplitude. There results a decrease in the amplitude of the D' wave, relative to the original D wave, as shown in Figs. 13-6a and 13-6b. When the D' and U' waves interfere at the image plane, the resultant P' wave then differs in *amplitude* relative to the original P wave, as shown by Figs. 13-6a and 13-6c.

For example, with an initial phase angle α of 45° between the P and S waves, the amplitude of the D wave, as stated above, is approximately 0.76 times that of the U wave. If the amplitude of the D' wave is reduced to approximately 0.38, the amplitude of the P' beam, as may be obtained by a calculation utilizing the law of cosines* is reduced to approximately 92% of the amplitude of the P or U beam. The observed intensity is then $(0.92)^2$ or about 85% as great as that of the surround.

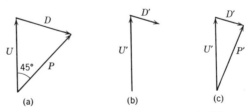

Fig. 13-6 Effect of low angular aperture on the D' wave.

*The law of cosines, which applies to any triangle, states that $A^2 = B^2 + C^2 - 2BC \cos \alpha$, where the sides and angles are as designated in the preceding footnote.

Figure 13-7 considers the case in which phase contrast arises because residual lens aberrations alter the phase rather than the amplitude of the resultant beam. An exaggerated case is shown in which the phase of the D' wave is increased 90° from that of the D wave. In practice the phase shift in an ordinary microscope, (as opposed to a phase contrast system) would be much smaller than this.

In Fig. 13-7, as in Fig. 13-6, a 45° phase shift between P and U waves is illustrated. The amplitude of the resultant P' wave, as shown in Fig. 13-7c, is appreciably reduced. Geometrical calculation, based on the law of cosines, shows that the amplitude of the P' wave in this example would be about 0.41 as great as that of the U' wave. The intensity of the image of the particle would thus be only $(0.41)^2$ or about 17% as great as the intensity of the surround.

Note that a different shift in the phase of the D' wave would result in increased amplitude of the P' wave. This is shown in Fig. 13-8, where a 90° advance in the phase of the D' wave, relative to the D wave, is represented. A geometrical calculation, utilizing the law of cosines, shows that the amplitude of the P' wave in this case is about 1.76 times as great as that of the U wave; the result is an image of the particle which is approximately three times as intense as that of the surround.

How does the shift in phase of the deviated wave as a whole come about in the ordinary light microscope? Portions of the deviated beam which pass through different zones of an aberrant lens travel different optical paths before reuniting at the image plane. The phase of the D' beam is then the resultant phase, at the image plane, of all the overlapping components which contribute to the wave.

PRINCIPLE OF THE PHASE CONTRAST MICROSCOPE

In the case of a (perfect) ordinary light microscope, all the rays leaving the specimen (i.e., all portions of the D wave) enter the objective aperture and travel through equal optical paths to the image plane. The phases of the final P' waves in different portions of the image may differ, but these variations are unobservable, since they do not affect the intensity.

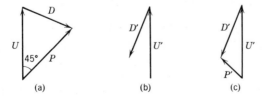

Fig. 13-7 Effect of lens aberrations on the D' wave.

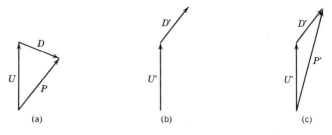

Fig. 13-8.

Information concerning specimen structure, though present at the image plane in the form of phase differences, is lost, and phase object specimens remain invisible. If the relative phases or the relative amplitudes, or both, of the U and D waves could be altered, differences in the amplitude and intensity of the P' waves would result, and the phase information provided by the specimen could be exploited. This is what is done, systematically, in the phase contrast microscope; the deviated and undeviated beams are physically separated, and one or the other is made to suffer a phase delay, a variation in amplitude, or both. The physical separation of the U and D beams is not in fact perfect, but it may be achieved to a reasonably high degree.

The magnitude of the changes of phase or amplitude or both which are produced in the phase contrast microscope is of the order diagramed by Figs. 13-6 through 13-8. Those diagrams illustrated the case of a 45° phase shift between P and S waves. Note that a phase shift of this magnitude represents an advance or delay of $\lambda/8$ or, for irradiation by light of $\lambda = 5000$ Å, a difference in optical path of 0.0625 μ. The latter could be achieved, for example, with an object 1.25 μ thick and of refractive index 1.35, viewed in a surround of refractive index 1.30. These are realistic values for representative biological specimens. The corresponding phase contrast effects shown in Figs. 13-7 and 13-8 are, respectively, a fivefold decrease or threefold increase in the intensity of the specimen relative to the surround.

MECHANISM OF PHASE CONTRAST MICROSCOPY

The shifts in relative amplitudes and phases of the U and P waves which are required for the production of phase contrast may be effected, as stated above, by physically separating these two elements of the beam and causing them to pass through absorption filters or regions of differential optical path. The physical separation, which at first thought might seem to be an impossible feat, in fact occurs at the rear focal plane of a microscope objective illuminated by parallel light (cf. p. 222). Figure 13-9 illustrates the illumination of a microscope specimen by a parallel beam. Part of the incident illumination falls upon the surround and enters the microscope objective without deviation; this light is brought to a focus at the central or zero order spot in the diffraction pattern formed in the back focal plane of the objective. The undeviated light diverges again to illuminate the image plane. Particles in the specimen scatter light in all directions. The resultant of all rays scattered out of the incident direction approximates the D wave shown in the preceding vector diagrams. (Note that the D wave is precisely that motion which would combine with unattenuated incident light to give the observed distribution of amplitudes leaving the specimen point.) Individual sets of rays scattered in any given direction contribute to points on the diffraction pattern in the rear focal plane; collectively, the (approximate) D wave occupies the entire focal plane. There thus exists a virtually complete separation of the U and D waves from the specimen in the rear focal plane.

In the conceptually simplest form of the phase contrast microscope a disc of at least partially transparent material of any refractive index greater than one, with a pinhole at the center, would be placed in the back focal plane of the objective lens. The pinhole, or *complementary area*, would transmit the U wave without alteration of phase, while the remainder of the disc, the *conjugate area*, would transmit the D wave with a relative delay in phase. If the disc is partially absorbing, the amplitude as well as the phase of the deviated light would be altered. An optical system of this sort is diagramed in Fig. 13-10. Experimental electron phase contrast systems have in fact been constructed according to this design. In practical light microscopes, however, the system is modified. The specimen is illuminated by convergent rather than collimated light. Since it is difficult to center the disc (known as a *phase plate*) in the focal plane exactly, the microscope is modified so that the plate can be placed at a more convenient position. An annular aperture controls the angle at which light from the condenser lens converges upon the specimen. The zero order beam then enters the objective lens as a hollow cone.

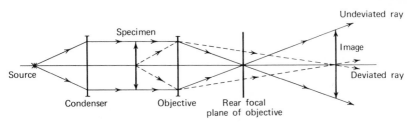

Fig. 13-9 Parallel illumination of a microscope specimen.

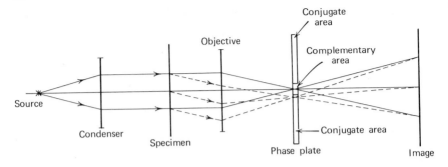

Fig. 13-10 Basic system for phase contrast microscopy.

The phase plate is placed beyond the objective at the position of the image of the annular condenser aperture and is itself of annular form. The complementary area of the phase plate thus coincides with the image of the open area of the condenser diaphragm. Careful matching and centering of the condenser aperture and phase plate are clearly of prime importance in obtaining satisfactory phase contrast.

The phase plate may be so constructed that either complementary or conjugate area represents the greater optical path, and that either of these areas may have greater absorption. A phase plate of appropriate characteristics is generally incorporated in the mounting of phase contrast objectives.

The optical system for phase microscopy is diagrammed in Fig. 13-11. Note that, whereas the specimen is illuminated by a hollow cone of light in both the phase contrast and the dark field microscopes (cf. p. 282), in the latter case the aperture of the objective is so limited as to exclude illumination by the direct beam. In phase contrast microscopy, however, the hollow cone of light must enter the objective lens.

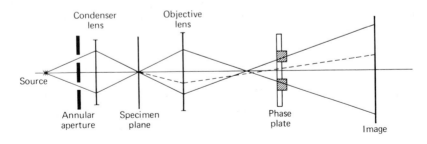

Fig. 13-11 Practical system for phase contrast microscopy.

POSITIVE AND NEGATIVE PHASE CONTRAST

Figure 13-12 summarizes the variation of the properties of the imaging system with variation of the characteristics of the phase plate. As before, the imaging of an object which retards the P wave by $\lambda/8$, or 45°, is considered. The resultant amplitude of the P' wave, that is, of the illumination forming the image of the specimen, is plotted as a function of the phase shift imposed upon the D wave by the phase plate. Two cases are considered: that of a nonabsorbing (or uniformly absorbing) phase plate, and that of a phase plate which reduces the relative amplitude of the D wave by half. It is evident that, for the specimen considered, optimum contrast would be provided by a plate which shifts the phase of the D wave by $-292°30'$ or $+67°30'$, giving a dark image of the object against the bright surround, or by a plate which shifts the phase of the D wave by $-112°30'$ or $+247°30'$, giving a bright image of the particle against a relatively dark background. These effects, for the example considered, are more pronounced in the case of the phase plate which produces no relative differences in amplitude.

The condition in which the specimen appears dark against a bright background is termed *positive phase contrast*; the reverse condition is termed *negative phase contrast*. Positive phase contrast is generally the more acceptable, psychologically, for biological observations, since the image then resembles the familiar appearance of an absorbing object observed in a transparent surround. From a theoretical point of view, either type of contrast is equally acceptable.

In principle, a phase plate would be chosen to give maximum contrast for a given specimen; in practice, the properties of many biological specimens are such that a phase shift of 90°, or $\lambda/4$, between deviated and undeviated beams is more or less optimal.

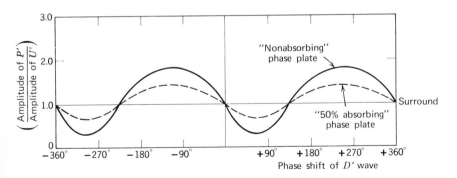

Fig. 13-12 Effect of variations in characteristics of the phase plate.

INTERPRETATION OF PHASE CONTRAST

Figures 13-13 and 13-14 illustrate the relative amplitudes and intensities, respectively, of the P' and U' waves as a function of the phase angle δ between the U and D waves. (The angle δ is indicated in Fig. 13-4.) A phase plate is considered which advances the D' wave 90° with respect to the U wave. The important conclusion to be drawn from these figures is that, within a total range of 360°, a given intensity may represent either of two phase shifts. Hence the optical thickness of the specimen is not indicated unequivocally by the appearance of the phase contrast image. For example, in Fig. 13-14 the condition of equal intensities of object and surround corresponds to phase shifts between the U and D waves of either 1.0λ or $\lambda/4$. Thus *intensity in the phase contrast image is not (even ideally) directly proportional to differences in optical path through the specimen.* It is for this reason that the phase microscope, as stated above, is not primarily a quantitative instrument.

In practice, bright or dark halos around the edges of particles contribute to the essentially qualitative nature of the phase contrast image. These halos, which appear in regions where there is a sharp discontinuity in refractive index, are a consequence of the inevitable imperfection of the separation of the U and D waves at the phase plate. Best phase contrast is observed for specimen points at which the D wave consists of light scattered through relatively large angles, a condition which occurs at refractive index boundaries. A cube of uniformly refractive material seen in positive phase contrast ought ideally to appear as shown in Fig. 13-15a, whereas in fact its appearance is rather more as shown in Fig. 13-15b. The ideal and observed densities measured across the image of such an object are of the form shown in Figs. 13-16a and 13-16b, respectively. The so-called *zone of action* of the phase plate is exceeded in the example

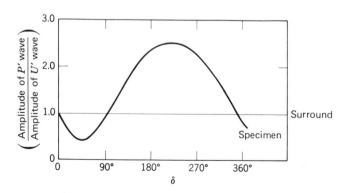

Fig. 13-13 Amplitudes of P' and U' waves.

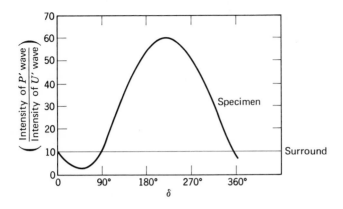

Fig. 13-14 Intensites of P' and U' waves.

shown in Figs. 13-15b and 13-16b, in which the image of the center of the refractive particle is of approximately the same intensity as the surround. The formation of such halos is not an obnoxious feature in an instrument used qualitatively for enhancement of contrast, since the halos may have the effect of highlighting details of the image.

In phase contrast microscopy, refractive index variations in the specimen are "seen," and the image must be interpreted rather differently from that of the ordinary light microscope. In fact, the refractive index of a medium is approximately proportional to the density of material in the medium. Apart from ambiguities of the type suggested by Fig. 13-15, a phase microscope thus provides an approximate map of the density of matter in the specimen. Since electron images also primarily depict

Fig. 13-15 (a) Object; (b) phase contrast image.

Fig. 13-16 (a) Densitometer tracing of object shown in Fig. 13-15a; (b) densitometer tracing of corresponding image.

specimen density, the interpretation of phase contrast images resembles that of electron micrographs more closely than does the interpretation of ordinary light microscope images.

QUANTITATIVE APPLICATIONS OF PHASE MICROSCOPY

Despite its fundamental suitability for qualitative observations, the phase contrast microscope may be applied in certain quantitative measurements, chiefly as an extension of the immersion method for determination of refractive indices (cf. 583). A protein solution, usually serum albumin of various concentrations, is used as the immersion medium for biological materials of unknown refractive index. Since proteins in aqueous solution give a *specific refractive increment* with concentration, a series of solutions of known refractive index can easily be prepared. The refractive index of an object is then measured by determining the protein concentration which causes the object to disappear from the phase microscope image. By using an immersion medium matched in refractive index to the bulk of the cytoplasm of a cell, it may be possible to view, clearly, the nucleus or other cellular organelles which otherwise may be obscured by overlying structures.

The refractive index of areas of the cell having been obtained, and a reasonable value for the specific refractive increment of cellular components assumed, the amount of matter present in various regions of intact, living, individual cells may be estimated.

SPECIAL TYPES OF PHASE CONTRAST MICROSCOPE

Most phase contrast microscopes employ fixed phase plate systems, since the optical path differences found in biological specimens are generally of the same order of magnitude, about $\frac{1}{4}\lambda$. A number of variable phase contrast systems have been developed, however; they allow variation of phase plate properties and, thus, the achievement of optimal contrast for any given specimen. One such system is the "polanret." In its simplest form, a so-called *zonal polarizer* serves as the phase plate. The zonal polarizer need introduce no differential absorption or retardation but acts to polarize light passing through the complementary area in one plane and that passing through the conjugate area in a plane perpendicular thereto. A zonal polarizer, shown in Fig. 13-17, can be constructed by piecing sheets of polaroid. In Fig. 13-17 the arrows represent the azimuths of vibration of the light transmitted by the device. The U and D waves passing through such a polarizer are polarized in mutually perpendicular directions and cannot interfere (except in the waveform sense of that term; cf. p. 79). Interference occurs when the directions of polarization are brought into complete or partial coincidence by additional polarizing devices. In the polanret instrument the analyzer is placed just in front of the image. Depending upon the orientation of the analyzer, the U- and D-wave components are transmitted to a variable extent. The orientation of the analyzer thus determines the type of phase contrast observed. For example, alignment of the direction of vibration transmitted by the analyzer with that transmitted by the con-

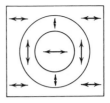

Fig. 13-17 Zonal polarizer.

jugate area of the polarizer results in complete transmission of the U wave and complete extinction of the D wave. The specimen then appears in dark or positive contrast against a bright background. Rotation of the analyzer through 90° from that position causes the D wave but not the U wave to be transmitted, and the object then appears bright against a dark surround. The intermediate orientation produces an equal relative reduction of the amplitudes of both waves and thus converts the system into an ordinary light microscope. The Polanret system is diagramed in Fig. 13-18.

Systems have also been devised for the production of phase color contrast. One approach to phase color contrast is the exploitation of the dispersive properties (cf. p. 508) of the materials used to construct the phase plate. In such a system there might be essentially no phase contrast for light of some wavelength λ, while for some other wavelength λ' the conjugate and complementary areas of the phase plate might differ in optical path by $\pm\frac{1}{4}\lambda'$. As a consequence, colored images are formed against a neutral background.

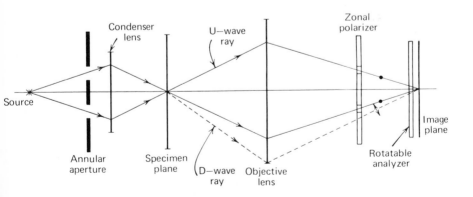

Fig. 13-18 Polanret system.

REFERENCES

H. Osterberg, in G. Oster and A. W. Pollister, Eds., *Physical Techniques in Biological Research*, Vol. I, *Optical Techniques*, Academic Press, New York, 1955, pp. 378–403. An excellent article on the theory of phase microscopy, in somewhat more detail than is given here.

R. Barer, in G. Oster and A. W. Pollister, *op. cit.*, Vol. III, *Cells and Tissues*. An equally excellent article on the theory of phase microscopy and its applications in biology.

F. Zernike, *Physica* **9**, 686, 974 (1942). "Phase Contrast, a New Method for the Microscopic Observation of Transparent Objects" (Parts I and II). These articles by the originator of the phase contrast microscope provide a clear discussion of diffraction theories of image formation and of the phase contrast principle.

A. H. B. Bennet, H. Jupnik, H. Osterberg, and O. W. Richards, *Phase Microscopy*, Wiley, New York, 1951. A classical and detailed treatise on this subject.

CHAPTER 14

The Interference Microscope

The interference microscope, like the phase contrast microscope, produces an image in which differences in optical path through the specimen appear as differences in intensity. It is also an instrument which can be used to observe living, unstained specimens. In the interference microscope, however, contrast is obtained in a manner which is adapted to direct *quantitative* measurement of differences in optical thickness of the specimen. This type of instrument is relatively unsuitable for general observations, since the field of view is normally traversed by a system of fringes. Careful, time-consuming adjustments are also required in order to obtain adequate viewing of the specimen in interference systems. In this chapter the principles of operation of several types of interference system are described, and applications of interference microscope measurements are summarized.

The discussion presented here is based upon the vectorial description of the light beams which contribute to the microscope image, as developed in Chapter 13 on pp. 289–294. The discussion of polarizing interference systems requires a general understanding of the properties of polarized light, as given in Chapter 4, while that of multiple beam interference systems is supplemented by the quantitative description of interference filters to be given in Chapter 22.

APPLICATIONS OF THE INTERFERENCE MICROSCOPE

Some quantitative applications of the phase microscope have been briefly described on p. 300. Similar measurements may be made in the interference microscope with generally greater ease and accuracy.

The optical path difference between light which has passed through a specimen point relative to that transmitted by the surround may be expressed as the *phase retardation*, Γ', such that

(14-a) $$\Gamma' = (n_{spec} - n_{med})t,$$

where t is the thickness of the specimen.

Note that Γ' is analogous to Γ, the relative retardation which arises between beams transmitted by a birefringent specimen, as defined in Chapter 6. Γ', like Γ, is a length.

Measurement of Γ' for a specimen of known thickness which is immersed in a medium of known refractive index yields the refractive index of the specimen. Thus,

(14-b) $$\boxed{n_{spec} = \frac{\Gamma'}{t} + n_{med}}.$$

The thickness of a specimen of known refractive index similarly may be obtained by measurement of Γ'.

If the thickness of a specimen of unknown refractive index cannot conveniently be measured but, if the specimen can be immersed successively in two media of (known) different refractive indices n_{med_1} and n_{med_2}, the refractive index of the specimen can be computed from the values of the two measured phase retardations Γ'_1 and Γ'_2. Thus,

(14-c) $$\boxed{n_{spec} = \Gamma'_2 \frac{n_{med_1} - n_{med_2}}{\Gamma'_2 - \Gamma'_1} n_{med_2}}.$$

The *dry weight concentration* and the *total dry weight* in an area of a specimen may be determined from measurements of Γ' if the *specific refractive increment* α is known. Although α usually cannot be measured precisely, it is found that, for most components of biological objects, a quite constant value, 1.8×10^{-3} refractive index units per gram of solute per 100 cc of solution, is appropriate. Thus the dry weight concentration C (expressed as grams dry weight per 100 cc volume) in an area of a specimen is as given by

(14-d) $$\boxed{C = \frac{n_{spec} - n_{H_2O}}{\alpha} = \frac{n_{spec} - 1.33}{1.8 \times 10^{-3}}},$$

noting that

(14-e) $$C = \frac{\text{g solids}}{100 \text{ cm}^3} = \frac{100 \times \text{g solids/cell}}{\text{area of cell} \times \text{thickness of cell}},$$

while

(14-f)
$$\Gamma' = \alpha C t = \frac{\alpha \times \text{g solids/cell} \times t}{\text{area of cell}},$$

where t is the thickness of the cell. The weight of solids per cell is then obtained as

(14-g)
$$\boxed{\text{g solids} = \frac{\Gamma' \times \text{cell area}}{100\alpha} = \frac{\Gamma' \times \text{area}}{0.18}}.$$

PRINCIPLE OF THE DOUBLE BEAM INTERFERENCE MICROSCOPE

Whereas, in the phase contrast microscope, contrast arises from interference between the deviated and undeviated portions of the light beam leaving the specimen, in the double beam interference system, interference between light from the specimen and a *reference* beam results in image contrast. Since the reference beam does not pass through the specimen at all, the physical separation of the two beams can be complete. Thus interference microscope images can be freed of halos and other defects which occur in phase contrast images as a consequence of the incomplete separation of deviated and undeviated light.

A numerical example may be considered in terms of the vectorial representation of phases and amplitudes developed in Chapter 13. (See, especially, pp. 289–294.) In Fig. 14-1 undeviated, particle, and reference waves (designated U, P, and R waves, respectively) are each of amplitude A_0. The R wave is shifted in phase by 90° and the P wave by 45° with respect to the U wave. As shown in Fig. 14-1a, the resultant of interference between the undeviated wave (from a background point) and the reference wave is the U' wave. The U' wave is of amplitude $\sqrt{2}$ $A_0 \cong 1.4 A_0$. Similarly, as shown in Fig. 14-1b, the resultant of interference between the particle wave (from a specimen point) is the P' wave. The amplitude of the P' wave (as found by application of the law of cosines) is approximately $1.8 A_0$. The specimen point thus appears in bright (negative) contrast with respect to its surround, the ratio of intensities of the two being $(1.8/1.4)^2$, or approximately $1.6\times$.

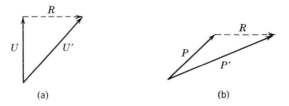

Fig. 14-1 Undeviated, particle, and reference wave vectors.

Choice of a different amplitude and phase for the R wave can produce dark (positive) contrasting of the specimen point. An example is considered in Fig. 14-2, where again the U and P waves are of equal amplitude A_0 but differ in phase by 45°. A reference beam of amplitude $\frac{1}{2}A_0$ leads the U wave in phase by 135° and is thus 180° out of phase with the P wave. The resultant amplitude of the U' wave is approximately $0.73A_0$ (Fig. 14-2a), while that of the P' wave is $0.5\,A_0$ (Fig. 14-2b). The relative intensity of the image of the specimen is thus $(0.50/0.73)^2$, or somewhat less than half that of the surround.

It should be clear that Figs. 14-1 and 14-2 illustrate but two of an infinite number of possible examples of interference with a reference beam.

For interference contrast to result from the overlapping of specimen and reference beams, the two beams must of course be *coherent*. (Note that this requirement need not be specified for phase microscopy, where interference occurs between portions of waves which, having arisen from the same point on the incident wavefront, are inevitably coherent.) All designs of interference microscope thus incorporate a device for splitting light from a source into two or more beams, a means of making fine adjustment in the relative phases of specimen and reference beams, and provision for the reunion of the beams at an appropriate position in the microscope system.

THE DYSON INTERFERENCE MICROSCOPE

The Dyson interference microscope uses a reflection lens system to form an image of the object below the objective lens. Reference and specimen beams unite at this "primary image." The subsequent lens system is essentially that of an ordinary light microscope.

The reflection lens system is diagramed in Fig. 14-3. Rays which represent the beam passing through the specimen are shown in the figure by dotted lines; rays which represent the reference beam are shown by solid lines. As shown, ray ① leaves the condenser lens in the direction of the specimen and primary image but is blocked by a fully silvered spot

(a)   (b)

Fig. 14-2.

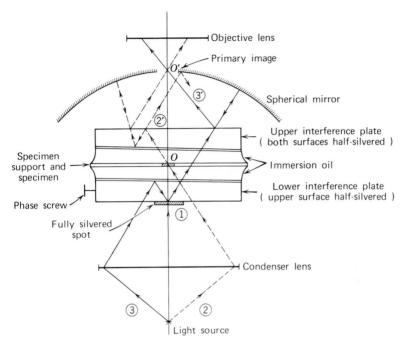

Fig. 14-3 Reflection lens system of the Dyson interference microscope.

(i.e., by a mirror) placed centrally on the bottom of the lower of two "interference plates." The function of the fully silvered spot is to prevent all *direct* illumination from reaching the plane of the primary image.

Rays ② and ③ leave the condenser lens and pass through the lower interference plate, proceeding in the direction of the specimen. (Note that in Fig. 14-3 refraction at the surface of the interference plates is ignored for the sake of clarity; the system is in fact so constructed as to compensate for refractive effects.) The upper surface of the lower interference plate is half-silvered; that is, it is coated lightly with reflecting material in such a manner that some (definite) fraction of incident intensity is transmitted and the remainder is reflected. Ray ② represents intensity which is transmitted by this half-silvered surface; ray ③ represents reflected intensity. Note that light traveling from the condenser lens in the direction of, for example, ray ② is split at the surface into two rays, one represented by the continuing course of ray ② and the other (not shown in Fig. 14-3) proceeding similarly to ray ③.

Beyond the lower plate, ray ② travels (through immersion oil matched in refractive index to that of the plates and specimen support slide) to and through the specimen, and thence to the upper plate, both surfaces of which are half-silvered. Ray ② is then successively transmitted by the lower surface of the upper plate, reflected by the upper surface, reflected by the lower surface, and transmitted by the upper surface. The ray is then incident upon the totally reflecting spherical mirror, from which it is reflected, via the upper surface of the upper half-silvered plate, to the primary image at O'. Note that components of the light beam from the specimen, such as ray ②, which follow a course different from ray ② at the half-silvered surfaces, are reflected in such a manner as to be excluded from the primary image. Note also that the system is so designed that the fully silvered spot lies at the center of curvature of the spherical mirror, so that rays reaching the spherical mirror are incident approximately along the normal to the surface of the mirror (i.e., along a radius) and hence are reflected approximately along the incident path, as shown for rays ② and ③.

Ray ③ is reflected at the upper surface of the lower interference plate, and again by the fully silvered spot before leaving the lower plate. It passes through immersion medium, specimen support, and both surfaces of the upper plate, reaching the spherical mirror, from which it is reflected, via the uppermost half-silvered surface, to the primary image at O'. Rays which undergo additional reflections at the upper plate travel paths which result in their exclusion from the primary image.

The primary image O' is thus formed by combination of rays such as ②, which have passed through the specimen, and rays such as ③ which have traversed only the surround. Clearly, components such as ② constitute the particle wave, while those such as ③ constitute the reference wave. Apart from any delay incurred at the specimen, rays ② and ③ differ in phase at O' if the additional optical path incurred by ray ② in making a double passage through the upper plate differs from the additional optical path incurred by ray ③ in making a double passage through the lower plate. The optical paths of the two beams may be adjusted in a controlled manner by constructing the interference plates as *wedges* and providing a means for their relative translation (i.e., movement in a direction perpendicular to the microscope axis). The wedging of the plates is illustrated in a much exaggerated manner in Fig. 14-3; in fact, the angle of the wedges is only about five minutes of arc, so that the variation from plane slabs is imperceptible on direct inspection. As indicated in the figure, the lower plate may be moved mechanically with respect to the upper by means of the so-called *phase screw*.

At some setting of the phase screw the total thickness through the two plates is equal for all zones of the field, and the optical paths of the particle and reference waves (rays ② and ③, respectively, in Fig. 14-3) are identical except for any optical path difference incurred by passage through the specimen. This setting of the phase screw is called the *white light* position, since the particle and reference beams from all portions of the background reunite in phase at the primary image, so that the specimen is observed against a uniformly bright background.

When the phase screw is turned away from the white light position, a series of fringes appears across the field of view. The fringes are alternately bright and dark if illumination is monochromatic, and colored if the system is illuminated by white light. These fringes represent progressive differences in optical path between the particle and reference waves across the field of view. Whenever a refractive object occurs in the field of view, the fringe system is displaced as a consequence of the corresponding variations in the optical path of the particle wave. Further discussion of the appearance of interference images is given on p. 314. Note that the outer zones of the field of view must be free of refractive material in order to avoid variations in the optical path of the reference wave.

Interference microscopes of the Dyson type require rather demanding and tedious adjustments; consequently, polarizing interference systems have tended to come into wider use.

POLARIZING INTERFERENCE MICROSCOPES

In polarizing interference microscopes, light which is plane polarized in mutually perpendicular azimuths is transmitted at *physically separate* positions of the specimen plane. Passage through specimen and surround, respectively, imposes a phase difference between the two. When the beams are caused to reunite (by a polarizing device which acts "in reverse"), the resultant vibration is, in general, elliptical in form. Differences in intensity are then observed by interposing a rotatable analyzer. Contrast is varied by illuminating the system with light from a rotatable polarizer; the relative intensities of specimen and reference beams may then be adjusted.

A polarizing interference system is illustrated in Fig. 14-4. This microscope is similar in general plan to that shown in Fig. 14-3, except that the interference wedges are replaced by birefringent plates. The plates are so cut that the optic axis lies in the plane of the plates, but the optic axes of the two lie in mutually perpendicular directions. In the instrument, light from the source is converged toward the specimen by the condenser lens, and forms separate O and E rays at the lower surface of plate A. (The plates shown in Fig. 14-4 are constructed from a negatively birefringent material, but positively birefringent plates could also be used.)

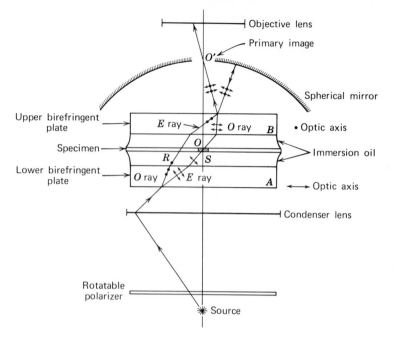

Fig. 14-4 Reflection lens system of the Polarizing interference microscope.

The space between plates and specimen is filled with an oil of refractive index intermediate between n_0 and n_e. Since the optic axis of plate B is rotated through 90° with respect to that of plate A, the O ray from plate A becomes an E ray in plate B, and vice versa. Consequently, the difference in refractive indices with respect to the two is *reversed* at the second birefringent plate. If A and B are of identical thickness, the two rays therefore reunite at the upper surface of plate B. From there they are reflected by the spherical mirror to the primary image at O'.

Interference contrast is not observable at O' but is produced by a rotatable analyzer placed at some subsequent position (in practice, at the eyepiece lens of the microscope). A numerical example may be considered in which the R and S beams are each of amplitude $\frac{1}{2}A_0$, the azimuth of the analyzer is set at 45° to each of these beams, and the path difference through O, relative to the surround, is $\lambda/8$.

Whatever the phase angle between the R and S beams, the analyzer transmits an amplitude of $(\sqrt{2}A_0/2)$ from each [cf. (15-b)]. The resultant intensity is determined by the relative phases of the two amplitudes. This may be computed with the use of the equation derived on p. 324.

(15-l) $$I = I_0[1 + \cos(\delta + \pi)],$$

where δ is the phase shift imposed by the specimen, and the term π represents an additional phase shift imposed at the analyzer.

If *both* beams traverse only the surround, $\delta = 0$, and

(14-h) $$I = I_0(1 + \cos \pi) = 0.$$

If the S beam traverses the specimen, δ (in this example) is 45°, so that

(14-i) $$I = I_0(1 + \cos 45°) = I_0\left(1 + \frac{\sqrt{2}}{2}\right) = 1.707I_0.$$

Thus the specimen appears bright against a dark surround.

Since the beam which traverses the specimen area is plane polarized, the phase shift with respect to the reference beam and thus the resultant intensity in the image are affected by any birefringence of the specimen. Rotation of the specimen reveals birefringence as variations in image intensity. In general, however, retardations originating from birefringence are quantitatively much less important than retardations arising from differences in refractive index between specimen and surround.

MULTIPLE BEAM INTERFEROMETRY

In multiple beam interference systems the specimen is placed between two surfaces of high reflectivity. In biological applications, where toxicity of the reflecting surface is a consideration, reflectances of the order of 0.90 are feasible. The light beam is reflected repeatedly between the two surfaces, so that components of the transmitted beam have undergone different numbers of reflections and differ from each other successively in physical path by twice the width of the gap between the surfaces. Differences in optical path between components of the transmitted beam are then determined by the refractive index of the material within the gap and by the effect of the phase changes which occur upon reflection.

Figure 14-5 illustrates the oblique illumination of a pair of parallel plates, the inner surfaces of which are of high reflectance. (Refraction of the beam upon entering and leaving the gap is ignored in Fig. 14-5 for

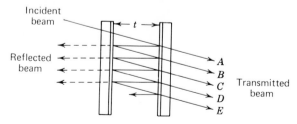

Fig. 14-5 Reflection and transmission from parallel plates.

the sake of clarity.) Note that, if the incident beam were *normal* to the pair of planes, the transmitted components *A*, *B*, *C*, . . . would all superpose.

For light incident normally, successive transmitted rays differ in physical path by the distance $2t$, where t is the separation of the plates. (Phase changes of 180° may also occur upon reflection at one or both sides, depending upon the refractive index of the reflecting surfaces and of the medium between them. As stated on p. 46, a phase change of π radians = 180° accompanies reflection from a medium of *higher* refractive index.) The beam transmitted by the plates accordingly is bright or dark as a function of the thickness t between a given set of plates. Note that an increase of plate separation such as to increase optical path by $\frac{1}{2}\lambda$ produces *no change* in the observed intensity of the transmitted beam, since the phase of each component ray is thereby increased by 2π, leaving the original phase relationships unaltered. The variation of transmitted intensity as a function of wavelength is considered quantitatively in the discussion of the interference filter in Chapter 22.

If a pair of plates is mutually inclined, as shown in exaggerated fashion by Fig. 14-6, t is no longer constant but increases linearly with y, while the optical path through the system increases at a corresponding rate. For a fixed refractive index of the medium between the plates there are successive positions of destructive interference along the y direction, with intermediate positions of maximum brightness. When the reflectance of the plate surfaces is high and illumination is by monochromatic light, a very sharp series of fringes is obtained.

The intensity pattern of the light transmitted by plates such as those illustrated in Fig. 14-6 might be as diagramed in Fig. 14-7a. The optical path between successive reflections could be diminished either by decreasing the refractive index of the medium between the plates or by moving the plates closer together. Either of these variations has the effect that the positions in which path lengths differ by one wavelength are

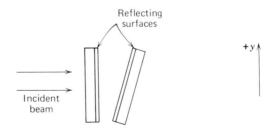

Fig. 14-6.

moved farther apart. The resulting fringe system might be as shown in Fig. 14-7b. Similarly, an increase in the refractive index of the medium between the plates, or an increase in the separation of the plates, produces an increase in the optical path difference between successive reflections. The consequent shift of fringes is as diagramed by Fig. 14-7c.

A refractive object placed between the inclined reflecting planes alters the path difference between successive reflections. Hence the distribution of fringes is changed in a manner which is related to the thickness and refractive index of the object.

Figure 14-8 illustrates the case of an object which is a *cube* of uniformly refractive material of higher index than its surround, and of such a thickness that the optical path through the cube is *one quarter wavelength* greater than that of the surround. Figure 14-8a illustrates the position of the cubic object between the inclined plates, and Fig. 14-8b shows the distribution of intensities in the image. The effect of the particle is a compression of the fringe system. Similarly, in the manner illustrated by Fig. 14-9, a particle which is of lower refractive index than its surround spreads out the fringe system. The particle considered in Fig. 14-9 is a cubic object of dimensions such that the optical path is one quarter wavelength less than that through the surround.

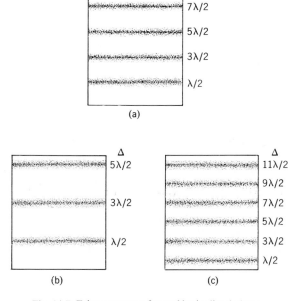

Fig. 14-7 Fringe systems formed by inclined plates.

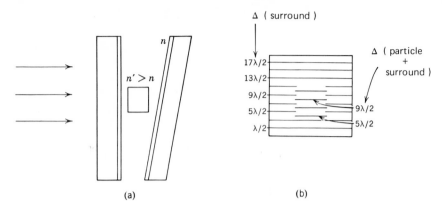

Fig. 14-8 Imaging by specimen placed between inclined plates; index of specimen greater than that of surround: (a) arrangement of specimen; (b) image.

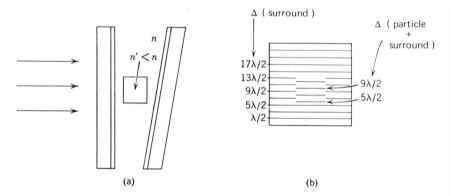

Fig. 14-9 Imaging of specimen placed between inclined plates; index of specimen less than that of surround: (a) arrangement of specimen; (b) image.

THE MULTIPLE BEAM INTERFERENCE MICROSCOPE

The patterns shown in Figs. 14-8b and 14-9b may be magnified by a compound microscope system. The optics of such a multiple beam interference microscope are diagramed in Fig. 14-10. As shown, the effective size of the source is reduced by interposing a pinhole aperture, thereby providing effectively coherent light with consequent production of sharp fringes. The condenser lens serves to collimate the incident beam.

THE INTERFERENCE MICROSCOPE IMAGE

Interference images of an idealized cubic object have been shown in

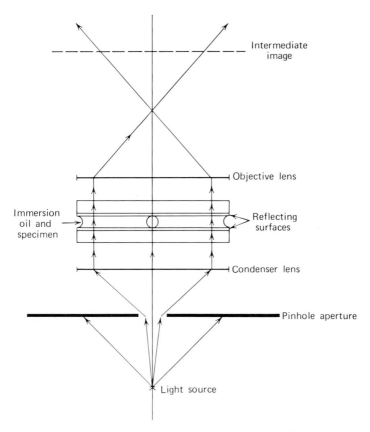

Fig. 14-10 Multiple beam interference microscope.

Figs. 14-8b and 14-9b. Biological specimens, however, generally lack such abrupt contours but are rounded in form. The appearance of the image of a *geometrically convex*, approximately egg-shaped object is considered in Fig. 14-11. If the refractive index of such a particle is greater than that of the surround, the particle is *optically convex* and its image has the form shown in Fig. 14-11a. The image of an *optically concave* particle (refractive index less than that of the surround) is diagramed in Fig. 14-11b. The refractive index of the particles shown in Fig. 14-11 is taken to be uniform, while their thickness increases uniformly from zero at the edge to some maximum value at the center. The thickness at the center of each particle corresponds to a path difference of *one wavelength* (i.e., between specimen and reference beams of a double beam instrument or between successive reflections in a multiple beam

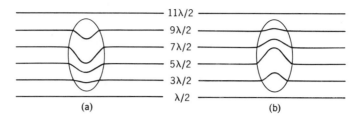

Fig. 14-11 Interference microscope images of egg-shaped objects: (a) index of specimen greater than that of surround; (b) index of specimen less than that of surround.

system). This implies a differential optical thickness of one wavelength in the case of observation in a double beam instrument, but one half wavelength in the case of observation in a multiple beam instrument.

In Fig. 14-11 the spacing of fringes is "advanced" or "retarded" by one full fringe only at the exact center of the particles, and the fringes are continuous and curved. The optical thickness of any point on the particle can be determined, at least in principle, by measuring the extent of fringe shift. (In practice, the finite width of fringes limits the precision of such measurements.) If the geometrical thickness of the particle is known from direct measurement, the refractive index of any portion of the particle can then be deduced. As noted on p. 304 the refractive index of a particle, such as a cell or subcellular inclusion, can be related to its weight.

COMPARISON OF DOUBLE AND MULTIPLE BEAM INTERFERENCE MICROSCOPES

Although the images formed by double and multiple beam interference systems are similar in appearance, the range of applicability is different for the two types of microscope. Fringes formed by the interference of two beams, only, are inevitably less sharp than those formed by the interference of many beams (cf. p. 59 ff.). The multiple beam interference microscope is thus more suitable for optical path measurements of extremely high precision. Differences in optical path must be at least 250 Å for measurement in double beam systems, whereas multiple beam instruments have been used to measure differences as small as 10 Å. Furthermore, the precision of double beam instruments is reduced if material at the periphery of the specimen changes the optical path of the reference beam.

In spite of these limitations, however, the double beam interference microscope is generally preferable for biological applications on account of its higher resolving power. In order to obtain sharp fringes, the aperture of multiple beam microscopes is severely restricted; useful numerical apertures cannot be much greater than about 0.1. The Dyson type of microscope, on the other hand, may have a numerical aperture (with oil immersion) as large as about 1.35, corresponding to resolving powers of the order of 0.4 μ.

PHASE MODULATED INTERFERENCE MICROSCOPY

The quantitative characterization of biological systems by interference microscopy, that is, the application of (14-b), (14-c), (14-d), and (14-g) is limited in practice by the difficulty of selecting fringe positions accurately. The reproducibility of visual selection of positions of minimum intensity is poor. Photographic measurements are somewhat more reliable but, at best, are time-consuming. Systems are currently undergoing development in which the limitations of convenience, accuracy, and reproducibility are much reduced by automatic electronic selection of match points. A microscope has been described in which the refractive index, bire-fringence and/or optical rotation of a specimen may be measured by a scanning system. The basic optical principles used in the design of such instruments are those of conventional interference microscopes, polarizing microscopes, or polarimeters. However, a continuously variable phase retardation (phase modulation) is imposed upon the signal produced by the object. The variable phase retardation originates from birefringence which is induced (in certain crystals) by applying a voltage parallel to the optic axis. (The observation of induced birefringence is known as the *Pockels effect*.)

REFERENCES

H. Osterberg, in G. Oster and A. W. Pollister, Eds. *Physical Techniques in Biological Research*, Vol. I, *Optical Techniques*, Academic Press, New York, 1955, pp. 404–437.

R. D. Allen, J. W. Brault, and R. M. Zeh, in R. Barer and V. E. Cosslett, Eds., *Advances in Optical and Electron Microscopy*, Vol. I, Academic Press, New York, 1966, pp. 77–114, "Image Contrast and Phase-modulated Light Methods in Polarization and Interference Microscopy."

CHAPTER 15

The Polarizing
Microscope

The polarizing microscope is used to reveal the birefringence of specimens. With the use of appropriate accessories the type of birefringence (form, intrinsic, etc.; cf. p. 109) may be determined, and the shape and orientation of the index ellipsoid (cf. p. 100) may be established. These measurements may be carried out on unstained or even living material.

Since the birefringence of materials is related to specimen structure at the molecular and macromolecular level, information obtained from polarization microscopy can be used to deduce the macromolecular structure of specimens, even though ultrastructure is not resolved directly. In more qualitative applications the instrument is used simply to locate structures of characteristic birefringence (e.g., fibers of striated muscle) and to observe their behavior.

A complete interpretation of polarization microscope images requires an understanding of the properties of polarized light beams and of the properties of birefringent materials, as presented in Chapters 4 and 6, respectively. Unfortunately, these topics seem almost invariably to present considerable difficulty to the biologist! While it must be emphasized that there is no "very simple explanation" for the effects observed in the polarizing microscope, these effects can be explained in an entirely straightforward manner. If the topics of polarization and birefringence are mastered, the properties of the polarizing microscope can easily be understood. Careful distinction between such terms as *direction of propagation*, *azimuth of polarization*, and *phase angle* is necessary, however, as is an appreciation of the significance of "circularly polarized light", "elliptically polarized light", and related concepts.

318

LAW OF MALUS; POSITIONS OF EXTINCTION

Figure 15-1 represents, vectorially, the interaction of a plane-polarized light beam, of amplitude A_0, with an analyzer so oriented as to transmit light vibrating in an azimuth which makes angle θ with the azimuth of vibration of the incident ray. (Note that it is often convenient to refer to such an azimuth of vibration as "the direction of the analyzer." etc. This phrase is in fact an abbreviation for "the azimuth of polarization of light transmitted by the analyzer" and must not be confused with references to directions of propagation.) As the geometry of Fig. 15-1 shows, the amplitude A which is transmitted by the analyzer is

(15-a) $A = A_0 \cos \theta.$

Since the square of amplitude is proportional to intensity, (15-a) may be squared to give

(15-b) $I = I_0 \cos^2 \theta.$

Equation 15-b, which describes the relative intensities of the incident and transmitted beams, is known as the *law of Malus.*

A polarizer transmits only that component of incident light which vibrates in a given azimuth, and light which vibrates in the plane at right angles thereto is excluded. (It is of course important to understand that, as developed in Chapter 3, a vibration in any azimuth may be analyzed as a pair of vibrations in any two azimuths which are mutually perpendicular, and perpendicular to the direction of propagation. See also the discussion of transmission of plane-polarized light given on p. 79.). If the analyzer is parallel to the polarizer, that is, if the angle θ between the transmission directions of polarizer and analyzer is 0°, then $\cos \theta = 1$, and $I = I_0$. That is, all the light passed by the polarizer is subsequently transmitted by the analyzer. Alternatively, if the transmission azimuths of polarizer and analyzer are set at 90°, then $\cos \theta = 0$ and none of the intensity passed by the polarizer is transmitted by the analyzer. In this position, polarizer and analyzer are said to be "crossed." At intermediate mutual orientations, intermediate levels of intensity are transmitted, as determined by (15-b).

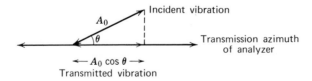

Fig. 15-1.

A consequence of these considerations is that, in varying the mutual orientation of a polarizer and analyzer through 360°, two positions of maximum brightness and two positions of darkness are obtained. The latter are termed *positions of extinction.* Similarly, a specimen so oriented between polarizer and analyzer as to form an image of negligible intensity is said to undergo extinction or to be in a position of extinction.

OBSERVATION OF BIREFRINGENT SPECIMENS BETWEEN CROSSED POLARIZER AND ANALYZER

If a polarizer and analyzer are mutually oriented in a position of extinction, light is transmitted by the analyzer only if a specimen placed in the space between the polarizer and analyzer has the effect of changing the azimuth of vibration of the light which it transmits. The azimuth of polarization of light from the polarizer is thereby shifted into an azimuth which is partially transmitted by the analyzer. A suitably oriented birefringent object has this effect; hence the incorporation of a polarizer and analyzer above and below the specimen plane of a microscope makes possible the observation of contrast in birefringent specimens. It is instructive to consider the effect of some specific orientations of the birefringent specimen.

A birefringent specimen might be oriented with its optic axis parallel to the axis of the microscope and at any arbitrary angle in the direction perpendicular to its optic axis. In this special case the two mutually perpendicularly polarized beams into which light is resolved within the specimen travel with equal velocity (cf. p. 102) and thus emerge from the specimen *in* phase. As shown on p. 81, the resultant of two in-phase mutually perpendicularly polarized components of vibration is itself a plane-polarized vibration. Thus the light beam, after passing through the specimen, is in the same state of polarization as before entering the specimen, and is not transmitted by the analyzer. The azimuths of vibration in this case are illustrated in Fig. 15-2. It is clear that the birefringence of the specimen is not revealed when it is viewed along its optic axis; that is, the specimen appears to be isotropic in this orientation.

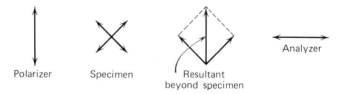

Polarizer   Specimen   Resultant
beyond specimen   Analyzer

Fig. 15-2 Azimuths of vibration.

In polarization microscopy it is usually desirable to view the specimen in directions which are perpendicular to the optic axis (i.e., the microscope axis and the optic axis are perpendicular). As explained on p. 108, the ordinary and extraordinary rays (i.e., the two mutually perpendicularly polarized components of vibration within the specimen) are then still coincident, but travel at different velocities, the ordinary ray at velocity c/n_o and the extraordinary ray at velocity c/n_e. Consequently the two components of vibration are shifted in phase when they emerge from the specimen.

Two sets of factors determine the intensity of the image of a specimen which is viewed between crossed polarizers and perpendicular to its optic axis. They are: the orientation of the azimuths of vibration within the specimen relative to those of the polarizer and analyzer (this, in turn, is determined by the physical orientation of the specimen); and the extent of retardation by the specimen (i.e., the phase difference between the mutually perpendicularly polarized rays which emerge from the specimen). The retardation is determined by the value of the birefringence B ($= n_e - n_o$) of the specimen and by its thickness. Confusion arises if it is not clear that both orientation and thickness affect the resultant intensity of the image.

EFFECT OF VARIATION OF THE ORIENTATION OF THE SPECIMEN

The effect of rotating a specimen of a given thickness and birefringence about a direction perpendicular to its optic axis may be considered. (As shown on p. 323 these variations are greatest if, for any given wavelength, retardation by the specimen is equal to half a wavelength.)

Light from a polarizer, of amplitude A_0, might vibrate in the azimuth represented in Fig. 15-3a and be incident upon a specimen in which the vibration azimuths of the O and E rays are as represented by Fig. 15-3b. If the azimuths of polarization of the incident light and that of the O ray are separated by any angle $\theta°$, the azimuth of the E ray must make an angle of $(90 - \theta)°$ with that of the polarizer. The amplitudes A_o and A_E of the O and E rays within the specimen are thus, as shown by Figs. 15-3c and 15-3d, respectively

(15-c) $A_O = A_0 \cos \theta$ and $A_E = A_0 \cos (90 - \theta) = A_0 \sin \theta.$

The analyzer, which is crossed with respect to the polarizer, transmits light which is polarized in the azimuth represented by Fig. 15-3e. The O- and E-ray vibration azimuth lie at $(90 - \theta)°$ and $\theta°$, respectively, with respect to that azimuth, as shown in Figs. 15-3f and 15-3g. Thus the

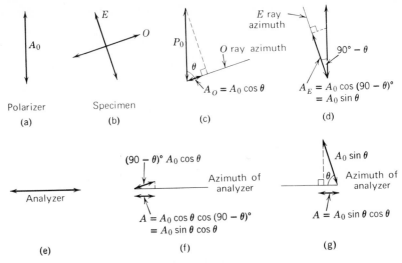

Fig. 15-3 Transmission of amplitude from a birefringent object.

amplitudes from the O and E rays passed by the analyzer are

(15-d) $\quad A = (A_0 \cos \theta) \cos (90 - \theta) = A_0 \cos \theta \sin \theta \quad$ for the O ray

and

(15-e) $\qquad\qquad A = (A_0 \sin \theta) \cos \theta \quad$ for the E ray.

Thus the O and E rays contribute equally to the amplitude passed by the analyzer, regardless of the orientation of the specimen. The total amplitude passed is the sum of the contributions by O and E rays and is thus

(15-f) $\qquad\qquad A = 2A_0 \sin \theta \cos \theta,$

while the transmitted intensity is

(15-g) $\qquad\qquad \boxed{I = 4I_0 \sin^2 \theta \cos^2 \theta}.$

Thus the transmitted intensity is zero wherever either $\sin \theta$ or $\cos \theta = 0$; that is, for $\theta = 0°, 90°, 180°, \ldots$. These are the positions in which the vibration directions of the specimen are parallel to those of polarizer and analyzer.

Transmitted intensity is maximal when $\sin \theta = \cos \theta$, as is the case for $\theta = 45°$. Then,

(15-h) $\qquad\qquad I = 4I_0 \left(\frac{1}{\sqrt{2}}\right)^2 \left(\frac{1}{\sqrt{2}}\right)^2 = I_0.$

These considerations show that rotation of a birefringent specimen between crossed polarizers produces *four positions of extinction* in the course of rotation through 360°. The positions of extinction correspond to $\theta = 0°, 90°, 180°$, and 270°, while four corresponding positions of maximum brightness occur for $\theta = 45°, 135°, 225°$, and 315°. Relative intensities at other orientations may be computed from (15-g). Note the difference between these positions of extinction and those of a mutually rotated polarizer and analyzer, which occur only twice during a rotation of 360°.

EFFECTS OF RETARDATION IN THE SPECIMEN

The effects of varying the retardation by a specimen which remains in a given orientation are now considered. (As shown above, maximum intensity is obtained in the image of a specimen of any given retardation when its vibration azimuths are oriented at 45° to those of the crossed polarizer and analyzer.) Monochromatic illumination is assumed.

O and E rays travel through the specimen at different velocities, so that their relative phase shifts continuously. The net phase shift upon emergence is, as given in Chapter 6,

(6-d)
$$\delta = \frac{2\pi\Gamma}{\lambda} = \frac{2\pi t(n_e - n_o)}{\lambda}.$$

In the absence of absorption the intensity of light leaving the specimen is the same as that incident upon it, since the O and E rays are polarized in different azimuths and thus, regardless of their relative phase, do not interfere in the sense of producing differences in intensity. (The significance of the term "interference" as applied to polarized light has been discussed on p. 79.)

At the analyzer a component of vibration from both O and E waves is transmitted. Each then vibrates in the *same* azimuth—that of the analyzer—and the two can interfere. The intensity observed beyond the analyzer is thus determined by the value of Γ.

The intensity of the image may be computed for any value of Γ. As derived on pp. 36–38, the interference of two beams of amplitudes A_1 and A_2 and of phases α_1 and α_2, respectively, yields a resultant amplitude A such that

(15-i)
$$A^2 = A_1^2 + A_2^2 + A_1 A_2 \cos(\alpha_1 - \alpha_2).$$

For a specimen oriented with its transmission azimuths at 45° to those of the polarizer and analyzer, and in which there is no dichroic absorption, $A_1 = A_2 = A_0$. Immediately beyond the specimen, the phase shift $(\alpha_1 - \alpha_2)$ between these amplitudes is the quantity, δ, as given by (6-d). However, as explained on pp. 94–96, the two beams undergo an additional mutual phase shift of 180° upon passage by the analyzer. Thus, beyond the analyzer,

(15-j) $$(\alpha_1 - \alpha_2) = \delta + \pi.$$

Accordingly (15-i) becomes

(15-k) $$A^2 = 2A_0^2 + 2A_0^2 \cos(\delta + \pi),$$

or, since I_0 is proportional to $2A_0^2$, the resultant intensity is

(15-l)
$$\boxed{\begin{aligned} I &= I_0[1 + \cos(\delta + \pi)] = I_0\left(1 + \cos\frac{2\pi\Gamma}{\lambda + \pi}\right) \\ &= I_0\left(1 + \cos\frac{2\pi t(n_e - n_0)}{\lambda + \pi}\right) \end{aligned}}.$$

The resultant amplitudes and intensities obtained for a series of values of δ, as computed from (15-l) are given in Table 15-1. (Note that the observed intensity before the light passes through the analyzer is always $I_0 = 2A_0^2$.)

The variation of the vibrations leaving the specimen may also be described as a variation of the *form* of the resultant wave motion (cf. Chapter 4). This is, in general, *elliptical* but becomes plane or circular at certain specific values of Γ. Representative waveforms are shown in Fig. 15-4, together with the graphical projection of these motions in the plane which is transmitted by the analyzer. Note that the lengths of these projections are proportional to the amplitudes computed in Table 15-1.

The *senses* of the elliptical and circular motions shown in Fig. 15-4

Table 15-1

δ	Amplitude Passed by Analyzer	Intensity Passed by Analyzer
0°	0.000 ($\times 2A_0$)	0.000I_0
45° = $\lambda/8$	0.270	0.073
90° = $\lambda/4$	0.500	0.250
135° = $3\lambda/8$	0.625	0.425
180° = $\lambda/2$	0.707	0.500
225° = $5\lambda/8$	0.625	0.425
270° = $3\lambda/4$	0.500	0.250
315° = $7\lambda/8$	0.270	0.073
360° = λ	0.000	0.000

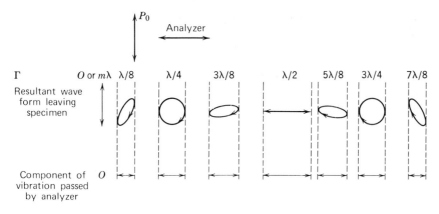

Fig. 15-4 Variation, with Γ, of resultant waveforms leaving the specimen.

apply to the case in which the fast ray in the birefringent specimen (i.e., the ray for which the refractive index is lower; this is the O ray for a positively birefringent specimen and the E ray for a negatively birefringent specimen) makes an angle of $+45°$ with the vibration azimuth of the *polarizer*. If the fast ray makes an angle of $-45°$ with the azimuth of the polarizer, the motions are of opposite sense (cf. p. 87).

In summary, the intensity of a birefringent specimen viewed perpendicular to its optic axis between crossed polarizers in monochromatic light varies cyclically with the rotational orientation of the specimen and according to the thickness (i.e., the relative retardation of the slow ray) of the specimen. Observed variations in intensity are described quantitatively by (15-m), in which the considerations of (15-g) and (15-l) are combined:

(15-m)
$$I = I_0[1 + \cos(\delta + \pi)]\sin^2\theta\cos^2\theta,$$

where, as before, δ is the phase shift between fast and slow beams, and θ is the angle between the vibration azimuths of polarizer and analyzer and those of the specimen.

If a birefringent specimen is oriented in a position of extinction (i.e., so oriented that the vibration azimuths of the specimen coincide with those of the polarizer and analyzer), or if the specimen is of such a thickness that the slow and fast rays emerge from the specimen *in phase*, no illumination is produced between crossed polarizers. Thus, although the observation of brightness between crossed polarizers is an indication of birefringence, it is not a necessary condition, since by chance the orientation or thickness of the specimen may be such as to produce extinction. The birefringence of biological specimens is usually small, however, so that specimens of a thickness suitable for microscopic examination are unlikely to produce retardations approaching one wavelength.

Some intensity may be observed between crossed polarizers as a result of polarization by the scattering of light from isotropic specimens. This condition may easily be distinguished from the effect of birefringence, since successive positions of extinction and brightness are not observed as the specimen is rotated. Note, however, that a specimen need not consist of a single birefringent unit. A randomly oriented group of crystallites, for example, may show brightness in every orientation, but individual members of the group may display characteristic positions of extinction. In the limit a specimen containing submicroscopic randomly oriented birefringent structures is indistinguishable from an isotropic specimen which polarizes by scattering.

RETARDATION COLORS

When a birefringent specimen observed at right angles to its optic axis is illuminated by *white* light (as opposed to monochromatic light), the retardation Γ represents a *different* phase shift δ for light of each wavelength. Consequently the image is colored. For any value of Γ, light for which $\Gamma = \lambda, 2\lambda, \ldots$ is removed from the image, and the observed color tends to be the complementary one. (The color actually observed is determined by the wavelength distribution of the source, however.) Conversely, if polarizer and analyzer are parallel rather than crossed, wavelengths for which $\Gamma = \lambda/2, 3\lambda/2, \ldots$ are removed from the image and correspondingly different colors are observed which, again, tend to be complementary to the one removed. A specimen for which $\Gamma = 600$ mμ may be considered as a specific example: Between crossed polarizers, *yellow-orange* light of $\lambda = 600$ mμ is *removed* from the beam, and the image tends to appear *blue-violet*. Between parallel polarizers, *violet* light of $\lambda = 400$ mμ ($\Gamma = 3\lambda/2$) is removed from the beam, and the image tends to appear *yellow*.

Table 15-2 lists the colors observed upon illumination by sunlight for a number of values of the retardation.

The table shows that colors recur in spectral sequence. Thus, for example, between crossed polarizers a "first order" yellow is observed at retardations of about 300 mμ, a "second order" yellow at retardations of about 900 mμ (and a third order at about 1500 mμ). The higher order colors are of successively lower spectral purity.

Note that, for retardations less than about 200 mμ, Γ is less than $\frac{1}{2}\lambda$ for all visible wavelengths. The effect of a birefringent specimen of corresponding thickness is thus to shift the azimuth of vibration of light of

Table 15-2   Retardation Colors

Retardation (Γ)	Crossed Polarizers	Parallel Polarizers
0	Black	White
100	Blue-gray	Yellowish white
200	Clear gray	Brownish yellow
280	Yellow-white	Violet
305	Light yellow	Blue-violet
330	Bright yellow	Blue
500	Orange	Blue-green
550	Red	Yellow-green
575	Violet	Greenish yellow
590	Blue-violet	Bright yellow
665	Blue	Orange
750	Green	Red
830	Yellow-green	Violet
910	Yellow	Blue-violet
1000	Orange	Blue-green
1100	Red-violet	Green
1150	Blue-green	Yellow
1375	Green	Violet

all wavelengths into a direction which can be transmitted by the analyzer. Consequently the specimen appears bright (i.e., white) but not colored when observed between crossed polarizers.

The order of colors produced by crossed polarizers is the same as that formed at thin films of increasing thickness, as described on p. 45.

RED I PLATE

The retardations produced by biological specimens are rarely larger than about 100 mμ; hence interference colors ordinarily are not observed directly in the polarizing microscope. However, with the use of a so-called "red I plate" brilliant interference colors can be observed in the images of these specimens.

The red I plate is a slab of highly birefringent material, such as mica, which is cut with the face parallel to its optic axis (i.e., the flat plate is observed at right angles to its optic axis). The thickness is such as to produce a retardation of about 550 mμ, so that a first order red color is obtained between crossed polarizers. A notation made on the plate indicates the azimuths of vibration of the slow and fast rays, i.e., the directions of maximum and minimum refractive index.

The red I plate can be inserted between polarizer and analyzer at a rotational orientation such that the azimuths of vibration make an angle of 45° with those of polarizer and analyzer. Relative orientations of the azimuths of vibration for the two possible positions of the red I plate are illustrated in Fig. 15-5. With the plate in either orientation the field of view appears uniformly red.

A birefringent specimen may be placed in either of two extreme positions with respect to the red I plate; the azimuth of vibration corresponding to maximum refractive index (conveniently referred to as the *slow direction*) of the specimen may be either parallel or perpendicular to the slow direction of the red I plate. If the two slow directions are parallel, the retardation produced in the red I plate is further increased in the specimen. For example, if the retardation by the specimen is 100 mμ, the total retardation in the specimen-plate combination is 650 mμ. As indicated by Table 15-2, a *blue* color results; this is termed an *addition color*. Alternatively, if the slow directions of plate and specimen are perpendicular, the slow ray of the plate partially "catches up" in the specimen. The net retardation is reduced to (550 − 100) = 450 mμ, giving a *yellow subtraction color*. Note that both addition and subtraction colors are observed in the same specimen upon rotation through 90°.

DESIGN OF THE POLARIZING MICROSCOPE

Basically, a polarizing microscope is an ordinary light microscope with certain accessories added. A special requirement is that optical components be constructed from glass which is strain-free, thus avoiding spurious sources of (strain) birefringence.

The characteristic accessories used in the polarizing microscope are: the *polarizer*, placed somewhere below the specimen; the *analyzer*, placed somewhere above the specimen; a *rotating specimen stage*; and *slots* for the insertion of the red I plate and other compensating devices. A means must be provided for the mutual rotation of polarizer and analyzer through not less than 180°. For example, the position of the analyzer might be fixed, while the polarizer might rotate through any angle. Calibrated scales should be provided for the measurement of rotations of both stage and polarizers.

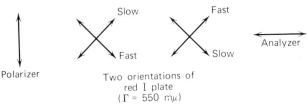

Fig. 15-5 Relative orientations of polarizer, analyzer, and red I plate.

The quality of a polarizing microscope is largely established by the completeness of extinction between crossed polarizers. This, in turn, is determined by the freedom of optical parts from strain birefringence and by the quality of polarizing devices.

The general plan of the polarizing microscope is shown in Fig. 15-6.

As already noted, the polarizing microscope is used to detect birefringence, to determine the sign of birefringence, to determine the type of birefringence, to locate the optic axis of birefringent structures, and, ultimately, to determine the shape and orientation of index ellipsoids.

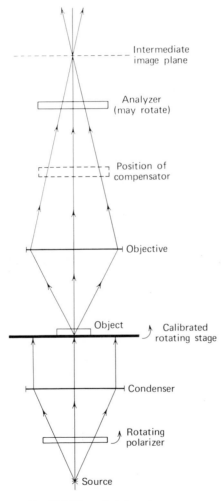

Fig. 15-6 The polarizing microscope.

The methods and accessories used for each of these determinations are now described.

DETECTION OF BIREFRINGENCE

The detection of birefringence implies the observation of *four* positions of extinction and four of maximum brightness as a specimen is rotated through 360° between crossed polarizers. Obviously, more than one orientation of apparently isotropic material should be examined, since a birefringent specimen may, by chance, be observed in a direction parallel to its optic axis. Some other misleading possibilities have been discussed on pp. 325 and 326.

The observation of birefringence in specimens other than fibers, etc., is an approximate test for crystallinity, since about 95% of all known crystals are birefringent. All crystals may be assigned to one of six structural systems (cf. Table 29-1); of these, only members of the highly symmetrical cubic system are optically isotropic.

LOCATION OF THE OPTIC AXIS

The optic axis of a specimen is located by finding that orientation of the specimen for which, regardless of thickness, extinction occurs at all rotational orientations (i.e., for which the specimen appears to be isotropic). When this condition is satisfied, the specimen is being observed in a direction parallel to its optic axis.

The approximate orientation of the optic axis may be determined very easily by inspection. Precise location of the optic axis requires careful adjustment and measurement of the orientation of the specimen. Examination of sections cut in a number of planes may be required.

Occasionally, radial, tangential, or helical orientations of the optic axis occur in fibers, as illustrated in Fig. 15-7. No view of such specimens appears to be isotropic. The characteristic orientation of the optic axis must be deduced in these cases by observations of sections cut in several different planes.

SIGN OF BIREFRINGENCE

The sign of birefringence can be determined unequivocally when the optic axis of a specimen has been located, thus permitting identification of the *O*- and *E*-ray vibration azimuths. If the optic axis has not been located, however, the presumed sign of birefringence may be specified with respect to some recognizable direction in the specimen, such as the long axis of a fiber.

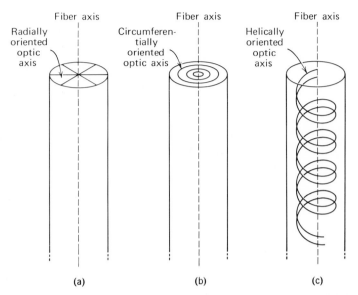

Fiber axis Fiber axis Fiber axis

Radially oriented optic axis

Circumferentially oriented optic axis

Helically oriented optic axis

(a) (b) (c)

Fig. 15-7 Special orientations of the optic axis.

In determining the sign of birefringence, a red I plate (p. 327) is placed in the position of the "compensator," as shown in Fig. 15-6. The azimuths of vibration of the red I plate are set at 45° to those of polarizer and analyzer, in either of the orientations diagramed in Fig. 15-5. The stage of the microscope is then rotated so that the optic axis (or reference axis) of the specimen is aligned with the slow ray azimuth of the red I plate. The image of the specimen then appears *blue* (shows an addition color) if the slow direction of the specimen is *parallel* to that of the red I plate, and *yellow* (shows a subtraction color) if the slow direction of the specimen is *perpendicular* to that of the red I plate. Rotation of the specimen through 90° reverses the observed blue or yellow color. The production of addition and subtraction colors is diagramed in Figs. 15-8a and 15-8b, respectively.

As explained on p. 103, the O ray always vibrates in an azimuth which is perpendicular to the optic axis, while the E ray always has a component of vibration parallel to the optic axis. Thus, when a specimen is observed in a direction perpendicular to the optic axis, the E ray must vibrate parallel to the optic axis. Accordingly the sign of birefringence may be deduced as follows: If a *blue* (addition) color is observed when the slow direction of the red I plate and the optic axis of the specimen are parallel, the E ray of the specimen, vibrating parallel to the axis, must be the *slow* ray. Accordingly n_e is greater than n_o, so that the birefringence of the

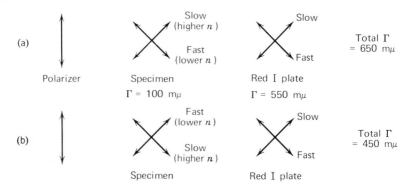

Fig. 15-8 (a) Formation of addition colors; (b) formation of subtraction colors.

specimen is *positive*. Conversely, if a yellow (subtraction) color is observed when specimen and red I plate are aligned, the E ray must be the fast ray in the specimen, n_e must be less than n_o, and the birefringence is negative.

Commonly the optic axes of fibers are at least approximately coincident with the longitudinal axis. In the exceptional case in which the optic axis lies at right angles to the selected reference direction, the sign of birefringence is, of course, estimated incorrectly.

QUANTITATIVE MEASUREMENT OF BIREFRINGENCE

The birefringence of a specimen is given by

$$(15\text{-}n) \qquad\qquad B = n_e - n_o = \frac{\Gamma}{t}.$$

The birefringence is thus determined by measuring the retardation Γ and the specimen thickness t.

Thickness t may be obtained directly by observation of the specimen in a plane at right angles to that in which measurements of retardation are made. Although straightforward in principle, measurements of specimen thickness may present practical difficulties; appropriate techniques vary with the nature of the specimen.

Γ may be obtained approximately by observation of retardation colors (cf. pp. 326 and 327). For small values of Γ the effect of summing retardation by the specimen with that of a device of known retardation (e.g., a red I plate) may be observed. Precise measurements of Γ can be made with the use of a compensator; four types of compensator are described below.

Note that, for accurate determination of the birefringence, which is the maximum difference between the (variable) *E*-ray refractive index and the (invariant) *O*-ray refractive index, the specimen must be so oriented that its optic axis is exactly perpendicular to the microscope axis.

WEDGE COMPENSATOR

In principle, the simplest type of compensating device is a calibrated wedge made of material of known birefringence. Compensators of this type are usually made of quartz. As shown in Fig. 15-9, the compensator is so cut that one of its azimuths of vibration is parallel to the sloping face of the wedge. The wedge is inserted into the tube of the polarizing microscope at the compensator position indicated in Fig. 15-6, with its slow direction parallel to the fast direction of the specimen. The wedge is translated past the specimen (i.e., in the direction indicated by *x* in Fig. 15-9) until a position of the compensator is located at which the image of the specimen undergoes extinction. At this position the retardations by specimen and compensator are *equal* and *opposite*. The corresponding value of the retardation is then obtained from the calibration factor of the wedge (i.e., the variation of Γ per unit length of wedge. This in turn may be obtained from the known value of the birefringence of quartz and the angle of the wedge.)

The precision of measurements of very small retardations is improved by the use of two opposing wedges, arranged in the manner shown in Fig. 15-10. A relative translation of the two wedges can be used to produce very small changes in net retardation.

Note that, if a (single) wedge compensator is so oriented that its slow direction coincides with that of the specimen, *no* position of compensation (extinction) can be found.

TILTING COMPENSATOR

The tilting compensator is a birefringent disc which is cut with its plane surfaces perpendicular to the optic axis. The disc may be tilted through a controlled angle with respect to the microscope axis. In the

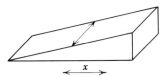

Fig. 15-9 Single wedge compensator.

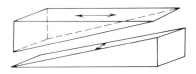

Fig. 15-10 Double wedge compensator.

neutral position, as shown in Fig. 15-11a, the optic axis of the compensator is parallel to the microscope axis, so that the compensator has no effect upon the retardation of the system. Rotation of the compensator through 90° to the position shown in Fig. 15-11b produces a maximum retardation by the compensator. Intermediate tilting, as diagramed in Fig. 15-11c, produces intermediate and variable retardations.

FIXED RETARDATION COMPENSATOR

A birefringent plate of known retardation, so cut that the optic axis is parallel to its surfaces, as shown in Fig. 15-12, may be rotated in the plane perpendicular to the microscope axis. As for a red I plate, the retardations of specimen and compensator add or subtract when the slow directions of the two are parallel or perpendicular, respectively. If the disc is rotated as the specimen is held stationary, the contribution by the disc to the total retardation varies continuously. Thus, for any specimen of retardation less than that of the compensator, an orientation of the disc exists at which the retardations of the two just cancel, and extinction of the image is observed. For specimen retardations which exceed that of the disc, no position of extinction exists.

Although fixed retardation compensators for which the value of Γ is low can thus be used only over a very limited range, this type of device provides a particularly accurate method for measuring very small values of the retardation.

QUARTER WAVE PLATE COMPENSATOR

The quarter wave plate, described on p. 112, is a birefringent disc for which the retardation is $\lambda/4$ for any chosen wavelength λ. A fixed quarter

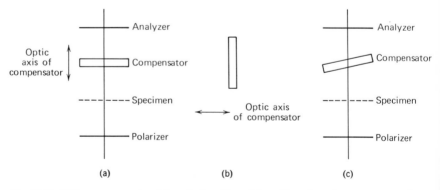

Fig. 15-11 Tilting compensator: (a) neutral position; (b) position of maximum compensation; (c) intermediate position.

Compensator

Optic axis

Fig. 15-12 Fixed retardation compensator.

wave plate is used in conjunction with a rotatable analyzer in the measurement of retardation, the arrangement and orientation of microscope
elements being as shown in Fig. 15-13a and 15-13b, respectively. Note
that the vibration azimuths of the specimen are set at 45° to those of
the polarizer and of the quarter wave plate. The analyzer is rotated from
its crossed position to one in which the image of the specimen is extin-
guished. A rotation of the analyzer through 180° is equivalent to a retardation of one wavelength by the specimen.

The action of the quarter wave plate compensating system may be
accounted for in the following manner. Two mutually perpendicularly
polarized beams travel through the specimen at different velocities. Each
of these beams, after traveling through the quarter wave plate, constitutes
a circularly polarized motion (i.e., each beam, within the quarter wave
plate, is resolved into two mutually perpendicularly polarized vibrations
of equal amplitude; a phase shift between them of 90° is then produced
upon passage through the quarter wave plate). These two wave motions
are circularly polarized in opposite senses (cf. p. 83). Their resultant is
a vibration which is polarized in a plane determined by the relative phase
of the circular motions (cf. pp. 87–89). The relative phase of the circular
motions is, in turn, determined by the relative phase of the beams leaving
the specimen, that is, by the retardation produced by the specimen.
Figure 15-13c, for example, illustrates the case of a retardation of $\lambda/8$ in
which the resultant azimuth of vibration is shifted through 22.5° relative

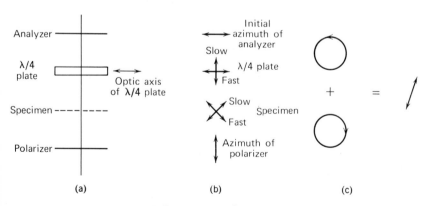

Fig. 15-13 Quarter wave plate compensator.

to that of the light incident upon the specimen. Extinction is then produced when the transmission azimuth of the analyzer is set at 90° to the plane of polarization of the resultant beam, that is, when the analyzer is rotated by 22.5° from the original crossed position.

TYPE OF BIREFRINGENCE

The existence of four types of birefringence—intrinsic, form, strain, and flow—has been noted on p. 109. The distinction between intrinsic and form birefringence in a static, unstrained specimen may serve to characterize its submicroscopic structure. Form birefringence derives from the micellar or paracrystalline arrangement of elements of structure, whereas intrinsic birefringence originates directly from the orientation of chemical bonds.

Form birefringence is observed when oriented units of substructure, such as macromolecules or macromolecular aggregates, differ in refractive index from the surrounding medium. *Positive* form birefringence occurs when substructural elements are in rodlike arrangements (with respect to the optic axis) in the manner suggested by Fig. 15-14a. *Negative* form birefringence is associated with a plateletlike substructure, as sketched in Fig. 15-14b. In either case, form birefringence is observed only if the substructural elements differ from the surround in refractive index. The greater the difference in refractive index, the larger is the amount of form birefringence, be it positive or negative. In addition, the substructural elements may or may not possess intrinsic birefringence.

The quantitative relationship between the properties of the specimen and the form birefringence has been derived by considering the passage of light through the specimen as the transmission of an electrical dis-

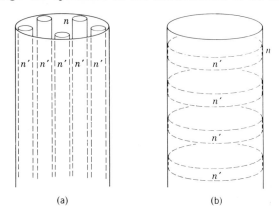

(a)   (b)

Fig. 15-14 Structures producing form birefringence: (a) positive; (b) negative.

turbance. The electrical disturbance is considered to be modified by the presence of substructural elements, which act as capacitors. This treatment, which is not presented in further detail here, leads, in the case of rodlet structures, to expressions such as

(15-o)
$$n_e^2 - n_o^2 = +\frac{\phi_p \phi_m (n_p^2 + n_m^2)^2}{(1 + \phi_p) n_m^2 + \phi_m n_p^2},$$

in which n_p and ϕ_p are the refractive index and volume fraction, respectively, of oriented substructural elements, while n_m and ϕ_m are the corresponding quantities of the medium.

For platelet structures, form birefringence is predicted from the corresponding equation,

(15-p)
$$n_e^2 - n_o^2 = -\frac{\phi_p \phi_m (n_p^2 - n_m^2)^2}{\phi_p n_p^2 + \phi_m n_m^2}.$$

(Derivations of (15-o) and (15-p) are given in[1]). Note that these equations) predict that positive or negative birefringence, respectively, *always* results from *rodlet* or *platelet* structure, regardless of specific values of the refractive indices. This is true because positive and negative signs, respectively, precede the (invariably positive) squared terms, while the volume fractions ϕ_p and ϕ_m must likewise always be greater than zero. Equations 15-o and 15-p apply strictly to idealized systems only, but provide a quite satisfactory description of the form birefringence of biological systems, in which substructural elements are never precisely of rodlet or platelet form. Note that, if $n_p = n_m$, absence of form birefringence is predicted, but isotropy of the specimen is not implied, since the elements of structure may themselves possess intrinsic birefringence.

Intrinsic and form birefringence of a specimen may be distinguished experimentally if the specimen can be infiltrated by media of different refractive indices, thus producing alterations in the observed birefringence of the specimen. While the replacement of the medium of a biological structure might seem to be a hazardous technique, application of this method to the study of a number of different biological systems has proved possible. Immersion media are, of course, limited to a number of inert solvents; for example, glycerol-water mixtures or sucrose solutions may be suitable media when the orientation of non-water-soluble structures is investigated. Total birefringence is then measured for a series of values of n_{med}; the resulting *form birefringence curves* are then as exemplified by Fig. 15-15. Figure 15-15a represents data obtained from a specimen of positive form birefringence and positive intrinsic birefringence. A minimum rather than zero value of birefringence is obtained. Figure 15-15b shows the form birefringence curve of a negatively

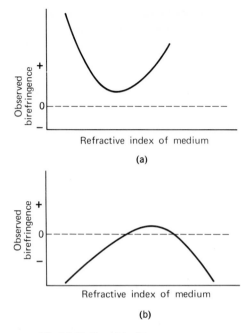

Fig. 15-15 Form birefringence curves.

form birefringent specimen in which the elements of structure are, as before, of positive intrinsic birefringence. Here the sign of the observed birefringence reverses as the refractive index of the immersion medium is changed.

DETERMINATION OF THE INDEX ELLIPSOID

Complete characterization of the birefringence of a specimen implies determination of the shape and orientation of its index ellipsoid. Stages in the characterization of the index ellipsoid of a uniaxial specimen, in which only intrinsic birefringence is present, may be summarized as follows:

1. The optic axis is located (cf. p. 330).
2. The sign of birefringence is determined with respect to the optic axis (cf. pp. 330–332).
3. Retardation is measured by means of a compensator (cf. pp. 332–336).
4. The thickness of the object is measured.
5. The birefringence of the specimen is obtained from (15-n).
6. The absolute value of n_0 is measured, usually by means of immersion

methods. The object must of course be viewed parallel to its optic axis during this procedure (cf. Chapter 26).

7. n_e is then evaluated:

(15-q) $$n_e = B + n_0.$$

The index ellipsoid is then, as defined in Chapter 6, an ellipsoid of rotation such that one axis is parallel to the optic axis and of length proportional to n_e, while the other two axes are of equal length proportional to n_0. (For biaxial structures, which are rarely encountered in biological systems, all three axes of the index ellipsoid are of different lengths and are inclined with respect to the optic axes.)

Figure 15-16 illustrates a fiber and the central section of its index ellipsoid. The optic axis is inclined at 30° to the fiber axis in this example; $n_e = 1.60$, $n_0 = 1.20$, and thus $B = +0.40$. On the scale used in Fig. 15-16 the three-dimensional ellipsoid would project 1.20 cm above and below the plane of the page.

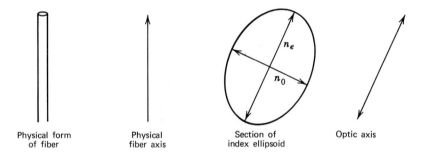

| Physical form of fiber | Physical fiber axis | Section of index ellipsoid | Optic axis |

Fig. 15-16.

REFERENCES

General References

H. S. Bennett, in R. M. Jones, Ed., *McClung's Handbook of Microscopical Technique*, 3rd ed., Hafner, New York, 1950, pp. 591–677. This chapter, "The Microscopical Investigation of Biological Materials with Polarized Light", is a clear and authoritative presentation of the theory and techniques of polarization microscopy.

F. Ruch, in G. Oster and A. W. Pollister, Eds., *Physical Techniques in Biological Research*, Vol. 3, *Cells and Tissues*, Academic Press, New York, 1955, pp. 149–174. This chapter, "Birefringence and Dichroism of Cells and Tissues" describes the application of polarization microscopy to biological specimens, with special emphasis on the study of botanical materials.

R. D. Allen, J. W. Brault, and R. M. Zeh, in R. Barer and V. E. Cosslett, Eds., *Advances in Optical and Electron Microscopy*, Vol. I, Academic Press, New York, 1966. An account of "phase modulated" polarization microscopy.

Specific Reference

[1] O. Wiener, Abhandl. Saechs. *Ges. Wiss., Math., Phys. Kl.* **32**, 509, (1912).

CHAPTER **16**

Electrons and Electron Beams

The wavelength of illuminating radiation is a factor which limits the resolution that can be achieved in the images formed by any lens system. As shown in Chapter 10, the resolution of optical microscopes can thus be no better than about $0.12 \, \mu$. When, during the 1920s, the work of both theoretical and experimental physicists showed that accelerated electron beams are associated with extremely short wavelengths, the possibility of achieving very high resolution with an "electron microscope" was at once evident, and the development of such instruments was soon undertaken.

In this chapter the concept of electrons as wave motions is considered, and some general features of "electron optics" are described. The physical processes whereby electron beams may be obtained are also summarized. Familiarity with these topics is helpful for further study of the electron microscope. It is interesting to note, however, that an understanding of the wave nature of the electron is scarcely required for understanding the operation (as opposed to the design) of electron microscopes. The physical interactions which take place in the "electron lens" and in the electron microscope column can be described, at least qualitatively, in terms of the classical concepts of particulate electrons. Only in predicting the ultimate resolving power of the instrument, in interpreting distributions of intensity in high resolution electron images, or in studying electron diffraction patterns is consideration of the wavelike nature of the electron essential.

PROPERTIES OF THE CLASSICAL ELECTRON

The electron is a fundamental particle bearing a unit negative charge e of 1.60×10^{-19} coulomb. Its rest mass is 9.11×10^{-28} g. In terms of classical theory it can be considered a charged sphere of a diameter of the order of 10^{-12} cm; more correctly, it can be said to represent a "probability distribution" for the location of negative charge. (The concept of electronic orbitals has been described on p. 126.)

THE DE BROGLIE ELECTRON

Prompted by an appreciation of the symmetries of the natural world, de Broglie considered the possibility that, just as light, which exhibits an unequivocal wavelike character, can be shown to be particulate in nature, the unmistakably particulate electron might be associated in some way with wave properties. Just as the existence of photons was unsuspected until appropriate experimental systems were studied, so also the wave properties of electron beams might be such as normally to escape detection. The wave properties of light beams are evident only when the beams studied, are *narrow* with respect to the wavelength. Should the electron wavelength be extremely small, de Broglie reasoned, previous failures to observe wavelike properties would be easy to explain. This in fact proved to be the case.

Assuming then that both photons and electrons (light beams and electron beams) possess both wave and particle properties, it should be possible to demonstrate parallels between the laws of particle mechanics and those of geometrical optics. It is instructive to compare the refraction of light with that of a stream of charged particles passing through two sucessive regions of different electrostatic potential. Figure 16-1 illustrates two such regions at potentials V_A and V_B separated by a plane interface. Electrons travel in region A at velocity v_A and in a direction which makes some angle θ_A with the normal to the interface. The motion of the particles may be resolved, vectorially, into lateral and perpendicular components of motion, as diagramed in Fig. 16-2. The corresponding values of the lateral and perpendicular components of velocity are:

$$(16\text{-a}) \qquad\qquad v_{A(\text{lat})} = v_A \sin \theta_A,$$

$$(16\text{-b}) \qquad\qquad v_{A(\text{perp})} = v_A \cos \theta_A.$$

Upon reaching region B, which is at some potential V_B different from V_A, the particles experience a different force in the perpendicular direction, while the lateral component of particle velocity is unaffected. Thus $v_{A(\text{lat})} = v_{B(\text{lat})}$, but $v_{A(\text{perp})} \neq v_{B(\text{perp})}$. A vector diagram of the components of

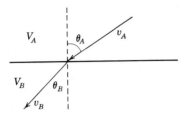

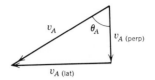

Fig. 16-1 Refraction of charged particles.

Fig. 16-2 Lateral and perpendicular components of velocity in region of potential V_A.

motion acting in region B is given in Fig. 16-3. The corresponding values of the perpendicular and lateral components of velocity are

(16-c) $\qquad v_{B(\text{perp})} = v_B \cos \theta_B,$

(16-d) $\qquad v_{B(\text{lat})} = v_B \sin \theta_B.$

Comparison of Figs. 16-2 and 16-3 shows that the particle trajectory is refracted as it enters a region of different electrostatic potential (i.e., $V_A \neq V_B$), just as a light beam undergoes refraction upon entering a medium of different refractive index.

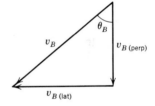

Fig. 16-3 Lateral and perpendicular components of velocity in region of potential V_B.

Since the lateral component of particle velocity is unaffected by the change of potential, (16-b) and (16-c) may be combined to yield

(16-e) $\qquad \sin \theta_A v_A = \sin \theta_B v_B \quad \text{or} \quad \dfrac{\sin \theta_A}{\sin \theta_B} = \dfrac{v_B}{v_A}.$

Equation 16-e appears similar to Snell's law for the refraction of light,

(16-f) $\qquad \dfrac{\sin i}{\sin r} = \dfrac{n_2}{n_1} = \dfrac{v_1}{v_2}.$

Equation 16-f, however, predicts a ratio of velocities which is the *inverse* of that given for particles in (16-e). Apparently electrons (or other particles) should travel faster in regions of higher "refractive index." Indeed, the proof, by direct measurement, that light travels more slowly in regions of higher refractive index and thus that its behavior is described by (16-f) rather than by (16-e), was accepted during the nineteenth century as conclusive proof of the wave rather than particulate character of light. Yet, if a symmetry of wave and particle properties of light and matter exists, the *same* equation would be expected to apply to both. The apparent obvious inconsistency of the two equations seems to preclude the evolution of a consistent view of electrons and their associated waves. The key to this dilemma lies in the fact that the velocities included in (16-e) and (16-f) are not identical. The particle velocity (electron velocity) of (16-e) is a *group* velocity; the wave velocity of (16-f) is a *phase* velocity. (The distinction between phase and group velocities has been discussed on pp. 73–75.) The particle velocity of electrons is correctly specified by the particle refraction equation (16-e) while (16-f) specifies the velocity of the associated "electron wave."

Comparison of (16-e) and (16-f) then shows that an inverse relationship must exist between phase and wave velocities. That is,

$$(16\text{-g}) \qquad \phi = \frac{K}{v}$$

where ϕ is the velocity of the electron wave, v the particle velocity of the electron, and K is a constant. Equation 16-g shows that a moving electron cannot remain in association with a *single* sinusoidal vibration since, in general, the two must travel at different velocities. The electron does, however, remain in association with a *group* of such motions. Similar behavior can be observed upon careful study of macroscopic mechanical wave trains in which individual wavelets outrun the pulse as a whole, while "new" wavelets reform at the rear of the pulse. The "explanation" of electron waves is directly analogous to that of the dual nature of light (cf. p. 22). Just as a large number of photons behave, collectively, as a wave, so a large number of electrons also behave collectively as a wave.

These considerations lead to the development of a remarkably simple expression for the associated wavelength (the de Broglie wavelength) of electron beams.

THE DE BROGLIE WAVELENGTH

The total energy E of an electron, as of a photon, is given by the expression

$$(16\text{-h}) \qquad E = h\nu.$$

Total energy must also be the sum of the kinetic energy (K.E.) and the potential energy (P.E.) of a particle. Thus,

$$(16\text{-i}) \qquad \text{K.E.} + \text{P.E.} = h\nu.$$

The kinetic energy of any particle of mass m moving with a "group" velocity v is

$$(16\text{-j}) \qquad \text{K.E.} = \tfrac{1}{2}mv^2$$

so that

(16-k) $$hv = \tfrac{1}{2}mv^2 + \text{P.E.}$$

Assuming that the potential energy remains constant, (16-k) can be differentiated with respect to λ:

(16-l) $$h\frac{\partial v}{\partial \lambda} = mv\frac{\partial v}{\partial \lambda}.$$

The group velocity $v(= v_g)$ and the phase velocity v_ϕ have been shown on p. 75 to be related by the expression

(3-k*) $$v = v_\phi - \lambda\frac{\partial v_\phi}{\partial \lambda},$$

which is equivalent to

(16-m) $$v = -\lambda^2\frac{\partial}{\partial \lambda}\left(\frac{v_\phi}{\lambda}\right),$$

or, since the frequency $\nu = v_\phi/\lambda$,

(16-n) $$v = -\lambda^2\frac{\partial \nu}{\partial \lambda}.$$

Substitution of (16-n) in (16-l) gives

(16-o) $$h\frac{\partial \nu}{\partial \lambda} = -m\lambda^2\left(\frac{\partial v}{\partial \lambda}\right)\left(\frac{\partial v}{\partial \lambda}\right) \quad \text{or} \quad \frac{\partial v}{\partial \lambda} = -\frac{h}{m\lambda^2}.$$

Integration of (16-o) yields

(16-p) $$v = \frac{h}{m\lambda} + C,$$

where C is a constant of integration. Assuming that $C = 0$, (16-p) becomes the de Broglie equation.

(16-q) $$\boxed{\lambda = \frac{h}{mv}} \quad \text{or,} \quad \lambda = \frac{h}{p},$$

where p is the *momentum* of the electron.

EQUIVALENT WAVELENGTHS OF ELECTRON BEAMS

Numerical values of the electron wavelength may be derived from (16-q). When relativistic corrections are negligible (cf. p. 347), an electron accelerated through V volts has a kinetic energy of (eV) electron volts or (eV/300) ergs. (Here e is the electronic charge, 4.80×10^{-10}

statcoulomb.) Thus,

(16-r) $$\tfrac{1}{2}mv^2 = \frac{eV}{300} \quad \text{or} \quad mv = \sqrt{\frac{MeV}{150}}.$$

Substitution of (16-r) in (16-q) then gives

(16-s) $$\lambda = \frac{h}{\sqrt{meV/150}}.$$

Numerical evaluation of (16-s) yields

(16-t) $$\lambda = \frac{12.27}{\sqrt{V}}$$

as the effective wavelength *in Angstrom units* of electrons accelerated through V volts. For example, an electron accelerated through 10 kilo-volts (kV) is associated with a wavelength of $\lambda = 12.27/\sqrt{10^4} = 0.1227 \text{Å}$.

THE ELECTRON OPTICAL ANALOG OF THE LAW OF REFRACTION

The refraction of charged beams at surfaces of changing electro-static potential (shown in Fig. 16-1) suggests an analogy between poten-tial and refractive index.

The law of refraction for light may be written

(16-u) $$\frac{\sin i}{\sin r} = \frac{\lambda_1}{\lambda_2},$$

in which the wavelengths λ_1 and λ_2 are the *physical* wavelengths in the media of incidence and refraction respectively (i.e., not invariant *in vacuo* wavelengths, $\lambda_{\text{vac}} = c/v$, as implied elsewhere in this book). Substitution of the electron wavelength, $\lambda = h/mv$, in (16-u) leads to

(16-v) $$\frac{\sin i}{\sin r} = \frac{m_2 v_2}{m_1 v_1}.$$

The momentum mv of the electron is thus seen to be the electron optical analog of refractive index.

A useful form of (16-v) may be obtained by substituting the value of mv obtained from (16-r):

(16-w) $$\frac{\sin i}{\sin r} = \sqrt{\frac{V_2}{V_1}}.$$

Equation 16-w, relating the refraction of an electron beam to the potential of the region through which the beam travels, is the practical electron optical analog of Snell's law. Equations 16-w and 16-v are but one example of electron optical expressions which parallel the equations of light optics.

LIMITATIONS OF PARALLELISM BETWEEN LIGHT AND ELECTRON OPTICS

The analogy between light and electron optics is subject to certain limitations. In geometrical optics the interactions of rays need not be considered, since photons lack both mass (rest mass) and charge. The negatively charged electrons in an electron beam, however, repel each other (as do the component particles of other charged particle beams). Consequently expressions analogous to those of ray optics apply only at relatively low electron densities; otherwise, correction for *space charge* effects is required.

Experimentally, different physical conditions apply during the passage of light and electron beams. For example, in light optics abrupt changes of refractive index occur, producing correspondingly abrupt deviations of the refracted rays. While, theoretically, an abrupt change of electrostatic potential may be considered (as on pp. 348–349), experimentally the refraction of electron beams occurs continuously as the electrons travel across a potential gradient.

Einstein's equation,

$$(16\text{-x}) \qquad\qquad m = \frac{m_0}{\sqrt{1 - v^2/c^2}},$$

shows that the mass of any particle is not a constant, but increases from its rest mass m_0, approaching infinity as the particle velocity approaches c. At "ordinary" velocities, v is an extremely small fraction of c, so that m is approximately equal to m_0 and is thus in effect constant. In the electron microscope, however, accelerated electrons travel with velocities of the order of $c/10$ so that, as found from (16-x), $m = 1.005 m_0$. The effect of this shift in mass is important when quantitative calculations are made, and it must be taken into account by microscope designers. Qualitatively, however, relativistic corrections are of secondary importance and can be ignored in subsequent discussions here.

The photon is massless at rest and invariably travels with velocity c. Thus no form of relativistic correction is applicable to the equations of light optics. (The relativistic mass of the moving photon is given by $h/\lambda c$, and is of the order of 10^{-33} g.)

ESTIMATION OF ELECTRON TRAJECTORIES

Computation of electron trajectories (i.e., of the path traveled by electrons) may be carried out by application of (16-w) and similar expressions. Although, as noted above, electrons in fact travel through continuous potential gradients, electron trajectories may be approximated by considering that changes in potential occur abruptly at each of a series of equipotential surfaces. The positions of equipotential surfaces within a given system may be located experimentally.

The procedure for computing trajectories may be illustrated by a simple example. An electron traveling initially with a velocity equivalent to acceleration across a potential of 10 volts, is incident at an angle of 60° (i.e., with respect to the normal) upon a linear potential gradient extending from 50 to 100 volts. Arbitrarily, the gradient is represented, as shown in Fig. 16-4, by a set of three equipotential surfaces at 50, 75, and 100 volts, respectively. The refraction of the electron trajectory is then computed at each successive surface. At the first surface,

$$(16\text{-y}) \qquad \frac{\sin 60°}{\sin \theta_1} = \sqrt{\frac{50}{10}}, \qquad \theta_1 = 23°;$$

at the second surface,

$$(16\text{-z}) \qquad \frac{\sin 23°}{\sin \theta_2} = \sqrt{\frac{75}{50}}, \qquad \theta_2 = 18°;$$

and, at the third,

$$(16\text{-aa}) \qquad \frac{\sin 18°}{\sin \theta_3} = \sqrt{\frac{100}{75}}, \qquad \theta_3 = 16°.$$

The approximate electron trajectory is thus that shown by the solid line in Fig. 16-4. The exact trajectory is shown by the dotted line in the same figure. Differences between the two are accounted for by the fact that the gradient (a) is continuous, and (b) extends some distance to the left of the 50-V equipotential line.

CONVENTION OF ELECTRON OPTICAL POTENTIALS

In the procedure just described, an electron at rest corresponds to an electron optical potential of zero. Similarly, any surface at which the net velocity of electrons is zero (i.e., at which there may be random motion of electrons, but no net translation) has an electron optical potential of zero, while at a surface at which the electron velocity is $\sqrt{eV/150m}$, the electron optical potential is, as given in (16-r), V volts. Note that this conven-

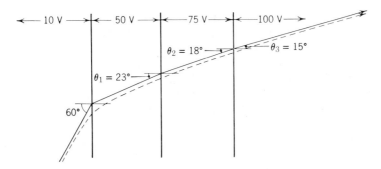

Fig. 16-4 Electron trajectory in a potential gradient.

tion of potentials is quite different from the usual one! For example, in the electron microscope, electrons leave a filament which is (conventionally) at a large negative potential, and they are accelerated toward an anode which is (conventionally) at ground potential. Electron optically, however, the filament is at a potential of zero volts, since the electrons are initially at rest there. The electrons of a 100 kV beam impinge upon the anode at a velocity of $\sqrt{100{,}000e/150m}$, and the electron optical potential of the anode is thus 100 kV. However, the *potential difference* is the same (100 kV) according to either convention.

SOURCES OF ELECTRON BEAMS

A number of physical processes produce electron beams. The radio-activity of β emitters is a spontaneous source of highly energetic electrons (β rays). Energy levels which are specifically characteristic of the emitter lie in the range from 100 to 10,000 kV (0.1–10 meV).

Electron emission from the surfaces of metals may be induced. Metals consist of fixed arrays of positive ions which share an equivalent number of valence electrons. The loosely bound valence electrons tend to migrate rather freely and can escape from the metal completely if sufficient additional energy is imparted to them. In *field emission* (cold emission) processes a strong electrostatic field "pulls" electrons out of the surface by the attraction of the external positive charge. In *photoemission* (cf. p. 20), electrons acquire excess energy by absorbing a photon. A certain level of energy, known as the "work function," must be supplied in order to allow the electron to escape from any given surface. Any energy which the photon imparts in excess of this amount appears in the form of electron velocity. Upon bombardment of metal surfaces by electron beams, the incident energy is partially converted into heat, but also reappears in the form of emitted x-rays (cf. p. 605) and *secondary electrons*. In gas discharge tubes, positively charged ions impinge upon a metal target, and a portion of their energy is imparted to the valence electrons of the target, inducing emission. In the process of *thermionic emission*, energy is supplied to loosely bound electrons by heating.

In electron microscopy a continuous, controllable beam of uniform velocity and of relatively high intensity is required. For the observation of biological specimens, accelerating voltages must be of the order of 50 to 100 kV (cf. p. 444). None of the emission processes mentioned above is capable of supplying such a beam directly (e.g., radioactivity is uncontrollable and produces excessive particle energies). However, *electron guns* have been developed which exploit the *thermionic* emission process in such a way as to provide reasonably satisfactory electron sources. In an uncontrolled system thermionic emission tends to cease as electron loss causes a net positive charge to build up on the surface of the emitter. In the electron gun this surface is maintained at a constant potential. The emitted electrons, leaving the surface at a variable but always low velocity, are immediately accelerated by a steep voltage gradient, acquiring an essentially uniform velocity. (The conventional design of electron guns is described in more detail on p. 379.)

REFERENCES

F. K. Richtmyer, E. H. Kennard, and T. Lauritsen, *Introduction to Modern Physics*, McGraw-Hill, New York, 1955, pp. 172–184. A short account of electron waves.

C. E. Hall, *Introduction to Electron Microscopy*, 2nd ed., McGraw-Hill, New York, 1966, Chapter 1 (Elements of Physical Theory) and Chapter 2 (Electrons in Electrostatic Fields). An account of the properties of electrons with special reference to electron microscopy.

V. K. Zworykin, G. A. Morton, E. G. Ramberg, J. Hillier, and A. W. Vance, *Electron Optics and the Electron Microscope*, Wiley, New York, 1945, pp. 1–8. A short summary of electron sources and of analogies between light and electron optics.

CHAPTER 17

Electron Lenses

Axially symmetric electrostatic and electromagnetic fields have the property of focusing electron beams, and thus of acting as electron lenses. In this chapter the general design and operating properties of such lenses are described, and the aberrations to which they are subject are discussed. Certain aberrations of electron lenses are analogous to the defects of glass lenses, while others originate from technical limitations in electron microscope design.

The treatment of electron lenses to be given here is necessarily more superficial and practical in nature than is the discussion of glass lenses presented in Chapter 8. Precise description of electron focusing action is of considerably greater mathematical complexity than the corresponding descriptions in geometrical optics. Furthermore the design and construction of electron lenses is an area of engineering practice with which the microscopist need have little concern, just as the lens designer, in turn, is rarely concerned by the interpretation of images. An excellent qualitative understanding of electron lens action can be obtained through appreciation of the many analogies between light and electron lenses, with due understanding of the modifications imposed by the physical character of electron focusing systems. Familiarity with the properties of optical lenses, as described in Chapters 8 to 11, is thus basic to an understanding of electron lenses. Parallels between light and electron waves have been discussed in Chapter 16. A quantitative discussion of certain features of electron lenses is included in Appendix F for the benefit of readers who are interested in a more mathematical approach.

While the quality of electron lens performance has improved rather steadily throughout the history of the electron microscope, the basic design of the lenses used in commercially available instruments has remained essentially unchanged. Axially symmetric magnetic fields have been employed as lenses in virtually all electron microscopes, and it is this type of focusing system which is considered almost exclusively in this chapter. However, further improvements in electron optical performance now seem to rely upon relatively radical innovations in lens design. The most promising approach seems to be the design of systems, both electrostatic and magnetic, in which there is a carefully controlled departure from complete axial symmetry. Such lens systems, which thus far remain experimental in nature, are described briefly. Otherwise the properties attributed to "electron lenses" in this chapter are those of the conventional axially symmetric systems.

LENS ACTION

An ideal electron lens causes the electron trajectories leaving any object point to cross at a single image point. Perfect lens action is completely analogous to that described for glass lenses on pp. 154–159, electron trajectories being analogous to the "rays" of geometrical optics. In Chapter 8 a proof was given (p. 159) which showed that a single refractive surface of spherical curvature approximates a perfect lens, provided that the aperture angle α is small. The crux of that proof was the derivation of the equation

(8-m)
$$\frac{n}{s} + \frac{n'}{s'} = \frac{n' - n}{R},$$

which gives the image position s' as a function of the object position s and of the refractive indices and the curvature of the surface. The important feature of (8-m) is the fact that s' is *independent* of the angle α, at which rays leave the object point (and of the related quantity h, which is the distance from the axis at which rays intersect the lens).

Proof that an axially symmetric field constitutes an electron lens is obtained in an analogous manner by deriving an equation for the location of the image. In this equation, also, the image position must be independent of the angle at which electrons leave the object, and dependent only upon the object position and the physical properties of the lens. The equation obtained is of the general form

(F-qq) $$sM(s') + N(s') = 0,$$

in which s and s' are the object and image positions, respectively, while M and N are functions of position along the lens axis. The values of M and N are determined by the distribution of the electrostatic or magnetic field but are independent of aperture angle.

The derivation of (F-qq) is considerably more detailed than that of (8-m) and is a matter of academic interest to the microscopist. In Appendix F a complete derivation is given for the case of an electrostatic lens, and the nature of the comparable derivation for magnetic lenses is outlined.

The important thing to understand about (F-qq) is that, like (8-m), it is *valid only when the aperture angle α is small*; that is, it is valid for *paraxial* electron trajectories. When α is appreciable, geometrical image defects occur which are analogous to those of light images, as described in Chapter 9. Either these aberrations or other limitations of technical origin may provide the ultimate limitation upon the performance of a given electron lens.

THE INEVITABLE NET CONVERGING ACTION OF ELECTRON LENSES

The conventional axially symmetric lens is always bounded by regions which are field-free. A consequence of this physical condition is that *the net action of electron lenses is inevitably convergent*. While limited portions of the field may be of diverging character, the same cannot be true of the lens as a whole. Unfortunately there seems to be no direct way of making this fact intuitively obvious, although a formal proof, given in Appendix F, establishes the property. Much attention has been devoted to attempts to design diverging electron lenses, but without avail. The serious consequence of this fact is that neither spherical nor chromatic aberrations can be corrected, as is done in light optics, by the use of "doublets" of positive and negative lenses. Aberrations can be controlled to some extent by the shape of the imaging field (cf. the process of "lens bending," as described on p. 193). However, whereas glass surfaces can be ground to high tolerances, magnetic fields cannot be precisely shaped at will. Thus certain field distributions which seem highly favorable on theoretical grounds are unattainable in practice.

The inability to correct geometrical aberrations fully, together with other limitations of a technical nature, precludes attainment of the full resolving power suggested by the electron wavelength. The nature of these limitations is discussed in later portions of this chapter.

ELECTRON TRAJECTORIES IN AN AXIALLY SYMMETRIC MAGNETIC FIELD

The motion of electrons in magnetic fields can be described by the *right-hand rule*, according to which the initial electron velocity is represented by the center finger of the right hand, the direction of the magnetic field by the forefinger, and the force exerted on the electron by the thumb. (Note that, if the center finger represents the direction of flow of *current*, rather than the movement of electrons, the right hand must be replaced by the left!)

The lines of magnetic flux in a plane containing the axis of an axially symmetric magnetic field are shown in Fig. 17-1a. (This figure illustrates the special case of a field which also is symmetrical about a perpendicular to the axis; the additional symmetry is not a requisite for lens action, however.) The flux density of the field is specified (in electron optics) by the vector **H**. Geometrically, the value of **H** at any point is represented by a vector tangent to the line of magnetic flux at that point. Thus, for the five points $A, \ldots, E$ as shown in Fig. 17-1b, **H**, like any other vector (cf. Appendix B) can be resolved into a set of three mutually perpendicular vectors. Specifically, the **H** vectors are the resultants of components $\mathbf{H}_z$, directed axially, and $\mathbf{H}_r$, directed perpendicular to the axis, and in the plane of Fig. 17-1. In general, there exists also a third component $\mathbf{H}_\theta$ which is perpendicular both to the axis and to the plane of Fig. 17-1; in axially symmetric systems, however, the forces directed into and out of the plane of Fig. 17-1 must be equal and opposite, so that the net value of $\mathbf{H}_\theta$ is always zero.

The resolution of **H** as the resultant of $\mathbf{H}_z$ and $\mathbf{H}_r$ at points A to E is illustrated in Fig. 17-2. At point A, $\mathbf{H}_r$ is appreciable in magnitude and is directed toward the axis. As the center of the field is approached, $\mathbf{H}_r$

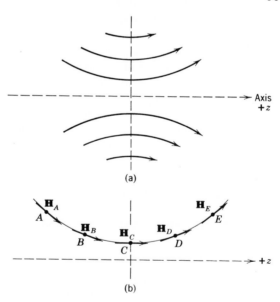

Fig. 17-1 (a) Lines of magnetic flux in an axially symmetric field; (b) flux densities at specific points.

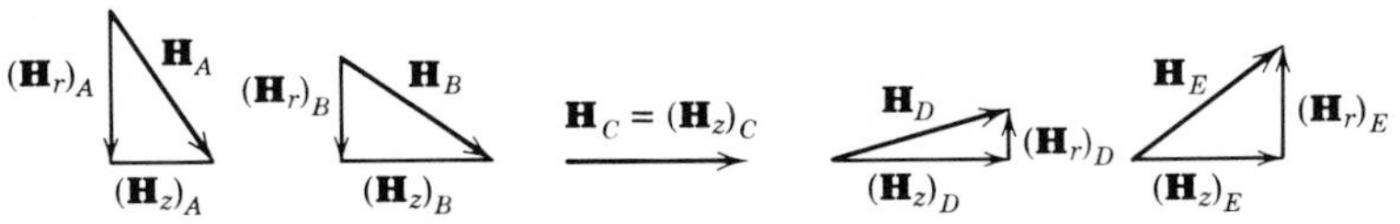

$(\mathbf{H}_r)_A$ $\mathbf{H}_A$ $(\mathbf{H}_r)_B$ $\mathbf{H}_B$ $\mathbf{H}_C = (\mathbf{H}_z)_C$ $\mathbf{H}_D$ $(\mathbf{H}_r)_D$ $\mathbf{H}_E$ $(\mathbf{H}_r)_E$

$(\mathbf{H}_z)_A$ $(\mathbf{H}_z)_B$ $(\mathbf{H}_z)_D$ $(\mathbf{H}_z)_E$

Fig. 17-2 Flux densities at points in an axially symmetric field as the resultants of their radial and axial components.

decreases in magnitude, while $\mathbf{H}_z$ increases; thus, at the center of the field, $\mathbf{H} = \mathbf{H}_z$ and $\mathbf{H}_r = 0$. Beyond this point, $\mathbf{H}_r$ again increases in magnitude but is now directed away from the axis. A continuous plot of the magnitudes of the radial and axial components of the magnetic field is shown in Fig. 17-3. Note that beyond the field the value of *both* components approaches zero.

The trajectory of an electron entering the field parallel to the axis may now be considered. If field strength essentially becomes effective at point A, the electron trajectory and $\mathbf{H}_z$ are parallel at A, so that *no* force is exerted upon the moving electron by the axial component of the field. $\mathbf{H}_r$, however, is perpendicular to the trajectory; application of the right-hand rule shows that a force is exerted to deflect the electron upward from the plane of Fig. 17-1. Once deflected, the electron no longer travels parallel to $\mathbf{H}_z$. Its velocity must then be represented as the resultant of three mutually perpendicular components v_z, v_r, and v_θ, as shown in Fig. 17-4. Each of these components of velocity interacts with *each* of the components of the magnetic field, producing a resultant force which is directed as determined by the right-hand rule. That is, the interactions between v_r and $\mathbf{H}_r$ and between v_z and $\mathbf{H}_z$ are both zero (i.e., since the fields and the respective components of velocity are parallel).

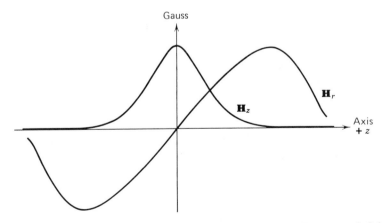

Fig. 17-3 Radial and axial components of the flux density of an axially symmetric field.

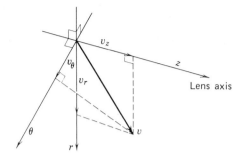

Fig. 17-4 Resolution of an electron velocity.

When the magnetic field is applied in the direction shown in Fig. 17-1, the interaction of v_θ with H_z produces a force in the $(-r)$ direction, that is, directed downward toward the axis. The interaction of v_θ with H_r produces a force in the $(-z)$ direction; that is, opposed to the incident velocity of the electron. The interaction of v_r with H_z produces a force directed into the plane of the page (i.e., in Figs. 17-1 to 17-4), while the interaction of v_z with H_r produces, as initially, a force out of the plane of the paper. These relationships between velocities, fields, and resultant forces are summarized in Table 17-1.

The net effect of these four forces may be estimated qualitatively. The interactions of v_r with H_z and of v_z with H_r produce forces which are exerted in opposite directions. Initially v_z and H_r are larger than v_r and H_z, respectively, so that movement *out* of the plane of the page predominates. Toward the center of the field H_r goes to zero and H_z to a maximum, while v_r increases and v_z diminishes. Accordingly, beyond some point in the field (as determined by the initial electron velocity and by the specific values of field strength) the electron trajectory rotates back *into* the plane of the page.

The force produced by the interaction of v_θ and H_r remains in the z direction throughout the lens since, beyond the center, where v_θ is reversed in direction, H_r is also reversed. The effect of this force is thus to de-

Table 17-1

Direction of Velocity Component	Component of Field	Direction of Resultant Force
$+\theta$	$+z$	$-r$
$+\theta$	$-r$	$-z$
$-r$	$+z$	$-\theta$
$+z$	$-r$	$+\theta$

celerate the electron. If the initial velocity is low, the electron can be trapped inside the magnetic field, where it executes a circular motion. Initial accelerations used in the electron microscope are large enough, however, that trapping is avoided.

The interaction of v_θ and H_z, which produces a force directed toward the axis, is the source of the focusing action of the magnetic lens. Note that magnetic lenses are highly "inefficient" in that only this minor portion of total field strength is actually effective in focusing the electron! The net trajectory of the electron is a spiral, since the convergence produced by the interaction of v_θ and H_z is accompanied by rotation resulting from the interaction of v_r and H_z. Thus, whereas the paths of individual light rays (or of electrons in electrostatic fields) always lie in a *plane*, electron trajectories in magnetic lenses are three-dimensional.

THE PHYSICAL ACTION OF THE MAGNETIC LENS

The preceding section shows that electrons traveling through axially symmetric magnetic fields experience a spiral trajectory of diminishing radius. The action of the magnetic lens is illustrated in Fig. 17-5. Figure 17-5a shows an electron trajectory in perspective. As shown, the electron leaves the object point O and travels parallel to the axis (z direction) until it enters the lens field at A. Within the lens the trajectory experiences both convergence and rotation, reaching point D at the lens limit. (Note that the electron does *not* cross the axis at point B; B is simply a position at which $y = 0$ but $x \neq 0$; similarly, at position C, $x = 0$ but $y \neq 0$.) Beyond D, the electron travels in a straight line through field-free space, cutting the lens axis at the second focal point F_2 and reaching the image plane at O'. As shown, the total rotation of the image is $(180 + \theta_i)°$; this rotation is a striking feature of the images formed by magnetic lenses. The value of θ_i is specified by (F-ww) (p. 727).

Planar projections of the electron trajectory, as seen from the lens limit and as seen from the side of the lens, are shown in Figs. 17-5b and 17-5c, respectively.

In a magnetic field applied in the opposite direction to that considered in Figs. 17-1 to 17-5, H_z and the electron velocity are antiparallel at the lens limit, producing, as before, no net force. Both v_θ and H_z are reversed in direction, so that the net radial force is directed *toward* the lens axis as before and the lens system is, again, convergent. However, since the force produced by the interaction of H_z and v_r is now reversed, rotation of the trajectory occurs in the opposite sense to that obtained previously. Relations between velocities, fields and resultant forces are summarized in Table 17-2, which may be compared with Table 17-1. The trajectory, as

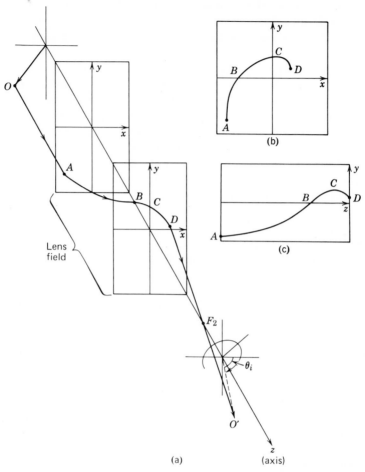

Fig. 17-5 Action of the magnetic lens: (a) in perspective; (b) electron trajectory in projection, along direction of propagation; (c) electron trajectory in projection, side view.

Table 17-2

Direction of Velocity Component	Component of Field	Direction of Resultant Force
$-\theta$	$-z$	$-r$
$-\theta$	$+r$	$-z$
$-r$	$-z$	$+\theta$
$+z$	$+r$	$-\theta$

seen in projection from the lens limits, is thus as shown in Fig. 17-6, which is analogous to Fig. 17-5.

THE DESIGN OF MAGNETIC LENSES

As can be deduced by observing the behavior of iron filings, a current flowing through a wire induces a magnetic field in the vicinity of the wire. The induced lines of magnetic force are oriented tangentially with respect to the wire, as illustrated in Fig. 17-7a. This magnetic field can be "concentrated" by twisting the wire into a coil, as shown in Fig. 17-7b. Induced field strength is then proportional to the number of turns of the wire, the strength of current flowing in the wire, and the magnetic permeability μ of the medium inside the coil. μ is unity for a vacuum but may be as large as several hundred thousand for ferromagnetic materials. Thus extremely strong magnetic fields can be induced within a ferromagnetic core surrounded by a current-carrying coil; for example, as shown in Fig. 17-7c.

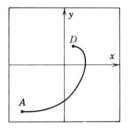

Fig. 17-6 Action of magnetic field applied in the reverse direction to that considered in Fig. 17-5. Electron trajectory in projection, along direction of propagation.

As Fig. 17-7b shows, the field induced by a coil tends to be axially symmetric; thus the device shown in Fig. 17-7c constitutes a simple type of electron lens. The focusing action of this type of lens is very weak, however, since the lines of flux are nearly parallel to the electron trajectories along most of the length of the coil. Only at the ends do the lines curve so as to be cut by the electron trajectories and thus exert a net force on the moving electrons.

Effective magnetic lenses are achieved by including a *gap* in the ferromagnetic core. Two *like* magnetic poles (i.e., both north or both south) are induced above and below the gap by causing current to flow in opposite directions through the respective windings. As illustrated in Fig. 17-8a, the lines of magnetic flux then curve around the edges of the ferromagnetic material and thus are concentrated within the gap in a direction perpendicular to the bore and to the electron trajectories. The field strength $\mathbf{H}_z$ induced within the gap is commonly plotted as a function of the distance z along the axis in the manner shown in Fig. 17-8b. Maximum field strengths are of the order of 20,000 gauss (G), although the availability of new types of magnetic materials now makes the development of lens fields of the order of 100,000 G seem feasible.

In practical magnetic lenses the (ferromagnetic) *pole pieces* shape the

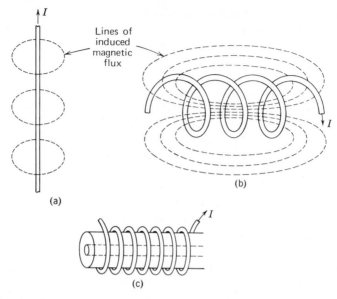

Lines of
induced
magnetic
flux

(a)

(b)

(c)

Fig. 17-7 Induced magnetic flux: (a) in the vicinity of a wire; (b) in the vicinity of a coil; (c) inclusion of a core within a coil.

bore and gap of the lens; a typical design is shown in Fig. 17-9. (Most lens systems also include one or more limiting apertures, which are not included in Fig. 17-9.) Lens strengths vary, as a function of the current flowing through the windings, up to a level of current at which magnetic

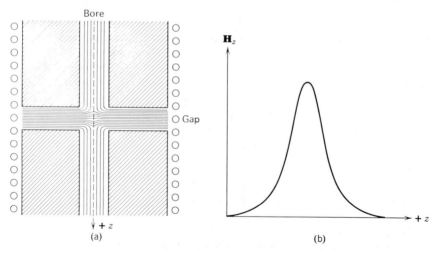

Bore

Gap

$+ z$

(a)

$\mathbf{H}_z$

$+ z$

(b)

Fig. 17-8 Magnetic lens: (a) basic design; (b) plot of magnetic flux.

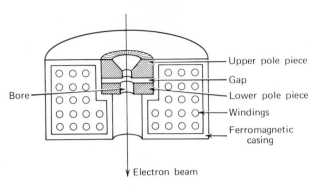

Fig. 17-9 Practical magnetic lens (cross-section).

saturation is produced. (For a discussion of magnetic saturation effects, see p. 362.) The machining of the pole pieces and the homogeneity of the ferromagnetic material are both critically important in determining the extent to which true axial symmetry of the system is achieved and, thus, the resolving power of the lens.

A practical problem in the design and operation of magnetic lenses is heating by the rather high levels of current which pass through the windings. Excessive heating may damage the lens, while all degrees of heating influence the level of current and thus the focal length, which is produced by a given level of applied voltage. (It will be recalled that the resistance of wires normally increases as temperature rises.) Many lenses are so constructed that they can be conduction-cooled by circulating cold water.

HYSTERESIS

Induced magnetic fields result from the orientation of "magnetic domains" within the ferromagnetic material, as shown schematically by Fig. 17-10 (N and S specify north and south magnetic poles, respectively). The induction of magnetization thus involves a *physical movement* within the magnetized material; this can be achieved only by overcoming some degree of inertia. A consequence of this fact is that magnetization,

Fig. 17-10 Magnetization.

that is, the production of magnetic flux H, tends to lag behind the magnet-izing force applied. Figure 17-11a illustrates the variation in H which is effected in a ferromagnetic core as the current passing through the windings is raised from zero to a high value and then returned to zero. The induced flux can be returned to zero only by application of current in the opposite direction, as indicated in Fig. 17-11b.

For a material which is once magnetized, variations of magnetization with current are described by the *hysteresis loop*, shown in Fig. 17-11c. (The dimensions of the hysteresis loop are characteristic of the particular magnetic material.) Demagnetization is achieved by subjecting the material to a reversing current of diminishing magnitude, as illustrated by Fig. 17-11d.

A practical consequence of the phenomenon of hysteresis is that the level of current used to energize a magnetic electron lens does not precisely specify the lens strength (i.e., focal length) achieved. For

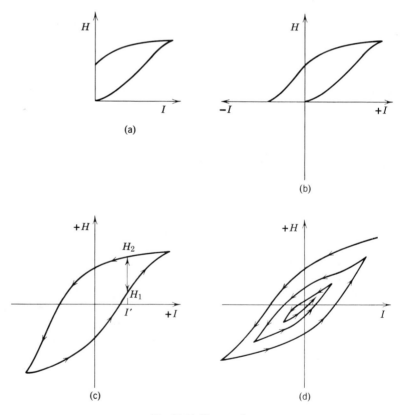

Fig. 17-11 Hysteresis.

example, as shown in Fig. 17-11c, a current I' may correspond to magnetic field strengths which range between H_1 and H_2, as determined by the previous condition of magnetization of the ferromagnetic core.

Electron microscope lenses may be *normalized* (i.e., returned to a standard level of magnetization) at any given setting by briefly reducing the applied current to zero some predetermined number of times. This procedure is essential when knowledge of the exact level of magnification is required.

ABERRATIONS IN ELECTRON LENSES

Axially symmetric fields, like glass surfaces of spherical curvature, act as lenses for paraxial rays, only; that is, under conditions for which the approximation $\sin \alpha \cong \alpha$ is valid. Thus, when electrons leave object points at appreciable angles with respect to the lens axis, geometrical aberrations occur in the image. These defects are essentially similar to those of glass lenses (cf. Chapter 9), although the conditions prevailing in the electron microscope influence their effects on the image as well as introducing certain additional types of aberration.

The problem of aberrations is a much more critical one in the electron microscope than in the light microscope, because systematic correction of spherical and chromatic aberrations by the construction of electron lens doublets is not possible. The use of quadrupole and octupole lens elements for correction of aberrations is under investigation at the present time (see p. 372), but the development of these systems has not yet proceeded to a point at which they would be useful in high resolution systems. Furthermore the ultimate resolution limit set by diffraction effects is only about 0.2μ (2000 Å) for glass lenses, a value large in comparison with the resolutions of 2 to 5 Å which are achieved by electron lenses of high quality. Thus defects at a level which would be undetectable in the light microscope are of major consequence in the electron microscope.

The nature and correction of aberrations in electron lenses is discussed in the following sections. Of these, spherical aberration may limit the ultimate performance of electron lenses. Coma, third order astigmatism and curvature of field also affect electron images, but are, fortunately, of minor importance. Distortion is often observed in electron images. In general, the effects of space charge, which produces a spreading out of electron trajectories by repulsion between these negatively charged particles, must be taken into account in the design of lenses.

The limitations imposed by aberrations of technical origin (i.e., imperfections in construction or regulation, or external disturbances) are vastly more important in high quality electron lenses than in their optical counterparts. The effects of lens asymmetry (so-called "astigmatism") are inevitably present in electron images at some level. Attempts to avoid chromatic aberration by the use of monochromatic electrons (i.e., electrons accelerated by a uniform voltage) are ultimately limited by the imperfect regulation of accelerating voltage. Also, lens current fluctuations introduce image defects.

SPHERICAL ABERRATION

The defect of spherical aberration has been discussed with respect to glass lenses on p. 188. Most of that discussion is relevant to electron lenses also; an exception is the unavailability of methods for correcting spherical aberration by the construction of electron lens doublets.

Spherical aberration causes *point* objects to be imaged as extended *discs*. The radius of the disc is the "lateral spherical aberration" defined by (9-c) and illustrated in Fig. 9-2. This radius, designated $(\Delta r_i)_{\text{S.A.}}$, is given for an image formed by a lens of aperture angle α as

$$(17\text{-a}) \qquad (\Delta r_i)_{\text{S.A.}} = M(C_3\alpha^3 + C_5\alpha^5 + C_7\alpha^7 + \dots),$$

where M is the magnification of the lens, and the terms $C_3\alpha^3, \dots, C_n\alpha^n$ represent the third to nth order components of aberration. An expression which is equivalent to (17-a) is

$$(17\text{-b}) \qquad (\Delta r_i)_{\text{S.A.}} = M(C_s f\alpha^3 + C'_s f\alpha^5 + \dots),$$

where f is the focal length of the lens, and $C_s = C_3/f$ is known as the *spherical aberration constant*.

The confusion of dimensions in object space which corresponds to the spherically aberrant image is $(\Delta r)_{\text{S.A.}}$, where

$$(17\text{-c}) \qquad (\Delta r)_{\text{S.A.}} = \frac{(\Delta r_i)_{\text{S.A.}}}{M}.$$

Since the aperture angle used in electron microscopy is always small, the spherical aberration is represented to a good approximation by the third order aberration term alone (although fifth order defects must also be considered in the design of highly corrected lenses). Thus

$$(17\text{-d}) \qquad (\Delta r_i)_{\text{S.A.}} \cong MC_3\alpha^3 = MfC_s\alpha^3.$$

The spherical aberration constant of a lens can be computed for hypothetical lens systems (i.e., systems for which a mathematical description of the field distribution is available). $(\Delta r_i)_{\text{S.A.}}$ can be *measured* in images formed by real lenses only if other types of defects do not produce larger radii of confusion about the images of points.

Minimal spherical aberration is obtained when minimal values of C_s are used at the shortest possible focal length. The focal length of the objective lenses presently used in electron microscopy is about 2 or 3 mm, although somewhat smaller values (ca. 1 mm) are feasible. Minimum values of C_3 are also about 1 mm for magnetic lenses. Minimum values of the spherical aberration constants for electrostatic lenses are substantially larger, a fact which has precluded the development of this type of electron lens for high resolution systems.

Spherical aberration is limited by the use of optimal magnetic field distributions (the factor which determines the value of C_s) and by *restricting the angular aperture of illumination*. The smaller the aperture of a lens, the less is the spherical aberration of the image. Ultimately, however, a limitation is imposed by the increase in the size of the Airy disc., $(\Delta r_i)_{\text{Diff}}$, which occurs as the aperture is reduced. As developed in Chapter 10,

$$(17\text{-e}) \qquad (\Delta r_i)_{\text{Diff}} = \frac{MK\lambda}{\sin \alpha},$$

where K is a constant of the order of unity.

An optimal value of the aperture angle thus exists at which the effects of spherical aberration and diffraction are exactly equal. This is the case when

$$(17\text{-f}) \qquad C_3\alpha^3 = \frac{K\lambda}{\sin \alpha}.$$

Or, since the value of α which satisfies (17-f) is *small*, the approximation $\sin \alpha_{\text{opt}} \cong \alpha_{\text{opt}}$ may be made. Taking $K = 1$, this gives

$$(17\text{-g}) \qquad \alpha_{\text{opt}}^4 = \frac{\lambda}{C_3} \quad \text{and} \quad \boxed{\alpha_{\text{opt}} = \sqrt[4]{\frac{\lambda}{C_s f}}}.$$

The corresponding value for the ultimate resolving power of the electron lens may then be found by substituting α_{opt} in (17-e) to obtain

$$(17\text{-h}) \qquad d_{\min} = (\Delta r)_{\text{Diff}} = (\Delta r)_{\text{S.A.}} = \frac{\lambda}{\sin \alpha_{\text{opt}}} \cong \frac{\lambda}{\alpha_{\text{opt}}}$$

or

$$\boxed{d_{\min} = \lambda^{3/4} C_3^{1/4} = \lambda^{3/4} C_s^{1/4} f^{1/4}}.$$

The treatment by which (17-h) is developed must be regarded as somewhat approximate, since the effective minimal size of Δr is affected by the three-dimensional distribution of intensities in the spherical aberration and diffraction figures. Nevertheless, evaluation of (17-h) yields values of $d_{\min}$ which are in good agreement with those obtained experimentally.

For example, if C_s and f are both equal to 1 mm (10^{-1} cm) and for an accelerating potential of 100 kV, combination of (16-t) and (17-h) yields

(17-i)
$$d_{min} = \left(\frac{12.27 \times 10^{-8}}{\sqrt{V}}\right)^{3/4} (10^{-2})^{1/4}$$

$$= 8.74 \times 10^{-8} \times 3.16 \times 10^{-1}$$

$$= 2.76 \text{ Å.}$$

Although "resolutions" slightly better than 2 Å have been obtained in the observation of extended lattices, the best values obtained for point-to-point resolution have in fact approached 3 Å.

Equation 17-h specifies the resolving power of an electron microscope in which spherical aberration is the limiting factor. Often other disturbances, including, principally, lens asymmetry, chromatic aberration, or faulty regulation of lens currents, establish a more severe limitation on the performance of lenses. That spherical aberration limits the resolving power of the electron microscope is rather well known; it is less generally appreciated that other factors are of equal practical importance. For further discussion of this point, see Chapter 20.

The *effective aperture* of an electron lens is that angle, α_{eff}, within which electrons contribute to image formation, rather than being so aberrantly focused that they contribute to a diffuse background intensity. The value of α_{eff} varies with the level of resolution sought in the image. The relation between effective aperture and image contrast is discussed on p. 426.

In principle, spherical aberration and other defects of electron lenses could be overcome by recording electron holograms (cf. p. 250 and [1]) and subsequently reconstructing the image by visible light with the use of corrected lenses. Unfortunately, however, the coherence of available electron sources is too low to permit the formation of holograms of adequate quality.

The spherically aberrant electron image is comparable in appearance to the spherically aberrant light image diagramed in Fig. 9-7. "Softness," which is observed in all image zones, becomes more pronounced at the periphery.

COMA, CURVATURE OF FIELD, AND THIRD ORDER ASTIGMATISM

The third order defects of coma, curvature of field, and astigmatism, which have been described on pp. 197, 209 and 204 respectively, are also present in electron images. However, since they are of consequence only at relatively large aperture angles and for relatively extended fields of view, their effects on resolution of electron images are negligible. Coma is a function of the square, and curvature of field and astigmatism are functions of the first power of the aperture angle, while coma is a function of the first power, and curvature of field and astigmatism are functions of the square of the distance of the object point from the lens axis. In the electron microscope, aperture angles are extremely small (10^{-2} to 10^{-3} rad) whereas, especially when high resolution is sought, magnifications are large, and the field of view correspondingly restricted.

DISTORTION

The third order defect of distortion has been described on p. 212, where it was shown that this aberration is observed in lens systems which contain apertures. Since it is true that distortion increases as a function of the cube of the distance of the object from the microscope axis, it might seem that the defect would be negligible in the very limited electron microscope field. Nevertheless, since extremely small apertures are used in electron lens systems, appreciable distortion of electron images may occur at magnifications less than about 10,000×. At higher magnification the field of view does become so limited that distortion is negligible. Thus the aberration is not a limitation in high resolution electron microscopy, but may be troublesome at low magnifications. When extended areas of cellular specimens are to be reconstructed by piecing together a series of overlapping micrographs, an exact flat fit is impossible if distortion of either type (pincushion or barrel) is present. Usually it is possible to eliminate distortion at any given level of magnification by adjustment of the spacings of lenses and apertures.

CHROMATIC ABERRATION

The focal lengths of electron lenses, like those of glass lenses, vary according to the wavelength of the incident illumination (i.e., according to the accelerating potential of the beam). The effect may be accounted for in particulate terms by noting that electrons which are accelerated through large potentials achieve high velocities. The greater the electron velocity, the larger is the momentum and thus the less susceptible is the trajectory to deflection by the lens field.

Variations in the focal length of a lens according to the acceleration potential of the electron beam produce chromatic aberration in the image. The effect is analogous to the chromatic aberration of glass lenses, as discussed on p. 214. Chromatic aberration, like spherical aberration, cannot be corrected in electron lenses; it is merely avoided through the use of uniformly accelerated electrons. However, although voltage supply is highly regulated in the electron microscope, variations in the acceleration of the electron beam must persist at some level, producing chromatic aberration of the image.

In the presence of chromatic aberration, point objects are imaged as a disc of radius $(\Delta r_i)_c$, such that

$$(17\text{-j}) \qquad\qquad (\Delta r_i)_c = M\alpha\Delta f.$$

As shown in Appendix F (equation F-zz), variations in accelerating voltage and focal length are related according to

$$(17\text{-k}) \qquad\qquad \frac{\Delta f}{f} = C_c \frac{\Delta V}{V},$$

where C_c, a constant characteristic of the lens, is known as the *chromatic aberration constant*. Substitution of (17-k) in (17-j) gives

$$(17\text{-l}) \qquad\qquad (\Delta r_i)_c = Mf\alpha C_c \frac{\Delta V}{V}.$$

The corresponding distance in object space, $(\Delta_r)_c$, establishes the resolution limit imposed by chromatic aberration

$$(17\text{-m}) \qquad\qquad \boxed{(\Delta r)_c = (d_{\min})_c = f\alpha C_c \frac{\Delta V}{V}}.$$

C_c is analogous to the quantity C_s, which describes the spherical aberration of a lens. Note, however, that the term "chromatic aberration constant" sometimes specifies, instead, the product fC_c and is then analogous to C_3 rather than to C_s. Unfortunately, the two forms of these two constants have not been consistently specified in the literature. They may be distinguished by noting that C_s and C_c as defined here are unitless quantities, while C_3 is given in units of length. Attainable values of C_c may be as low as 1.0.

Three sources of variation in electron velocity (i.e., in effective acceleration potential) may be distinguished. As explained on p. 350, electron beams are obtained by acceleration of the electrons freed from a heated metal surface. Nonuniformity of electron velocities may thus originate at the electron source by two mechanisms: *Variations in thermal velocity* and *fluctuations in accelerating potential*. A third factor, *energy loss at the specimen*, is not directly related to the characteristics of the imaging system; its discussion is postponed to p. 425.

The thermal velocities of electrons range from zero, through a preferred value which is characteristic of the temperature of the emitter, to somewhat higher values. Variations in velocity are thus superimposed upon the much greater velocity acquired by acceleration. The most probable thermal velocity, as expressed in statvolts ($\doteq$ practical volts/300), can be obtained from

$$(17\text{-n}) \qquad\qquad v = \frac{kT}{e},$$

where k is the Boltzmann constant (1.38×10^{-16} erg/degree), T is the absolute temperature of the emitter (about 2400°K for the tungsten filaments used in electron microscopes), and e is the electronic charge (4.803×10^{-10} statcolumb). Evaluation of (17-n) shows that the preferred velocity of thermal electrons is equivalent to acceleration through about 0.2 volt. While some electron velocities are greater than this, virtually all the electrons leaving the tungsten surface have a velocity of less than 1 volt. Thus, for $\Delta V = 1$ volt, $V = 100$ kV, $f = 3$ mm, $\alpha = 5 \times 10^{-3}$ rad, and $C_c = 1.0$, (17-m) gives the approximate thermal limit on resolution as

$$(17\text{-o}) \quad (d_{\min})_c, \text{thermal} = \frac{3 \times 10^7 \text{Å} \times 5 \times 10^{-3} \times 1.0 \times 1.0 \text{ volt}}{10^5 \text{ volt}} = 1.5 \text{ Å}.$$

The effect is not limiting in present electron microscopes, but it could easily become so in systems of somewhat higher resolving power.

Variations in the level of accelerating voltage are an important source of chromatic aberration, and they may be the factor which limits the resolving power of a given electron microscope. Equation 17-m shows that if the minimum resolvable distance is to be reduced to 2 Å, the quantity $\Delta V/V$, which here represents the "ripple voltage" remaining after rectification of the ac power supplied to the microscope, must be no greater than one part in 10^5. Thus the ripple voltage level must be reduced to one volt for 100 kV operation, or to 0.5 volt for 50 kV operation. These specifications can be met in present systems under optimum operating conditions, although a somewhat higher ripple voltage level (about 5 volts) is often tolerated. Future improvement of power supplies to give $\Delta V/V = 10^{-6}$ seems feasible. (It should be appreciated that the control of voltage at these levels is an impressive engineering achievement!) While even further improvements are probably possible, the expense and complexity of such systems would not be justified as long as factors other than chromatic aberration continued to impose substantial limitations upon $d_{\min}$.

Electron images which suffer from chromatic aberration may be thought of as being produced by the superposition of a series of images, each of which is formed by electrons of a different wavelength. Since the focal length of the lens is different for each wavelength, the superposed images are of different magnifications. Furthermore, since images are rotated through a different angle for each level of magnification by the magnetic lens, the superposed images are also rotated with respect to each other. An object and its (much exaggerated) chromatically aberrant image are as illustrated in Fig. 17-12. (Note that Fig. 17-12 shows, principally, the effects of the electron optical analog of *lateral* chromatic aberration.) As a consequence of the accompanying *longitudinal* chromatic aberration, only one of the series of superposed images shown in Fig. 17-12b is in exact focus.

Images of the type shown in Fig. 17-12b are generally described as suffering from a "high voltage smear."

LENS CURRENT FLUCTUATIONS

Lens current levels, like accelerating potentials, vary at the level of about 10^{-5}; that is, $\Delta I/I \cong 10^{-5}$. Corresponding fluctuations in lens strength (i.e., focal length) are induced. Since these fluctuations are rapid, images of different magnifications are superimposed. The effect on the observed image is similar to that shown in Fig. 17-12, although the image defects resulting from current instability may differ in detail from those produced by voltage instability (i.e., the relative magnitude of the defects may be different at different positions in the image).

A possible method of reducing effects of both current and voltage instability is the electronic coupling of voltage and current supplies in such a way that instantaneous variations can be made to nullify each other. Another approach to the elimination of current fluctuations is the use of "superconducting lenses." At temperatures close to absolute zero the phenomenon of superconductivity is observed in certain materials; that is, application of a single pulse of voltages causes current to flow for an indefinite time. The level of currents produced in this way is thus free of any variations in the level of voltage supply. Liquid helium is used for the cooling of superconducting lenses.

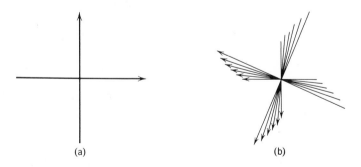

(a)   (b)

Fig. 17-12 Chromatic aberration by the magnetic lens: (a) object; (b) image.

Whereas rapid fluctuations of lens current (over a period of seconds or milliseconds) affect image quality, slow variations (over periods of hours or weeks) affect the level of magnification. The same is true of fluctuations in accelerating voltage. Electron microscope exposures are usually made over a period of one to three sec; stability for this length of time is essential if full resolution of image details is to be obtained.

The effect of lens current fluctuations is often described in the literature as "chromatic aberration." Strictly speaking, this is incorrect, since variations in focusing action occur as a function of *time* and are independent of electron wavelength.

LENS ASYMMETRY

The technology of producing homogeneous optical glasses and of grinding spherical glass surfaces has long since reached a high degree of perfection. Techniques for grinding aspheric surfaces also have advanced in recent years. Thus defects arising in the construction of glass lenses can be reduced to a level well below the resolving power of these systems. The situation is quite different in electron lenses, for which criteria are more stringent in proportion to the higher resolving powers sought. Neither the homogeneity of available magnetic materials nor the accuracy of machining these metals into lens pole pieces is adequate for the direct production of lenses capable of displaying the theoretical resolving power established by the spherical aberration-diffraction limit. Fortunately, corrections may be superposed upon the lens field during the operation of the electron microscope; if the correction is adequate, lens asymmetry does not limit the resolving power of the instrument.

Asymmetry has the effect of producing images of which the focal level varies with direction. (The analogous defect of glass lenses has been discussed on p. 202.) To a first approximation the directions of maximum and minimum focus are separated by 90°, so that the system is equivalent to the combination of a cylindrical lens with one of spherical curvature (as shown by Fig. 9-19). Correction of the defect (termed *compensation of asymmetry*) is attained by imposing a second cylindrical lens field of the same magnitude as that already present, but oriented at right angles thereto.

Lens asymmetry may be introduced by the build-up of layers of contamination on lens elements; thus compensation of asymmetry must be carried out periodically. A residual lens asymmetry can sometimes be observed when the correction procedure is completed. This defect is known as *quadrature*, since a single quadrant of the image then differs in focal level from the other three.

The appearance of images formed by asymmetric electron lenses and the nature of the correction procedure are discussed on p. 409 after the description of focus effects in electron images given on p. 402. There it is shown that, in the absence of other aberrations, resolving power is limited by lens asymmetry approximately to the extent predicted in Chapter 18:

(18-f) $$d_{\min} = \sqrt{\lambda \, \Delta f},$$

where Δf is the maximum difference in the focal length of the asymmetric lens, and λ is the electron wavelength. Evaluation of (18-f) shows that Δf must be reduced to about 0.3 μ in order to obtain resolution at the 10 Å level, and to about 0.01 μ for resolution of 2 Å. Correction even to the 0.3 μ level is difficult either to distinguish or to maintain over any usefully long period. Thus compensation of asymmetry becomes, in the limit, somewhat of a process of trial and error. While, in principle, asymmetry may be corrected to any desired level, in practice it is very often the factor which limits the resolution achieved by electron lenses.

INNOVATIONS IN ELECTRON LENS DESIGN

Possibilities for further improvement of the conventional axially symmetric magnetic electron lens are limited, since there seems to be little hope of substantially reducing spherical aberration beyond the level now attained. Therefore designers are exploring the use of other types of lens systems, principally of systems which lack complete rotational symmetry. They include combinations of *quadrupoles* and/or *octupoles*, that is, of four or eight symmetrically positioned electrodes. (Quadrupole and octupole systems have 4- and 8-fold axes of symmetry, respectively, but not complete axial symmetry; that is, rotation of the systems about their axes through an arbitrary angle does not necessarily generate an orientation which is indistinguishable from the original one.)

Any electrode ("pole") is surrounded by an electrostatic field, which may be represented by a series of equipotential surfaces. Similarly, a magnetic pole is surrounded by a magnetic field, which may be represented by lines of magnetic flux. A negatively charged electrostatic dipole and the surrounding equipotentials are shown in Fig. 17-13. Electrons are repelled by the negatively charged poles, thus being deflected in the x direction toward the axis. However, no comparable deflection is experienced in the y direction (i.e., perpendicular to the plane of the figure). Therefore electron trajectories which are inclined with respect to the plane of Fig. 17-13 are focused at points on a line image which extends in the y direction (i.e., perpendicular to the z axis and to the line of the dipole). The action of the dipole may be compared to that of a cylindrical

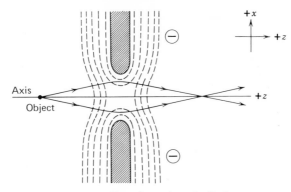

Fig. 17-13 Focusing action of a dipole.

lens, which forms a line image of a point illuminated by light. A dipole like that shown in Fig. 17-13, but oriented perpendicular to the plane of the figure, similarly forms a line image which extends in the x direction. The action of a mutually perpendicular pair of dipoles is thus to focus rays originating from a point object with respect to two mutually perpendicular directions, as shown in Fig. 17-14. Images of the type shown in Fig. 17-14c are also formed when the pairs of dipoles lie in the same plane, constituting a quadrupole.

The action of magnetic dipoles is similar to that shown in Figs. 17-13 and 17-14, except that the line images are rotated with respect to the plane of the dipoles. The general design of a magnetic quadrupole lens is illustrated in Fig. 17-15.

Further "compression" of the image shown in Fig. 17-14c may be achieved by the action of additional partially symmetrical lens elements. It might seem that the net effect of a number of multipole lenses would be, at best, the formation of an image with resolution approximating that of a lens of full axial symmetry. However, correction of aberrations is

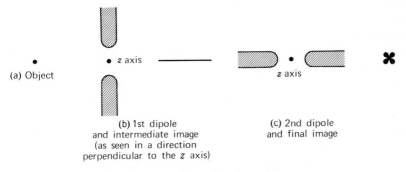

Fig. 17-14 Focusing action of a quadrupole.

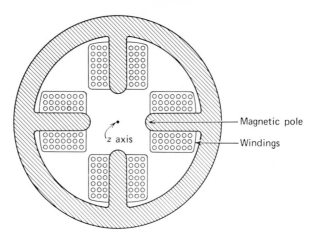

Fig. 17-15 Practical magnetic quadrupole lens.

possible in these systems, at least in principle, because the strength of individual poles may be varied independently. Encouraging results have been obtained in pilot systems of relatively low resolution, where it has been possible to eliminate third order spherical aberration from the image, leaving only fifth and higher order defects. Thus far, however, the ultimate resolving power of conventional lenses has not been matched in quadrupole systems.

A quite different method for the possible elimination of spherical aberration from electron images involves exploitation of the wavelike properties of the electron beam. Spherical aberration may be thought of as arising from interference between portions of the electron wavefront which are out of phase with each other (i.e., as a consequence of having passed through different zones of the lens). An aperture containing a series of appropriately spaced concentric openings could, in principle, be employed to remove the out-of-phase elements of the wavefront. As sketched in Fig. 17-16, rays OBO' and $OB'O'$ reach point O' half a wavelength out of phase with rays OAO', OCO', etc. Removal of rays such as OBO' by the aperture would result in the formation of the image exclusively by rays which are, approximately, *in* phase. It appears that construction of suitable apertures, containing rings not more than about $100\,\mu$ in diameter, might be technically possible. However, although elegant in principle, this method for lens correction would be seriously limited in practice by the accumulation of contamination on the edges of the rings.

The design of lens systems in which variations in lens current levels are avoided has been discussed on p. 370.

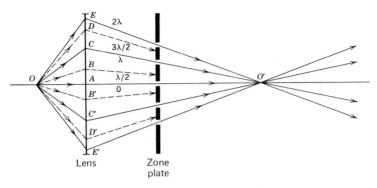

Fig. 17-16 Zone plate for correction of aberrations.

REFERENCES

General References

C. E. Hall, *Introduction to Electron Microscopy*, 2nd ed., McGraw-Hill, New York, 1966, Chapter 4 (The Electrostatic Lens), Chapter 5 (The Magnetic Lens), and Chapter 6 (Aberrations and Diffraction). This book is a classical and authoritative work on the electron microscope.

M. E. Haine, *The Electron Microscope*; *The Present State of the Art*, Interscience, New York, 1961.

E. Ruska, in R. Barer and V. E. Cosslett, *Advances in Optical and Electron Microscopy*, Vol. I, Academic Press, New York, 1966, pp. 116–179, "Past and Present Attempts to Attain the Resolution Limit of the Transmission Electron Microscope."

A. Septier, in Barer and Cosslett, *op. cit.*, pp. 204–274, "The Struggle to Overcome Spherical Aberration in Electron Optics." This article, together with Ruska's, provides an up-to-date description of the status of electron lens design.

V. K. Zworykin, C. E. Hall, G. A. Morton, E. G. Ramberg, J. Hillier, and A. W. Vance, *Electron Optics and the Electron Microscope*, Wiley, New York, 1945, Chapter 1 (Electron Optics), Chapter 6 (Aberrations and Tolerances in the Electron Microscope), Chapter 11 (Determination of Potential Distribution), Chapter 12 (Electron Trajectory Tracing), Chapter 13 (Gaussian Dioptrics of Electrostatic Lenses), Chapter 14 (Magnetic Fields), Chapter 15 (Electron Motion in Magnetic Fields and Magnetic Lenses), Chapter 17 (Magnitude and Correction of Electron Lens Defects). While decidedly out of date on the subject of electron *microscopy* (which is discussed in the first nine chapters), the book is a valuable and classical reference on electron *optics*. Many of the topics discussed in the present chapter are developed in detail in the latter half of the book.

F. Lenz, in G. F. Bahr, and E. H. Zeitler, Eds., *Quantitative Electron Microscopy*, Armed Forces Institute of Pathology, Washington, D.C., 1965. (This book also appeared as part of Vol. 14 of the journal; *Laboratory Investigation*, pp. 70/808-80/818.) This article, "The Influence of Lens Imperfections on Image Formation," includes a discussion of the zone plate.

B. M. Siegel, in B. M. Siegel, Ed. *Modern Developments in Electron Microscopy*, Academic Press, New York, 1964, Chapter 1 (The Physics of the Electron Microscope).

R. Uyeda, Ed., *Electron Microscopy, 1966* (Proceedings of the Sixth International Congress for Electron Microscopy), Maruzen Co., Ltd. Tokyo, 1966, Vol. I, pp. 19–284. This section, "Electron Optics and Instrument Design," includes abstracts of papers describing many aspects of current work on electron microscope design.

Specific Reference

[1] D. Gabor, *Nature*, **161**, 777 (1948).

CHAPTER 18

The Electron Microscope

In the electron microscope as in the light microscope, radiation is incident upon a specimen, interacts with the specimen, and is subsequently focused by a series of lenses which form a magnified image. The two types of instrument are broadly analogous in terms of optical principles. They differ markedly, however, in details of construction and operation. In this chapter the general features of the electron microscope are described, with emphasis on comparisons and contrasts to light microscopy. Certain procedures used in the operation and maintenance of the electron microscope, and upon which the attainment of high resolution depends critically, are also described and explained. These descriptions are necessarily rather general in nature, since the design and method of operation of different models of electron microscopes varies much more radically than for light microscopes. Manufacturers' literature must be consulted for instructions for carrying out a particular procedure, such as, for example, lens alignment, with a given type of instrument. The discussion presented here is intended only to explain the rationale of such procedures.

Understanding of the material presented in this chapter can be aided by familiarity with the properties of the corresponding optical systems as described in Chapters 8, 12, and elsewhere in this book, and with those of electron lenses, as discussed in Chapter 17.

GENERAL PLAN OF THE ELECTRON MICROSCOPE

The optical elements of the electron microscope, as of the light microscope, consist of source, condenser lens, specimen, objective lens, and projector lens(es). Mechanical and other features of the electron microscope

377

differ appreciably from those of the light microscope, as a consequence of technical problems encountered in obtaining electron images. The optical systems of light and electron microscopes are compared in Fig. 18-1.

The source (electron gun) is usually located at the top of the microscope column, rather than at the bottom, as is standard in light optical systems. The electron gun is the source of an electron beam of high, and usually to some degree, adjustable intensity consisting of electrons which travel with uniform velocity.

Since the mean free path of the electron in air at atmospheric pressure is only of the order of one centimeter, the electron microscope must be operated at moderately high vacuum. Vacuum operation not only permits transmission of the electron beam through the microscope column, but also avoids deflections of electron trajectories otherwise than by inter-actions with the specimen or lens fields. Pressures of the order of 10^{-4}

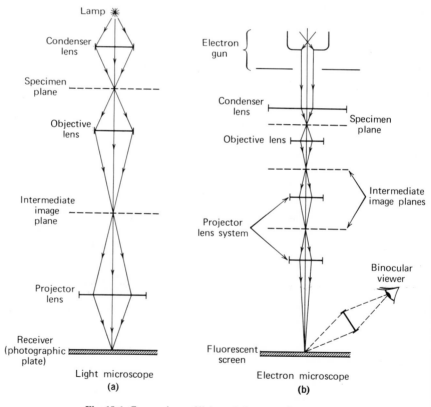

Fig. 18-1 Comparison of light and electron microscopes.

or 10^{-5} mm of mercury are used. The vacuum system must also be workably fast, since delays incurred in restoring an operating vacuum after the exchange of specimens or photographic materials should not be unduly prolonged.

The illumination supplied by the electron gun is focused at or near the specimen plane by means of a condenser lens, or pair of condenser lenses. In a few systems, focusing action of the electron gun itself has been used in place of a separate condenser system.

The electron microscope specimen is so mounted that it can easily be exchanged and can be translated reproducibly through extremely small distances. Requirements for mechanical stability of the specimen and specimen stage are high; the specimen must also be stable with respect to irradiation and heating by the electron beam.

Electrons which leave the specimen are focused by the objective lens. The objective, like its counterpart in the light microscope, is the most critical element of the optical system, since the image formed by this lens must undergo further magnification.

The counterpart of the ocular or projector lens of the light microscope is usually a *pair* of lenses, termed the "intermediate" and "projector." Three-stage imaging makes it possible to achieve a wide range of magnifications in a microscope column of convenient length. Total magnification by electron microscopes ranges from about 1000× to as much as 300,000×.

The receptor in light microscopy is usually the eye; sometimes, as in Fig. 18-1a, it is a photographic plate. The electron image is recorded photographically, but may be observed visually (although with impaired resolution because of the appreciable grain size) on a fluorescent screen. Glass monocular or binocular lenses are used for convenient observation of the image formed on the screen. A means for convenient exchange of photographic materials from the evacuated column must be provided.

Each of the elements of the electron microscope is now discussed in some detail.

THE ELECTRON GUN

Thermionic emission (cf. p. 350) is the source of the electron beams used in microscopy. In this process, which occurs to an appreciable extent only at very high temperatures, electrons are literally boiled off from a thin filament of tungsten. Significant emission from tungsten occurs above about 2200°K; at higher temperatures the thermionic current increases rapidly but, as the melting point (3410°K) is approached, evaporation of the atoms of the filament also increases. Thus an operating temperature must be selected at which adequate electron flux is provided without eroding the filament in an impracticably short time. The requirement is best met by temperatures of about 2400°K, a temperature which can be produced by resistance heating of a thin filament. While some evaporation of tungsten atoms occurs at this level, the operating lifetimes of standard filaments range from about 20 to 40 hours.

The replaceable portion of the standard filament assembly consists of a V-shaped tungsten wire mounted on an insulating base as shown (in perspective) by Fig. 18-2. In the region surrounding the tip of the "V", heating is greatest, and emission is correspondingly most intense. Nevertheless, the maximum intensity of the electron beam is limited because emission from this simple type of filament is spread over a relatively wide region. In "pointed" filaments the V-shape is modified in order to produce more intense emission from a more limited region. Figures 18-2b and 18-2c show front and side views of a tungsten "hairpin" filament which is mechanically ground to form a pointed tip. In the device shown in Fig. 18-2d a wire is attached to a hairpin base and subsequently etched to form an extremely thin filament.

The electrons emitted by the filament travel with variable but always low velocities, which are equivalent to acceleration through potentials of the order of 0.2 volt (cf. p. 369). In the electron gun they are then accelerated across a potential which is of the order of 100 kV. The accelerating voltage ranges from 40 to 100 kV in standard electron microscopes, but voltages as low as about 10 kV or as high as 1000 kV have also been used. The choice of accelerating potential is largely determined by the nature of the specimen (cf. p. 444).

In the *biased electron gun*, a diagram of which is given in Fig. 18-3, the region around the heated filament is maintained at a high negative potential. Acceleration is achieved by providing a region of positive potential beyond, toward which the electrons are attracted. A shield (the "gun cap") is maintained by the "bias resistor" at a potential slightly negative with respect to that of the filament, while the anode is at ground potential. Figure 18-3 describes the basic design of practical electron guns

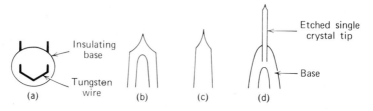

Fig. 18-2 Filaments: (a) standard; (b) pointed (front view); (c) pointed (side view); (d) etched.

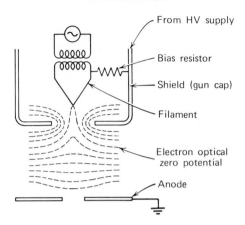

From HV supply

Bias resistor

Shield (gun cap)

Filament

Electron optical
zero potential

Anode

Fig. 18-3 Biased electron gun.

but ignores the electronic circuitry used to control both the accelerating voltage and the filament heater current.

In terms of the electron optical convention of potentials (cf. p. 348), the highly negative filament is at zero potential (i.e., corresponding to the initial zero velocity of the electrons), the shield is at a slight negative potential, and the grounded anode is at a high positive potential. The resulting distribution of equipotentials is of the form shown by dotted lines in Fig. 18-3. Electrons which are accelerated from the filament toward the anode are thus repelled by the negative shield, so that they tend to funnel through the center of the aperture in the shield. In this way the gun exerts a focusing action on the emitted beam.

The shield also plays a role in controlling the level of beam current. As the filament heater current is first raised from zero, *no* beam current is produced, since no significant electron emission occurs at filament temperatures less than about 2200°K. Beam current then appears, increasing with increases in filament current in a manner which is, at first, characteristic of emission by the tungsten. As the beam current increases further, however, emission of electrons causes the surface of the filament to become relatively positive with respect to the shield. The emitted electrons thus tend to be repelled back onto the filament by the negative equipotential surfaces surrounding the shield. Above a certain level, described as *saturation*, increases of filament current cause the filament to become hotter but produce no further net increase of emission. These features of the operation of the gun are illustrated in Fig. 18-4. The levels of filament and beam current at saturation are determined by the value of the bias resistor, which in some systems is variable, and by the distance between the filament and shield.

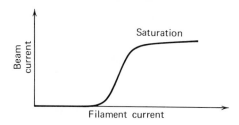

Fig. 18-4 Beam current characteristic.

These characteristics explain certain operational features of the electron microscope. Thus, when the high voltage is turned on but no filament current is applied, no electron intensity is observed, since the cold filament emits no electrons. As filament current is increased, the screen of the microscope remains dark until a level is reached at which emission becomes appreciable. Beyond this point the intensity of illumination increases rather rapidly with filament current until the maximum, saturation intensity is reached. The microscope is operated just *at* the saturation level, thus combining maximum electron intensity with maximum filament life. The intensity of the image is controlled by varying the level of condenser focus (as explained on p. 388); it should *not* be adjusted by means of variations in the beam current.

Older filaments reach saturation more rapidly than ones which have recently been replaced. This is a consequence of the evaporation of tungsten from the filament during use; filaments become thinner, and thus are heated to a higher temperature by the passage of current at a given level. It should be clear that the only effect of oversaturating the filament is to increase the rate of evaporation of tungsten atoms, and thus to reduce the operating lifetime.

After the filament burns out, no intensity is observed when the high voltage is applied, but a small value of beam current is obtained. The level of this current, which is known as the *leakage current*, is insensitive to increases of the filament current setting. Leakage current is produced by the direct flow of electrons and ions between the negative shield and the (electron optically) highly positive anode. This condition is diagnostic of a discontinuous filament. In other failures of either electrical elements or the vacuum system, high voltage is automatically interrupted by safety relays, so that no beam current reading whatsoever is obtained. Failure to observe image intensity when high beam current levels prevail is due to obstruction of the electron beam. Because of the restricted apertures used in the electron microscope, this can occur rather easily.

The intensity level at saturation may be increased either by decreasing the value of the bias resistor of the gun (in systems where this is adjustable) or by screwing the gun cap (shield) upwards so that it is closer to the filament. While high electron intensity is often desirable, especially for work at high magnification, the beam current must be limited to a level which (a) is rated as safe in terms of the emission of x-rays which occurs as the beam is absorbed (i.e., by the microscope screen or photographic plate, by limiting apertures, and by other elements of the column.), and which (b) avoids destruction of the specimen.

A means for *modulating* the high voltage (i.e., for changing the voltage level temporarily by approximately 5%) is provided in some instruments. This control is useful in the process of alignment. (See p. 399.)

VACUUM SYSTEMS

A short account of vacuum systems and gauges is given in this and the following sections. This discussion is devoted exclusively to types of system employed in electron microscopes.

The level of vacuum is specified in units of *torr*. one torr being defined as a pressure of one mm of mercury. Thus atmospheric pressure is approximately 760 torr. Vacuum regions are defined as

atmospheric pressure to 1 torr = "rough vacuum",
1 to 10^{-3} torr = "medium vacuum" (or "industrial vacuum"),
10^{-3} to 10^{-6} torr = "high vacuum",
10^{-6} to 10^{-9} torr = "very high vacuum",
less than 10^{-9} torr = "ultrahigh vacuum".

Standard electron microscopes operate in the high vacuum range, at about 10^{-4} torr. A particular requirement for these systems is high speed, since prolonged periods of outgassing interfere with effective use of the microscope.

Preliminary outgassing of the electron microscope column, to a pressure of about 10^{-3} torr, is achieved by a *mechanical pump* (the fore pump). In this type of pump an eccentric rotor traps and compresses gas from the chamber to be evacuated, subsequently allowing it to escape through a vent. Diagrams of the successive stages in mechanical pumping are shown in Fig. 18-5.

Evacuation of the microscope column is completed and maintained by a *diffusion pump*. In this type of pump, which is shown in Fig. 18-6, a stream of vapor traps gas molecules from the chamber to be evacuated, concentrating them at a position where they can be removed by mechanical pumping. Diffusion pumps produce pressures in the high vacuum range but require the backing action of a mechanical pump. As shown in

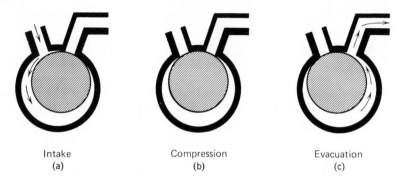

Intake	Compression	Evacuation
(a)	(b)	(c)

Fig. 18-5 Action of a mechanical pump.

Fig. 18-6a, the diffusion pump consists of an electrically heated resevoir of pump fluid, a chimney, and a water-cooled casing. The system is connected at the top to the chamber to be outgassed, and near the base to the backing pump. The nature of pumping action is illustrated by Fig. 18-6b. As shown, the pump fluid is vaporized by heating. Streams of molecules then travel upward and escape from vents in the chimney as jets of vapor which are directed downward. The streams of vapor return to the resevoir after condensing on the cold casing. In the process they trap the volatile outgassed molecules, which accumulate near the base of the pump. If the chimney is appropriately designed, the current formed by the jets of vapor effectively prevents back-diffusion of the trapped gases. The latter are then removed by the mechanical pump.

The efficiency of the diffusion pumping process is critically dependent upon the design of the chimney but is also affected by the properties of the pump fluid. If the fluid has a low vapor pressure and is of relatively high molecular weight, diffusion away from the streams of vapor (as well as possible back-diffusion into the outgassed chamber) is minimized. The fluid should be of low viscosity to provide efficient return of the condensed fluid to the resevoir. Stability is also important; heated pump fluid must not be exposed to atmospheric pressure, since oxidation can then cause the pumping properties to deteriorate.

Silicone oils are the type of pump fluid most commonly used in electron microscopes; they have replaced the less stable hydrocarbon oils. Mercury diffusion pumps are sometimes used instead, since it is then possible to tolerate a relatively high fore pressure at the mechanical pump, allowing the latter to be shut off for extended periods. Exhaust from the mechanical pump used to back a mercury diffusion pump must be allowed to escape through a hood in order to avoid accumulation of toxic mercury vapor.

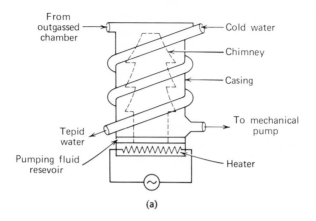

From
outgassed — chamber
Cold water
Chimney
Casing
To mechanical
pump
Tepid
water
Pumping fluid
resevoir
Heater

(a)

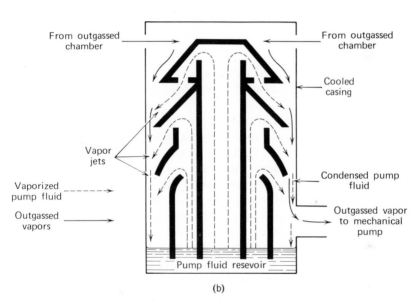

From outgassed
chamber
From outgassed
chamber
Cooled
casing
Vapor
jets
Condensed pump
fluid
Vaporized
pump fluid
Outgassed
vapors
Outgassed vapor
to mechanical
pump
Pump fluid reservoir

(b)

Fig. 18-6 Diffusion pump: (a) external view; (b) cross section.

Satisfactory operation of a diffusion pump depends upon a correct rate of flow of the cooling water. If flow is inadequate, pump oil fails to condense and tends to diffuse into the chamber undergoing evacuation. Apart from the fact that operating pressures cannot then be achieved, accumulation of pump oil deposits in the microscope column is highly undesirable. On the other hand, evaporation of pump oil is restricted if cooling action is too efficient; pumping action then fails.

Joints between elements of the microscope column are usually fitted with rubber gaskets. These gaskets are lightly coated by a grease of low vapor pressure. A recent innovation is the use of viton gaskets, which require no greasing. In this way the rate of contamination of column elements tends to be reduced. Viton is a plastic which tends to flow into the crevices of joints; it is thus more useful for semipermanent fittings than for those which must be demounted often.

Somewhat higher vacuums than 10^{-5} torr have been employed in experimental electron microscopes for the purpose of reducing the rate of contamination of specimens and other column elements. Ion pumps operating at pressures of the order of 10^{-9} have proved effective for this purpose (e.g., [1]. (In ion pumps a strong electric field ionizes the particles to be outgassed and subsequently causes these ions to be trapped at a surface.) Unfortunately, the speed of ion pumping is so low as to be incompatible with the normal operation of electron microscopes.

VACUUM GAUGES

Vacuum gauges are used in the electron microscope to measure the "fore pressure" at the intake of the mechanical pump and to measure the pressure in the column.

Fore pressure, which ranges between atmospheric pressure and about 10^{-3} torr, is measured by a *thermocouple gauge*. In this type of gauge a heated wire is in thermal contact with a thermocouple. Within the medium vacuum range, heat loss from the wire occurs primarily by the mechanism of conduction and thus is proportional to the pressure of the surrounding gas. As the pressure decreases, less heat can be lost from the wire, so that its temperature (and that of the thermocouple) increases. The rise in temperature is accompanied by an increased flow of current through the thermocouple circuit. The characteristic of this type of gauge is thus of the form shown by Fig. 18-7.

Column pressure is read by an *ion gauge* (discharge gauge) for which a typical characteristic is shown in Fig. 18-8. In the ion gauge a stream of

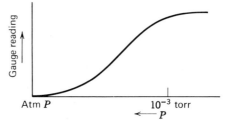

Fig. 18-7 Thermocouple gauge characteristic.

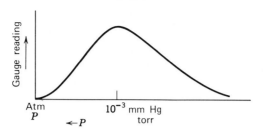

Fig. 18-8 Ion gauge characteristic.

electrons, produced by thermal emission from a tungsten filament, is accelerated through about 150 volts toward a positively charged grid. Collisions between gas particles and the stream of electrons result in the formation of positively charged ions; these ions are attracted toward a negatively charged collector electrode, thus forming an *ion current*. At relatively high pressures the mean free path of the ions is small, so that the intensity of the ion current is low. As pressure decreases, the mean free path increases, and ion current rises to a maximum at pressures of the order of 10^{-3}. At still lower pressures, the current decreases, owing to the unavailability of ionizable particles. In the descending portion of the characteristic curve the level of the ion current is a reliable index of pressure. Ion gauges used in the electron microscope read pressures down to about 10^{-5} torr. However, modifications of the standard ion gauge permit reading of pressures as low as 10^{-10} torr.

CONTAMINATION

Even at operating vacuum levels, small amounts of gaseous hydrocarbons and other materials are present in the electron microscope column. These substances originate from pump oils, vacuum grease, from parts of the specimen itself, or are present in the air which must be admitted to the column from time to time. Upon irradiation by the electron beam, various chemical reactions (ionization, polymerization, reduction, etc.) may occur which produce nonvolatile deposits of these materials on any surface, including that of the specimen, which is irradiated by the beam. The rate at which contamination accumulates on surfaces may be appreciable; for example, small holes in an electron-transparent film can be observed to fill in completely in as little as a minute. In this way small details of structures are buried, and the contrast of the specimen as a whole is reduced by the deposition of a uniform layer of material. Since an appreciable length of time necessarily elapses during the focusing and recording of images, contamination tends to limit the resolution obtained.

Fortunately, the problem of specimen contamination can be virtually eliminated by the use of the "cold finger" devices which are now commercially available as accessories for some types of microscope. The cold finger, which is cooled by liquid nitrogen, maintains a region near the specimen at a temperature below −100°C, but the specimen itself is not cooled. Potential contaminants in the vicinity of the specimen then condense on the surface of the cold finger. The device is found to reduce contamination during several minutes of irradiation by the electron beam to a negligible level.

The deposition of contamination on column elements may also be a source of difficulty, although the presence of thin symmetrical layers of contamination on optical elements is without significant effect on performance. Very heavy deposits must be avoided, however, since they may flake off, distorting the electron beam; such deposits normally form only on the shield of the electron gun. More serious problems arise from *asymmetric* contamination, particularly that of the limiting objective aperture (as discussed on p. 429). Contamination may be reduced by maintaining this aperture at a *high* temperature.

THE CONDENSER LENS

The condenser lens, which converges the electron beam to a focus at or near the plane of the specimen, is used to control the intensity of the image. Focusing of the beam *in* the plane of the specimen causes an area of minimum size to be illuminated with maximum intensity, a condition termed "cross-over". Focusing of the beam above or below the plane of the specimen produces an out-of-focus image of the source in the plane of the image; illumination of the specimen is then less intense but is spread over a wider area. These possibilities are summarized by Fig. 18-9.

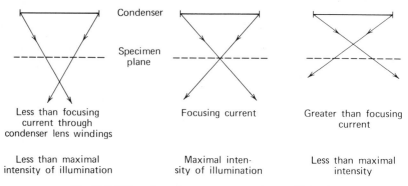

Fig. 18-9 Effect of condenser current on observed intensity.

Variation of the level of condenser focus, as of electron lenses in general, is effected by varying the lens strength (i.e., by varying the level of current) rather than by changing the position of the lens.

The image of the source which is formed by the condenser system is only a few microns in diameter, so that minor deflections of the beam readily cause the focused spot to disappear from the axial region. This condition is illustrated in Fig. 18-10. The position of illumination in the plane of the specimen is determined by the mechanical alignment of the condenser lens and its limiting aperture, and by magnets (electronic deflection coils) which may be placed in the column near the level of the lens.

The limiting aperture which is included in the condenser lens assembly blocks the passage of excessive radiation from the electron gun and thus controls the total electron intensity within the microscope column. (Electron intensity at the specimen is of the order of one amp/cm^2.) Electrons which pass through the column are ultimately absorbed at some point, where they may excite emission of x-rays. While the housing of the microscope column is designed to absorb this inevitable radiation, shielding is adequate only up to some specified intensity level. The maximum size of condenser aperture is thus of the order of 300μ in diameter. Accidental burning out or displacement of the aperture is usually detected at once, since an excessive flow of electrons may heat the metal specimen support above its melting point. Under this condition, the electron microscope can become a radiological hazard. Under normal operating conditions, however, the level of x-radiation escaping from the microscope is extremely low; exposure of the operator is negligible.

Double condenser lens systems are often used. A crossover spot as small as 2μ in diameter may then be obtained, as compared with 20μ or larger for single condenser systems. Use of the double condenser provides several advantages; from a practical point of view the most im-

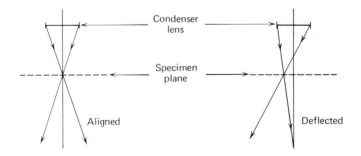

Fig. 18-10 Effect of condenser alignment.

portant one is that the illumination of smaller areas reduces the total accumulation of contamination of the specimen. (This is particularly important if the cold finger device described above is not used.) Improvement of image contrast, which is due to greater effective coherence of the electron source, is associated with the use of a small illuminating spot.

Illumination by an extended (i.e., single condenser) spot is diagramed in Fig. 18-11a. The illumination from point A, which is centrally located with respect to the microscope axis, is symmetrically included in the imaging system, but illumination from the asymmetrically located point B is only partially and asymmetrically included. If the action of the objective lens were ideal, the consequence of this condition would simply be that B would be imaged at lower resolution than A (i.e., in accordance with the ideas of Abbe; cf. p. 237). The image of A would be entirely unaffected by illumination of the peripheral point B. In fact, however, the action of the objective lens is nonideal to the extent that the contrast of *all* portions of the image is reduced by a background of inadequately focused electrons from *all* parts of the specimen. Illumination originating from points such as B increases the background intensity in the vicinity of the image of A, and the contrast of A correspondingly deteriorates. As shown in Fig. 18-11b, B is no longer illuminated when the size of the crossover spot is reduced. No electrons from B then contribute to the image at or near A, so that the contrast of the image of A is correspondingly improved.

The minimum resolved distance may be reduced by half when *periodic* structures are observed under *tilted, coherent* illumination (cf. p. 241 and especially Fig. 10-14). The coherence associated with the small size of the crossover spot thus becomes important when structures at the limit of resolution are to be studied. The successful observation, with the use of tilted double condenser illumination, of crystal spacings of 2 Å or even less confirms the predictions made by the Abbe theory of resolution.

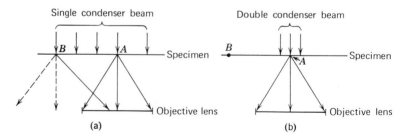

Fig. 18-11 Comparison of the effect of single and double condensers on image contrast.

SPECIMEN, SPECIMEN SUPPORT, AND STAGE

Electron microscope specimens must be thin. Electrons are scattered by all types of atoms; thus even the thinnest (ca. $1\,\mu$) of specimens used in light microscopy produces an electron image which is so hazy as to be useless. At the same time the specimen must be thick enough that it interacts with the beam to produce an image of adequate contrast. (The nature of interactions between the specimen and beam is discussed in Chapter 19.) The specimen must also be mechanically and thermally stable and must not accumulate electrostatic charge to an extent which deflects the imaging electrons.

Whether the specimen is a tissue section or a suspension of particles, it is supported by a thin "electron-transparent" film. Layers of evaporated carbon 200–500 Å thick are almost universally used for this purpose. Carbon films are the support of choice because of their relative transparency, limited granularity, and stability to irradiation. Silicon monoxide and various plastics are also used for the purpose.

A discussion of the nature and preparation of specimens is beyond the scope of this book; for treatises on this subject see [2], [3]. The relation of the specimen to image contrast is discussed on pp. 430 and 441.

The specimen and supporting film require the mechanical support of a metal grid which is necessarily electron opaque. A portion of the grid is illustrated in Fig. 18-12a, and a cross section of grid, film, and a particulate specimen is shown in Fig. 18-12b. The grid may have about 200 openings per linear inch, with an open area $100\,\mu$ or somewhat less on a side. However, a larger percentage of open area is required when extended tissue sections are studied. (Note that there is only about 25% open area in the grid shown in Fig. 18-12.) Reasonable mechanical stability is combined with a large percentage of open area in individually cut grids with a continuous rim. The design of this type of grid is suggested by Fig. 18-13.

The metal from which the grid is constructed may distort the magnetic

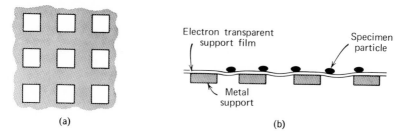

(a)   (b)

Fig. 18-12 Grid for electron microscope specimen.

The Electron Microscope

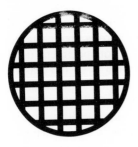

Fig. 18-13 Design of grid with large percent open area.

field of the objective lens. Copper is an ideal material from this point of view, although nickel, which is more easily handled, is often used when the highest resolution is not sought. At high resolution, slight deterioration can sometimes be observed in images of portions of the specimen which are very close to the support bars (even when these bars are made of copper). The grid is placed in the holder, which is the object actually moved by the specimen stage controls.

The magnification achieved by the objective lens is, as given in Chapter 8,

(8-mm)
$$M = \frac{s_2''}{s},$$

where s_2'' and s are the image and object distances, as measured from the second and first principal planes, respectively. Thus the location of the specimen along the microscope axis determines the magnification. While major changes in the level of magnification are effected in the electron microscope by changing lens current levels (possibly with the use of different pole pieces at different levels of magnification), adjustment of the position of the specimen may also be used to vary magnification. *High magnification specimen holders*, which are commercially available as accessories, place the specimen as close as possible to the objective lens field. Accidental variations in the axial position of the specimen may produce unsuspected variations in magnification.

The specimen stage supports the specimen holder, and is provided with external controls for positioning the specimen in the beam. Movement of the specimen is controlled to within a few hundred angstrom units. Vibratory and thermal motions, mechanical drift, and movements of electrostatic origin must be reduced to such a level that the specimen moves through less than the minimum resolved distance during exposure of the image.

Various modifications of the specimen stage assembly are used for special purposes. The specimen may be tilted through a controlled angle in order to record stereoscopic pairs of exposures in which an impression of depth is given by binocular viewing. The stereo angle corresponding to normal binocular vision is about 7°, but larger stereo angles may also be useful in detailed analyses of three-dimensional structures. Because the depth of field of the electron microscope is very great (cf. p. 414), the observed electron image is actually a superposition of images of many planes in the specimen. Three-dimensional relationships of specimen structure are accordingly difficult to assess. In principle, stereomicroscopy should be a useful method of overcoming this difficulty. In fact, little such work has been done, largely because specimen techniques have been unreliable from the point of view of the preservation of three-dimensional structure at the 100 Å level. (Three-dimensional relationships of sectioned materials may be deduced, as in light microscopy, by the observation of serial sections.)

Hot stages, in which specimen temperature may be increased in a controlled manner, are used to study structural details of phase changes which occur at high temperatures. The stages are primarily useful in metallurgical electron microscopy.

Temperatures usually rise to about 100°C during irradiation of the specimen by the electron beam. Thermal damage can be avoided by the use of *cold stages*, which are cooled by liquid nitrogen. The rate of contamination may be reduced also, although the cold stage is less effective for this purpose than is the cold finger (cf. p. 388). The variations of contamination rate with temperature may seem paradoxical: it is found that the rate is reduced when both stage and specimen are held at temperatures of −50°C, while contamination is virtually *eliminated* for a hot specimen placed near a surface which is at a temperature below about −100°C. The usefulness of the cold stage is limited by the necessity to warm the specimen to room temperature before bringing it to atmospheric pressure. (Otherwise the specimen may be destroyed by condensation of moisture from the air.) Cold stage operation thus is time consuming.

THE OBJECTIVE LENS

In the electron microscope as in the light microscope, the objective lens is the most critical optical element, since it performs the first stage of imaging. Aberrations which are present in the intermediate image formed by the objective lens are subject to further magnification, both by the projector system and photographically.

Complete axial symmetry ("roundness") of the objective lens is extremely important. The lens assembly includes a device by means of which lens asymmetry may be corrected (compensated). Electron lens asymmetry and its correction are discussed on pp. 409 and 412.

The objective lens assembly may include a limiting aperture at or close to the rear focal plane of the lens. The diameter of the aperture ranges from about 30 to 100 μ; its function is to intercept electrons which have been scattered by the specimen through excessively large angles.

The contribution of the objective aperture to image contrast is discussed on p. 428.

Variation of objective lens strength is the primary means of effecting focus in the electron microscope. Other means of focusing the image also exist; they are discussed on p. 408.

THE PROJECTOR LENS SYSTEM

A two-element projector lens system is the usual electron optical counterpart of the ocular of the light microscope. The two lenses are called the *intermediate* (or intermediate projector) and *projector* (or final projector), respectively. Use of two lenses is convenient because it makes a wide range of magnifications available with the use of a microscope column of fixed length. The projector system forms a *real* image which is recorded either by the fluorescent screen or by a photographic emulsion.

In a compound light microscope the level of magnification is determined by the choice of ocular and objective lenses; thus only those fixed values of magnification corresponding to available combinations of lenses are available. In the electron microscope, on the other hand, magnification can be varied continuously over a wide range by changing the currents applied to the projector lenses. Low magnifications are obtained by using the intermediate as a demagnifying lens; high magnifications result from the use of both intermediate and projector as magnifying lenses. In some microscopes different pole pieces may be used for different ranges of magnification. The choice of the level of magnification is discussed below.

Magnification is determined by the level of objective current as well as that of projector current. However, only slight variations in objective current level are used in effecting focus. Thus, at close-to-focus settings, magnification does not change significantly with focal level.

Owing to the effects of *hysteresis* (cf. p. 361), the strength of lens fields, and thus the magnification, depend in part upon the level of current previously applied to the lens windings. If precise levels of magnification are important, the effects of hysteresis must be eliminated by normalizing the projector lenses in the manner described on p. 363.

During operation the lens currents which correspond to any given instrumental setting may vary detectably. During a day, currents decrease as heating increases the resistance of the lens windings. These changes, which occur principally during the first hour of operation, may be by as much as 10% in systems which are not water-cooled. Lens current levels are maintained with respect to the voltage established by a reference battery. They may thus change slowly, over a period of months, as the batteries age; replacement of reference batteries often produces abrupt changes in instrumental magnification which are of the order of 5%.

CHOICE OF MAGNIFICATION LEVEL

The availability of a range of magnifications is important in electron microscopy. The light microscopist usually begins with some idea of the size and general plan of an object but, in electron microscopy, "orientation" with respect to the size of the specimen is often difficult to achieve. Thus it is helpful first to view the object at a relatively low magnification and then to proceed to progressively higher magnifications.

Maximum values of instrumental magnification range up to as much as $300,000\times$, as compared with about $1200\times$ in the light microscope. However, it should be appreciated that it is the *resolution* achieved by the electron microscope, not its magnifying power, which is truly impressive (cf. p. 234).

A number of considerations set both lower and upper limits on the magnification which should be used with any given specimen. The magnification should be sufficient to reveal detail which is resolved by the electron beam. Since photographic grain limits resolution at enlargements of more than about $10\times$ (cf. p. 467), instrumental magnification should be at least one tenth of total final magnification. (Note, however, that "detail which is resolved" in any given image may be considerably larger than the minimum distance resolvable by the instrument.) Also, high magnification aids accurate compensation and focusing of the image.

From other points of view, minimum feasible magnifications are desirable. Even at the lowest maginfications, the area observed on the microscope screen is not more than about $100\,\mu$ on a side. In order to study a reasonably large number of particles or cells, it is convenient to record images of relatively large areas; regions of particular interest may then be enlarged photographically.

Since magnification is a linear measure, the area occupied by the image of a given specimen area increases as the *square* of the magnification. For a given level of illumination at the specimen, the intensity of the image thus decreases as the square of magnification [cf. p. 267 and (12-a)]. Accordingly, very intense illumination is needed to produce usable image intensity at high magnification. The possibility of damage to the specimen then tends to make very high magnification undesirable.

The appropriate levels of magnification used in observation of various types of specimen are indicated in Table 18-1.

Calibration of magnification is discussed on p. 416.

Table 18-1

Structure to be Observed	Approximate Instrumental Magnification
Full area of single opening in the specimen support	1000 × (i.e., minimum setting)
Cellular structure	About 10,000×
Shadow cast specimens (in which resolution is limited to about 20 Å by metal grain)	About 20,000×
Subcellular or individual stained particles	20,000–300,000×
Spacings at the limit of resolution	300,000× (i.e., maximum feasible setting)

VIEWING AND RECORDING OF THE IMAGE

The final magnified electron image is formed on a fluorescent screen at the base of the microscope column. Detail in this image is conveniently observed through an external monocular or binocular. (The novice should distinguish between the requirement for focusing the glass lens(es) on the fluorescent screen, and that for obtaining an in-focus electron image by adjustment of the microscope controls!)

Impact of the electron beam on the fluorescent screen (or photographic plate) produces emission of x-rays. This is normally not a hazard, since the viewing chamber is constructed with panels of leaded glass which are sufficiently thick to absorb this radiation.

Whereas much light microscopy is done by direct observation, the fluorescent screen of the electron microscope generally serves only as a means of locating and focusing on areas which are then recorded photographically. This is necessarily the case, since the image formed on the screen lacks the high resolution which is inherent in the electron beam. The fluorescent screen is constructed by coating a surface with crystallites of a material (the phosphor) which has the property of partially dissipating the energy of incident photons as quanta of visible light. In the fluorescent emulsions presently used, the grain size is of the order of 100 μ, so that, to be resolved on the screen, image detail must be somewhat larger than that size. For example, it is found that magnification of about 230,000× is required in order to observe 6 Å spacings directly. The resolving power of the fluorescent screen may be less than that predicted on the basis of grain size if the coating of phosphor is not extremely uniform.

The *image intensifier* is a fairly recent development in instrumentation for electron microscopy. In this device electronic scanning of the image produces a signal which is amplified before being converted to a visual display on a second fluorescent screen. The latter may be placed outside the column.

The gain in contrast of the intensified image is comparable to that obtained by photographic processing; ultimate resolution is *not* improved by the use of the device. Nevertheless, the image intensifier may be of practical importance in obtaining very high resolution. Both compensation and focusing of the microscope are so critical at limiting resolution as to become, to some degree, processes of trial and error. Under these conditions, bright (photopic) vision is more advantageous than dark adapted (scotopic) vision. Thus, if the image intensifier can provide a generally higher level of intensity than is available on the microscope screen, recording of micrographs of high quality on a more routine basis should be possible. The importance of image intensification would vary according to the quality of the microscopist's eyesight.

Image intensifiers are useful for displaying images simultaneously to groups. Also, if the scanning device preserves the resolution which is inherent in the electron beam, photographic recording can be carried on outside the microscope column. Delays due to the outgassing of photographic materials are thus avoided.

In the recording of light photomicrographs the location of the plate is precisely determined by the focus; if the plate is misplaced even slightly, the recorded image is out of focus. Owing to its very large depth of focus, the electron microscope is much more convenient in this respect. Exact focus may be selected at the plane of the fluorescent screen, and the image then recorded several centimeters above or below this position. Mechanisms for photographic recording vary between instruments; either the screen is tilted to expose an emulsion lying below, or the film is interposed between projector lens and screen. Mechanically timed shutters are often used. An important consideration in the design of the recording system is that it must not produce mechanical vibrations during operation.

Exposures, which are determined on the basis of experience, may either be estimated visually or determined from a photometer. Since the brightness of the image formed on the screen varies with the age and type of the phosphor, "experience" must be related to each instrument individually. Usually the exposed plate is used to make a print, so that the level of the initial exposure is relatively uncritical. Photometer readings must be interpreted cautiously, since the sensitivity of photographic emulsions is found to increase after prolonged outgassing.

Photographic processing of images is discussed in Chapter 21.

OBSERVATION OF ELECTRON DIFFRACTION PATTERNS

In the light microscope a focused diffraction pattern is formed in the rear focal plane of the objective lens (cf. p. 222). This pattern can be observed with the use of a specially designed ocular lens (or with the naked eye, after removal of the ocular). In the electron microscope, diffraction patterns can be observed in an analogous manner by changing lens current levels in such a way that *a magnified image of the rear focal plane of the objective lens is formed on the fluorescent screen.* The optical system is then as shown in Fig. 18-14. In some microscopes a switch can be used to change the lens currents abruptly to a level which is approximately suitable for diffraction; in others, lens current are simply adjusted manually.

Electron diffraction patterns are usually "powder patterns" similar to the x-ray powder patterns described on p. 648 and illustrated by Figs. 30-7 and 30-8. Single crystal patterns may also be recorded. In fact, a particular advantage of electron diffraction (as opposed to x-ray diffraction) is the possibility of isolating the pattern formed by specific

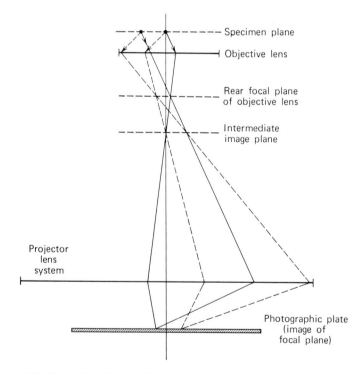

Specimen plane

Objective lens

Rear focal plane
of objective lens

Intermediate
image plane

Projector
lens
system

Photographic plate
(image of
focal plane)

Fig. 18-14 Optical system for recording electron diffraction patterns.

areas of the specimen. In *selected area diffraction* a microscope image is obtained first. An aperture of variable dimensions, located within the projector lens assembly, is then so adjusted as to mask off the area to be diffracted. In this way the diffraction pattern from single crystals of precisely known orientation may be observed. Alternatively, typical areas may be selected for the formation of powder patterns. After adjustment of the aperture, lens currents are altered to produce the diffraction image. The interpretation of the electron diffraction patterns is analogous to that of the corresponding types of x-ray patterns, as described in Chapter 30.

Electron diffraction patterns, as commonly recorded, are suited to the study of maxima corresponding to spacings in the size range from about 1 to 25 Å. Larger spacings, which are of greater biological interest, tend to be confused with the central spot of unscattered electrons. Recently, however, *low angle electron diffraction* has been studied in highly magnified patterns; only the area close to the central spot is then included in the field of view. Such patterns may be recorded by reducing the strength of the objective lens; that is, by increasing its focal length. The rear focal plane then lies *closer* to the projector system, so that a *larger* magnification of the final diffraction image results [i.e., according to (8-mm)]. To obtain such patterns *in* focus it is practical to place the specimen relatively far from the objective lens field; a special "low magnification" holder is used. Low angle electron diffraction has been used to study spacings of the order of a few hundred angstroms; for example, in virus particles.

ALIGNMENT OF THE MICROSCOPE COLUMN

Ideally the optical elements of any microscope system must be coaxial, that is, the axes of symmetry of each lens must be exactly coincident. In the light microscope, alignment of the lenses to satisfy this condition is achieved and maintained quite easily. In the electron microscope, however, aperture angles are small, and fields of view only a few microns in diameter. Thus optical alignment is much more critical. Frequent adjustments, also, are required to maintain alignment at a level suitable for high resolution microscopy.

Alignment must be effected with respect to the electron gun, condenser lens(es), and imaging lenses. The aim of gun alignment (source alignment) is the centering of an image of the emitting filament on the viewing screen. Condenser alignment ensures centering of illumination at all levels of condenser current. Under ideal conditions the position of the illuminating spot at under-, in-, and overfocus settings of the condenser lens is as shown in Fig. 18-15a. In practice, some degree of "condenser sweep" is observed, as illustrated in Fig. 18-15b. When the condenser lens is badly misaligned, the illumination can sweep entirely out of the field of view as the level of current is altered. Apart from any effect upon image quality, this can be so annoying as to hamper effective use of the microscope.

Alignment of the electron gun and condenser lens is achieved by translation of the gun, the condenser lens, or the condenser lens aperture, by tilting the gun, and/or by deflections of the beam by electromagnets. (Not all of these adjustments are available on all models of microscope.) Major adjustments of gun and condenser alignment normally are required only after disassembly of the microscope column.

Because of the substantial aberrations of electron lenses, an image of acceptable quality is produced only within a limited paraxial region. Hence it is important to view only this part of the image. Alignment of the imaging lenses causes the optical center of the image to coincide with the physical center of the viewing area.

The effect of fluctuations in high voltage level (chromatic aberration) is to superimpose a continuum of images of different magnifications and focal levels. This results in sharp definition of the image only near an axis or voltage center, as illustrated by Fig. 17-12. Fluctuations of lens current levels likewise cause images of different magnifications to super-impose, also resulting in a composite image which is sharp only at an axis, the current center. Ideally, the voltage center and current center of the image should coincide, but in practice they do not do so exactly.

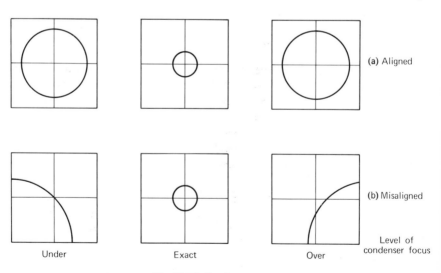

Fig. 18-15 Condenser sweep.

Lens alignment consists of the physical translation of the lenses so that either the voltage or current center coincides with the physical microscope axis. If the regulation of lens current levels is superior to that of the high voltage, voltage alignment is recommended by the manufacturer, and vice versa.

The effects of misalignment and of the failure of high voltage regulation (presence of excessive "ripple voltage") are compared in Fig. 18-16. Figure 18-16a shows the image of an object of random structure (e.g., the background of a shadow cast specimen) as formed by an aligned, well regulated microscope. The axis is at the center of the image, and smearing is undetectable at the periphery (but would be observed if the field were extended). Figure 18-16b is an (exaggerated) example of the effects of failure of high voltage regulation. The optical center remains at the center of the screen, and the image is sharp close to the axis, but smearing is observed at the edges of the field of view. (A smear with both radial and tangential components is illustrated; in any given system, either of these directions of smear might predominate.) Figure 18-16c illustrates (again with exaggeration) the effect of misalignment. Here the optical center has been displaced to the lower right. As a result, image quality is satisfactory at the lower right-hand corner of the image, but deteriorates progressively toward the upper left. In the process of voltage alignment the image is observed as the voltage level is modulated (i.e., as the voltage level is varied). All parts of the image except the voltage center then move; in an aligned system the position of no movement coincides with the center of the image.

If a microscope is voltage aligned, and if the voltage and current centers do not coincide, changes in the objective lens current produce a movement or "sweep" of the image, as shown in Fig. 18-17. Objective sweep is annoying, but it is unlikely to affect image quality significantly so long as the objective current regulation remains satisfactory.

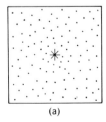
(a)

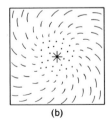
(b)

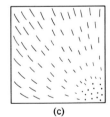
(c)

Fig. 18-16 Effects of misalignment and high voltage ripple: (a) ideal image; (b) image formed when regulation of high voltage is defective; (c) misaligned image.

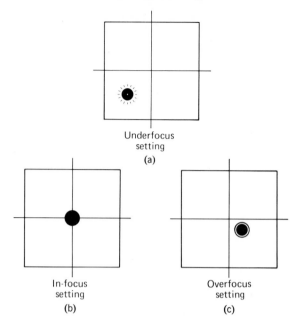

Underfocus
setting
(a)

In-focus
setting
(b)

Overfocus
setting
(c)

Fig. 18-17 Objective sweep.

"Perfect" alignment of electron microscopes is generally impossible, since lens axes are usually found to be very slightly tilted with respect to each other. Nevertheless, the best possible alignment is of critical importance in obtaining high resolution. Procedures for effecting alignment vary so widely that no general discussion can be attempted here.

FOCUS FRINGES

Characteristic fringes appear in electron images according to the level of focus. When well defined edges are observed, the appearance of these fringes is striking. The effect is best observed in images of small *holes* in a thin film, as shown in Fig. 18-18 and Plate II. At exact focus, the fringes coincide with the geometrical edge of the object (Plate II, center). At underfocus a halo of high intensity is formed on the electron transparent side of the edge (Plate II, left); at overfocus a fringe of high intensity is formed on the electron opaque side (Plate II, right). Figure 18-18a illustrates the images of a hole as observed on the fluorescent screen of the microscope, where high electron intensity produces a *bright* image. Figure 18-18b shows the images formed on a photographic plate, where the reversal of contrast causes regions of high electron

Plate II Electron microscope images of a hole at different focal levels. Magnification 60,000 ×. Left, slightly underfocus; center, close to exact focus; right, slightly overfocus.

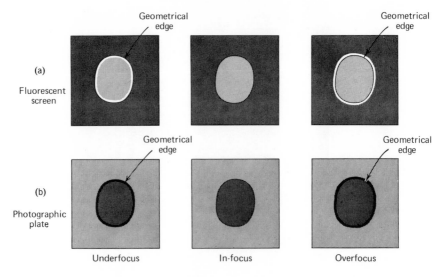

Fig. 18-18 Focus fringes in the image of a hole in a thin film: (a) as observed on fluorescent screen; (b) as recorded on photographic plate.

intensity to appear dark. Similar fringes are formed in images of specimens of more complex structure. The overall effect is then a "hypersharp" appearance of the image at slightly underfocus settings of the microscope, while even very slight overfocus produces a hazy image. At exact focus, boundaries are precisely defined, but image contrast is minimal.

The fringes may be considered to be the electron optical analog of the Fresnel diffraction effects of light optics (cf. p. 65). Diffraction at a straight edge, upon illumination by a source placed a finite distance from the edge, produces a fluctuating pattern of intensities. This pattern, which has been shown in Fig. 3-28, is reproduced as Fig. 18-19a. These Fresnel fringes are emphasized in the electron image because the microscope is a system of restricted aperture in which the widths of beams are correspondingly narrow. Figure 18-19a approximately represents the intensity distribution in electron images taken rather far from exact focus. At slight underfocus, in-focus, and slight overfocus settings, intensity distributions are of the forms shown in Figs. 18-19b, 18-19c, 18-19d, respectively. At these slightly out-of-focus settings the intensity of diffraction maxima other than the first is so slight as to be virtually undetectable.

The focus fringes may also be accounted for in terms of the particulate behavior of electrons. According to this view, two beams of anomalously high intensity are scattered at any interface in the specimen at which an abrupt change in the potential of the medium occurs. That is, electrons

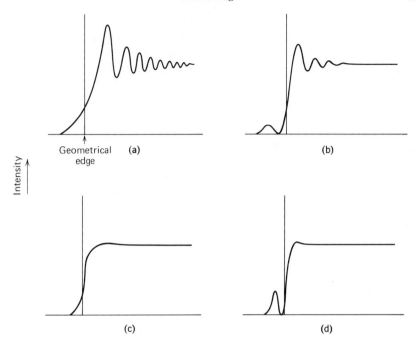

Intensity

Geometrical **(a)**
edge

(b)

(c)

(d)

Fig. 18-19 Densitometer tracings of fringes.

are scattered, as by all matter, through a continuum of angles, but scattering is particularly intense in the directions ① and ② shown in Fig. 18-20. Of these two beams, ① is the more intense and thus the more readily observed.

Each of the rays shown in Fig. 18-20 originates essentially at a single point, and thus can be recombined by the lens at a single point in the image. If the image is viewed in focus, the rays will have recombined, so that the existence of two directions in which scattering is of high intensity is not revealed. If, however, the objective lens is too weak to focus the

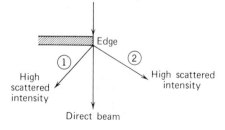

Edge

① ②

High
scattered
intensity

High scattered
intensity

Direct beam

Fig. 18-20 Electron scattering at an edge.

image at the fluorescent screen, the condition of focus is achieved at some point below (i.e., beyond) the screen, as illustrated by Fig. 18-21. (Recollect, in considering Fig. 18-21 and also Figs. 18-22 and 18-23, that the real image formed by a single convergent lens is inverted.) In the plane of observation (i.e., as opposed to the plane of focus) the high intensity ray ① then lies outside the geometrical image of the edge, while ray ② lies inside the edge. Since ray ① is the more intense, the observed effect at underfocus is one of increased electron intensity just outside the edge, as represented in Fig. 18-19b.

The underfocused image may also be considered from the point of view illustrated by Fig. 18-22. The distance s_2'' from the lens to the in-focus image of the edge is conjugate to the distance s_1, from the edge to the lens; that is, s_1 and s_2'' satisfy the relationship

(8-z)
$$\frac{1}{s_1} + \frac{1}{s_2''} = \frac{1}{f}.$$

(Strictly speaking, s_1 and s_2'' are measured from the corresponding principal planes; however, the same argument applies when these distances are so measured.) Similarly, for the distance s_u'' between the lens and the plane of observation, there exists a conjugate object distance s_u for which (8-z) also is satisfied. For s_u'' less than s_2'', s_u must be greater than s_1. Thus the plane of observation can be said to contain an image of some plane which lies *above* the object. The effective object includes the virtual projections ①' and ②' of rays ① and ②. Since the projection ①' lies outside the edge of the object, the bright fringe lies beyond the geometrical edge in the image.

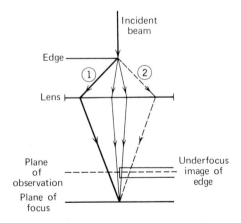

Fig. 18-21 Observation of an underfocused image.

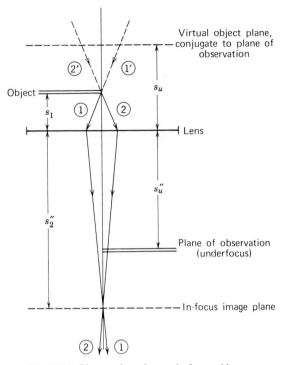

Fig. 18-22 Observation of an underfocused image.

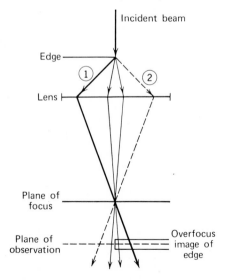

Fig. 18-23 Observation of an overfocused image.

The formation of an overfocused image is considered in Fig. 18-23. Here the high intensity ray ① forms a fringe inside the geometrical edge of the image (as represented by Fig. 18-19d). From the point of view of conjugate distances, the overfocused image can be said to be a real image of some plane which lies between the specimen and the lens.

Note that, since any plane in the vicinity of the specimen can be imaged at a corresponding focal setting, the observed spacing of over- or underfocus fringes can be varied at will. This property makes the focus fringes formed at holes a useful resolution test "object" (see p. 417).

METHODS FOR FOCUSING THE IMAGE

Focusing of the image is effected by variations in the objective lens current. However, in microscopes which are aligned with respect to voltage, objective "sweep" (cf. p. 401) may hinder the process of focusing. In extreme cases the object can sweep entirely out of the field of view during attempts to set the objective lens current level. This difficulty is avoided if variations in high voltage level are used as the means of effecting fine focus. Changes of the high voltage level change the wavelength of the electron beam, and thus the effective focal length of the lenses. (Equivalently, voltage changes can be said to vary the velocity of electrons, and thus their susceptibility to deflection by magnetic fields.) Since these changes affect magnification also, alterations of accelerating voltage are used only for fine adjustment of focus, while approximate focus is established by the objective current level.

It is of interest to calculate the magnitude of high voltage change which corresponds to a change of 1 μ in focal length. As given on p. 368,

(17-k) $$\frac{\Delta V}{V} = C_c \frac{\Delta f}{f},$$

where C_c is the chromatic aberration constant characteristic of the objective lens. C_c and the focal length are of the same order of magnitude; thus, at $V = 100\ \text{kV}$,

(18-a) $$\Delta V = C_c V \frac{\Delta f}{f} \cong V \Delta f = 10^5 \times 10^{-4} = 10 \text{ volts.}$$

A second method of 'voltage focusing' makes use of variations in the operating temperature of the filament. The operating temperature can be varied by changing the level of the electron gun bias resistor and, thus, the level of filament current which is required for "saturation" of the emission. (cf. p. 382.) The most probable initial (thermal) electron velocity, expressed in terms of the equivalent accelerating voltages, V is, as given by

(17-r)
$$V = \frac{kT}{e},$$

where k is Boltzmann's constant, T is the absolute temperature, and e is the electronic charge. Thus:

(18-b)
$$\Delta V = k\frac{\Delta T}{e}.$$

Combination of (18-b) and (17-o) then gives

(18-c)
$$\Delta f_{thermal} = \frac{k\Delta Tf}{C_c eV}.$$

Changes in the operating temperature by about 500° are feasible. The corresponding change in focal level for 100 kV operation, and for equal values of f and C_c is

(18-d)
$$f = \frac{1.38 \times 10^{-16} \times 5 \times 10^2}{1.6 \times 10^{-19} \times 10^5} = 0.43\ \mu.$$

Filament temperature adjustment is thus a sensitive method for fine level variation of image focus.

LENS ASYMMETRY AND COMPENSATION

The effect of lens asymmetry is described in Chapters 9 and 17 as a variation of focal length with direction. This effect can be observed very graphically in the images of holes formed by an asymmetric lens, as shown in Fig. 18-24 and Plate III. These illustrations show the asymmetric image at three different focal levels. It is evident that the directions of maximum and minimum focus remain separated by 90°, and thus that *no true in-focus image can be formed by an asymmetric lens.*

The magnitude of asymmetry is expressed as the difference Δf in focal length between the directions of extreme focus. If, at some setting of the objective lens current, the most overfocus direction of the image (e.g., the x direction in Fig. 18-24) is at a given focal level (e.g., exact focus in Fig. 18-24a), then Δf is the change (decrease) in focal length required to produce the same focal level in the most underfocused direction of the image. In the example considered, exact focus is achieved in the y direction in Fig. 18-24b. Thus, if the focal length used to record the images has been changed by, for example, 2 μ between Figs. 18-24a and 18-24b, the lens is said to suffer from 2 μ of "astigmatism."

The distance between the geometrical edge and the center of the under- or overfocus fringe in the direction of minimum or maximum focus, when the direction at right angles is at exact focus, may be designated $(\Delta r_i)_A$.

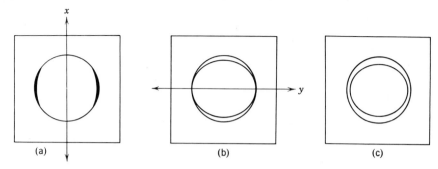

Fig. 18-24 Asymmetric images of a hole in a thin film.

This distance is illustrated by Fig. 18-25. The corresponding distance in object space, $\Delta r_A = (\Delta r_i)_A/M$, is then the resolution limit imposed by lens asymmetry, or by defocus of a symmetrical image. The magnitude is estimated by regarding the focus fringes, whether symmetric or asymmetric, as Fresnel diffraction effects. Various approximate expressions which describe the spacing of Fresnel fringes differ in detail but are similar in general form. One of them is

$$(18\text{-}e) \qquad \Delta r = \sqrt{\Delta f \lambda (2n-1)}, \qquad \Delta r_i = M \sqrt{\Delta f \lambda (2n-1)},$$

where Δf is the amount of defocus, and n is the number of the diffraction maximum. In electron microscope images, only the *first* Fresnel fringe is normally observed, so that the limitation on resolution imposed by either defocus or lens asymmetry is

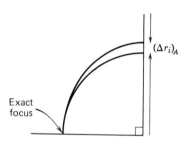

Fig. 18-25 Measurement of asymmetry.

$$(18\text{-}f) \qquad (\Delta r)_A \cong \sqrt{\lambda \Delta f}.$$

An asymmetric lens, as shown in Fig. 9-19, is equivalent to the combination of a cylindrical lens with the "spherical" or "round" (i.e., axially symmetric) field of the basic lens. Compensation is achieved by imposing a second cylindrical field of equal magnitude but opposite direction. This is frequently supplied by an electrostatic multipole lens, that is, by a set of symmetrically arranged electrodes to which a variable set of voltages may be applied. The net *magnitude* and the *azimuth* (direction) of the field imposed by the compensator may be varied independently. The process of compensation consists of the selection of correct values for both azimuth and magnitude.

Plate III Astigmatic electron microscope images of a hole. The asymmetry shown here was produced by the use of a misaligned limiting objective aperture. Magnification 60,000×. Left, relatively low focal setting; center, closest possible approach to an in-focus image; right, overfocus setting.

While compensator devices and recommended compensation procedures vary in different makes of microscope, a rational procedure for effecting compensation may be described as follows:

1. The relative azimuth of astigmatism in the image formed by a completely uncorrected lens is noted. Since this is the asymmetry obtained in the absence of any compensating field, it is termed the *residual asymmetry*. In Fig. 18-26a this azimuth is designated θ with respect to the arbitrarily selected direction z (z is usually the direction of a side of the photographic plate). Note that $(180 - \theta)°$ might equally well specify the residual azimuth, should angles be measured in a clockwise rather than a counterclockwise direction from the z axis. Also, the direction of maximum underfocus rather than that of overfocus might have been selected for reference. Clearly, the designation of the angle of asymmetry is arbitrary, but further specifications of azimuths must be consistent with the convention selected at this point.

2. The maximum possible magnitude of compensation is then applied. The azimuth of this *applied asymmetry* is then adjusted so that the fringes lie at an angle of $(\theta \pm 90)°$ according to the convention chosen as explained above. This condition is illustrated in Fig. 18-26b. The maximum magnitude of asymmetry is considerably in excess of the residual asymmetry, so that effects of residual asymmetry upon the azimuth of applied asymmetry are negligible. The quality of the image after completion of this step is, of course, far less satisfactory than it was initially.

3. Maintaining the *azimuth* selected in step 2, the *magnitude* of applied asymmetry is reduced to the point at which it just equals the magnitude of the residual asymmetry. If the azimuth is once determined correctly, the extent of correction is determined by the accuracy of the applied magnitude. As the condition of compensation is approached, the observed

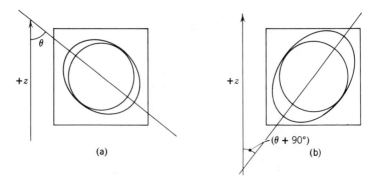

(a)   (b)

Fig. 18-26 Stages in the compensation of asymmetry: (a) residual asymmetry; (b) applied asymmetry.

azimuth of *resultant asymmetry* shifts from values close to those illustrated in Fig. 18-26 toward a direction 45° from either. At the same time the magnitude of the resultant astigmatism approaches zero. Figure 18-27 illustrates the variations in magnitude and azimuth of the resultant asymmetry which occur as the applied magnitude is reduced past the point of compensation.

The process of compensation may be carried out visually or, more tediously but with greater accuracy, photographically. The level of compensation achieved varies considerably with both operator and instrument. Compensation can be set and maintained at levels of about $0.10\,\mu$, provided that asymmetric contamination of objective apertures is avoided. Reduction of asymmetry to the low levels required for ultimate resolution cannot be routinely achieved or maintained in present microscopes.

DEPTHS OF FIELD AND FOCUS

The approximate depth of field for light microscopy was estimated on p. 278 by equating the diameter of the Airy disc to that of the out-of-focus cone of illumination from the image. Similarly, in the electron

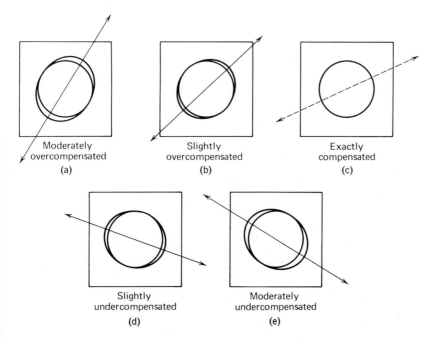

Fig. 18-27 Effect of varying the magnitude of applied asymmetry.

microscope the depth of field is that distance for which the diameter y of the cone of illumination shown in Fig. 18-28 is no greater than the minimum resolvable distance. Here d_{min} is the value determined by the spherical aberration of the objective, or by other characteristics of the microscope, rather than by the size of the Airy disc. With reference to Fig. 18-28,

(18-g)
$$\tan \alpha = \frac{\frac{1}{2}y}{\frac{1}{2}D} = \frac{d_{min}}{D},$$

where D is the depth of field, and α is the aperture angle of the microscope. Since α is small, $\tan \alpha \cong \alpha$, and thus

(18-h)
$$\boxed{D = \frac{d_{min}}{\alpha}}.$$

For example, if $d_{min} = 5\,\text{Å}$, and $\alpha = 5 \times 10^{-3}$ rad, then

(18-i)
$$D = \frac{5 \times 10^{-8}\ \text{cm}}{5 \times 10^{-3}} = 0.1\ \mu.$$

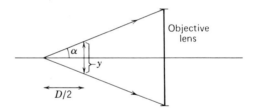

Fig. 18-28 Depth of field.

The value of D is of the same order of magnitude as in light microscopy, despite the fact that the level of structural detail resolved is very much smaller. The depth of field is in fact as great as, or greater than, the thickness of the specimen, so that *all* planes of the specimen are focused simultaneously in the image. This condition precludes the use of continuous adjustments of fine focus to indicate the three-dimensional relationships of the specimen, as is done in light microscopy. Electron images must be interpreted with due regard to the fact that the three-dimensional organization of the specimen is not directly revealed. Note that the relatively very large depth of field does not eliminate the requirement for very careful focusing of the image.

In light microscopy, depths of field and focus are approximately equal. In the electron microscope, however, the depth of focus is found not to be a fraction of a micron, but at least several *centimeters*. Images focused on the fluorescent screen are recorded without loss of focus on emulsions placed some distance above or below the screen. This can be accounted for by the fact that, at the aperture angles used in the electron microscope, angular and lateral magnifications are inversely related (cf. p. 185). Thus the angular and total magnifications of the systems may not be equated, as was done in deriving an expression for the depth of focus of the light microscope.

The depth of focus of the electron microscope may be estimated, noting first, with reference to Fig. 18-29, that

$$(18\text{-j}) \qquad \tan \alpha' \cong \alpha' \cong \frac{\frac{1}{2}z}{\frac{1}{2}D'}, \qquad D' = \frac{M_{lat}d_{min}}{\alpha'},$$

where D' is the depth of focus.

From the definition of angular magnification and the inverse relationship of angular and lateral magnifications at small values of α,

$$(18\text{-k}) \qquad \alpha' = \alpha M_{ang} \cong \frac{\alpha}{M_{lat}}.$$

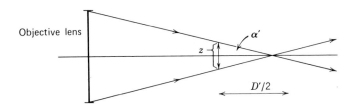

Fig. 18-29 Depth of focus.

Thus substitution of (18-k) in (18-j) gives

$$(18\text{-l}) \qquad \boxed{D' = M_{lat}^2 \frac{d_{min}}{\alpha}}$$

as the expression for the depth of focus of the electron microscope. Evaluation of (18-1) for $d_{min} = 5 \text{ Å}$, $\alpha = 5 \times 10^{-3}$ rad, and $M = 10,000\times$ yields

$$(18\text{-m}) \qquad D' = \frac{10^8 \times 5 \times 10^{-8} \text{ cm}}{5 \times 10^{-3} \text{ cm}} \qquad \text{or} \quad \text{one meter.}$$

To the extent that the paraxial condition is not strictly satisfied, even in the electron microscope, (18-l) and (18-m) exaggerate the depth of focus. Nevertheless, these equations correctly indicate that the depth of focus is very large.

CALIBRATION

The approximate magnifications corresponding to certain instrumental settings of an electron microscope may be provided by the manufacturer, but they are accurate only to within about 10%. Also, magnification at any given set of lens settings changes with time, as a consequence of both the heating of resistors and the aging or replacement of reference batteries or other circuit elements. Direct determinations of magnification must be made when accurate measurements are required.

Calibration procedures depend upon the availability of a specimen containing structures of suitable size which can be measured accurately by some independent means.

Relatively few calibration objects have been obtainable in the size range studied in the electron microscope. Replicas of diffraction gratings are the standard calibration specimen; they are useful at magnifications up to about 50,000×. Gratings (cf. Chapters 3 and 23) may be ruled with as many as about 50,000 lines per inch. Interlinear spacings are thus of the order of one micron, so that, at magnifications of the order of 10,000×, images of individual spacings are about one cm across. The exact value of the spacing may be obtained by measurements of the angular distribution of orders diffracted from the grating [i.e., according to (3-d*): $m\lambda = d \sin \theta$]. The spacings of a series of rulings must be measured, since the image of part of any individual ruling, as seen at high magnification, may be irregular or distorted. As indicated by Fig. 18-30, any spacing which is characteristic of the grating may be measured.

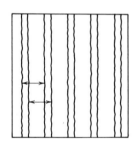

Fig. 18-30 Measurement of rulings in the image of a diffraction grating.

The average value $\bar{d}$ for a reasonably large number of spacings is determined:

(18-n)
$$M = \bar{d}N,$$

where N is the number of lines per unit length in the grating replica.

At very high magnifications, where images of grating spacings are inconveniently large, certain crystal spacings may be used as the calibration object. The size of these spacings may be determined independently by measurements of either x-ray or electron diffraction patterns. Unfortunately, crystals with spacings of the order of 100 Å are virtually unknown; therefore the distances to be observed may be inconveniently small. Also, crystals may decompose, or alter in structure, during irradiation by the electron beam. A number of different crystalline substances have been used for calibration purposes, including, principally, copper pthalocyanin, which has a crystal spacing of 12.4 Å.

RESOLUTION TESTS

No standard criterion has been formulated for determination of the resolving power of the electron microscope. However, three types of specimen have proved useful for the determination of resolution, even though each is somewhat limited with respect to its general applicability. These specimens are: focus fringes, crystal spacings, and point-to-point resolution specimens. Each type of specimen is discussed here.

FOCUS FRINGES AS RESOLUTION TEST OBJECTS

The overfocus fringes in images of holes provide a particularly convenient resolution test specimen. As explained on p. 404, the focus fringes result from the formation, at boundaries within the specimen, of scattered electron beams of anomalously high electron density. These beams form a so-called "selfless" object, which is imaged by the lens system. By changing focal level, the spacing between the geometrical edge of an object and the fringe can be varied at will. Specifically, the separation can be made as small as desired.

Resolving power is measured by recording a number of images of a hole very close to focus. The most closely spaced overfocus fringe which can be distinguished is then measured. It has been customary to measure fringe width from the center of the image of the geometrical edge to the center of the overfocus fringe, in the manner shown in Fig. 18-31b. However, careful observations at limiting resolution suggest that the resolving power estimated in this way is too great (i.e., that d_{min} is unrealistically small). The criterion $d_{min} = 2\Delta r$ appears to be more realistic.

In a system of inferior resolving power to that used to record Fig. 18-31,

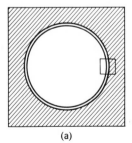
(a)
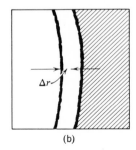
(b)

Fig. 18-31 Measurement of fringe widths.

the images of fringe and edge are "softer." Because they tend to overlap, the minimum separation which can just be distinguished is larger.

CRYSTAL SPACINGS AS RESOLUTION TEST OBJECTS

The smallest spacings thus far resolved by electron microscopy are those of certain lattices. Copper pthalocyanine (12.4 Å), MoO_3 (6.9 Å), tremolite (3.2 Å), and gold (ca. 2 Å) spacings are among those used as resolution test objects at the corresponding levels of performance.

The resolution of point objects may differ from that obtained by the same instrument in the imaging of crystal lattices; in general, it is poorer. This is due partly to the greater ease of distinguishing *rows* of molecules in the crystal as compared with separate points in a random structure. Also, contrast in images of crystals is primarily a *phase* effect. Phase contrast results from interference between beams scattered in different directions by the crystal planes. The contrast of point objects, however, is primarily an amplitude effect which can be related to the total scattering power of the object. It is to be expected that the resolutions achieved by two such entirely different mechanisms of contrast may differ.

POINT-TO-POINT RESOLUTION TESTS

Specimens for the determination of point-to-point resolution contain a variety of spacings of different sizes. A suitable specimen is a thin film on which a metal of minimal grain size, usually platinum, is evaporated very lightly. Two identical micrographs of an area of the specimen are then recorded at exact focus. Enlarged prints of these micrographs are then studied in order to locate minimum spacings between particles of metal which appear in both exposures. In this way, random effects of photographic grain can be distinguished from bona fide resolution of object points.

The point-to-point resolution test provides a realistic estimate of the performance of the electron microscope. However, since resolution may be limited by lack of contrast in the test specimen, and also since spacings just equal to d_{min} could be absent from the object or overlooked in the image, the value of d_{min} estimated by this method tends to be an upper limit.

REFERENCES

General References

C. E. Hall, *Introduction to Electron Microscopy*, 2nd ed., McGraw-Hill, New York, 1966.

M. E. Haine, *The Electron Microscope; The Present State of the Art*, Interscience, New York, 1961.

A. W. Agar, in D. H. Kaye, Ed., *Techniques for Electron Microscopy*, 2nd ed., Blackwood and F. A. Davis, Philadelphia, 1965, pp. 1–41. A good general article on "The Operation of the Electron Microscope."

V. K. Zworykin, G. A. Morton, E. G. Ramberg, J. Hillier, and A. W. Vance, *Electron Optics and the Electron Microscope*, Wiley, New York, 1945, pp. 33–42. A section on "The Electron Gun."

H. Fernandez-Moran, *J. Appl. Phys.* **31**, 1840 (1960). An abstract describing "Improved Pointed Filaments of Tungsten, Rhenium, and Tantalum for High Resolution Electron Microscopy and Electron Diffraction."

H. A. Steinherz, *Handbook of High Vacuum Engineering*, Reinhold, New York, 1963; especially Chapter 3 (Behavior of Gasses at Low Pressure), Chapter 4 (Methods of Obtaining Vacuum), and Chapter 5 (Gauges). An excellent collection of information about vacuum systems, with reference both to industrial and to scientific applications.

H. G. Heide, in S. S. Breese, Jr., Ed., *Electron Microscopy* (Proceedings of the Fifth International Congress for Electron Microscopy, Philadelphia, 1962), Academic Press, New York, Paper A4. This short paper, "The Prevention of Contamination without Beam Damage to the Specimen," presents the results of a systematic study of contamination.

R. H. Alderson and J. S. Halliday, in Kaye, *op. cit.*, pp. 478–527. This chapter, "Electron Diffraction," is a good general discussion of the subject even though not specifically related to the diffraction of biological specimens.

B. M. Siegel in B. M. Siegel, Ed. *Modern Developments in Electron Microscopy*, Academic Press, New York, 1964, pp. 17–27 (a general description of light and electron microscopes), pp. 36–45 (The Resolving Power of the Electron Microscope), and pp. 51–78 (Properties of High Resolution Electron Microscopes). Other chapters in this book (by various authors) describe certain techniques for specimen preparation and discuss applications of the electron microscope in histology and cytology, in bacteriology, and in studies of biological macromolecules.

G. F. Bahr and E. Zeitler, in G. F. Bahr and E. Zeitler, Eds., *Quantitative Electron Microscopy*, Armed Forces Institute of Pathology, Washington, D.C., 1965 *Lab. Invest.* **14** 142/880–152/890 (1965). An article on "Means for the Determination of Magnification."

Specific References

[1] A. W. Crewe, *Science* **154**, 593 (1966).
[2] D. C. Pease, *Histological Techniques for Electron Microscopy*, 2nd ed., Academic Press, New York, 1964.
[3] Kaye, *op. cit.*, especially Chapter 3 (The Preparation of Support Films, by D. E. Bradley), Chapter 4 (Techniques for Mounting, Dispersing, and Disintegrating Specimens, by D. E. Bradley), Chapter 5 (Replica and Shadowing Techniques, by D. E. Bradley), Chapter 6 (Techniques for Optical Selection of Particles, by D. E. Bradley), Chapter 7 (The Fixation and Embedding of Biological Specimens, by A. M. Glauert), Chapter 8 (The Preparation of Thin Sections, by A. M. Glauert and R. Philips), Chapter 9 (Section Staining, Cytology, Autoradiography and Immunochemistry for Biological Specimens, by A. M. Glauert), Chapter 10 (The Examination of Small Particles, by R. W. Horne), and Chapter 11 (Negative Staining Methods, by R. W. Horne).

CHAPTER 19

Contrast and Resolution in the Electron Image

The resolution achieved in electron images is limited, almost invariably, by lack of contrast rather than by lack of resolving power. Although ultimate resolving powers are of the order of 2 Å, there is at present very little prospect of resolving detail in biological structures at that level. The best resolution obtained in images of biological objects has been of the order of 8 Å. While much attention has rightly been devoted to possibilities for extending the resolving power of the electron microscope, it should be appreciated that the extension of image contrast is an equally demanding problem.

The contrast of electron images is determined by the nature and extent of interactions between the electron beam and the specimen. Properties both of the specimen (inherent contrast) and of the microscope system (instrumental contrast) are of importance in this respect. These factors and their effects upon resolution are discussed in this chapter; methods for enhancing contrast are also described. The relative importance and even the nature of the contrast mechanisms which operate at limiting resolution in the electron microscope are still not fully understood. Thus the discussion of this aspect should be regarded more as an outline of current thinking than as the "last word" on electron contrast.

A general knowledge of the design of electron microscopes and of the properties of electron lenses, as covered in Chapters 17 and 18, respectively, is required for appreciation of the factors which affect image quality. It is also helpful to understand the nature of the contrast mechanisms which are operative in light microscopy, as discussed in Chapter 11.

SUMMARY OF MECHANISMS OF ELECTRON CONTRAST

The contrast of the electron image, as of the light microscope image, can arise from both amplitude and phase effects. *Amplitude contrast* is produced by the loss of amplitude (i.e., of electrons) from the beam; *phase contrast* originates from shifts in the relative phases of the portions of the beam which contribute to the image (cf. p. 254).

Electron images must be interpreted with awareness of the fact that the relative importance of the various contrast mechanisms is quite different in electron microscopy from what it is in light microscopy. For the majority of specimens, *differential electron scattering*, an amplitude effect, is the primary source of electron contrast. While amplitude effects are also predominant in ordinary light microscopy, the important mechanism is one of differential absorption. Absorption by usefully thin electron microscope specimens is completely negligible, whereas scattering contrast is only occasionally of importance in light microscopy.

As the limit of resolution of the electron microscope is approached, the importance of phase contrast effects increases. It should be understood that electron phase contrast is *not* analogous to that of the high quality, in-focus images formed in the phase and interference microscopes (cf. Chapters 13 and 14), but to the (limited) phase contrast which is sometimes obtained in the ordinary light microscope (cf. p. 257). Electron phase contrast originates from: defocusing of the image; or from interference between rays which are misfocused by spherically aberrant lenses. The second mechanism is of consequence only very close to the limit of resolution. Some attempts have been made to develop in-focus electron phase contrast systems (see p. 446), but these systems are not used in standard electron microscopes at the present time. Nevertheless, electron *images* are more analogous to phase contrast microscope images than to ordinary light images. The reason for this is that both the differential scattering power for electrons and refractive index for light are related to the *mass density* of the specimen. On the other hand, the absorption contrast of ordinary light microscope images is largely related to the chemical bonding of the specimen, a factor which has no direct effect on electron contrast.

ELECTRON SCATTERING; SCATTERING CROSS SECTIONS OF ATOMS

Electrons which are incident upon matter are scattered in a manner determined by the quantity and the atomic number Z of the atoms present. A quantum mechanical treatment is required for accurate prediction of the angular distribution of scattered intensities. Here only a simple classical treatment of electron scattering is outlined. This treatment provides a pictorial description of the events occurring during the process of scattering. Predictions based on this approach are approximately correct; limitations of the treatment are indicated.

The *scattering cross section* (cross section for electron scattering) of an atom, designated σ_e, is the *effective area* of the atom for scattering through an angle greater than any specified value. That is, the scattering cross section gives the diameter of a particle which would deflect the observed number of electrons through the specified angle if all collisions resulted in scattering. σ_e is, in fact, a measure of probability. For example, σ_e for scattering through more than 10^{-2} rad might be 10^{-18} cm², while the value for scattering through more than 10^{-3} rad might be 10^{-16} cm². This means that scattering through 10^{-3} rad or less is one hundred times more likely than scattering through 10^{-2} rad or more. It does *not* mean that the atom is literally 10^{-16} or 10^{-18} cm in diameter. From a classical point of view, however, the resulting distribution of electron intensities is *as if* electrons incident upon the central 10^{-18} cm² of a particle of well defined size were scattered through at least 10^{-2} rad, while those incident beyond the central 10^{-16} cm² were scattered through 10^{-3} rad or less. The concept of cross sections is also applied to a number of other events for which probability distributions exist; for example, macromolecules, virus particles, and bacteria may be described in terms of their cross sections for inactivation by radiations of various types (see Setlow and Pollard [1]).

The nature of the scattering process may be explained by noting that an electron passing near a nucleus at some distance r_{n_1} is attracted toward the positive charge, thus describing a hyperbolic path in the region of the nucleus. As the electron passes beyond the nucleus, it again travels on a linear path, but one which is directed at some angle θ_1 with respect to the original trajectory. An electron passing closer to the nucleus (i.e., at $r_{n_2} < r_{n_1}$) is more strongly attracted by the positive charge and is therefore deflected through a larger angle. Two such electron trajectories are shown in Fig. 19-1. In the limit the negatively charged electron could be captured by the positively charged nucleus; here it is considered that the momentum of electrons is sufficient for them to escape capture.

Comparable electron trajectories in the vicinity of a stationary electron are illustrated in Fig. 19-2. The force exerted between the similarly charged moving and stationary particles now becomes one of repulsion. As before, the closer the approach, r_e, the larger is the deviation θ of the electron trajectory. Since electrons approach the stationary particle, whether nucleus or electron, from all sides, the latter behaves approximately as a point source of scattered electrons. Thus nuclei and indivi-

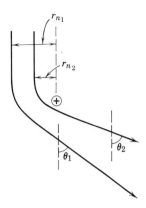

Fig. 19-1 Electron trajectories in the vicinity of a nucleus.

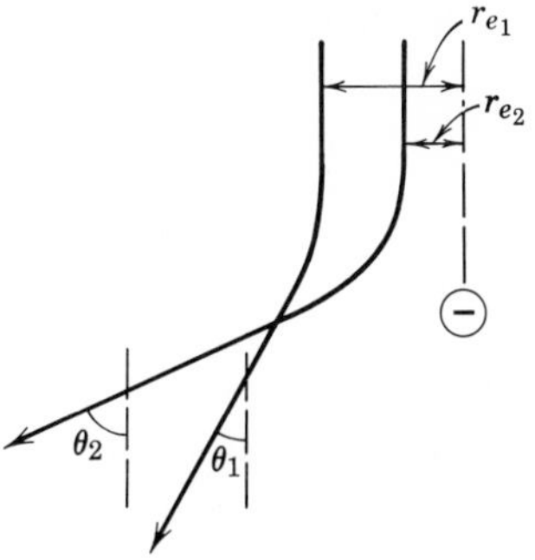

Fig. 19-2 Electron trajectories in the vicinity of a stationary electron.

dual electrons are comparable to the "self-luminous points" considered in light optics.

A shortcoming of the classical view of electron scattering may be noted. As described here, the attraction or repulsion of electrons varies continuously as a function of distance from the stationary charged particle. According to this view, electrons which pass anywhere through a specimen should experience some electrostatic force and thus some deflection, however slight. But this prediction is not confirmed by experiment. In fact, when a thin layer of matter is irradiated, part of the beam is scattered in all directions, but *many electrons are not deflected at all.*

An expression may be derived for the fraction of incident electrons scattered by a specimen through any angle θ. The effective radii of isolated, stationary nuclei and electrons are

(19-a)
$$r_n = \frac{Ze}{2V \tan \left(\frac{1}{2}\theta\right)},$$

(19-b)
$$r_e = \frac{e}{V \tan \theta},$$

where Z is the atomic number of the nucleus, e the electronic charge, V the acceleration potential of the moving electron, and θ the angle of scattering. Equations (19-a) and (19-b) are not derived here; for a derivation see [2].

In the electron microscope appreciable scattering occurs only through small angles. The approximation $\tan \theta \cong \theta$ is then valid, so that (19-a) and (19-b) become

(19-c)
$$r_n = \frac{ZE}{V\theta},$$

(19-d)
$$r_e = \frac{e}{V\theta}.$$

$\sigma_{e.atom.\theta}$, the cross section for scattering of electrons through $\theta°$ by the complete atom, may then be estimated. It is assumed, as a first approximation, that the atom consists of independent electrons and nucleus, that is, that the nuclear charge is not shielded by the surrounding electrons. Since the atom contains one nucleus and a corresponding number Z of electrons, the cross section is

(19-e)
$$\sigma_{e,atom,\theta} = (r_{atm,\theta})^2 = \frac{Z^2 e^2}{V^2\theta^2} + \frac{Ze^2}{V^2\theta^2}$$

$$= Z\left(\frac{e}{V\theta}\right)^2 (Z+1).$$

The probability for scattering of an incident electron through $\theta°$ by a *specimen* is then the *scattering cross section per atom*, as given by (19-e), times the *total number of atoms in the specimen* divided by the *area* of the specimen. The total number of atoms, N, is

(19-f)
$$N = \frac{dN_0 ta}{A},$$

where d is the density of the specimen, N_0 is Avogadro's number (i.e., the number of atoms per mole), t is the thickness of the specimen, a is its area, and A is the average atomic weight of the specimen.

Combination of (19-e) and (19-f) thus yields an expression giving the probability, $N(> \theta)/N$, for scattering of electrons by a *specimen* through an angle greater than θ. A thin layer, of thickness Δt, must be considered, so that multiple deflections of the same electron can be ignored. This expression is

(19-g)
$$\frac{N(> \theta)}{N} = \frac{dN_0 Z^2 e^2}{A\ V^2\theta^2}\left(1 + \frac{1}{Z}\right) \Delta t.$$

For heavy atoms, Z is large, so that $(1 + 1/Z) \cong 1$, and (19-g) reduces to

(19-h)
$$\frac{N(> \theta)}{N} = \frac{dN_0 Z^2 e^2}{A\ V^2\theta^2} \Delta t.$$

The simplification made in obtaining (19-h) (i.e., the use of the assumption $1 + 1/Z \cong 1$), is equivalent to assuming that *only* scattering from *nuclei* is important; scattering from atomic electrons is ignored. Since, as already noted, shielding of the nuclear charge is also ignored, it is evident that (19-h) must tend to exaggerate the nuclear component of electron scattering. A more reasonable expression for the atomic scattering cross section thus is

(19-i) $$\frac{N\,(>\theta)}{N} \cong KZ,$$

where $K = (e/V\theta)^2(dN_0/A)\Delta t$. Experimentally, it is found that (19-i) is roughly correct for elements of atomic number less than about 40; for elements of high atomic number, scattering is even less strongly dependent upon Z.

ELASTIC AND INELASTIC COLLISIONS OF ELECTRONS; CONTRIBUTION OF INTERACTIONS WITH THE SPECIMEN TO CHROMATIC ABERRATION

The way in which the atoms of an electron microscope specimen act as "point sources" of electrons has been explained above, and the intensity of electron scattering has been shown to be a function of the atomic number of the scatterer. A third feature of the interactions of electrons with the specimen is the possibility for change of the electron wavelength upon collision, that is, in particulate terms, of possible energy loss with a corresponding reduction of electron velocity.

Interactions which produce no change in the wavelength of the incident electron are termed *elastic collisions*. According to the classical picture, elastically scattered electrons are those which "bounce off" the massive and essentially stationary nuclei of the scattering atoms. *Inelastic collisions* occur when energy is transferred by the beam to the light, mobile electrons of the specimen.

The proportions of inelastic and elastic collisions are determined by the accelerating voltage and by the nature of the specimen. For a carbon film about 500 Å thick, which is illuminated by 50 kV electrons, it has been estimated that about 34% of incident electrons are undeflected, 11% are elastically scattered, and 55% inelastically scattered. Elastic scattering deflects the incident electrons through angles of the same order of magnitude as those used in practical electron microscopy; only a very small proportion of total scattering is through angles greater than about 10^{-2} rad. Elastic scattering occurs almost exclusively within the lens aperture; few electrons are elastically scattered through more than about 5×10^{-4} rad.

The proportional energy loss due to inelastic collisions is found to be, approximately, a function of the *square* of θ, the angle of scattering. Thus,

(19-j)
$$\frac{\Delta E}{E} = \frac{\Delta V}{V} = \sin^2 \theta \cong \theta^2,$$

so that, for $\theta = 5 \times 10^{-4}$ rad, $\Delta V/V = 2.5 \times 10^{-7}$. This figure may be compared with the fluctuations in the level of accelerating voltage which occur in present electron microscopes at levels of 10^{-4} or 10^{-5}. The change in wavelength produced by single electron collisions is seen to be relatively insignificant. However, *multiple* elastic collisions, which may occur in relatively thick specimens, can be a serious source of chromatic aberrations in the image.

SCATTERING CONTRAST IN THE ELECTRON IMAGE

The angular distribution of scattered intensities varies as a function of the atomic composition and density of the object. "Electron opaque" object points produce appreciable inelastic scattering through relatively large angles. Many of the electrons which are incident on such points are thus excluded from the lens aperture. The intensity of the images of these points is correspondingly low. Conversely, the "electron transparent" support or surround, which is of lower average atomic number or thickness or both, produces very little scattering beyond the lens aperture. The intensity of images of these points is correspondingly high. The production of contrast by these effects is illustrated in Fig. 19-3. Further increase of scattering contrast is obtained with the use of a limiting objective aperture, as discussed on p. 428.

The contrast brought about as shown in Fig. 19-3 is analogous to the contrast observed at reduced apertures in certain types of light microscope specimens (cf. p. 255). In electron images, additional contrast is produced as a consequence of the aberrations of peripheral lens zones. The *effective aperture* of electron lenses, within which electrons are focused essentially at the corresponding Gaussian image points, is often smaller than the physical lens aperture. Electrons which are scattered into the lens margins can be considered to contribute to a continuous *background intensity*, rather than to the correct image points. The net effect may seem somewhat paradoxical: overall image contrast deteriorates, but certain points (which correspond to objects of low scattering power) acquire contrast *because* of the aberration. This is the case because the loss of electrons from these image points to background reduces the amplitude of the image just as much as would the total exclusion of the same electrons from the lens aperture. The effect may be of considerable quantitative importance as a source of contrast in images of biological specimens. Correction of spherical aberration of the peripheral lens

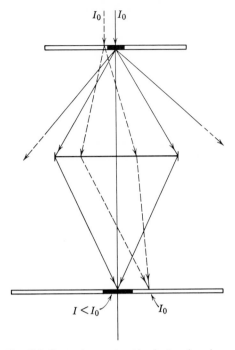

I_0 I_0

$I < I_0$ I_0

Fig. 19-3 Scattering contrast in electron imaging.

zones would therefore produce a loss of contrast from images of this type of specimen!

In addition to the amplitude contrast discussed here, lens aberrations also may produce phase contrast in the image, as discussed on p. 437.

The mechanisms just discussed have thus far been used exclusively as the source of scattering contrast in high resolution electron microscopes. However, other scattering contrast mechanisms are likely to be exploited in the future development of the instrument. For example, differential analysis of inelastically scattered electrons has proved effective in experimental systems of moderate resolving power (see [3]). In this method, images are formed only by those scattered electrons which travel with velocities within some selected range. The distribution of velocities (*energy loss spectrum*) is much more characteristic of the scattering element than is the total scattering within any angular range. Furthermore, since inelastic collisions are primarily interactions with the valence electrons of the specimen, the distribution of energies is determined to some extent by the chemical combinations of the atoms in the specimen. Recently, the same principal has been exploited successfully at resolutions comparable to those of standard high resolution microscopes.

ENHANCEMENT OF CONTRAST BY A LIMITING OBJECTIVE APERTURE

The background intensity produced by spherical aberration can be curtailed by limiting the angular aperture of the objective lens. The proportion of electrons which are totally excluded from the lens aperture is correspondingly increased. In practice, this is done by placing an aperture at or near the rear focal plane of the objective lens. The action of this limiting aperture is shown in Fig. 19-4. Figure 19-4a shows rays leaving two object points Q and P in three directions Ⓐ, Ⓑ, and Ⓒ. The rays which leave Q and P in direction Ⓐ, parallel to the lens axis, cross on the axis in the focal plane; rays leaving the object points in directions Ⓑ and Ⓒ, inclined to the axis, intersect at other positions in the focal plane. The rays which travel from each point in directions Ⓐ and Ⓑ intersect the

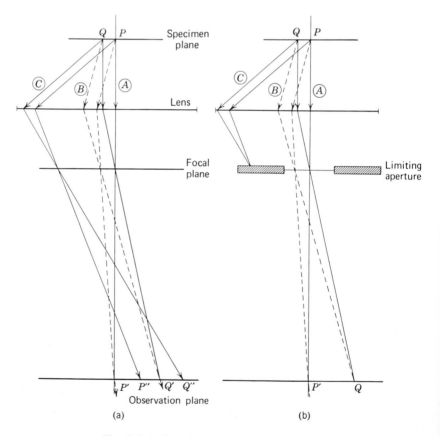

Fig. 19-4 Action of the limiting objective aperture.

lens field close to its axis and reunite at the corresponding image points P' and Q'. However, rays traveling in direction © intersect the lens close to its margin, and thus are converged excessively, reaching the image plane at points P'' and Q'', where they contribute to background intentity.

Figure 19-4b shows the same lens as does Fig. 19-4a, but with an aperture placed in the rear focal plane; rays traveling in direction © no longer reach the image plane, and background intensity is correspondingly reduced.

These considerations seem to imply that the minimal aperture consistent with optimal resolution should be used in the objective lens at all times. In fact, however, use of the aperture is limited for practical reasons. The limiting aperture is very easily contaminated by the electron beam; if this contamination is *asymmetric*, lens asymmetry results, with corresponding deterioration of resolving power. Furthermore, the rate of contamination may easily be so great that the asymmetry shifts much too rapidly for effective correction by changes in the compensation of the lens. The asymmetry introduced by a contaminated aperture is by far the most important source of lens asymmetry in the electron microscope. Use of the limiting aperture is thus profitable only if great care is taken to avoid its contamination, since *improvement in contrast is absolutely worthless if accompanying asymmetry leads to a serious loss of resolution*. The effect upon resolution may, unfortunately, be far from subtle.

Thus, provisions are always made for easy removal of the limiting aperture from the objective lens field. The extent to which the aperture is used depends in part upon the nature of the specimen. Contrast is improved most effectively for specimens which scatter principally into the peripheral zones of the lens (i.e., beyond the "effective aperture" rather than for specimens containing a high proportion of heavy atoms, which scatter principally through larger angles *beyond* the physical aperture of the lens. A limiting aperture thus may not be required for observation of particles shadow cast by heavy metal atoms, but is virtually essential in recording images of biological material which is either stained or untreated. Surveying of the general features of areas of a specimen may also be carried out without an aperture in the lens.

Procedures for the use of the aperture tend to vary also with the level of resolution sought. Traditionally, semipermanent platinum apertures have been used. The contamination rate is held to a minimum, and the aperture is cleaned periodically. Some degree of lens asymmetry may be removed by compensation. For more demanding work these procedures are inadequate: once contaminated, the apertures never regain the symmetry required for very high resolution. Series of apertures have then been used which are prepared by etching meshes of metal (usually copper). Entirely fresh apertures can then be shifted into the field as required. At ultimate resolution, continuous use of a limiting aperture is usually essential, since contrast is limiting at this level of operation.

Inspection of Fig. 19-4 suggests that the aperture need not be placed at the rear focal plane, but could be located at any other level reasonably close to the lens. The choice (in most electron microscopes) of the focal plane location is the most suitable for two reasons. At this position the aperture does not restrict the field of view, that is, the aperture is then purely an aperture stop and not a field stop also (cf. p. 181). Even more important is the fact that contamination is most easily controlled at this position. The greatest part of the intensity leaving the specimen is that of the direct beam. The unscattered electrons are brought, at least approximately, to a point focus in the focal plane, whereas at other levels they are distributed over a finite radius. Thus it is easier to avoid intersecting the contaminating beam at the focal plane.

Procedures must be devised for centering the aperture about the focused zero order beam without contaminating it in the process. The most desirable method of doing so varies from instrument to instrument. The physical dimensions of apertures used at the focal plane vary from about $20\,\mu$ to about $100\,\mu$. Aperture diameters of about 20 or $30\,\mu$ limit the angular aperture to its optimal value (cf. p. 365), but apertures of this size are very difficult to keep free of contamination. The upper range of diameters of the limiting aperture is thus more useful.

ENHANCEMENT OF THE INHERENT CONTRAST OF THE SPECIMEN

The level of electron amplitude contrast is determined by $\bar{Z}$, the average atomic number of the specimen, as (19-i) suggests. For biological specimens, which consist of carbon, hydrogen, nitrogen, oxygen, and minor amounts of only somewhat heavier elements, $\bar{Z}$ is low; therefore lack of inherent contrast (i.e., of electron scattering power) is decidedly a limiting factor. Inherent contrast may be increased by the preferential addition of materials of high atomic number during specimen preparation.

The nature of methods for the enhancement of inherent contrast is summarized briefly (see also[4], [5]). In the method of *shadow casting*, heavy atoms (usually platinum) are evaporated at an angle to the surface of the specimen. Thus metal atoms accumulate on the surface contours which may be, for example, virus particles supported by a carbon film. Electron *staining*, a method analogous to the staining techniques used in optical microscopy, is used principally for the study of tissue sections. Appropriate stains are substances which contain heavy atoms and/or are of high density. In *negative contrast* techniques (also referred to as "negative staining") relatively electron transparent particles are surrounded by a continuous film of electron opaque "stain." In each of the methods the dimensions of the material used to increase the effective atomic number of specimen details tend to be of the same order as the size of the structure resolved. Possibilities for enhancement of contrast by altering the composition of the specimen are thus inherently much more limited than in light microscopy, where individual stain molecules are much smaller than the minimum resolvable distance.

The effect of the specimen support upon image contrast is described on p. 441.

PHASE CONTRAST EFFECTS IN ELECTRON IMAGES

Electron scattering, which has been considered in particulate terms thus far, may also be described as the diffraction of electron waves. If the interference of diffracted waves produces differences in intensity at the image, phase contrast is effected. A general description of electron scattering, in terms of electron waves, is given in this section. (Although specifically phrased in terms of electrons, the following account requires only minor modifications to apply to the scattering and interference of light.)

All specimens contain a distribution of atoms and thus are associated with a corresponding distribution of inner potential. For example, a carbon film has an average inner potential of 11 volts, with local variations of ± 3 volts. Potential (i.e., $\sqrt{V}$) is the electron optical analog of refractive index (cf. p. 346). Therefore electron waves which impinge upon a specimen are retarded according to the level of inner potential which they encounter, just as light waves are retarded by specimens according to the value of the refractive index. In this way spatial fluctuations of the inner potential of the specimen produce distortions of the transmitted electron wavefront.

An idealized electron microscope specimen is considered in Fig. 19-5. This specimen consists of a matrix (surround) of uniform inner potential, V, in which spherical atoms of uniform inner potential $V' < V$ are embedded. An incident plane wavefront AA' is distorted by transmission through the specimen, forming the wavefront BB'. The corresponding electron trajectories (rays) are found by applying Huygens' principle; that is, by drawing normals to BB' (cf. p. 50). As shown, electrons transmitted by regions of uniform potential (solid lines) are undeflected; electrons transmitted by regions in which a *gradient* of inner potential

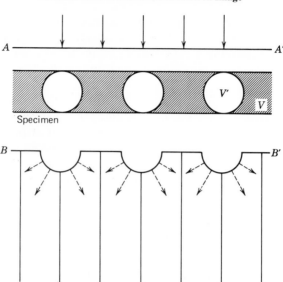

Fig. 19-5 Distortion of the electron wavefront by the varying inner potential of a specimen.

exists are scattered through a range of angles. The electron trajectories are most strongly deflected by regions of very steep potential gradient. (The analogous effects in the case of visible light are described on p. 683.)

Electrons scattered through large angles (corresponding to the most distorted portions of the wavefront) may fall outside the lens aperture, thus giving rise to amplitude contrast. Ideally, all other scattered electrons are focused by the lens at the corresponding image points, at which they arrive *in* phase. At planes above or below the ideal image plane, however, *out*-of-phase electron trajectories intersect. The resulting interference produces differences in intensity. Specifically, the intensity differences consist of the focus fringes (Fresnel fringes, contour phenomena) already described on p. 402.

Because of the effects of lens aberrations, entirely "pure" focus fringes are not obtained in the image. The optical paths through paraxial and marginal zones of spherically aberrant lenses differ. Consequently, the *lens* imposes distortions upon the electron wavefront which add to those already imprinted by the specimen. A slight phase contrast then persists at exact focus, since the optical paths of the electrons which contribute to each in-focus image point are no longer equal. At other focal settings the effects of out-of-focus and aberration phase contrast superpose.

The two sources of phase contrast — defocusing and lens aberrations — and the effects of phase contrast on the imaging of the specimen support are discussed in detail in the following sections.

OUT-OF-FOCUS PHASE CONTRAST

The appearance of out-of-focus fringes has been illustrated in Fig. 18-18 and Plate II. Even slight overfocusing tends to confuse the image, but at underfocus the geometrical edge of the object is pleasingly high-lighted. Similarly, in images of objects of more complex structure than the holes shown in Plate II, contrast is minimal at exact focus but is enhanced at overfocus. This is illustrated by Plate IV, which shows a preparation of myosin molecules, shadow cast by platinum atoms. The upper picture is recorded very close to exact focus. The lower picture shows the same area at an underfocus setting. As these illustrations suggest, underfocusing is an important source of contrast in electron images. It is essential to understand, however, that, *to the extent that an image is defocused, resolving power is sacrificed.* In Plate IV it is particularly obvious that the particles appear to be larger when recorded at an underfocus setting.

Qualitatively it can be said that, if two closely spaced points are imaged at an underfocus setting, the halo surrounding the geometrical image of one overlaps the image of the other, as indicated in Fig. 19-6. Consequently, it may be that the two points cannot be distinguished as separate; that is, that they cannot be resolved. Careful comparison of the micrographs shown in Plate III shows that loss of resolution has occurred in the underfocused micrograph, even though this image is apparently sharper.

Fig. 19-6 Confusion of the image by over-lapping of under-focus haloes.

Quantitatively the loss of resolving power due to defocus is given by the expression obtained on p. 410

$$(18\text{-}f) \qquad\qquad \Delta r = \sqrt{\lambda \Delta f}.$$

where λ is the electron wavelength, Δf the extent of defocus, and Δr the corresponding confusion in the size of the object.

An optimal level of phase contrast can be produced by defocusing the image to the extent that resolution of the structure to be observed is *just* not impaired. Evaluation of (18-f) shows that defocusing by about 0.5 μ (a condition which may be selected with relative ease) is suitable for the enhancement of contrast at the level of about 15 Å. For resolution of

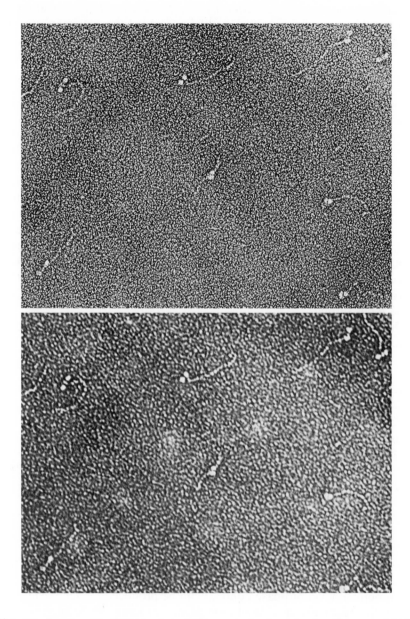

Plate IV   Images of individual rotary shadow cast myosin molecules. The same field is shown in both parts of the figure at a magnification of 92,000×. Top, image recorded very close to exact focus; bottom, image recorded at a level of underfocus which is typical of published electron micrographs.

2 Å, however, focusing to within 0.01 μ is required. So close an approach to exact focus cannot be achieved *systematically* in present microscopes, but depends upon trial and error; that is, of a number of micrographs recorded as close as possible to focus, some may fall within the required tolerance. Improvement of the focusing procedure depends upon the availability of high intensity at very high magnifications, possibly with the aid of image intensifier systems.

A slight defocus thus produces an optimum combination of contrast and resolving power. It remains debatable whether such focal settings are in fact ideal from the point of view of obtaining reliable information about the structure of the specimen. It has been argued that, since phase contrast is not directly related to the structure of the specimen, it is preferable to record images precisely at exact focus. This point is discussed at length on p. 441 with reference to the contrast introduced by the substrate film.

Whether images are recorded at exact focus or at the optimal defocus, it is critically important to record high resolution electron micrographs very close to focus. This fact does not appear to be generally appreciated; micrographs have often appeared in the literature which are defocused to a level that is obviously inconsistent with the claimed resolution. On the other hand, underfocusing is a desirable source of contrast at moderate levels of resolution, that is, for d_{min} greater than about 20 Å. Since lack of scattering contrast limits resolution to values of this order in the vast majority of biological specimens, a moderate extent of defocus usually can only improve image quality.

It is of interest to obtain a quantitative estimate of the path and phase differences introduced by defocusing the lens. Figure 19-7 illustrates the focusing of a ray from an axial point P at P' by a lens of focal length f. Overfocusing of this lens reduces the focal length to $(f - \Delta f)$ and shifts the image plane through a distance $-\Delta s_2''$, such that the axial image of P is formed at X, while the ray from P through X intersects the in-focus image plane at Q'. The in-focus and overfocus rays differ in path by a distance $Q'Y (= \Delta_F)$, as shown in Fig. 19-8.

The distance $\Delta s_2''$, may first be evaluated. From the thin lens equation (8-z),

(19-k)
$$\frac{1}{s_2''} = \frac{1}{f} + \frac{1}{s}.$$

Equation 19-k may be differentiated, noting that, since the specimen position s is held constant, only the image position s_2'' varies as a function of f. (Variations in s which result from movement of the first principal plane of the lens are assumed to be negligible for small values of Δf.) Thus,

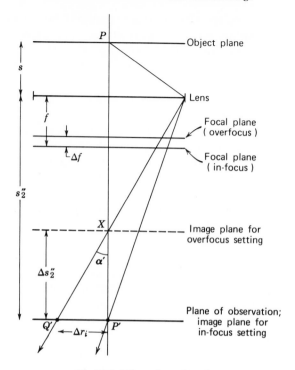

Fig. 19-7 Effect of overfocusing.

$$(19\text{-}l) \qquad \frac{ds_2''}{(s_2'')^2} = \frac{df}{f^2},$$

or, equating differentials to small but finite changes,

$$(19\text{-}m) \qquad \Delta s_2'' = \Delta f \left(\frac{s_2''}{f}\right)^2.$$

In the electron microscope the specimen is always placed very close to the first focal plane, so that $f \cong s$. Since the (lateral) magnification is defined by (8-mm) as the ratio s_2''/s, (19-m) becomes

$$(19\text{-}n) \qquad \Delta s_2'' = M^2 \Delta f.$$

Δr_i, the image spread produced by defocus, can then be evaluated. In triangle $XQ'P'$ of Fig. 19-7,

$$(19\text{-}o) \qquad \frac{\Delta r_i}{\Delta s_2''} = \tan \alpha' \cong \alpha'.$$

From (8-rr) and (8-ss),

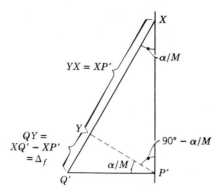

Fig. 19-8 Path difference between in-focus and overfocus rays.

(19-p)
$$\alpha' \cong \frac{\alpha}{M_{\text{lat}}}.$$

Thus, substituting (19-n) and (19-p) in (19-o) yields

(19-q)
$$\Delta r_i = \frac{\alpha}{M} \Delta f M^2 = \alpha M \Delta f.$$

$YQ' = \Delta_F$ may now be evaluated. With reference to Fig. 19-8,

(19-r)
$$\sin \frac{\alpha}{M} \cong \frac{\alpha}{M} = \frac{Q'Y}{\Delta r_i} = \frac{\Delta_F}{\Delta r_i} \quad \text{and} \quad \Delta_F = \frac{\alpha}{M} \alpha M \Delta f,$$

or
(19-s)
$$\boxed{\Delta_F = \pm \alpha^2 \Delta f}.$$

The choice of sign in (19-s) refers to the possibility of overfocusing *or* underfocusing the lens.

The corresponding phase shifts ϕ_F are

(19-t)
$$\boxed{\phi_F = \pm \frac{2\pi}{\lambda} \alpha^2 \Delta f}.$$

Numerical evaluations of (19-s) and (19-t) are given in Table 19-1. Note that, while the rays PLP' and PLQ' do not overlap, point Q' in the overfocused images is also illuminated by rays from specimen points in the vicinity of P. The resultant of *all* rays which overlap at Q' determines the intensity observed at this point in the image.

ABERRATION PHASE CONTRAST

Beams which pass through different zones of an aberrant lens experience different optical paths, and thus interfere to produce differences in intensity, even at the in-focus image plane. If all points of the specimen were to produce identical angular distributions of scattered electrons, interference between the scattered beams would be unrelated to the structure of the specimen. However, as has been shown by Fig. 19-5, regions of the specimen in which there is a gradient of inner potential scatter electrons through relatively large angles. A correspondingly high proportion of electrons from these regions is transmitted by peripheral lens zones; unscattered electrons, from regions of uniform inner potential, are transmitted by paraxial zones. As a result, electrons from different parts of the specimen tend to be segregated in different zones of the lens. Consequently, the path and phase differences imposed by the lens are to some degree related to specimen structure.

A quantitative estimate of aberration phase contrast may be made by comparing the optical paths of rays focused by the margins of real (i.e., aberrant) and ideal lenses, respectively. As shown in Fig. 19-9a, rays originating at the axial object point P would ideally be focused in the (Gaussian) image plane at P'. In fact, electrons which pass through the lens margin at L are deflected too strongly, so that they intersect the image plane at Q'. Q' is also the position which corresponds to imaging of a specimen point Q by an ideal lens. Thus, in Fig. 19-9, PLP' and QLQ' are the ideal or paraxial trajectories, while PLQ' is a real trajectory. An aberration figure of radius $\Delta r_i = Q'P'$ is the corresponding image of point P. At exact focus in a fully compensated microscope, Δr_i may be equated to the spherical aberration radius $(\Delta r_i)_{\text{S.A.}}$, the value of which was given in Chapter 17:

(17-d) $$(\Delta r_i)_{\text{S.A.}} \cong MC_3\alpha_3$$

where M is the lateral magnification of the microscope, $C_3(= C_sf)$ is the spherical aberration constant of the objective lens, and α is the angular aperture. (Chromatic aberration, though present, is unlikely to increase the value of Δr_i significantly.)

As shown in Fig. 19-9b, the path PL exceeds QL by a distance PR in object space, whereas, in image space, LQ' exceeds LP' by a distance $Q'R'$. If $PR = Q'R'$, the total optical paths PLP' and QLQ' through the ideal lens are equal. In triangle PQR,

(19-u) $$\cos(90-\alpha) = \frac{PR}{C_s\alpha^3},$$

or, since $\cos(90-\alpha) \equiv \sin\alpha$ and since α is small,

(19-v) $$PR = C_s\alpha^3 \sin\alpha = C_s\alpha^4.$$

Similarly, in triangle $P'Q'R'$,

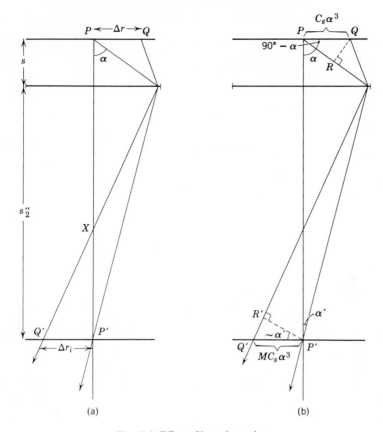

Fig. 19-9 Effect of lens aberrations.

$$(19\text{-}w) \qquad \langle (Q'P'R') = \frac{Q'R'}{MC_s\alpha^3}.$$

If PQ and thus $P'Q'$ are small, LP' and LQ' are very nearly parallel; thus $R'P'$ is perpendicular to $Q'P$. (These restrictions are applicable for the nearly paraxial imaging by electron lenses.) It then follows that $\angle(Q'R'P') \cong \alpha'$, so that

$$(19\text{-}x) \qquad Q'R' = MC_3\alpha^3 \sin\alpha'.$$

Since α' is small, $\sin\alpha' \cong \alpha'$, while, from (8-qq) and (8-ss),

$$(19\text{-}y) \qquad \frac{\sin\alpha'}{\sin\alpha} = M_{\text{ang}} = \frac{1}{M_{\text{lat}}} \quad \text{or} \quad \sin\alpha' = \frac{\sin\alpha}{M}.$$

Substitution of (19-y) in (19-x) thus gives

$$(19\text{-z}) \qquad Q'R' = C_3\alpha^3 \sin \alpha = C_3\alpha^4 = PR.$$

The optical paths through the ideal lens from closely spaced points P and Q to the corresponding image points are thus shown to be equal. The margin of a *non*ideal lens, however, deflects the rays from P to Q'. Δ_A, the phase difference so introduced between marginal and paraxial rays from, P is therefore the distance $Q'R'$. That is,

$$(19\text{-aa}) \qquad \boxed{\Delta_A = C_3\alpha^4},$$

and the corresponding phase shift ϕ_A is

$$(19\text{-bb}) \qquad \boxed{\phi_A = \frac{2\pi}{\lambda} C_3\alpha^4}.$$

These values may be compared with the corresponding quantities Δ_F and ϕ_F, which were derived on p. 437.

$$(19\text{-s}) \qquad \Delta_F = \pm\,\alpha^2 \Delta f, \qquad\qquad (19\text{-t}) \qquad \phi_F = \pm\frac{2\pi}{\lambda}\,\alpha^2\,\Delta f.$$

While the path and phase differences introduced by defocus are proportional to the *square* of the aperture angle, those introduced by spherical aberration are proportional to the *fourth power* of that angle. Since α is always much smaller than one, this means that the spherical aberration effect is of relative importance only when Δf is reduced to small values. Some realistic numerical values are compared in Table 19-1. (Aperture angles of 10^{-2} and 10^{-3} rad represent the approximate upper and lower limits, respectively, of apertures used in normal microscopy.) The specific values given in Table 19-1 must not be taken too seriously, however, since a number of assumptions have been made in deriving (19-t) and

Table 19-1 Path and Phase Shifts Due to Spherical Aberration or Defocus
(For 100 kv electrons, and $C_s \cong 1.0$ cm)

α	Δf and assoc. d_{min}	Δ_A	ϕ_A	Δ_F	ϕ_F
10^{-3} rad	$1\mu = 10^{-4}$cm (15 Å)	10^{-12}cm	$\lambda/400$	10^{-10}cm	$\lambda/4$
10^{-3}	$0.1\,\mu = 10^{-5}$ (6 Å)	10^{-2}	$\lambda/400$	10^{-11}	$\lambda/40$
10^{-2}	$1\mu = 10^{-4}$ (15 Å)	10^{-8}	25λ	10^{-8}	25λ
10^{-2}	$0.1\,\mu = 10^{-5}$ (6 Å)	10^{-8}	25λ	10^{-9}	2.5λ

(19-aa). (For example, only electrons scattered into the lens margins, as shown in Fig. 19-8, have been considered. Equation 19-aa correspondingly applies to the imaging of a periodic object at the limit of resolution. More generally, scattering into extended lens zones should be considered.) Although merely approximate, the data of Table 19-1 nevertheless show that aberration phase contrast usually is of relative importance only very close to the limit of resolution.

The contrast which results from the phase shifts computed in Table 19-1 depends upon the relative intensities of the electron beams scattered into the different zones of the lens. These intensities are determined by the level of accelerating voltage and by the nature of the specimen.

EFFECT OF THE SUBSTRATE FILM ON IMAGE CONTRAST

The continuous film which provides mechanical support for the electron microscope specimen is certainly indispensable (for the great majority of biological specimens), yet it imposes severe limitations on observations at very high resolution. In light microscopy the specimen can be mounted on a transparent slide (which is also of uniform optical path), but in electron microscopy the contrast introduced by the support may be comparable to that of the specimen itself. The situation may be likened to using stained glass to make slides for light microscopy!

The contrast of the specimen support originates from both amplitude and phase effects, although only the former are important when resolutions poorer than about 20 Å are considered. Amplitude contrast is minimized by using very *thin* films constructed of a material of low atomic number. Standard evaporated carbon films, which are about 200–500 Å thick, are satisfactory at moderate levels of resolution. However, when these films are observed at resolutions better than about 20 Å, "structure" is evident in the image. As resolution is improved (i.e., by more exact focusing of a microscope of high resolving power), this structure persists, becoming progressively finer. As exact focus is approached, limiting point-to-point distances of from 3 to 6 Å may be measured in the image; beyond this level the structure of the film tends to disappear. The image of the film, which is superimposed on that of the specimen itself, may often create a misleading impression of substructure in the specimen.

The observed structure may be accounted for as a phase contrast image of the individual, randomly arranged carbon atoms of the film; actual interatomic distances average about 2.7 Å. The random distribution of atoms distorts the transmitted electron wavefront in the same way as does the regular distribution of idealized atoms considered in Fig. 19-5. The distribution of atoms in the film may be regarded as the *summation* of a large number of regular periodic spacings of atoms. Correspondingly, the distorted wavefront represents the superposition of many sinusoidal wave surfaces of different periodicities (and different amplitudes). This point of view is illustrated by Fig. 19-10.

In the space beyond the specimen the components of the transmitted wavefront propagate in the manner described by Huygens' principle (cf. p. 50). *Reinforcement* occurs at positions where the secondary wavelets which originate from successive maxima of the wavefront differ in path by exact multiples of one electron wavelength. (Note that the maxima referred to here are positions on electron wavefronts, not crests of the instantaneous amplitude of the electron waves!) As shown by Fig. 19-11, reinforcement occurs along a line at a series of points which corresponds to the positions of crests on the transmitted wavefront. In Fig. 19-11a, for example, the path length BA' is exactly one wavelength longer than AA', while CA' is *two* wavelengths longer.

Components of the transmitted wavefront which correspond to different periodicities in the specimen form series of interference maxima *at different distances from the specimen*. This is illustrated in Fig. 19-11b for a larger spacing (lower "spatial frequency") than that considered in part Fig. 19-11a; the line of reinforcement is farther from the specimen. Any plane in the vicinity of the specimen may be observed by appropriate defocusing of the lens. Therefore *any* spacing can be observed at the corresponding level of defocus, up to the limit of about 20 Å, above which well defined periodicities do not exist. Very small spacings exist but are unobservable because of limitations in resolving power.

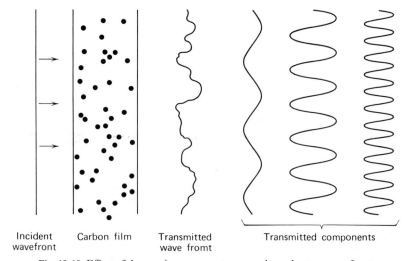

Incident   Carbon film   Transmitted   Transmitted components
wavefront   wave fromt

Fig. 19-10 Effect of the specimen support upon a plane electron wavefront.

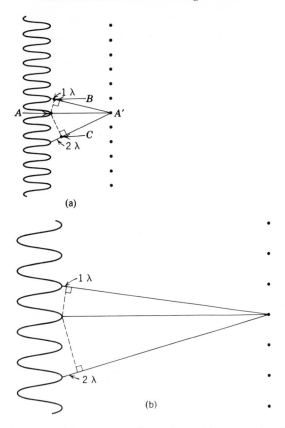

(a)

(b)

Fig. 19-11 Interference maxima corresponding to "spatial frequencies" of the specimen.

Both experiment and theory thus show that the spacing of phase contrast detail in the image varies with the level of focus. The same is not true of amplitude contrast. For this reason the observation of amplitude contrast, only, should provide the most reliable indication of the structure of the specimen. The elimination of substrate phase contrast at very close-to-focus settings also improves the real contrast of the image of the specimen itself. Note that this is an exception to the generality that improvements of contrast and resolving power tend to be incompatible!

In order to record micrographs in which phase contrast imaging of substrate structure is avoided, optimum recording conditions (i.e., very high magnifications and intensities) are necessary. The microscope must also be compensated to within $0.1\ \mu$. Phase and amplitude contrast may be distinguished by the fact that the spacing of amplitude-contrasted structures is independent of the focal setting.

From the point of view of phase effects, thick substrate films are the most desirable, since optical path differences between adjacent areas of the film then tend to average out. Carbon films which are 150 or 200 Å thick should represent an optimum compromise between phase and amplitude effects; attempts to reduce substrate noise by producing films thinner than this are probably misguided.

EFFECT OF THE HIGH VOLTAGE LEVEL ON CONTRAST AND RESOLUTION

Relatively slow moving electrons, that is, electrons accelerated through relatively low voltages, are more susceptible to scattering by the specimen than are higher voltage electrons. Quantitatively, (19-h) indicates that scattering beyond any given angle, Θ, is proportional to the inverse square of the accelerating potential, V. That is

$$(19\text{-cc}) \qquad\qquad \frac{dN\,(>\theta)}{N} = \frac{K'}{V^2},$$

where K' is a constant characteristic of the specimen. Thus, optimum contrast tends to be obtained at low voltages. The improvement of contrast is limited, however, by the accumulation of scattered background intensity.

To a more limited degree, resolving power, also, is determined by the accelerating potential. As given in chapter 17, the minimum resolvable distance is

$$(17\text{-h}) \qquad\qquad d_{min} = \lambda^{3/4} C_s^{1/4},$$

while, from chapter 16

$$(16\text{-t}) \qquad\qquad \lambda = \frac{12.27}{\sqrt{V}}.$$

Combination of these expressions gives

$$(19\text{-dd}) \qquad\qquad \boxed{d_{min} = \text{constant} \times V^{-3/8}}\,.$$

Accelerating voltages available in conventional electron microscopes are usually in the range from 40 to 100 kV. In principle the use of the lower range of voltages is desirable in the study of biological specimens, in order to take advantage of the noticeably superior contrast so obtained.

For the study of materials such as shadow cast specimens, in which the average atomic number is substantially increased, the greater penetrating power and somewhat superior resolution of operating at higher voltages tends to be more desirable. In practice, however, it is usually most important to choose that accelerating potential for which a given instrument is optimally tuned.

Recently, electron microscopes have been developed which operate at voltages of the order of 1000 kV, and early results in the study of biological specimens with these instruments have proved encouraging. The particular advantages of high voltage electron microscopy are: (1) Improved ultimate resolving power associated with the shorter electron wavelength. (Resolutions of at least 1 Å have been demonstrated). (2) Increased penetrating power of the electron beam, which makes it possible to study specimens as much as 5μ thick. Apart from the convenience of being able to work with thicker specimens, this opens the way for extensive applications of stereomicroscopy in the determination of three dimensional structure. (3) Gross specimen damage is avoided. Although equation (19-cc) predicts minimal amplitude contrast in the million volt range, substantial phase contrast is in fact obtained in images of biological specimens.

It should be noted that instrumentation for million volt electron microscopes differs in practical detail from that of standard instruments. Because of the greatly increased level of x-irradiation produced when the highly energetic electron beams impinge upon matter, voluminous shielding is required. Also, regulation of the voltage level to a degree consistent with the theoretical resolving power is more difficult than at lower levels.

MAXIMIZING CONTRAST IN ELECTRON IMAGES

The methods of obtaining maximum contrast in electron images are summarized here. The inherent scattering contrast of the specimen may be improved by preferential increase of the atomic number. Improvement so achieved is limited but may be of considerable importance. It is essential to observe specimens which are sufficiently thin that multiple scattering is negligible. Otherwise, improvement of contrast is related to instrumental factors. These are:

1. Scattering contrast is enhanced by the use of a *limiting objective aperture*, which frees the image from the background fog contributed by electrons scattered into the lens margins (cf. p. 428). Improvement of contrast by this method is especially striking for specimens of *low* scattering power.

2. An intense, focused electron beam of minimal size is used to illuminate the specimen. The manner in which a reduced size of crossover spot aids image contrast has been explained on p. 390. This condition requires the use of a *pointed filament* in the electron gun (cf. p. 380) and of a *double condenser lens*.

3. A relatively *low accelerating voltage* is used, 40–50 kV is suitable for many types of biological specimen (cf. p. 444).

4. The image may be *defocused* by the maximum amount which does not produce a deterioration of resolution (cf. p. 433).

5. Image contrast may be enhanced by the judicious use of photographic processing (cf. p. 478). In the special case of images of *periodic* objects the method of *optical filtering* (described below) may be applied.

Certain modifications of the electron microscope have been studied in attempts to improve image contrast. They are the *dark field* and the *electron phase contrast* microscopes. Interesting results have been obtained in experimental instruments, but neither method has produced improvement of contrast sufficiently great for resolution to exceed that of conventional electron microscopes.

THE ELECTRON PHASE CONTRAST MICROSCOPE

Electron phase contrast microscopes make use of the same principles as the optical phase contrast microscope (cf. Chapter 13) to produce *in-focus* phase contrast. The phase contrast so produced must be distinguished from the out-of-focus and aberration effects discussed above.

The electrons which leave the specimen form focused diffraction patterns at the rear focal plane of the objective and again at the rear focal plane of the intermediate lens. In these patterns, unscattered electrons are focused on the axis, while scattered electrons intersect other positions on the focal plane. Deviated and undeviated rays are thus physically separated, just as in the light phase contrast microscope. The phase plate of the electron system is a carbon film of a thickness such as to increase the optical path of the electron beam by one quarter wavelength. A small central aperture transmits the unscattered beam without imposing a phase delay. The electron phase microscope is thus equivalent to the system shown in Fig. 13-10 (which shows a phase plate located at the focal plane) rather than to the system shown in Fig. 13-11 which is actually used in practical optical phase contrast microscopes.

The principal difficulty in devising practical electron phase contrast microscopes is that contamination accumulates on the phase plate. Rapid and unpredictable increases in optical thickness and thus in the resulting phase shift may easily result. Contamination may be minimized by heating the phase plate (to about 250°C) and by locating it in the rear focal plane of the intermediate rather than of the objective. Also, contamination by the intense unscattered beam is avoided, since this component is transmitted by the central aperture rather than by the carbon film.

Definite improvements in contrast have been demonstrated with the use of electron phase optics, but resolving powers better than about 20 Å have not yet been demonstrated in these instruments.

Lenses of wider angular aperture may result from current work in electron optics. However, since the scattering of electrons through angles larger than the aperture angles of present lenses is negligible (for most specimens), the use of superior lenses would tend to eliminate amplitude contrast. Thus the development of high quality phase systems, which utilize scattering within the lens aperture, seems essential.

DARK FIELD ELECTRON MICROSCOPY

Dark field electron microscopy is analogous to dark field light microscopy, as described on p. 282 and in Fig. 12-16. Dark field effects can be obtained quite easily (or even accidentally) in the electron microscope by misaligning the objective aperture (or any other optical component) so that it just blocks the unscattered electron beam. Scattered electrons then enter the lens field and are focused to form a bright image against a dark background. The images so formed are generally of poor quality, however, since electrons which contribute to the image have passed through the outer, highly aberrant zones of the objective lens. High resolution dark field images can be obtained with the use of tilted illumination. In this case the direct beam passes beyond the objective aperture, while scattered electrons are focused by paraxial zones of the objective (e.g., see van Dorsten).

A serious limitation of dark field electron microscopy (to which dark field light microscopy is not subject) is the intensity contributed to the image by the substantial scattering from points of the specimen support. This has the effect of illuminating the background and thus of reducing contrast.

OPTICAL FILTERING

When *periodic* specimens are observed in the electron microscope, *random* variations in intensity, introduced by the substrate, may be removed from the image by the technique of *optical filtering*. This method is illustrated in Fig. 19-12. As shown in Fig. 19-12a, the transparency obtained in the electron microscope (plate or film) is illuminated by a collimated light beam. (The collimated beam must originate from a coherent

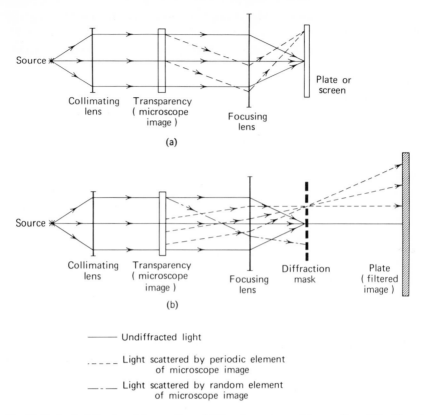

Fig. 19-12 Optical filtering: (a) recording of diffraction pattern; (b) recording of the filtered image.

source.) Transmitted intensity is then focused by a lens. The diffraction pattern of the transparency is formed in the rear focal plane of the second (focusing) lens and is recorded. Portions (in general, sets of spots) of this diffraction pattern may be identified as arising from the regular structure of the specimen, while the remainder of the pattern consists of light diffracted by random features of the image. If the origins of the diffracted intensities can be deduced, those portions of the diffraction pattern which originate from the regular structure can be isolated by means of a suitable mask.

As shown in Fig. 19-12b, the transparency is replaced in its original position. The mask, which is placed at the second focal plane of the focusing lens, transmits *only* intensities which correspond to the structure of the periodic image. The filtered image is then recorded on a photographic plate placed at the image plane of the second lens.

Considerable skill is required for the identification of those features of the diffraction pattern which correspond to the structure of the periodic specimen. If parts of the diffraction pattern are excluded by the mask, or if additional intensities are included erroneously, the appearance of the filtered image can be completely misleading. Also, the mask itself must be cut with considerable care. When all steps in the procedure are correctly carried out, however, the clarification of detail in the final image may be very striking.

The instrument in which the diffraction pattern and the filtered image are recorded is known as an *optical diffractometer*. The light source used in the instrument is, preferably, a laser (cf. Chapter 22), since this type of source is sufficiently intense to form a directly observable diffraction pattern. However, aperture beams from a mercury arc lamp may also be used; in this case prolonged photographic exposures are required.

REFERENCES

General References

C. E. Hall, *Introduction to Electron Microscopy*, 2nd ed., McGraw-Hill, New York, 1966, Chapter 9 (Scattering Phenomena) and Chapter 10 (Image Characteristics).

O. Scherzer, in G. F. Bahr and E. H. Zeitler, *Quantitative Electron Microscopy*, Armed Forces Institute of Pathology, Washington, D.C. *Lab. Invest.* **14**, (1965), 59/797–62/800. This article, "Image Formation as a Problem of Wave Theory," provides a short general discussion of contrast mechanisms in the electron microscope; Fig. 4 shows an extreme example of spurious image detail produced by diffraction effects.

R. D. Heidenreich, *Fundamentals of Transmission Electron Microscopy*, Interscience, New York, 1964. Despite its title, this book is devoted almost exclusively to considerations of the image-formation process. Considerable familiarity with diffraction theory is assumed.

E. Zeitler, in Bahr and Zeitler, *op. cit.* pp. 36/774–48/786. This article, "Theory of Elastic Scattering of Electrons," discusses the methods by which electron scattering is computed, and includes considerable numerical data.

A. J. Glick, in Bahr and Zeitler, *op. cit.* pp. 49/787–57/795. An article on "Theory of Inelastic Scattering of Electrons."

A. C. van Dorsten, in Bahr and Zeitler, *op. cit.*, pp. 81/819–86/824. An article on "The Role of Acceleration Voltage in Image Formation."

A. Klug and J. E. Berger, *J. Mol. Biol.* **10**, 565 (1964). This article, "An Optical Method for the Analysis of Periodicities in Electron Micrographs and Some Observations on the Mechanism of Negative Staining," describes the optical filtering technique and certain of its applications.

D. J. De Rosier and A. Klug, *Nature* **217**, 130 (1968). This article, "Reconstruction of Three Dimensional Structures from Electron Micrographs," describes the mathematical analog of the optical filtering technique.

H. M. Johnson and D. F. Parsons, in C. J. Arsenault, Ed., *Proceedings of the Electron Microscopy Society of America*; *25th Anniversary Meeting*, Claytor's Bookstore, Baton Rouge, La., 1967, p. 236. This abstract, "Four-Stage Electron Microscope for Phase and

Strioscopic Microscopy," briefly describes some recent experiments with an electron phase contrast system.

B. M. Siegel, *Modern Developments in Electron Microscopy*, Academic Press, New York, 1964, pp. 27–35. A discussion of wave optical effects in electron image formation and contrast.

Specific References

[1] R. B. Setlow and E. C. Pollard, *Molecular Biophysics*, Addison-Wesley, Reading Mass., 1962, pp. 323–334.

[2] V. K. Zworykin, G. A. Morton, E. G. Ramberg, J. Hillier, and A. W. Vance, *Electron Optics and the Electron Microscope*, Wiley, New York, 1945, pp. 674–691. A section on "Scattering and Absorption Processes in the Object."

[3] A. W. Crewe, *Science* **154**, 593 (1966).

[4] D. C. Pease, *Histological Techniques for Electron Microscopy*, 2nd ed., Academic Press, New York, 1964, Chapter 7 (Staining) and pp. 331–334.

[5] D. H. Kaye, Ed., *Techniques for Electron Microscopy*, 2nd ed., Blackwood and F. A. Davis, Philadelphia, 1965, Chapter 5 (Replica and Shadowing Techniques, by D. E. Bradley), Chapter 9 (Section Staining, Cytology, Autoradiography, and Immuno-chemistry for Biological Specimens, by A. M. Glauert), and Chapter 11 (Negative Staining Methods, by R. W. Horne).

CHAPTER 20

High Resolution Electron Microscopy

Operation of the electron microscope at ultimate resolution is a topic of lively interest, in terms both of present accomplishments and of future prospects. This chapter is a review in which material discussed in Chapters 16 to 19 is summarized from the point of view of high resolution. The relationship of resolving power, contrast, and resolution is explained. The factors which limit resolution are discussed, and possibilities for improving ultimate resolution are considered. The practical operating conditions required for optimum resolution are also listed.

High resolution electron microscopy may be defined as the study of spacings smaller than 20 Å. It should be understood that very little current work in electron microscopy is, actually, high resolution microscopy. This statement is by no means a criticism of current ultrastructural studies. A vast amount of information remains to be collected concerning biological structure at a level which is beyond the resolution limit of the optical microscope, but well within the resolution limit of the electron microscope. Specimens of interest frequently lack the inherent contrast essential for resolution of structural detail smaller than 20 Å. In studying such specimens there is no justification for the frequent, time consuming adjustments of operating conditions which are essential for work at ultimate resolution. Nevertheless, extension of resolution below 20 Å is important in studies of the structure of subcellular particles or of individual macromolecules.

It is not always understood that, although 2 Å crystal spacings have been resolved in the electron microscope, resolution of individual

451

atoms cannot be expected, at least at present. The resolution of spacings is not the same thing as point-to-point resolution; resolving power is defined in terms of the latter (cf. Chapter 10). It is always easier to discern closely spaced pairs of linear structures than pairs of adjacent points.

Ultimately, both instrumental resolving power *and* image contrast might be improved to the extent that atomic resolution would become possible, thereby making the electron microscope a tool for direct chemical analysis. However, such developments are by no means foreseeable now.

RESOLVING POWER, CONTRAST, AND RESOLUTION

The nature of the relationships between resolving power, contrast, and resolution may be summarized. *Resolving power* is the property of an *instrument* which specifies its ability to resolve structure in a theoretical specimen of ideal properties. Resolving power is thus determined, *without reference to the nature of the specimen*, by the electron wavelength, the aperture angle of illumination, and other instrumental features.

Contrast is the property of images which specifies the ability to distinguish structures because of differences in intensity. The contrast of any given image is determined both by the properties of the instrument (instrumental contrast) (i.e., electron wavelength, aperture angle, focal level, etc.) and by the properties of the specimen (inherent contrast) (i.e., thickness, distribution of substructure, electron scattering power, etc.)

Resolution also is a property of *images*; it specifies the minimum separation of points which may be discerned in the image formed of a given specimen by a given instrument. High resolution requires both high instrumental resolving power and high image contrast. High resolution cannot be obtained unless *both* of these factors are adequate. Yet, since instrumental factors which favor image contrast tend to cause resolving power to deteriorate, and vice versa, resolution may sometimes be improved in practice, by improving *either* resolving power *or* contrast at the expense of the other. If it is understood that this is so, the significance of some apparently contradictory statements concerning resolution should become clear. High resolution electron microscopy requires a delicate balance of instrumental factors which favor contrast with those that favor resolving power. On the other hand, improvement of the inherent contrast of the specimen is without effect on resolving power, and thus is entirely favorable to high resolution.

LIMITATION OF ULTIMATE RESOLVING POWER BY INHERENT ELECTRON LENS PARAMETERS.

Electron lenses, like glass lenses, provide "perfect" lens action only "very close" to the axis. Thus peripheral lens zones suffer from aberrations, of which spherical aberration (cf. p. 364) and chromatic aberration (cf. p. 367) are particularly important. These defects can be restricted but, at least in conventional designs of electron lens, they cannot be substantially corrected.

Spherical aberration of electron images is minimized by using only a restricted paraxial region of the lens. Yet, as in optical lenses, diffraction effects limit resolving power when aperture angles are small. Ultimate resolving power is thus obtained at angular apertures of the order of 10^{-3} rad, for which losses of resolving power due to spherical aberration and to diffraction are just balanced. In this way an ultimate point-to-point resolving power just below 3 Å can be predicted (cf. p. 366).

Decrease of spherical aberration could result from the improvement of conventional lens designs, that is, from reduction of the spherical aberration constant. It has been estimated that the resolution limit might be extended in this way to about 1 Å. An alternative approach is the design of entirely different types of lens system, including quadrupole and octupole (as opposed to axially symmetric) lens elements, in which spherical aberration is in fact corrected. These systems seem promising, but much further work remains to be done before they can even match the resolving powers of conventional lenses of high quality.

That the resolving power of the electron microscope is limited by the spherical aberration of electron lenses is rather well known. It is much less commonly appreciated that this is by no means the only source of limitation of resolving power. Improvement of spherical aberration alone would only cause other types of image defect to limit resolution at about the same level. Also, development of high quality electron lenses of relatively wide angular aperture would eliminate the contrast now obtained in electron images *because* of aberrant focusing by peripheral lens zones. Since lack of contrast often limits resolution in practice, the higher resolving power of such lenses, if unaccompanied by other developments, might fail to effect any general improvement of resolution. It would appear that the solution of this dilemma should be the development of effective phase contrast electron microscopes, since phase imaging effects contrast by exploiting scattering within the effective lens aperture.

Chromatic aberration in electron images is minimized by the use of uniformly accelerated (i.e., monochromatic) electrons. Residual variations in electron velocity nevertheless tend to limit resolving power.

Chromatic aberration is reduced by minimizing the chromatic aberration constant of the lens. It is generally considered that improvement of the chromatic aberration constant could keep pace with improvements in the spherical aberration constant as electron lens design evolves.

Three sources of fluctuations in electron velocity (wavelength) must be considered. One of these, the level of *accelerating voltage regulation*, tends to limit resolving power in present systems. $\Delta V/V$ is of the order of 10^{-5}, a condition which reduces the effects of chromatic aberration to the level of the spherical aberration limit. High voltage regulation (and also that of lens currents) can presumably be improved to almost any desired level; the problem is a technological one. However, the additional cost of improving regulation beyond the 10^{-5} level would be considerable and would be justified only if resolution were not otherwise limited.

Ultimately, improvements in high voltage regulation could cause the other sources of variation in electron velocity to become limiting. They are: variations in the initial thermal velocities of electrons emitted by the hot filament of the electron gun (cf. p. 380); and energy losses resulting from inelastic collisions with the valence electrons of the specimen (cf. p. 425). Improved control of initial velocities could probably be achieved only by radical innovations in the design of electron guns. Energy losses at the specimen can be minimized by observing only very thin specimens. However, the degree of interaction with the specimen (i.e., the extent of both elastic and inelastic collisions) is also minimal when the specimen is very thin; hence image contrast is correspondingly reduced. Fortunately, variation of electron velocities resulting from either of these sources remains, at present, considerably less important than that produced by fluctuations in the accelerating voltage.

A related source of limitation of the resolving power is the *fluctuation of lens current levels*. $\Delta I/I$ is held at about one part in 10^5 in present systems, so that resolving power is limited in this way to about the level set either by spherical aberration or by regulation of accelerating voltages. Improvement of lens current regulation, as of accelerating voltage regulation, should be feasible if this source of image defects were to become limiting. The use of superconducting lenses (cf. p. 370) is a possible method for avoiding fluctuations in lens currents.

DISTURBANCES WHICH LIMIT INSTRUMENTAL RESOLVING POWER

The limits of performance of the light microscope can be estimated accurately on theoretical grounds. In electron microscopy, however, the more stringent technological requirements easily become limiting. These factors are discussed below, roughly in the order of their present importance with respect to ultimate resolving power. The first three items discussed (image movements, lens asymmetry, and focusing) are ones which can limit the ultimage resolving power of an instrument. The others need not impair resolving power if appropriate precautions are taken.

It is important to realize that problems which seem trivial in terms of macroscopic dimensions may become profound when resolutions of a few Angstrom units are sought. The limitation to their present levels of the disturbances discussed here is an impressive achievement.

(1) Movement of the Image

Exposure times of the order of one second are required for the photographic recording of electron images. Thus movements of the specimen through only a few Angstrom units per second limits resolution. (It is instructive to calculate that movement through 3 Å/sec is equivalent to less than one cm per *year*!)

Movements may result from instabilities of the specimen holder and stage assembly, or of the specimen itself. Problems with the specimen are minimized by the use of intact support films of high stability. Evaporated carbon, the standard substrate material, has favorable properties from this point of view. Copper support grids are good conductors of heat, and they help to minimize thermal movements of the specimen.

Vibrations of the microscope column as a whole must also be avoided; appropriate location and mounting of the instrument are essential. The fore pump of the vacuum system may produce vibrations which interfere with the recording of high resolution images. Some types of microscope can be operated temporarily with this pump shut off.

Apart from improvements in mechanical design, limitation of resolution by residual movements of the specimen or image is reduced with the use of illumination of the highest possible intensity. It may then be feasible to reduce exposure times to as little as 0.5 sec, even at maximum instrumental magnifications. Vibrations caused by operation of the photographic shutter mechanism must be avoided if the reduction of the exposure time is to be profitable. The development of faster photographic emulsions which would also be capable of recording fine detail in electron images seems unlikely.

(2) Lens Asymmetry (cf. pp. 371 and 409)

Realization of the full resolving power of an electron microscope is dependent upon compensation of lens asymmetry to a level consistent with that resolving power. In present instruments this is generally possible, but compensation to the level required for ultimate resolution is so difficult to achieve or to maintain that asymmetry tends to be the defect which limits resolution in practice.

The greatest magnitude of asymmetry compatible with resolution of any spacing, r, is given in Chapter 18 (18-f) as

$$(20\text{-a}) \qquad\qquad \Delta f = \frac{(\Delta r)^2}{\lambda}.$$

For operation at $100\,\mathrm{kV}$, $\lambda = 0.04\mathrm{\AA}$; hence asymmetry must be corrected to about $0.02\,\mu$ for resolutions of 3 Å. Systematic compensation of asymmetry is difficult even at the level of $0.10\,\mu$, which corresponds to a resolution of about 6 Å.

Improvements in resolving power will thus call for improvements in the *ease* of effecting compensation. Apart from reliable fine level control of the compensation mechanism, high intensity of illumination at high magnification is required simply for the operator to be able to see very narrow asymmetric fringes. The use of image intensifier systems may become essential with microscopes of superior resolving power. Development of essentially contamination-free systems would also make it possible to maintain a high degree of symmetry over a workable length of time.

(3) Focusing of the Image (cf. pp. 402 and 408)

The effect of defocusing the image, like that of lens asymmetry, can be described by (20-a). Thus, defocusing of the image by more than about $0.02\,\mu$ precludes resolution at the 3 Å level, while resolution at 6 Å requires focusing to within about $0.10\,\mu$. Although these requirements pose no *theoretical* limit, they are so critical that difficulty in selecting focus is often a practical limitation on resolution. Note, for example, that the minimum variation of focal length which can be imposed in one model of high resolution electron microscope is of the order of $0.03\,\mu$.

Fine focusing, like compensation, is aided by operation at high magnification and intensity and, possibly, by the use of an image intensifier.

(4) Electrical and Magnetic Disturbances

Electron microscope columns are constructed from materials which shield the electron beam from external fields. While certain levels of disturbance may thus be tolerated in the vicinity of the microscope, stronger stray fields can cause marked deterioration of image quality. Limitation of resolution is avoided by suitable location of the instrument, away from power lines, magnets, etc.

Distortion of the objective lens field by the metal grid supporting the specimen must be avoided. Nickel grids should not be used at high resolution. Even with nonferromagnetic grids (i.e., copper), images of regions of the specimen which are very close to the wire supports should not be recorded.

(5) Alignment (cf. p. 399)

Correct alignment of the electron microscope column must be rigorously maintained during operation at high resolution. Alignment conditions which are compatible with limiting resolution are not difficult to obtain in microscopes of high quality.

EFFECTS OF IMAGE CONTRAST UPON RESOLUTION

Whatever the resolving power of an electron microscope, resolution is not obtained unless the interactions of the electron beam with the specimen are sufficiently extensive to produce a detectable level of contrast. Limitations imposed by lack of contrast are particularly serious in biological electron microscopy since, while electron contrast originates from scattering processes, carbon and other elements of biological importance are relatively weak scatterers of electrons. Progress in electron microscopy depends as much upon the development of means for improving contrast (either instrumentally or by means of specimen preparation techniques) as upon the improvement of resolving power.

The methods by which electron image contrast can be maximized have been summarized on p. 445. As is generally true in light microscopy also, most of these factors tend to improve contrast *at the expense of resolving power*. Thus, for example, shadow casting of particles creates excellent contrast but limits resolution to about 20 Å, which is the grain size of the metal deposits. The resolution of stained structures is also limited, ultimately, by the size of the stain molecules (or atoms) themselves. Defocusing of the image enhances contrast but, beyond a restricted range, causes resolving power to deteriorate. Reduction of the objective lens aperture to less than α_{opt} would tend to improve contrast but would also cause diffraction effects to limit resolving power. (In practice, the objective aperture is always larger than α_{opt}, since uncontrollable lens asymmetry produces an extreme deterioration of resolving power when very small limiting apertures are used.) Low voltage operation favors contrast, but has (slight) unfavorable effects upon resolving power. The phase contrast produced by spherical aberration is also reduced in systems of high resolving power.

While resolving power and contrast thus tend to be incompatible, some exceptions to this general picture exist. As explained on p. 443, the phase contrast "grain" size of support films varies according to the level of focus but disappears at exact focus. Thus at moderately high resolution an image of the support is unavoidably superimposed upon that of the specimen itself. Granularity at the 15–20 Å level confuses images of specimen substructure and thus reduces effective contrast. At exact focus, however, where *resolving power* is optimal, *effective* contrast (i.e., as opposed to variations in intensity which are *not* related to the structure of the specimen) is also improved by the absence of background grain.

A second condition which favors *both* resolving power and contrast is the use of focused electron beams of high intensity and minimum spot size (cf. p. 390).

SUMMARY OF OPERATING CONDITIONS WHICH PROVIDE OPTIMAL RESOLVING POWER

Instrumental features which characterize electron microscopes of high resolving power are: lens design and construction such as to result in minimum values of the spherical and chromatic aberration constants (C_s and C_c), and in low values of the residual asymmetry of the uncompensated lens; regulation of accelerating voltage supply to values of the order of one part in 10^5; regulation of lens current supplies to values of the order of one part in 10^5; intense electron source; double condenser illumination; fine level compensation mechanism; fine level focusing mechanism; mechanical design and construction such as to minimize drifting or vibration of microscope components, effective shielding of the microscope column; availability of anticontamination device; photographic shutter which does not disturb the image during its operation.

Requirements for operation at high resolution are for the most part similar to those for moderate resolution, but correspondingly more demanding. They are:

1. Maintenance of column *alignment*.
2. *Compensation* of the asymmetry of the objective lens to a level consistent with the resolution sought.
3. *Focusing* of the image at the level consistent with optimum resolution.
4. Detection of any failure in the *regulation* of accelerating voltage or lens current supplies.
5. Although contrast of detail at limiting resolution often originates primarily from phase effects, use of a *limiting objective aperture* is required in order to enhance amplitude contrast by removal of background scattering (cf. p. 428). The aperture must be uncontaminated, and properly centered about the microscope axis, so that it introduces no asymmetry

into the objective lens field. Attempts to compensate for any asymmetry of the aperture, or to use apertures which have been contaminated and subsequently cleaned, are generally unsatisfactory at limiting resolutions.

6. Use of a *pointed filament* in the electron gun. (cf. p. 380). This type of filament provides a smaller, more intense spot of illumination than does the conventional type of filament. The effect of small spot size is to improve image contrast; the effect of high intensity is both to improve visibility during instrumental adjustments (focusing and compensation) and to minimize exposure times.

7. Use of an *anticontamination* device (cf. p. 388). This is essential for optimal resolution, although it may be inconvenient. Much longer times for outgassing of photographic emulsions are required when the "cold finger" is used.

8. Loose or torn *specimen support films* tend to drift when irradiated by the electron beam, and must be avoided.

9. *Copper* grids should be used for the support of specimens. Areas of the specimen which are recorded should be far from the wires of the grid in order to prevent spurious deflections of electron trajectories.

10. *Photographic processing* of electron micrographs should be carried out with particular care. (Photographic methods are discussed in Chapter 21.)

REFERENCE

E. M. Slayter, in F. J. Leach, Ed. *Physical Theory and Techniques of Protein Chemistry*, Academic Press, New York (1969). This chapter on "The Study of Globular Protein Molecules by Electron Microscopy" emphasizes practical techniques in high resolution microscopy. Recent work on some representative molecular systems is described.

CHAPTER **21**

Photographic Methods in Optical Instrumentation

Photographic measurements are of central importance in many areas of biological instrumentation. Two categories of application may be distinguished: the *recording* of images, and the *quantitation* of light intensities.

Both light and electron images may be recorded photographically. The recording of electron images is of particular importance, since the resolution inherent in the electron beam is often not revealed in the image formed on the fluorescent screen of the microscope, and since the specimen may be destroyed by prolonged irradiation. Thus the image is normally studied in the form of a photographic record, which is a more suitable item for permanent storage than is the original specimen.

Quantitation of light intensities is often carried out by densitometry of photographic records of microscope images, x-ray or electron diffraction patterns, emission spectrographs, ultracentrifuge records, and other types of light pattern. While precise and accurate results can be obtained in this way, the exposure, development, and densitometry of each photograph is time consuming, so that, when many intensities must be measured, photoelectric systems, when available, are more convenient. The latter may scan the intensity of the beam which emerges from an instrument, providing, after appropriate electronic processing, an immediate record of intensity (or some related parameter) as a function of position. The operations of development and densitometry are thus replaced by electrical interactions which provide numerical results within a matter of seconds. For example, photoelectric methods have long since replaced photography in routine spectrophotometry, while sophisticated electronic systems have been developed for the recording and processing of diffraction data. Nevertheless, the expense and complexity of these systems may be formidable; thus photographic recording remains the method of choice for many types of intensity measurement.

The aim of scientific photography is the complete and faithful reproduction of object detail. Image intensities should invariably be related to the structure of the object. This aim is, of course, quite different from that of pictorial photography, which seeks to provide a pleasing and perhaps impressionistic image. Techniques, including those of processing the photographic record, differ correspondingly.

This chapter provides an account of considerations relevant to practical scientific photography. More theoretical aspects, such as the nature of the latent image (p. 467) are discussed briefly.

GENERAL NATURE OF THE PHOTOGRAPHIC PROCESS

The light-sensitive photographic emulsion consists of small *grains* (crystallites) of silver halide (usually silver bromide) suspended in a gelatin matrix and supported by a glass plate, paper, or plastic film. Upon exposure to light, some of the grains are so altered as to form a *latent image*; chemical *processing* is required to render the latent image visible and stable.

In the first stage of photographic processing, developer reduces the exposed grains, forming a deposit of metallic silver:

$$(21\text{-}a) \qquad\qquad Ag^+ + e \rightarrow Ag^0.$$

Grains which have formed a latent image are reduced to metallic silver *as a whole*, thus producing an observable blackening of the emulsion; unexposed grains ideally are unaffected by the developer. A *stop bath* may or may not be used to terminate development abruptly. A *fixative* is then used to remove the unexposed, undeveloped silver halide, and, finally, *washing* removes the excess of fixative. The image so obtained may then be observed directly or may be used to form a magnified print. If quantitative measurements are required, the photographic density of the image is measured by means of a densitometer.

Stages in exposure and processing are illustrated in Fig. 21-1. Figure 21-1a shows an object which consists of an opaque central band surrounded by areas which transmit light to different extents. Figure 21-1b represents the initial distribution of grains in a photographic emulsion; however, grain size is greatly exaggerated for the purpose of illustration.

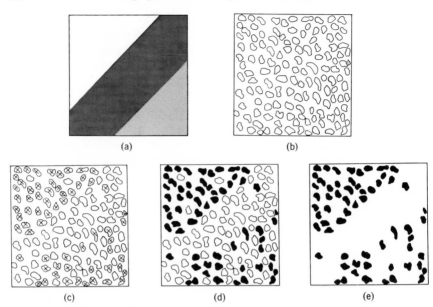

(a)   (b)

(c)   (d)   (e)

Fig. 21-1 Stages in photographic processing: (a) object; (b) unexposed emulsion; (c) exposed but unprocessed emulsion; (d) developed but unfixed emulsion; (e) final image after fixation.

(Practical grain diameters range from about 0.1 to several microns.) Figure 21-1c is a diagram of the emulsion after suitable exposure to light transmitted by the object; grains which have interacted with the light to form the (invisible) latent image are marked by an x. (No indication of the physical nature of the latent image is implied in Fig. 21-1c; for a discussion of this topic, see p. 467.) Note that most, but not all, grains in the areas corresponding to the partially transmitting regions of the specimen have formed a latent image, the proportion being higher in the area which corresponds to the relatively more transparent upper corner of the specimen. Figure 21-1d represents the emulsion after an optimum time of contact with the developer. As shown, virtually all the grains which had interacted with the illumination to form a latent image have been reduced to metallic silver, while virtually none of the unreacted grains have been reduced. Finally, the image which remains after the unexposed halide has been removed by fixative is shown in Fig. 21-1e.

PHOTOGRAPHIC DENSITY AND EXPOSURE

Photographic density D is a quantitative measure of the blackening of a photographic emulsion. It is defined by

(21-b)
$$D \equiv \log \frac{I_0}{I} = -\log T = \log O,$$

where I_0 and I are the incident and transmitted intensities, respectively, T $(\equiv I/I_0)$ is the *transparency* of the emulsion, and O $(\equiv I_0/I)$ is the *opacity*.

Photographic density is analogous to optical density, the measure of light absorption by solutions (cf. pp. 531–532). Photographic density, like optical density, has the property that the combined value for two superposed emulsions (solutions) is the *sum* of values for each separately. Useful exposures produce photographic densities which range up to or slightly beyond 2.0 (1% transmission), while a photographic density of unity corresponds to a moderately light exposure.

The properties of photographic emulsions are commonly specified by means of the *characteristic curve* ("H and D" or Hurter and Driffield curve) which is a plot of photographic density as a function of the logarithm of exposure. Features of characteristic curves are discussed in the following section.

Exposure is determined by the *reciprocity law*,

(21-c)
$$E = It,$$

where I is the intensity of illumination and t its duration. Confusion may be avoided by distinguishing between the loose use of the term "exposure" to refer to the exposure time and its correct use to specify the product (It).

Experimentally, (21-c) is found to apply over a wide range of intensities and exposure times. Thus exposure of a given type of emulsion to intensity I for one second (for a fixed set of processing conditions) generally results in a photographic density which is identical with that produced by exposure to an intensity of $\frac{1}{2}I$ for two seconds. However, when exposures are either extremely long or extremely short, failures of reciprocity may be observed; in both cases the photographic densities produced are lower than the expected values.

PROPERTIES OF PHOTOGRAPHIC EMULSIONS

The photographic emulsion consists of silver bromide (or chloride) grains embedded in a gelatin matrix and supported on a paper, plate, or film. (Nuclear emulsions may be supplied in the form of a paste which is applied directly to the specimen.) As well as providing a mechanical binder for the light-sensitive grains, the matrix prevents aggregation of the crystallites and tends to protect unexposed grains from the action of developer. The gelatin also supplies trace amounts of sulfur-containing compounds which aid in the sensitization of the halide grains to the action of light.

Emulsions are characterized by their *spectral sensitivity, contrast properties,* and *speed and grain size.*

Direct photochemical reduction of silver halide is effected by light of wavelengths which are absorbed. Blue and violet visible light and ultraviolet and shorter wavelength electromagnetic radiations are effective. High velocity electrons and other energized particles also serve to expose emulsions.

Sensitization to the longer visible wavelengths and to the near infrared can be produced by coupling the grains to an appropriate absorbing dye. Emulsions thus sensitized to green light are termed *orthochromatic*; those sensitized to all visible wavelengths are termed *panchromatic.* Where feasible (as for measurement of electron, X-ray, or ultraviolet intensities), the use of unsensitized emulsions is convenient, since an appreciable intensity of yellow or red light may then be used for illumination during processing.

The speed and contrast of an emulsion *for exposure by a given light source and upon development under a fixed set of conditions* are specified by the *characteristic curve.* As explained above, this curve is a plot of photographic density as a function of log (exposure). The general form of the characteristic curve is shown in Fig. 21-2. Several features of these curves may be noted. Some degree of blackening always occurs at zero exposure (point *A*); this is described as the *fog density* of the emulsion. Fog density arises from the development (reduction) of grains that have not interacted with the exposing beam, or which have previously formed a latent image due to exposure by stray radiation or to pressure on the surface of the emulsion.

The photographic density is relatively insensitive to exposure over the

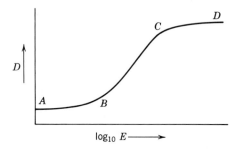

Fig. 21-2 Characteristic ("H and D") curve of a photographic emulsion.

"toe" of the characteristic (region AB). This portion of the curve is one of underexposure. Between points B and C, photographic density increases linearly with log E. If all portions of the object produce an exposure within this range, all density levels in the image are proportional to those of the object. With certain exceptions, this is a desirable condition for the inspection of images, and it is essential when quantitative measurements are to be made. (However, relatively faint features of an image may be removed photographically, by means of underexposure, in order to provide a clearer image of more dense regions.) The "shoulder" of the curve (beyond point C) is a region of more or less uniform blackening, corresponding to overexposure.

The *speed* of an emulsion is a measure of the extent of the toe of the characteristic curve. A number of related but not strictly comparable operational measures of emulsion speed have been used but need not be considered in detail here. A simple measure of speed is the exposure required to produce some specified value of photographic density, such as 1.0.

Speed is related to the physical size of light-sensitive grains. Absorption of only a few quanta per grain is sufficient to produce a latent image and, thus, the reduction of an entire grain upon development. Accordingly, larger grains produce more extensive blackening of the image for absorption of the same number of quanta. The relation of speed and grain size to the observed "graininess" of photographic images is discussed below.

The contrast of an emulsion (i.e., as distinct from image contrast related to the inherent contrast of the object) is expressed by its gamma (γ), which is the slope of the linear portion of the characteristic curve (region BC of Fig. 21-2). The steeper the characteristic, the greater is the difference in photographic density between portions of the image which are exposed within the linear range of the emulsion. Hence, provided that the range of exposure is suitable, greater contrast is provided by emulsions of larger γ values.

In Fig. 21-3 the emulsion characterized by the curve shown in Fig. 21-2 is compared with two other emulsions of identical speed, but of greater and lesser contrast, respectively. Note that higher contrast can be obtained only by a reduction in the range of exposures at which contrast is produced. Hence, when high contrast emulsions are used, exposures must be selected with greater care in order not to miss the relatively limited range in which contrast varies linearly with the exposure.

The use of different developer solutions or development times may change the speed or contrast of an emulsion. The characteristic curves shown in Fig. 21-3 could thus represent, also, the properties of a single type of emulsion when developed under different conditions.

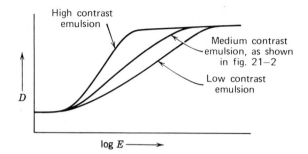

Fig. 21-3 Characteristic curves of emulsions of different contrast properties.

Photographic emulsions are exposed by electrons which have been accelerated through voltages greater than about 0.5 kV. Properties of emulsions with respect to electron exposure can be represented by typical characteristic curves. It is found, however, that for the accelerating voltages used in electron microscopy (50–100 kV) a linear plot can be obtained for photographic density as a function of exposure (i.e., rather than of the logarithm of exposure). This indicates that the exposure of emulsion grains by electrons is a "one-hit" process (i.e., a single electron is sufficient to expose a single grain). Photon exposure is a "multiple-hit" process: two or more quanta are required for exposure of each grain. The difference is accounted for by the fact that accelerated electrons produce *tracks* of ionizations as they pass through an emulsion. On the average, a sufficient number of ionizations is produced within any one grain to ensure formation of a latent image.

EMULSION SPEED AND GRAIN SIZE, AND IMAGE GRAININESS

Graininess may limit resolution in micrographs which have undergone excessive photographic magnification. The origin of this effect is easily misunderstood; image graininess is *not*, as might be supposed, an image of the individual grains of the original photographic emulsion. (Note that grains 1μ in diameter would require a magnification of at least $100 \times$ to become directly visible, yet graininess is observed at magnifications not much greater than $10 \times$.) The observed graininess in fact results from the *random spatial distribution of quantum or electron arrivals at the emulsion*. The apparent grains seen in the magnified image are a photographic record of the randomly clustered exposure of the original emulsion. Nevertheless, graininess of this type is related to physical grain size.

The signal carried by a particulate beam is proportional to N, the total number of particles arriving at a detector. However, the behavior of individual particles, such as photons or electrons, can be accurately predicted only in terms of probabilities (cf. p. 23). Thus any signal is accompanied by randomness in the arrival of photons or electrons. The randomness, which is known as "noise," can be shown to be proportional to $\sqrt{N}$. The *signal-to-noise ratio*, which is the measure of information transfer by a beam, is proportional to $N/\sqrt{N} = \sqrt{N}$ and thus increases with the size of the signal. A larger signal contains relatively less random noise than a smaller one and thus carries more information.

Only a small fraction (in the range from 1% to 10%) of the light quanta which impinge upon a photographic emulsion is effective in producing exposure. In order to obtain useful photographic densities, therefore, a large signal (in terms of numbers of quanta) is required, so that the signal-to-noise ratio is correspondingly high. For electrons, however, the quantum efficiency of exposure (sometimes referred to as the DQE, or detector quantum efficiency) is very nearly unity. Thus the number of electron arrivals required to produce any given level of photographic density is relatively small. The signal-to-noise ratio is correspondingly low for electron exposures; that is, the spatial distribution of electron arrivals is highly random. Consequently, exposures of halide crystallites tend to be clustered, and thus the image is grainy. The effect is particularly obvious if coarse grained emulsions are used, since the number of electron arrivals which produces a given photographic density is then minimal (i.e., because single electrons expose relatively large grains). Conversely, in a fine grained emulsion, many halide crystallites are present per unit area, and a considerably larger total number of electrons is required to produce a given level of overall photographic density. The larger number of incident electrons corresponds to a less random distribution of electron arrivals, so that the resulting image is less grainy. In short, fine grained emulsions are suitable for recording high resolution images *because they interact with many electrons*.

The use of fine grained, low speed emulsions in high resolution electron microscopy is limited by the instability of the microscope itself, since mechanical movements and voltage or current fluctuations preclude the use of extended exposure times. Nevertheless, the graininess of photographic images need not limit electron microscope resolution, since the availability of high beam intensities makes possible the use of instrumental magnifications which are large enough to render photographic enlargements of more than $10 \times$ unnecessary.

FORMATION OF THE LATENT IMAGE

Although the nature of the processes which contribute to the formation of the latent image has not been fully established, a consistent description of the general features of latent image formation has been evolved, and it is presented briefly here.

First, it may be noted that the term "latent image," as used in the current literature, applies specifically to the change effected by exposure in individual silver grains, rather than to the invisible "picture" as a whole.

The illumination of silver halides slowly effects their decomposition to form metallic silver. Correspondingly, the latent image consists of discrete clumps of reduced silver atoms within the silver halide crystal. This may be demonstrated by direct analysis for exposures which correspond to the high density regions of characteristic curves; the amount of reduced silver present after exposures corresponding to the toe of the curve is so small as to escape detection.

Centers of development, known as *sensitivity specks*, appear to exist before exposure, and subsequently act as nuclei for the deposition of silver atoms. After exposure the sensitivity specks are visible in the electron microscope; they increase in both size and number with time of exposure.

During exposure the absorption of quanta by the halide crystallites raises electronic energies to the conduction band (cf. Fig. 7-2), allowing the energized electrons to migrate freely through the grain. These electrons are preferentially trapped in localized regions of structural imperfection, while the resulting space charge is neutralized by migration of silver ions. The latter are reduced by the trapped electrons to metallic silver. As the cycle of migration and neutralization is repeated, finite numbers of silver atoms are deposited at the condensation nuclei.

Electron migration leaves positive halide "holes" within the crystallite lattice. Holes, like energized electrons, are highly mobile (by a mechanism of electron exchange) and thus tend to combine with the liberated electrons. This "dead end" recombination of electrons and holes is prevented by "trapping" the holes through interaction with a sensitizing agent. The sensitizing agent consists of preexisting silver atoms, or of silver sulfide molecules formed by interaction of the halide grain with the gelatin matrix.

During chemical development the condensed specks of silver, which constitute the latent image, serve as centers for the deposition of more massive amounts of metallic silver.

According to this account of latent image formation, the diffusion of the relatively massive silver ions should be rate limiting. More rapid formation of latent images is in fact observed at high temperatures, at which the diffusion rate is increased.

PHOTOGRAPHIC PROCESSING

Development, fixation, and *washing* are the essential states of photographic processing.

In the process of development the sensitized halide grains must be reduced to metallic silver; unexposed grains should remain unaffected. Under appropriate conditions this specification can be met adequately, although not precisely. The *developing agent* is a reducing agent; the two substances used for the purpose at present are hydroquinone and *p*-methylaminophenol sulfate, structural formulas for which are shown in Figs 21-4a and 21-4b, respectively. Since these compounds are active reducing agents in their ionized forms, the speed of development is controlled by the pH of the solution; the more alkaline the developer, the more rapid is its action. The *accelerator* component of developer solutions thus consists of alkali and buffer salts. Speed of development is also quite highly temperature dependent.

Sodium sulfite acts as a *preservative* in developer solutions. This substance not only protects the developing agent from air oxidation, but also interacts with the oxidized developer to form colorless compounds. In the absence of preservative, the developed emulsion may be stained by the highly colored products of the oxidized developer.

Some developer solutions also contain potassium bromide, which serves as a *restrainer*. The restrainer inhibits reduction of unexposed halide grains, thus limiting the development of fog density.

In order to obtain satisfactory development, all portions of the emulsion must remain in contact with an excess of unexpended developing agent throughout the process of development. Thus agitation of the solution is essential. Insufficient agitation, or the use of a spent developer solution, results in the formation of streaky and faded images. For most purposes, continuous movement of the emulsion through the tank or tray of developer is adequate. For critical work, however, the solution may be agitated at regular intervals by bursts of nitrogen gas, or the surface of the emulsion may be stroked by a very soft, camel's hair brush.

Developer solutions require rather frequent replacement. The concentration of active developing agent is reduced both by interaction with

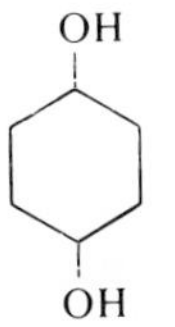

OH
OH

OH
CH_3 N   H   H: $\frac{1}{2}$ H$_2$SO$_4$

Fig. 21-4(a). 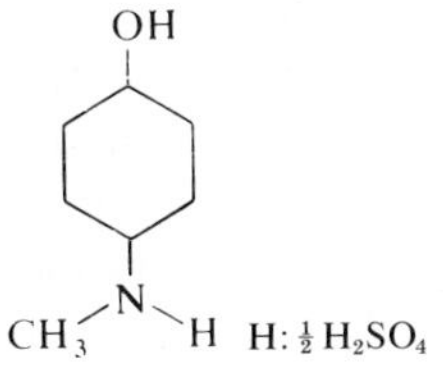   **Fig. 21-4(b).**

the emulsion and by air oxidation. Even more important is the fact that accumulated reaction products inhibit the rate of development by means of mass action effect.

The *time, temperature,* and *extent of agitation* during development all affect the characteristic curve actually followed by any given type of emulsion. When quantitative measurements are to be made, a standard development procedure must therefore be adhered to exactly. Development at excessively high temperatures is to be avoided, since the developer then tends to be spent rapidly by air oxidation. Also, loosening or "reticulation" (formation of a network of wrinkles in the emulsion) may occur at high temperatures.

The development of an emulsion may be followed by immersion in a stop bath. This is simply an acidic solution which quickly lowers the pH to a level at which the extent of ionization of the developing agent is negligible, thus rendering the latter ineffective. The stop bath must consist of a weak acid; strong acids would cause swelling and softening of the gelatin matrix of the emulsion. Dilute acetic acid is commonly used. If a stop bath is not employed, the fixer solution, which also is acidic, serves to terminate development.

Fixation causes the removal of unexposed silver halide from the emulsion and also hardens (tans) the gelatin matrix. The *fixing agent* consists of sodium or ammonium thiosulfate (hypo), which reacts with unreduced ions to form a soluble complex ion:

$$(21\text{-}d) \qquad Ag^+ + 2(S_2O_3)^{2-} \rightarrow Ag(S_2O_3)^{3-}.$$

The hardener (tanning agent) consists of potassium alum. Acetate and borate salts are included in order to provide favorable conditions for the hardening reaction. In addition, sodium sulfite is added as a *preservative.* In fixing solutions the effect of this substance is to supply an excess of bisulfite ion, thus suppressing the reversible decomposition of thiosulfate ion, which occurs as shown by

$$(21\text{-}e) \qquad S_2O_3^{2-} + H^+ \rightleftharpoons HSO_3^- + S.$$

Adequate duration of fixing is essential, since any residual silver halide remains light sensitive, and therefore improperly fixed images gradually blacken upon exposure to light.

Washing, serving to remove completely all excess fixative, is also an essential stage in photographic processing. Discoloration and, ultimately, disappearance of the image may slowly be effected by residual fixative according to the reactions

$$Na_2S_2O_3 + 2\ Ag \longrightarrow Ag_2S + Na_2SO_3,$$

(21-f)

$$Ag_2S + 2\ O_2 \xrightarrow[\text{light}]{\text{humidity}} Ag_2SO_4.$$

Silver sulfate, the final product of this sequence, is a colorless soluble substance.

Excessive fixative is relatively easy to remove from glass plates, but prints require prolonged washing (about 1 hour in running water) for complete removal of adsorbed fixative. The inconvenience can be obviated by the use of special papers which have a resistant plastic coating.

Recommended processing procedures for each type of emulsion are specified by manufacturers.

DENSITOMETRY

Densitometry is the quantitative measurement of the blackening of exposed photographic emulsions. (A related term often found in the photographic literature is *sensitometry*; it refers to the densitometric determination of characteristic curves.) As defined earlier, the photographic density D is

(21-b)
$$\boxed{D \equiv \log \frac{I_0}{I}}.$$

Practical densitometry thus consists of comparison of the light intensity incident upon and transmitted by the developed emulsion.

Experimentally, two types of measurement of photographic density are obtained, the *diffuse* and *specular* densities. Diffuse densities are obtained when all the light leaving an emulsion is collected and measured; specular densities are obtained when illumination and measurement are confined to a narrow beam. The difference between the two arises because light is scattered from the emulsion as well as being absorbed by the developed deposits of metallic silver. Because scattered intensity is always distributed over a wide angle, specular densities are always at least slightly higher than diffuse densities. The difference between the two varies according to the physical properties of the emulsion and its support, and is characterized for any given photographic material by the Callier Q Factor, which is the ratio of the specular and diffuse densities; that is,

(21-g)
$$Q \equiv \frac{\text{specular density}}{\text{diffuse density}}.$$

The extent of light scattering is unrelated to the exposure originally received by an emulsion. Accordingly, specular densities are sought when determinations of photographic density are made for the purpose of determining the intensity of the exposing light. For certain other purposes, such as the determination of the net light stopping power of an absorption filter, measurements of diffuse density are appropriate.

The type of density measured is determined by the geometry of the densitometer; that is, by the angular distribution of transmitted radiation which is focused upon the detector. The latter is a photoelectric device which produces an electrical signal. The signal may be directly proportional to incident intensity, or it may compare the intensity transmitted by an emulsion with that of a reference beam. In practice, the measured densities are only approximately specular or diffuse, since some scattering always occurs within the aperture of the beam transmitted by a "specular" densitometer or beyond the aperture of the beam collected in a "diffuse" densitometer. For this reason the photographic density readings obtained on different models of densitometer are not strictly comparable.

The *microdensitometer* is used to make quantitative measurements of the photographic density of finely structured patterns, such as images of diffraction patterns. In this instrument a focused spot or line of light (the probe) scans the surface of the emulsion. The intensity transmitted at each position is compared with that transmitted by a standard of known optical density. A recording system then produces an automatic read-out of density as a function of position. In Fig. 21-5 two photographic images and their microdensitometer tracings are shown. The image shown in Fig. 21-5a can be scanned by a *line* probe, that is, by a line of light extending in the y direction, and moving along the x direction. The line probe tracing (Fig. 21-5b) is similar to that which would be formed by a *spot* probe moving along any line parallel to the x axis. However, random variations in the photographic density of the image tend to be averaged along a line, so that a smoother trace is produced by that type of probe. The image shown in Fig. 21-5c must be scanned by a spot probe if the intensity of the *individual* image points is of interest.

When densitometer sensitivity is high, noise may also be introduced as a consequence of the random arrival of photons at the photocell. Figures 21-6a, 21-6b, and 21-6c compare densitometer tracings of a given specimen as recorded at low, optimal, and excessive sensitivity, respectively.

The spectrophotometer (Chapter 24) can be used to measure the photographic densities of extended specimens. The optical density of the exposed emulsion is measured using a physically identical developed but unexposed emulsion as a blank.

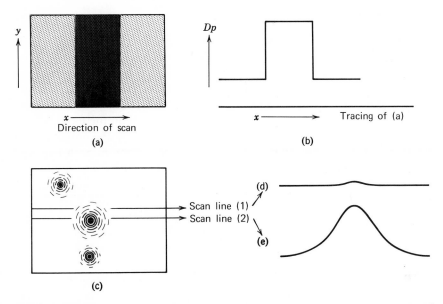

Fig. 21-5 Densitometry: (a) object; (b) corresponding line probe tracing; (c) object; (d) and (e) two corresponding spot probe tracings.

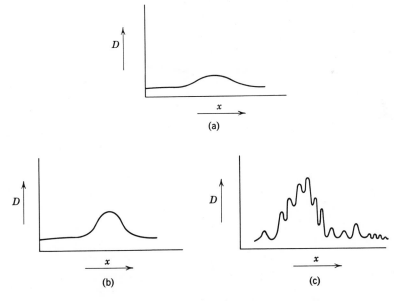

Fig. 21-6 Sensitivity: (a) low; (b) optimal; (c) excessive.

ENLARGEMENT

A diagram of a typical photographic enlarger is shown in Fig. 21-7. The *light source* of this instrument provides a beam of high and even intensity. The level of intensity may be controlled by adjustment of the energizing voltage. Light from the source is collected in the plane of the photographic plate (the negative) by means of a *condenser lens*, while the *enlarger lens* focuses an image of this emulsion on the easel below.

The contrast of the resulting image is influenced by the conditions of illumination. Intensity tends to be scattered equally from both exposed and unexposed regions of the negative; thus, scattered light, if admitted to the aperture of the enlarger lens, reduces the overall contrast of the image. (In terms of the discussion of the preceding section, a more contrasty image is obtained if specular rather than diffuse densities are recorded by the enlarger.) Thus a high contrast illumination system provides a light beam of narrow aperture angle and requires the use of a lamp of correspondingly high intensity.

Focus and *magnification* are controlled by physical movements of the enlarger components. Typically, source, condenser lens, negative, and

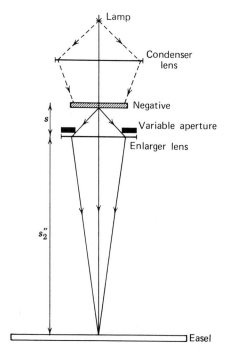

Fig. 21-7 Photographic enlarger.

enlarger lens may be raised or lowered as a unit, thus changing the distance s_2'' between lens and easel (see Fig. 21-9), while the magnification $(=s_2''/s)$ varies accordingly. Focus is achieved by adjustment of the distance s between negative and enlarger lens. While obvious differences in magnification can be observed by defocusing, recording of the image is always *in* focus, so that a given vertical position of the enlarger lens with respect to the easel always corresponds to a definite value of magnification.

Enlargers can be calibrated (for a given enlarger lens) by noting the lens positions which correspond to known magnifications. A transparent ruling of known dimensions, such as a clear plastic ruler, is used as the calibration object. In practice, these measurements are used to determine the approximate level of magnification, while very accurate measurements of the magnification of prints are made by comparison of the separations of identifiable objects on print and negative.

The question whether focusing of an enlarger can be used to compensate for misfocusing of the negative which is copied sometimes arises. The answer is that it can *not*. The action of the enlarger lens, as of any other lens, is to reunite, at the image, rays which originate from any point on the object (i.e., in this case, from any point on the negative). In an out-of-focus negative the illumination which originates from any single point on the original specimen is distributed over a finite area. Rays from different points on the negative intersect at various positions beyond the lens but do not do so systematically in any one plane. (This is illustrated in Fig. 21-8, in which A and C represent the edge of the disc of light produced by a specimen point which would be in focus on

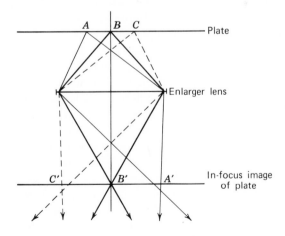

Fig. 21-8 .

the negative at point B. The action of the enlarger lens is simply to magnify this disc of confusion; there is no plane at which all rays from points A, B, and C unite.) To state these conclusions in other words: information lost from a negative by defocus cannot be restored by the enlarger.

As suggested by the reciprocity relationship (21-c), exposures used in printing may be varied by adjustment of either time or intensity. The latter is controlled principally by varying the *aperture* of the enlarger lens (although the maximum available intensity may be adjusted by variation of the power supplied to the lamp). Enlarger lens aperture is controlled by means of a variable diaphragm included in the lens assembly; this either may be continuously variable or may be set at any one of a series of fixed f numbers.

The f number has been defined on p. 178 as the ratio of the focal length of a lens to its aperture diameter; that is,

$$\text{(21-h)} \qquad\qquad\qquad f\text{ number} \equiv \frac{f}{a}.$$

The intensity of light passing through a lens is proportional to the illuminated area and thus to a^2. Accordingly, for a lens of fixed focal length, *the transmitted intensity is inversely proportional to the square of the f number.*

In practice, a series of f numbers is used for which squares are related in the proportion $1:2:4:8:\ldots$ Each successive f number setting then corresponds to *half* the intensity of illumination provided by its predecessor. The series of f numbers used in enlarger (and camera) lenses includes the values $\ldots, 4.0, 5.6, 8, 11, 16, 22, \ldots$ For example, a lens of 10 cm focal length has an aperture diameter, when set at f/5.6, of $10/5.6 = 1.8$ cm, corresponding to an illuminated area of $\pi (1.8/2)^2 = 2.54 \text{ cm}^2$. When the f number is changed to f/8, a is reduced to $10/8 = 1.25$ cm, and the illuminated area to $\pi (1.25/2)^2 = 1.23 \text{ cm}^2$. The ratio of transmitted intensities is $1.23/2.54$ or approximately one half.

A high degree of correction of lens aberrations is more difficult to achieve in enlarger lenses than in microscope lenses, since the enlarger must be used to form images of *extended* fields. (The negatives copied in the enlarger may often be two to four inches across, while, even at low magnifications, microscope fields are only a few millimeters in diameter.) In particular, the defects of third order astigmatism, curvature of field, and distortion (cf. pp. 204–214) are significant in enlarger systems. Since these aberrations are controlled by aperture diameter as well as by the extent of the field, it is desirable in high resolution imaging to use the *smallest* aperture consistent with a reasonable duration of exposure. While the paraxial zones of the images formed by an enlarger may be of high quality, deterioration of the image is often quite obvious in peripheral zones.

An ideal exposure for any given emulsion (and processing procedure) is determined initially by a process of trial and error. Variations in f number and exposure time can then be made in accordance with the reciprocity law. The change in exposure required to compensate for a change in magnification can be obtained by noting that intensity of illumination is inversely proportional to the square of distance from the source. As shown in Fig. 21-7, the distance from the negative (the effective light source) to the easel is $(s_2'' + s)$ or, approximately, s_2''. Since s varies only slightly when focus is adjusted for different levels of magnification, the magnification $(-s_2''/s)$ also varies, to a first approximation, with s_2''. Thus suitable exposures at different magnifications are related as the square of the ratio of magnifications; that is,

$$(21\text{-i}) \qquad \boxed{\frac{\text{exposure at } M_2}{\text{exposure at } M_1} = \left(\frac{M_2}{M_1}\right)^2.}$$

For example, the exposure required for printing at $8\times$ is $(8/4)^2$ or four times as great as that required for printing at $4\times$.

Correct exposures may be obtained directly from the reading of a photometer (exposure meter) which has initially been calibrated with respect to an ideal exposure. The calibration consists in determining by trial and error the ideal exposure for a typical negative at any given level of magnification. The light-sensitive element of the meter is exposed to the illuminations transmitted by the negative under these conditions. The meter reading so obtained is then used to set scales which specify equivalent f number and exposure time combinations. The way in which subsequent exposures are determined varies in detail with the nature of the photometer; manufacturers' instructions must be consulted. In general, a standard reading equivalent to that of the ideal exposure is obtained by (a) adjusting the sensitivity of the photometer by means of a calibrated dial (which specifies appropriate combinations of f numbers and exposure times) or (b) adjusting the illumination which reaches the meter by changing the f number of the enlarger. (In this way the f number to be used with a standard exposure time is obtained directly.) In either case the specified combinations of exposure time and f number may be varied by making use of the fact that successive f number values require successive doublings of the exposure time. For example, if a photometer specifies a one sec exposure at f/8, a four second exposure at f/16 may be used instead.

The selection of exposure is accomplished automatically in some systems; they are particularly convenient when many prints must be made on a routine basis. However, critical photographic printing often demands individual determinations of exposure, since it may be desirable to emphasize a feature which is of a different photographic density than is characteristic of the negative as a whole. In so doing, a "test strip" may be made along which exposure time is increased successively. Alternatively, masks are available commercially which absorb specified fractions of the light reaching various portions of the test print.

PHOTOGRAPHIC ENHANCEMENT OF CONTRAST

The contrast of electron micrographs and other images can be enhanced by the appropriate use of photographic techniques. It is important to realize, however, that *photographic technique cannot supply information which is not inherent in the original image.* That is, photography serves only to emphasize features which are already evident in the original image. More ambitious use of photographic methods can easily lead to mistaken conclusions; for example, excessively contrasted shadowing metal particles are easily mistaken for macromolecular substructure.

Excessive photographic contrast is undesirable, since background may become obvious to an extent which diverts attention from the specimen. For this reason the printing of electron micrographs should, ideally, be such as to reduce the contrast of shadowing metal grain and of the structural detail of supporting films. In this way the contrast of the specimen itself is enhanced. The imaging of an object with insufficient, ideal, and excessive contrast is suggested by Figs. 21-9a, 21-9b, and 21-9c, respectively.

A high level of background noise must generally be tolerated in electron micrographs, even when optimal printing contrast is obtained. Considerable improvement by the method of noise filtering or optical

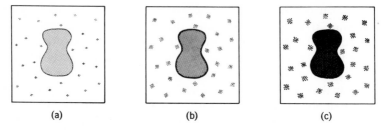

    (a)   (b)   (c)

Fig. 21-9 Contrast: (a) low; (b) optimal; (c) excessive.

filtering (cf. p. 447) has been demonstrated in images of objects of *periodic* structure.

Adjustment of contrast level can be carried out with the use of a series of emulsions which differ in γ value but are otherwise similar. Alternatively, a series of developer solutions is available which vary the γ of a single type of emulsion.

REFERENCES

P. Hansell and R. Ollerenshaw, *Longmore's Medical Photography*, 7th rev. ed., The Focal Press, London, 1962. The book gives an excellent general account of scientific photography, as well as of specifically medical applications.

"Some Things Every Electron Microscopist Ought to Know," *Kodak Tech Bits* **2**, 3 (1961). A discussion of graininess and other limitations on the photographic recording of electron images. Short descriptions and characteristic curves of emulsions used in electron microscopy are given, and a list of practical hints for processing and handling is included.

R. C. Valentine, in R. Barer and V. E. Cosslett, *Advances in Optical and Electron Microscopy*, Vol. I, Academic Press, New York, 1966, pp. 180–201. An article on "The Response of Photographic Emulsions to Electrons."

"Processing Chemicals and Formulas," Kodak Professional Data Book J-1, 6th ed., 1966.

"Some Chemical Reactions in Photography," Kodak Pamphlet J-15, 1965.

"Kodak Notes on Practical Densitometry," Kodak Pamphlet E-59, 1967.

C. E. Mees, *The Theory of the Photographic Process*, 2nd ed., Macmillan, New York, 1954. A comprehensive treatise which traces the history of present understanding of the photographic process.

"Photography through the Microscope," Kodak Scientific and Technical Data Book P-2, 4th ed., 1966.

E. Zeitler and J. R. Hayes, in G. F. Bahr and E. H. Zeitler, *Quantitative Electron Microscopy*, Armed Forces Institute of Pathology, Washington, D.C., 1965 *Lab. Invest.* **14**, 1965, 586/1324–595/1333. This article, "Electrography," discusses scientific photography from the point of view of communications theory.

R. G. Hart, *Science* 159, 1464 (1968). This article, "Electron Microscopy of Unstained Biological Material: The Polytropic Montage," describes a photographic method for contrast enhancement.

R. C. Jones, *Sci. Am.* **219**, 111 (1968). This article, "How Images Are Detected," emphasizes the concept of *DQE*.

CHAPTER 22

Sources of
Monochromatic Light

A number of optical instruments, including, in certain special cases, microscopes, but principally spectrophotometers, fluorimeters, and similar spectroscopic devices (see Chapters 24 and 25) require the use of a source of monochromatic illumination. Monochromatic illumination may be provided at a "single" wavelength by a device which produces, essentially, *only* light of that wavelength. Certain conventional types of lamp fit into this category, of which the *laser* is also an example. Light of a single wavelength may also be isolated from polychromatic emission by a filter of suitable properties. Both *absorption filters* and *interference filters* are used for this purpose.

Single wavelength sources provide relatively highly monochromatic illumination but are unsuitable for use in instruments in which the effect of a range of wavelengths is to be studied. In the latter case, mono-chromators are used to isolate any chosen wavelength from the light produced by a source which emits a more or less continuous spectrum. Generally the degree of monochromaticity of light passed by a mono-chromator is considerably less than that of light from devices which emit or transmit a single wavelength only. Prisms, diffraction gratings, and the use of these elements in monochromators are discussed in Chapter 23.

The nature of conventional monochromatic light sources, absorption and interference filters, and lasers is described in this chapter. Aspects of topics discussed in Chapters 1, 3, 4, and 7 are basic to this material.

ABSORPTION FILTERS

Absorption filters consist of materials which selectively absorb light of certain wavelength ranges. Appropriate glasses or solutions are used

480

for this purpose. Figure 22-1a shows the absorption spectrum of an idealized absorption filter which strongly absorbs all light within some wavelength range from λ_1 to λ_2 but is completely transparent in the wavelength ranges λ_0 to λ_1 and λ_2 to λ_3. Such transmission properties are only approximated in reality; absorption spectra are of the general form shown by Fig. 22-1b. Absorption filters are thus unsuitable for the production of highly monochromatic light from a white beam, since they tend to remove or to pass light of rather wide wavelength ranges, with partial transmission of intermediate wavelengths.

An example of the use of absorption filters to remove a broad wavelength band is the inclusion of cells containing water as a "heat filter." Certain wavelengths in the infrared and far ultraviolet may be removed from the beam in this way, so that their potential destructive effects are avoided.

The effects of reflection and scattering at small crystals can be exploited to produce a type of filter which transmits a restricted range of wavelengths. Powdered quartz is suspended in a mixture of benzene and carbon disulfide. If the refractive index of the solvent mixture exactly matches that of the quartz, light is transmitted, as if by a homogeneous solution. If there is even a very slight mismatch of indices, however, light intensity is lost by reflection and scattering from the many small quartz crystals. The variation of refractive index with wavelength (i.e., the *dispersion*; cf. p. 508) is *different* for quartz from that of the solvent mixture. Thus the wavelength which is transmitted without attenuation is determined by the exact composition of the mixture of solvents. Also, since the *temperature* dependence of the refractive index of the solvents is much greater than that of the quartz, the transmitted wavelength may be varied by changing the temperature of the suspension. Ground glass has also been used to make filters of this type.

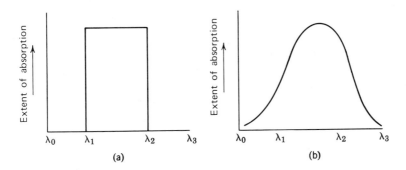

(a)

(b)

Fig. 22-1 Characteristics of absorption filters; (a) ideal; (b) real.

COMBINATIONS OF CONVENTIONAL LIGHT SOURCES AND ABSORPTION FILTERS

As explained in Chapter 7, elements in their gaseous states may be excited to high electronic energy levels with subsequent emission of light at characteristic wavelengths as the atoms return to the ground energy level. The wavelengths which are emitted correspond in quantum energies to the difference in total energy between the ground state and the respective excited states. Since atoms may exist only in a discrete number of energy states, only a finite number of different wavelengths are emitted; that is, a *line* spectrum is formed. One or a few of the wavelengths present in an atomic line spectrum may be of high intensity and may differ appreciably in wavelength from other lines of significant intensity. An absorption filter may then be used to remove undesired wave-

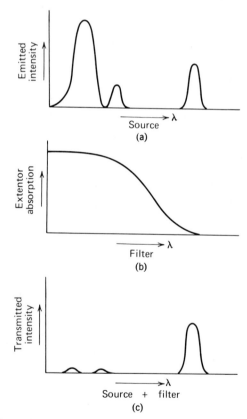

Fig. 22-2 Monochromatization by an absorption filter: (a) emission spectrum of source; (b) absorption spectrum of filter; (c) spectrum of intensities transmitted by filter.

lengths, resulting in passage of an approximately monochromatic beam by the source-filter combination. The respective spectra are shown in Fig. 22-2.

Two atomic vapors commonly used in this way are sodium and mercury. Among other wavelengths, sodium vapor lamps emit an intense, closely spaced doublet at 589.0 and 589.6 mμ (the "D line"), with weaker but appreciable emission in the violet and ultraviolet. A glass lamp housing serves as a filter for the removal of the latter. Mercury discharge lamps emit strongly in the green at 546 mμ, in the ultraviolet at 254 mμ, and appreciably at a number of other wavelengths in the blue, violet, near ultraviolet, and infrared. While the relative intensities of the emission lines vary with the manner in which the gaseous mercury is excited, appropriate filters may be used to isolate any one of the emission maxima.

INTERFERENCE FILTERS

The interference filter may consist of a layer of (dielectric) material that usually is substantially nonabsorbing, and which is coated on both surfaces with highly but not totally reflecting layers of metal. Alternatively, this type of filter may be constructed from multilayer dielectric stacks. Interference between successive reflections from the upper and lower surfaces of the dielectric results in the transmission of light of the desired wavelength, while light of other wavelengths is reflected. On casual inspection, an interference filter thus appears to be a mirror. It may be noted that the interference filter is a special case of the parallel plate interferometer.

The wavelength of the maximally transmitted amplitude is critically determined by the thickness of the dielectric layer, while the reflectivity of the surface coatings determines the degree of selectivity of the filter. The more highly reflective the coatings, the narrower is the range of wavelengths transmitted with appreciable intensity. The relationship of reflectivities and dielectric thickness to the properties of the filter is developed quantitatively in the succeeding sections.

The standard type of interference filter can produce bandwidths (spectral ranges over which transmitted intensity is equal to or greater than half the maximum transmitted intensity) down to about 100 Å, with a peak transmittance of about 30%. By replacing the metallic reflecting layers with multiple dielectric layers, the bandwidth can be reduced to about 10 Å and the peak transmittance increased to about 80%.

While the action of the interference filter is usually diagramed in the manner shown in Fig. 22-3a (since this type of diagram illustrates the

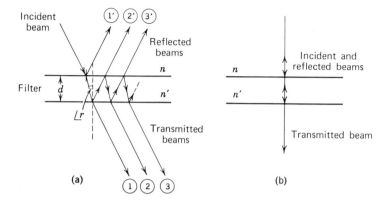

Fig. 22-3 Successive reflected and transmitted beams at a pair of parallel plane surfaces: (a) oblique illumination; (b) normal illumination.

relationship between individual reflected rays), these filters are usually illuminated normally, as shown by Fig. 22-3b. The discussion on p. 486 and especially (22-e) make clear that the wavelength of maximum transmission by a given interference filter varies somewhat according to the angle of illumination; the transmitted wavelength becomes *shorter* as the angle of incidence increases.

RELATIONS GOVERNING REFLECTIVITY AND TRANSMISSIVITY

The principle of reversibility of light rays (cf. p. 3) may be applied to the incidence of light upon a partially reflecting, partially transmitting surface. Certain quantitative relationships which apply to the interference filter may thus be obtained. In the following discussion it is assumed that no absorption of light occurs on either side of the interface.

Figure 22-4 shows the paths of incident, reflected, and refracted rays, designated ①, ②, and ③, respectively, at the interface. If the direction of the reflected ray ② is reversed, a new pair of reflected and refracted rays, ⑥ and ⑤, respectively, results as shown in Fig. 22-4b. In accordance with the principle of reversibility, the reflected ray ⑥ must retrace the path of the original incident ray ①, shown in Fig. 22-4a.

If the amplitude of the reversed ray ④ is Ar, then those of the reflected and transmitted rays ⑥ and ⑤ must be Ar^2 and Art, respectively, where r and t are, respectively, the fractional amplitudes *reflected* and *transmitted* at the surface.

Reversal of the original transmitted ray ③ to form ray ⑦ similarly produces reflected and refracted components of amplitude, ⑧ and ⑨,

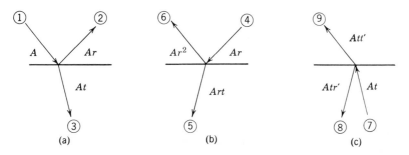

Fig. 22-4 Incident, reflected and refracted amplitudes.

respectively, as shown in Fig. 22-4c. Since in this case reflection and re-fraction occur at the opposite side of the interface to that considered in Figs. 22-4a and 22-4b, the fractional amplitudes transmitted and re-flected may differ from r and t, and thus are designated t' and r', res-pectively. Reflected ray ⑧ is therefore of amplitude Atr', and refracted ray ⑨ is of amplitude Att'.

Conservation of energy demands that simultaneous reversal of both rays ② and ③ should have the net effect of reversing ray ① without change in its amplitude. Furthermore, *no* net amplitudes should travel in the directions of rays ⑤ or ⑧, since no amplitude is observed to travel in these directions upon incidence of light in the direction of ray ①. These considerations are expressed quantitatively by

(22-a) $$A = Ar^2 + Att' \quad \text{or} \quad tt' = (1 - r^2),$$

(22-b) $$Art + Ar't = 0 \quad \text{or} \quad r = -r'.$$

Equation 22-b states that the fractional amplitude reflected is the same on both sides of the interface. The minus sign indicates that a *reversal of phase* occurs upon reflection from *one* of the two sides of the interface. (This reversal of phase has been discussed on p. 46).

It might appear that t and t' should be equal also. This is *not* in fact so; as shown on p. 151, the law of conservation of energy requires that

(7-q) $$r^2 + \frac{n_2 \cos r'}{n_1 \cos i} t^2 = 1,$$

where r' and i are the angles of refraction and incidence, respectively. The numerical relationship between t and t' is therefore dependent upon the specific values of the refractive indices of the media and upon the angle of incidence.

QUANTITATION OF PROPERTIES OF THE INTERFERENCE FILTER

As shown in Figs. 22-3a and 22-5, the light transmitted by an interference filter consists of the successive beams ①, ②, ③, . . . , each of which, relative to the preceding component, has made *two* additional passages through the dielectric layer. The optical path difference Δ between successive components of the transmitted beam is thus

$$\text{(22-c)} \qquad \Delta = 2nd \cos r',$$

where n is the refractive index, d the thickness of the dielectric layer, and r' is the angle of refraction within the layer.

In the special case of normal incidence, illustrated in Fig. 22-3b, for which $r' = 0°$ and $\cos r = 1$,

$$\text{(22-d)} \qquad \Delta = 2nd.$$

If the quantity Δ is an integral number of wavelengths, successive components of the transmitted beam emerge from the filter *in phase* and reinforce each other; light of a wavelength such that this condition holds is thus transmitted with maximum intensity by the filter. That is,

$$\text{(22-e)} \qquad \lambda_{\max} = \frac{2nd \cos r'}{m}$$

or, for normal incidence,

$$\text{(22-f)} \qquad \boxed{\lambda_{\max} = \frac{2nd}{m}},$$

where m is any integer.

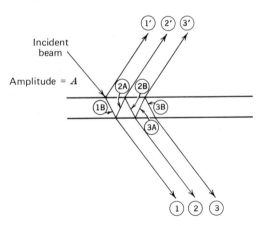

Fig. 22-5 Successive reflected and transmitted beams.

Similarly, for light of the same wavelength, the reflected beams ②, ③, ④, . . . are also *in* phase with each other. Thus it might seem that light of the wavelength transmitted with maximum intensity would also be reflected with maximum intensity. However, the *first* component ① of the reflected beam undergoes reflection from the *air* side of the air-dielectric interface, whereas the subsequent reflected beams undergo 1, 3, 5, . . . reflections from the *dielectric* side of that interface. ① thus differs in phase by 180° from all subsequent reflected components [cf. (22-b) and the related discussion above]. As shown in the following paragraph, the amplitude of beam ① is exactly equal to the sum of the amplitudes of all the subsequent reflected beams; hence, when the latter are exactly in phase with each other, there is total destructive interference with beam ①, and *no* intensity is reflected.

The amplitudes of the successive reflected components and also of the successive transmitted components may each be calculated in terms of the incident amplitude A and the fractional amplitudes t, t', r, and r' which have been defined above. The results of such a calculation are summarized in Table 22-1, in which intermediate portions of the light paths are designated in the manner indicated in Fig. 22-5.

At an appropriate wavelength, all components are in phase, and the sum of the second and subsequent reflected amplitudes (②, ③, ④, . .). is the

(22-g) $$Att'r'[1+(r')^2+(r')^4+(r')^6+\cdots) = \frac{Attr'}{1-r^2} \qquad \text{for all } r < 1.$$

Since r is a fractional amplitude, and thus always less than one, the sum of these reflected amplitudes converges for all physically possible cases [i.e., the infinite series can be expressed as the quantity given by the right side of (22-g)]. Therefore,

(22-h)   summed amplitudes of beam components ②, ③, ④ $= \dfrac{Att'r'}{1-r^2}.$

Substitution for tt' and r' from (22-a) and (22-b) then gives

(22-i)   summed amplitudes $= \dfrac{-A(1-r^2)r}{1-r^2} = -Ar.$

Table 22-1   Amplitudes of Portions of Incident Light Reflected or Transmitted by an Interference Filter

① Att'	①B At	②A Atr'	①' Ar
② $Att'(r')^2$	②B $At(r')^2$	③A $At(r')^3$	②' $Att'r'$
③ $Att'(r')^4$	③B $At(r.)^4, \ldots$ etc.		③' $Att'(r')^3$

Thus it is established, as stated above, that the summed amplitude of all reflected beams but the first is equal in magnitude to the amplitude of the first reflected component. The phases of these two beams are reversed, leading to complete destructive interference of the reflected beam at λ_{max}. Note that this result could also be predicted on the basis of conservation of energy: if all intensity at λ_{max} is transmitted, none can remain to be reflected.

The extent to which wavelengths other than the maximum are transmitted determines the selectivity or "bandwidth" of the interference filter. The extent of transmission of light of any given wavelength may be calculated by adding the amplitudes of all the components of the transmitted beam, with due regard to phase relationships. The numerical examples to be given demonstrate that the selectivity of the filter is a function of the fractional reflectivity of the surface layers.

When the dielectric is coated with layers of low reflectivity, most of the incident intensity appears in the first transmitted beam, whereas only some small or negligible fraction of total intensity is contained in the subsequent transmitted or reflected beams. There is then limited or negligible opportunity for interference between successive reflected or transmitted beams; light of all wavelengths is transmitted without significant alteration of intensity, and the device fails to function as a filter. Consider a dielectric layer with surfaces of 5% reflectivity; that is, a reflectivity of the same order as uncoated glass in air. The amplitudes of the first and second transmitted beams are found, by substitution in (22-a), to be

(22-j)   amplitude of 1st transmitted beam $= Att' = A(1 - r^2)$
$$= A(1 - 0.0025) = 0.9975 \text{ Å.}$$

(22-k)   amplitude of 2nd transmitted beam $= Att'(r')^2$
$$= (0.9975 \times 0.0025)A$$
$$= 0.00249\ A$$

Subsequent components of the transmitted beam are further reduced in amplitude by successive factors of (0.0025). Clearly, the transmitted intensity is composed overwhelmingly by the first transmitted beam. The relative phases of the subsequent beams are of little consequence, since the resultant amplitude can vary, at most, by less than 1% in accordance with these phases.

As the reflectivity of the surface layers is increased, destructive interference between beams becomes quantitatively more important. In the limit, films of very high reflectivity ($> 95\%$) may produce transmission bandwidths as narrow as 15 Å.

Surfaces of 60% reflectivity may be considered as a case of intermediate reflectivity. The amplitude of the first transmitted beam is given,
again by substitution in (22-a), as

(22-l)   amplitude of 1st transmitted beam $= A(1-r^2)$
$$= A(1-0.36) = 0.64\ A$$

while subsequent amplitudes are obtained through successive multiplication by the factor $(r')^2 = 0.36$. These values are listed in Table 22-2.
The sixth and all subsequent transmitted components are of an amplitude
less than 0.5% as great as the incident beam, and their contributions to
resultant intensity are small. If the first six components should be exactly
in phase, the resultant amplitude would be $(0.640 + 0.230 + 0.083 + 0.030 +$
$0.011 + 0.004) = 0.998A_0$ or, approximately, the same as that of the incident beam. If the successive components should be exactly *out* of phase,
the resultant transmitted amplitude (neglecting components beyond the
sixth) would be $(0.640 - 0.230 + 0.083 - 0.030 + 0.011 - 0.004)A_0$, or
$0.470A_0$. The transmitted intensity would be $(0.470\ A)^2$, or about 22% as
large as for the *in*-phase case. For example, if the filter thickness were
chosen for maximal transmission of violet light of $\lambda = 4000\ \text{Å}$, near
infrared radiation of $\lambda = 8000\ \text{Å}$ would be transmitted only to the extent
of about one fifth. (Note that these estimates are based on the assumption that the reflectivity r of the surface does not vary with the wavelength of light. In fact, variations do occur, reflectivity being, in
general, higher at longer wavelengths. The extent of the changes in r
across the visible spectrum varies greatly between coatings of different
compositions.) Similarly, for green light of $\lambda =$
5333 Å, successive transmitted beams differ in
phase by one quarter wavelength. The resultant
amplitude can be computed graphically by the
method described on p. 39. As Fig. 22-6 shows,
this amplitude is about 0.6 Å. The corresponding
resultant intensity is proportional to $(0.6A_0)^2$, or
about 36% of the incident intensity. The device
thus displays a certain selectivity in the transmission of light; transmissivity varies by a factor of
about four across the visible spectrum. This
selectivity is inferior to that of absorption filters;
much higher surface reflectivity is in fact required
to produce a useful interference filter.

A surface reflectivity of 90% may be considered
as an example of moderately high reflectivity.
Amplitudes of the individual transmitted

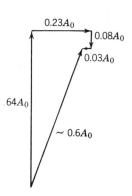

Fig. 22-6 Vector diagram
for determination of
amplitude transmitted
by an interference filter
of 60% reflectivity.

Table 22-2   Successive Amplitudes Transmitted by an Interference Filter of 60% Reflectance

Component	Amplitude
1	$0.640A_0$
2	$0.230A_0$
3	$0.083A_0$
4	$0.030A_0$
5	$0.011A_0$
6	$0.004A_0$

components can be computed for this case, as before, by evaluation of the terms listed in Table 22-1. Twenty sets of reflections are now required to reduce the amplitude to less than $0.005A_0$; the values of these first twenty components are listed in Table 22-3. The graphic method of computing resultant amplitude becomes somewhat cumbersome when as many as twenty components must be considered, but may nonetheless be used to deduce approximate resultant intensities. This is done in Fig. 22-7 for the case of transmitted beams which differ successively in phase by 90°. The resultant amplitude is seen to be about $0.146A_0$, while the corresponding intensity is proportional to $(0.146A_0)^2$ or about 2% of I_0. Figure 22-7 might represent, for example, the case already considered of the passage of light of $\lambda = 5333$ Å by a filter designed to transmit maximally at 4000 Å. While such a device still cannot be described as a highly selective filter, it is clear that further increases in the reflectivity of the filter surfaces, producing a still larger number of components of

Table 22-3   Successive Amplitudes Transmitted by an Interference Filter of 90% Reflectivity

Component	Amplitude	Component	Amplitude
1	$[1-(0.9)^2] = 0.190A_0$	11	$0.023A_0$
2	0.154	12	0.019
3	0.125	13	0.015
4	0.101	14	0.012
5	0.082	15	0.010
6	0.066	16	0.008
7	0.054	17	0.007
8	0.043	18	0.008
9	0.035	19	0.005
10	0.028	20	0.004

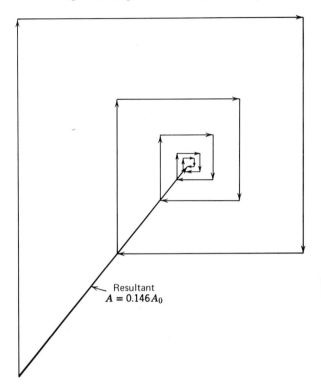

Fig. 22-7 Vector diagram for determination of amplitude transmitted by an interference filter of 90% reflectivity.

transmitted intensity of significant amplitude, would result in more complete destructive interference of transmitted amplitudes at wavelengths relatively close to the wavelength of maximum transmission.

ALGEBRAIC COMPUTATIONS OF RESULTANT INTENSITIES PRODUCED BY INTERFERENCE FILTERS

The vector diagrams in Figs. 22-6 and 22-7 are an interesting graphic illustration of the contributions of successive transmitted amplitudes to the resultant. However, an *algebraic* method for the summation of amplitudes is less cumbersome, and it allows for inclusion of the complete infinite series of reflections. In so doing, the complex amplitude of the resultant is obtained as the sum of the complex amplitudes of the individual components. Properties of the complex form of the wave equation, and the process by which the sum of three amplitudes may be obtained, are described in Appendix C. That discussion may be referred to in considering the following derivation of the infinite sum of amplitudes which is applicable to the interference filter.

In general, the complex amplitude of the resultant of an infinite series of amplitudes may be written

$$(22\text{-m}) \qquad A e^{i\theta} = a_1 + a_2 e^{i\delta} + a_3 e^{2i\delta} + a_4 e^{3i\delta} + \dots ,$$

where θ and A designate the phase and amplitude, respectively, of the resultant; δ is the *phase shift* between successive components, the phase of the first transmitted component being selected, arbitrarily, as zero. $a_1, a_2, a_3, \dots$ are the individual amplitudes of the transmitted components, evaluation of which, in the manner given by Table 22-1 yields, from (22-m),

$$(22\text{-n}) \qquad A e^{i\theta} = A_0(1 - r^2)(1 + r^2 e^{i\delta} + r^4 e^{2i\delta} + r^6 e^{3i\delta} + \dots),$$

where A_0 is the incident amplitude. r, as noted above, must always be less than one, so that the infinite sum of (22-n), like that of (22-g), is convergent. The sum may thus be expressed as

$$(22\text{-o}) \qquad (1 + r^2 e^{i\delta} + r^4 e^{2i\delta} + r^6 e^{3i\delta} + \dots) = \frac{1}{1 - r^2 e^{i\delta}};$$

This gives the complex amplitude of the resultant as

$$(22\text{-p}) \qquad \frac{A_0(1 - r^2)}{1 - r^2 e^{i\delta}} = A e^{i\theta}.$$

Appropriate algebraic rearrangement could lead to the separation of the quantity given by (22-p) into real and complex portions; more simply, however, the *intensity* of the resultant may be obtained directly by multiplication of the complex amplitude with its complex conjugate:

$$(22\text{-q}) \qquad (A e^{i\theta})(A e^{-i\theta}) = A^2 = \left[\frac{A_0(1 - r^2)}{1 - r^2 e^{i\delta}}\right]\left[\frac{A_0(1 - r^2)}{1 - r^2 e^{-i\delta}}\right]$$

$$= \frac{A_0^2(1 - r^2)^2}{1 - r^2(e^{i\delta} + e^{-i\delta}) + r^4}$$

noting, from (C-e), that

$$(22\text{-r}) \qquad \cos \delta = \tfrac{1}{2}(e^{i\delta} + e^{-i\delta}),$$

this gives

$$(22\text{-s}) \qquad I = \frac{I_0(1 - r^2)^2}{1 - 2r^2 \cos \delta + r^4},$$

in which I_0 is the incident intensity, r the fractional amplitude reflected on a single reflection from the surface of the filter and δ the phase angle between successive transmitted beams.

Equation 22-s may be applied, for example, to the case of surfaces with 99% reflectance, at which an extremely large number of component beams of comparable amplitude are formed. For $\delta = 90°$ (as for light of $\lambda = 5333$ Å transmitted by a filter designed to pass light of $\lambda = 4000$ Å), $\cos \delta = 0$. Then,

(22-t) $$I = \frac{[1 - (0.99)^2]^2}{1 + (0.99)^4} I_0 = 2.02 \times 10^{-4} I_0.$$

For light of $\lambda = 4050$ Å passing through the same filter, $\delta = 4°30'$, which gives

(22-u) $$I = \frac{[1 - (0.99)^2]^2}{1 - 2 \cos (4°30')(0.99)^2 + (0.99)^4} I_0 = 5.15 \times 10^{-4} I_0.$$

Note that the intensity transmitted at a wavelength 50 Å distant from that of maximum transmitted intensity is only slightly greater than that at a wavelength 1333 Å distant.

Calculations similar to those given in (22-t) and (22-u) show that in the case of surfaces of 60% reflectivity, as shown by Fig. 22-6, there is 35.72% transmission of light for which successive beams vary in phase by 90°. For surfaces of 90% reflectivity, as shown by Fig. 22-7, beams differing in phase by 90° are transmitted to the extent of 2.18%. In practice, these transmission values are reduced in proportion to the extent of absorption by the filter.

THE LASER

The word "*laser*" is an acronym for "Light Amplification by Stimulated Emission of Radiation." The term *optical maser* is used synonomously with laser.

The special properties of laser light are: (1) high monochromaticity and coherence (2) high intensity, and (3) high degree of collimation. These properties are possessed by laser light to a degree which generally differs in order of magnitude in comparison to light from conventional sources. It is sometimes said that the light produced by lasers is "a different kind" from that resulting from other sources; such statements may be misleading, since the special properties of laser light can in fact be analyzed as differences in the *degree* of monochromaticity, intensity, and parallelism. The individual quanta emitted by a laser are not different from "ordinary" quanta. However, the intensity of laser beams may be so great that a significant proportion of irradiated atoms absorb *two* quanta. In this case the resulting photochemical or luminescent effects are quite different from those induced by ordinary light beams. This aspect of laser beams is discussed on p. 503.

The nature of monochromaticity and coherence has already been discussed in Chapters 1 and 3 but is briefly recapitulated here. The principles of laser action, and the prototype of practical lasers — the ruby — are described. Understanding of the energy level diagrams discussed in Chapter 7 is basic to a discussion of laser action. Applications of lasers which are of interest to biologists are described briefly.

COHERENCE AND MONOCHROMATICITY

As explained in Chapter 1, a pure monochromatic beam can be represented by a sine wave of infinite duration, but any real emission of light corresponds to a sine wave of *finite* duration. Such a pulse is in fact equivalent to the superposition of infinitely extended sine waves of a *range* of frequencies and of various relative amplitudes. A finite pulse is always equivalent to a mixture of pure wavelengths; thus, since emission times must always be finite, *there is no such thing as a pure monochromatic source*. The minimum emission time is that required for the emission of a quantum by a single atom, about 10^{-15} second. In fact, however, emission by neighboring atoms within a conventional source tends to be coupled, so that the duration of emitted pulses is effectively prolonged. Accordingly, the emission from a limited region of the source may be quite highly monochromatic. Such emission may also be described as coherent. That is, a definite phase relation exists between different portions of the wavefront (lateral coherence) and between the successive phases at a given position over an extended period of time (time coherence). The terms "coherence" and "monochromaticity" describe the same property of electromagnetic radiation, although from different points of view; they are "two sides of the same coin".

It is possible to generate radio waves, in the long wavelength, low energy portion of the electromagnetic spectrum, with a high degree of monochromaticity and coherence. This property permits the formation of very sharp interference patterns and is the basis for the usefulness of radio waves in communications. In other terms, it may be said that radio waves exhibit *wave* properties to a high degree. For radiations of shorter wavelength, *quantum* properties, especially *randomness* in the emission of photons are more evident (cf. p. 30). In order to produce visible (or infrared) light of a high degree of coherence and monochromaticity, it is necessary to cause the atoms of the emitter to emit in phase with each other over an extended period of time. It is this which the laser achieves.

LASER ACTION

The principle of laser action depends upon the phenomenon of *stimulated emission*. As discussed in Chapter 7, atoms which are raised to an excited energy level by the absorption of a quantum of energy may return to lower energy levels by the emission of a quantum of somewhat lower energy content than that absorbed, that is, by the process of fluorescence. The diagramatic representation of fluorescence has been given in Fig. 7-16 and is shown in simplified form in Fig. 22-8. In that diagram, energy sublevels $(1, 2, 3, \ldots)$ are shown only for the excited state.

An electron, gaining energy $h\nu_1$ by absorption, loses part of its excess energy by internal conversion (i.e., by heat exchanges with adjacent atoms), falling rapidly from higher to lower energy sublevels of the excited state. It then emits a (fluoresced) quantum of energy $h\nu_2$. Fluorescence, which here may be termed *spontaneous emission*, commonly occurs after an excited lifetime of the order of 10^{-8} sec.

It can be shown that an alternative to the normal process of spontaneous emission is that of *stimulated emission*. If a quantum of energy of frequency identical with that which would be emitted spontaneously impinges upon the excited atom *during the lifetime of the excited state,* both quanta are emitted immediately and are in phase with each other. Each of the two quanta is then capable of stimulating emission by other excited atoms, producing four quanta in phase, and so forth.

The process of stimulated emission is in fact just as "normal" as that of spontaneous emission, but is normally not detected since (a) very few of the atoms of a medium normally exist in the excited state at any one time, and (b) the lifetime of the excited state is generally so short as to permit few interactions to precede spontaneous decay. Ideal conditions for stimulated emission occur when the emitting atoms are relatively well separated.

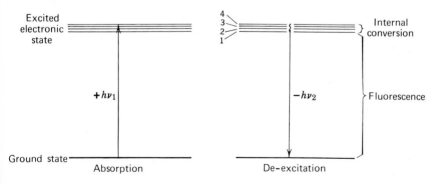

Fig. 22-8 Energy level diagrams.

Laser action depends upon the creation of a medium in which a substantial proportion of the active atoms exist in an excited state, and in which the lifetime of the excited state is sufficiently prolonged that the process of stimulated emission can become predominant. An initial spontaneous emission can then result in an amplified "cascade" of stimulated emission. The laser can be described as an amplifier in which spontaneous emission provides the driving signal.

An excited state of prolonged lifetime is called a *metastable* state. In quantum mechanical terms a metastable state is one from which transitions to the ground state, or to any other state of lower energy, are forbidden (i.e., are of low probability). Thus the lowest triplet states are metastable in atoms for which the ground state is singlet.

An appreciable or predominant proportion of the atoms on which laser action depends must exist in the metastable state simultaneously. This condition, termed a *population inversion*, may be contrasted to the ordinary distribution of the energy levels of atoms, in which the ground state predominates. The production of laser action thus depends on: (a) the availability of a medium in which a transition from a metastable state corresponds to quantum energy of the desired wavelength; (b) production of a condition of population inversion so that the active atoms of the medium can exist simultaneously at the metastable level. The process of applying energy to the medium in order to produce population inversion is termed *pumping*; (c) mechanical construction of a suitable *optical cavity* which favors amplification of the cascading beam of emitted photons. Typical construction of the optical cavity is that described for the ruby laser below. The term refers to the analogy between the optical reflecting system and the resonant cavities employed in radio wave engineering. The optical cavity differs from radio wave systems in that many cycles of vibration occur during one passage through the system.

The availability of suitable energy level transitions seemed at first to be a severely limiting factor in the development of lasers for visible and ultraviolet light. In recent years, however, a large number of possible transitions have been studied, and laser emission has been obtained at wavelengths as short as about 3000 Å. Further extension is provided by the generation of optical harmonics, as described below. Certain variations in the active medium of a laser (temperature, concentration, composition) may be used to produce variations in the emitted wavelength. Combination of all these factors opens the way, at least in principle, for the availability of laser radiation at any desired wavelength.

Three general categories of laser have been widely developed; they are the *solid*, *gas*, and *semiconductor junction* types. Possibilities for *chemical* lasers are being explored at present. The (solid) ruby laser is described in detail below, and the other types of laser are described more briefly.

THE RUBY LASER

Rubies are aluminum oxide crystals which contain varying proportions of chromium atoms as impurities, the characteristic red colors resulting from the chromium content. After the absorption of light quanta by the chromium atoms, a part of the excess energy is lost by internal conversion, the atoms decaying thereby to a metastable level which persists for more than 10^{-3} seconds or about 100,000 times as long as usual excited states. In the absence of a condition of population inversion, decay from the metastable level produces ordinary (i.e., not highly monochromatic) red fluorescent light at 6943 Å (at room temperature and for rubies of 0.05% chromium content). When these crystals are illuminated by light of only moderate intensity, population inversion is not achieved, and fluorescence occurs, with the spread of wavelengths implied by diagrams such as Fig. 7-16.

When the ruby is illuminated at extremely high intensities (i.e., by an electronic flash tube), population inversion occurs, and the process of stimulated emission becomes predominant. If the illuminated ruby is an ordinary crystal, however, the emitted quanta escape before a cascade of high intensity can accumulate. The ruby must therefore be so fashioned that the light makes many passages through the crystal, the intensity building up through stimulated emission during each passage. As shown in Fig. 22-9, the ruby crystal is machined into a rod which has polished plane ends. These ends are coated so as to be partially reflecting. (Or, one end is made totally reflecting and the other partially reflecting.) The machining of the rod is highly critical; plane surfaces must be optical flats, parallel to each other. Dimensions are as shown in Fig. 22-9.

After pumping by the intense flash, which is provided by a xenon lamp, some quanta are emitted spontaneously from excited atoms (after periods shorter than the average lifetime of the metastable excited state).

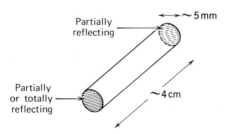

Fig. 22-9 Optical cavity of ruby laser.

Normally; that is, in the absence of a state of population inversion, these quanta would either be emitted as fluorescent light or would undergo reabsorption by atoms in the *ground* state. Here, however, the process of reabsorption is replaced by stimulated emission, the quanta so produced being extremely closely matched in frequency to the incident quanta. The quanta produced by stimulated emission stimulate further emissions from other excited atoms, and a beam of high intensity accumulates as the process continues. The intensity of the beam which escapes from the only partially silvered end(s) of the laser rod eventually becomes extremely high. The light path is greatly increased as a consequence of successive reflections, within the optical cavity, thus allowing the beam to continue traversing the rod until all the chromium atoms are returned to the ground state. The silvered surfaces also serve to collimate the beam, since quanta which do not travel parallel to the axis of the rod are removed from the beam after a number of reflections, as illustrated diagramatically by Fig. 22-10.

These considerations account for the high intensity and high degree of collimation of the laser beam. Monochromaticity and coherence arise from the fact that quanta produced by the process of stimulated emission must be very precisely in phase with each other. Thus all additions to the laser beam, as a whole, are at constant relative phase. (In practice, some departures from this condition are observed, but they are extremely slight in comparison to conventional light sources.) The high coherence of light from ruby and other lasers is easily demonstrated by pasting a pair of parallel slits over the face of the laser. Whereas double slit diffraction patterns can be obtained from conventional sources only by taking great care in the positioning of slits, etc., very sharp fringes are easily formed by extended laser beams.

The overall features of the electronic transition upon which ruby laser action depends are summarized by Fig. 22-11. The chromium atoms

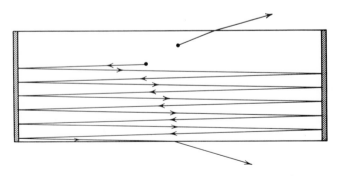

Fig. 22-10 Collimation of laser beam by successive reflections.

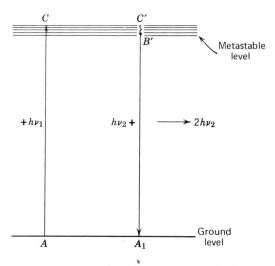

Fig. 22-11 Energy level transitions in the ruby laser.

in the ground level (A, A') absorb energy ($h\nu_1$) during pumping, thus making the transition to the excited state (CC'). Line $C'B'$ then represents a rapid heat exchange (internal conversion) within the crystal lattice, leading to the metastable excited state (B'). Whereas in ordinary fluorescent systems the decay $B'A'$ occurs rapidly, this process occurs spontaneously in the chromium atoms of the ruby only after a delay of a few milliseconds. The ruby laser uses the phenomenon of stimulated emission to precipitate *simultaneously* the decay $B'A'$ for all excited atoms.

Q-switching is a modification of the ruby laser which is frequently used to increase the peak power output. In this method a shutter is placed in front of the half-silvered exit surface of the laser and is held in position until the pumping flash from the xenon lamp is almost complete. While the shutter is in position, no amplified cascade can accumulate by successive passages through the rod and, thus, relatively little radiation is lost from the system. Accordingly, the metastable state is more fully occupied at the termination of pumping. Upon removal of the shutter, the cascade of stimulated emission builds up very rapidly, since essentially *all* the chromium atoms can now emit over a very short period of time. A time sequence representation of Q-switching is shown in Fig. 22-12. (In Fig. 22-12, atoms occupying the metastable state are represented by solid circles, and atoms in the ground state by open circles.) Note that it is the *peak* intensity rather than the *total power* of the laser beam which is increased by Q-switching.

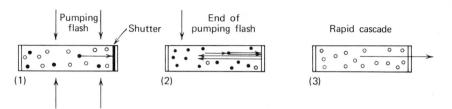

Fig. 22-12 Q-switching.

The ruby laser can be modified to produce continuous emission but, as the foregoing description implies, this device is primarily a source of pulses of illumination. Certain other types of laser are more suited to continuous operation.

OTHER TYPES OF LASERS

Solid phase lasers, of which the ruby is an example, have been developed which operate over the wavelength range from about 3100 Å to 2.5 μ. One of them is neodymium-containing glass, which produces radiation in the near infrared. This and the ruby are presently the best sources of highly intense radiation. Further amplification of laser intensity can be produced by causing the laser light to travel through a laserlike rod which is unsilvered. The second rod is pumped by an appropriately synchronized flash, and adds the emission stimulated from its atoms to the cascade produced by the original laser.

As stated above, the most favorable conditions for stimulated emission occur when the emitting atoms are well separated. Ruby and neodymium glass satisfy this requirement, since the active atoms are present only at low concentrations as impurities in media which do not interact with radiation of the wavelengths used for pumping. (In quantum mechanical terms, the aluminum oxide or glass matrices possess no energy level differences in the quantum energy range of the pumping light.) An alternative means of achieving suitable atomic separations is to use a *gas* as the active medium. This is done not only in the helium-neon gas laser, but also in the chemical lasers reported to date. Similarly, in the semiconductor junction type of laser, positions of active sites correspond to the distribution of trace impurities.

The helium-neon gas laser is pumped by means of electron bombardment. Thereby helium atoms are raised to an excited state from which energy cannot be lost radiatively, but only by collisional exchanges. The absorbed energy is thus transferred to neon atoms by collision. The neon atoms lose part of the energy by internal conversion, reaching a metastable energy level. Stimulated emission can then occur, causing the atoms to fall from the metastable level part way to the ground level. The final fall to ground energy level occurs in a rapid transition which does not contribute to laser action. The energy level transitions undergone by the two types of atom are diagramed in Fig. 22-13.

The laser wavelength of the He-Ne system is much more variable than that of emission by the ruby, being significantly altered according to both temperature and composition of the medium; thus these lasers can be "tuned" to a desired wavelength. Gas lasers produce a number of wavelengths in the range from 0.24 to 300 μ.

The semiconductor junction (*pn* junction) laser utilizes the energy changes associated with electron exchanges. It may be recalled that, whereas the electrons of insulating substances occur at rather well defined positions in a fixed matrix, electrical conductors contain many very loosely bound electrons. Semiconductors are an intermediate category of material in which a relatively small number of free electrons exists. Semiconductors can be described essentially as insulating crystals in which impurities produce either an excess or a deficit of electrons relative to the crystal lattice as a whole. For example, germanium, the prototype of semiconducting materials, has four valence electrons per atom; electronic forces are precisely balanced in a perfectly pure Ge crystal. When a trace of boron, which has only three valence electrons, is present in the germanium lattice, a relative deficit in electrons is introduced; the electron deficit is called a *hole*. Electrons may be attracted to the relatively positively charged boron nuclei, leaving behind a partial positive charge on the germanium nuclei. In this way, holes are said to migrate through the medium. Similarly, traces of arsenic atoms, with five valence electrons, produce *extra* electrons which are free to migrate through the lattice. The electron-deficient germanium crystal is called a *positive* or *p-type* semiconductor; the electron-rich crystal is a *negative*

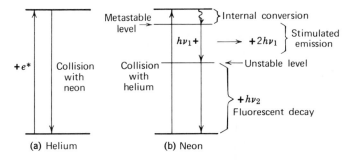

Fig. 22-13 Energy level transitions in the helium–neon gas laser.

or *n-type* semiconductor. Note that the spacing of the holes or free electrons in the semiconductor is that of the boron, arsenic, or other atoms present as impurities, and is thus in the range suitable for optimal stimulated emission.

The energy of free electrons is greater than that of electrons associated with atoms. The energy difference is released when a free electron is captured by a hole and must be absorbed for an electron to escape from an atom. In terms of energy level diagrams, free electrons of semiconductors occupy the conduction band level, as shown in Fig. 7-2.

In some types of semiconductor, of which gallium arsenide and gallium arsenide-gallium phosphide are examples, the energy released upon loss of a free electron can be emitted as a light photon. (In other types of semiconductor, including germanium crystals, the energy is absorbed by the crystal lattice in the form of heat.) When *p* and *n* types of semiconductor meet at a junction, and when a source of electrons is applied to the *n* side of the junction (i.e., when a suitably directed voltage is applied across the junction; in practice, the voltage must be sufficient to produce very high currents), recombination of electrons and holes proceeds at a high rate in the area of the junction. Illumination of the junction by a light beam of low intensity then triggers stimulated emission of laser radiation at the same wavelength.

A schematic diagram of a *pn* junction laser is given in Fig. 22-14. In this system, mirrors can be formed simply by grinding and polishing parallel plane surfaces of the semiconductor crystal. Depending upon composition and temperature, semiconductor junction lasers can be made to produce wavelengths in the range from about 4800 Å to 22 μ.

Although their exploitation is only beginning, chemical reactions provide, at least in principle, a very desirable source of laser action. Theoretically, the mixing of suitable reactants could result in the formation of products in which excited states could persist over an appreciable

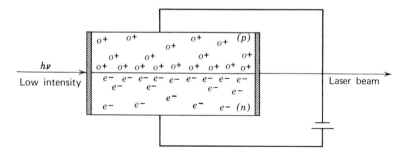

Fig. 22-14 The semiconductor junction laser.

period of time. A condition of complete population inversion would prevail initially, energy loss from the excited state then proceeding by the process of stimulated emission. In other words, the pumping energy would be provided by the energy content of the reactant. Emission would persist over a prolonged interval provided that reactant could be pumped into the system at an adequate rate. This hypothetical scheme is outlined by 22-u:

$$A + B \rightarrow (AB)^*,$$
(22-v)
$$(AB)^* + h\nu \rightarrow AB + 2h\nu.$$

As yet, chemical laser action has been achieved only in systems pulsed by flash photolysis. The first such system utilized the photodissociation of CF_3I, a reaction in which a metastable activated iodine atom is formed:

(22-w)
$$CF_3I \xrightarrow{h\nu} CF_3 + I^*.$$

The iodine subsequently radiates intensely at $1.3\,\mu$ by stimulated emission. Build-up of the amplified cascade is achieved by multiple reflections in the usual manner. This particular chemical laser reaction utilizes the energy of an electronic transition, but energies in the vibrational-rotational and pure rotational ranges might also be achieved in suitable systems. Chemical lasers thus may be applicable to a wavelength range extending from the ultraviolet far into the infrared.

NON-LINEAR EFFECTS INDUCED BY LASERS

An inherent assumption in the discussion of the properties of light presented thus far has been that the properties of transmitting media (refractive index, etc.) are not a function of the intensity of the transmitted light. That is, it has been assumed that optical effects are *linear*. The assumption is in fact correct at the intensities produced by conventional sources. Illumination with the most intense available laser beams, however, reveals a detectable degree of nonlinearity.

Nonlinear effects of the passage of light have been predicted in the past in view of the fact that optical properties of materials can be altered by imposing an extremely strong electric or magnetic field (cf. p. 117). Since light is itself a coupled oscillation of the electric and magnetic fields, it is to be expected that passage of beams of high energy content would induce temporary alterations in the properties of the medium. In fact, the refractive index is found to be a function of intensity at high intensities, and there results a distortion of transmitted sine wave forms. Application of the methods of Fourier analysis reveals that the distorted pulses can be analyzed as the resultant of the original (fundamental) frequency and its harmonics (wave motions whose frequencies are integral multiples of the fundamental frequency). The second harmonic (twice the frequency and half the wavelength) and third harmonic (three times the frequency and one third the wavelength) of the fundamental frequency have in fact been detected in beams originating from a ruby laser. That is, in addition to light of wavelength 6942 Å, radiation of 3471 Å and 2314 Å can be detected. Normally the intensity of the second harmonic is very low, and that of the third harmonic much smaller yet. Under optimal conditions, however, it is possible to convert about one fifth of the incident intensity from a laser into its second harmonic. The production of harmonics thus provides a means for the extension of laser operations to shorter wavelengths.

Addition and subtraction frequencies can be observed when intense monochromatic beams of different wavelengths pass through a medium in opposite directions. These frequencies are analogous to the "beat" frequencies which occur in the interactions of sound waves. Beat wavelengths of 305 mμ and 2.56 μ are produced, for example, by superposition of ruby laser light with the quite highly monochromatic mercury green line. Note that:

(22-x)
$$\nu_{Hg} = \frac{c}{5461 \text{ Å}} = 5.49 \times 10^{14} \quad \text{cycles per second,}$$

$$\nu_{ruby} = \frac{c}{6943 \text{ Å}} = 4.32 \times 10^{14} \quad \text{cycles per second,}$$

so that

(22-y) $\qquad \nu_{Hg} + \nu_{ruby} = (5.49 + 4.32) \times 10^{14} = 9.81 \times 10^{14} \quad \text{cps,}$

and thus

(22-z) $$\lambda_{addn} = \frac{c}{9.81} \times 10^{14} = 305 \text{ m}\mu.$$

Likewise,

(22-aa) $\qquad \nu_{Hg} - \nu_{ruby} = (5.49 - 4.32) \times 10^{14} = 1.17 \times 10^{14} \quad \text{cps,}$

so that

(22-bb) $$\lambda_{subtr} = \frac{c}{1.17} \times 10^{14} = 2.56 \text{ } \mu.$$

The generation of these and other beat wavelengths provide another means of extending the number of available laser wavelengths.

APPLICATIONS OF THE LASER

Until recently the greatest interest of the laser from the scientific point of view has been in the field of physics, since understanding of the propagation of light and the theory of the interactions of light with matter must now be refined in order to describe the properties of extremely intense and coherent beams.

Technological applications, most of them still potential, are numerous. In particular, it would be very desirable to be able to use the wide range of frequencies contained in the visible spectrum in the field of communications (telephone, television, etc.), since the amount of information which can be carried by a beam of radiation is determined by the range of frequencies it contains. Hitherto, available light beams have been too incoherent for use in this way. Laser beams possess the necessary degree of coherence and it may be expected that they will be applied to the problem of communications once a number of technical problems have been overcome.

The exploitation of the laser in biological studies is only beginning. The most interesting applications of this device are to diffraction experiments which, while previously feasible in principle, were rendered impracticable by the relatively low intensity and coherence of conventional sources. Thus, for example, the recording of optical diffraction patterns from electron micrographs generally required exposures of many hours, whereas the corresponding patterns formed by laser beams can be observed directly. Consequently, holographic (cf. p. 249) and optical filtering (cf. p. 447) methods are now undergoing rapid development. In particular, laserlike x-ray or electron beams would make possible the extension of the resolution limit of microscopes by holographic methods. At present, such devices do not exist, but the development of an "x-ray laser" is considered possible. The *destructive* effects of focused laser beams, in which high intensity is achieved over a very limited area, have been utilized; for example, for the microdissection of cellular organelles, the destruction of experimental tumors, and in retinal surgery. Focused laser beams could also be used for the evaporation of high melting point metals in shadow casting for electron microscopy. The particular advantage of all these applications is that the small size of the focused spot of light and the brief duration of the irradiation reduce damage to nearby structures.

PHOTOMETRIC UNITS

From the point of view of the scientist, as opposed to that of the illumination engineer, the intensity of light sources is usually expressed in terms of *power* output, that is, of energy per unit time. Thus a power output of one watt is equal to one joule per second $= 10^7$ ergs per second $= 0.239$ calories per second. Other units in which light intensity is described are summarized here. This system of units of illumination tends to be confusing because it is based on the psychological sensation of brightness as perceived by a hypothetical "standard" observed. Units of brightness may be converted into energy units, however, at a reference wavelength, for which 556 mμ has been chosen.

Power output, defined as above, is sometimes referred to as *radiant flux*. Radiant flux is distinguished from *luminous flux*, which is a measure of properties related to the sensation of brightness. The luminous flux at any wavelength is equivalent to the radiant flux times the efficiency of the eye at that wavelength relative to 556 mμ.

The unit of luminous flux is the *lumen*, which is correctly defined as the flux through a unit solid angle (i.e., $\frac{1}{4}\pi$ of the total flux) proceeding from a uniform point source with an intensity of one candle. The *candle*, in turn, is defined on an absolute basis as 1/60 of the intensity proceeding from 10 cm^2 of surface of an ideal black body radiator at a temperature of 2046°K. The lumen was formerly defined as 1/685 watt at 556 mμ or its equivalent (in terms of the standard curve of efficiency of perception) at any other wavelength. A flux of one lumen/steradian (i.e., per unit solid angle) was formerly defined as one candle; this quantity is now termed one *spherical candlepower*.

The *lambert* is a unit of brightness and is defined as $1/\pi$ candles per cm^2 surface of the source. *Illumination* is a measure of luminous flux incident per unit area on a surface and is usually expressed in lumens per square meter. Note the distinction between *brightness*, a property of the source, and *illumination*, a property of the illuminated surface.

REFERENCES

R. G. Greenler, in J. Strong, *Concepts of Classical Optics*, Freeman, San Francisco, 1958, pp. 580–595. This article, "Optical Filters," is informative and well illustrated. Filters for visible light, ultraviolet, and infrared are described.

F. A. Jenkins and H. E. White, *Fundamentals of Optics*, 3rd ed., McGraw-Hill, New York, 1957, Chapter 14 (Interference Involving Multiple Reflections).

J. M. Stone, *Radiation and Optics*, McGraw-Hill, New York, 1963, pp. 486–488. A section on "Amplification by Stimulated Emission. Masers and Lasers."

W. H. Miller, G. D. Bernard and J. L. Allen, *Science* **162** 760 (1968) "The Optics of Insect Compound Eyes." This article not only describes biological structures which act as interference filters, but also provides an excellent example of the application of several optical techniques to a biological problem of current interest.

A number of articles have appeared in *Scientific American* which describe various aspects of the design and applications of lasers. The following are of particular interest:

A. L. Schawlow, **209**, 34 (1963), "Advances in Optical Masers."
G. C. Pimentel, **214**, 23 (1966), "Chemical Lasers."

A. Javan, **217**, 238 (1967), "The Optical Properties of Materials."

A. Lempicke and H. Samelson, **216**, 80 (1967), "Liquid Lasers."

A. L. Schawlow, **219**, 120 (1968), "Laser Light."

D. R. Herriott, **219**, 141 (1968), "Applications of Laser Light."

E. N. Leith and J. Upatnieks, **212**, 24 (1965).

K. S. Pennington, **218**, 40 (1968).

CHAPTER 23

Monochromators

Monochromators isolate light of any desired wavelength from an illuminating beam which may contain a wide distribution of wavelengths. While the extreme spectral purity of light from monochromatic sources is rarely matched by the monochromator output, these devices have the advantage of versatility. The two basic types of dispersing element used in monochromators are the *prism* and the *diffraction grating*. Prisms separate light as a consequence of the variation of their refractive behavior as a function of wavelength, while diffraction gratings, as their name implies, make use of interference effects to separate light of different wavelengths. The properties of both types of dispersing elements are discussed in this chapter, and their application to biological emission spectroscopy is described. The material to be discussed in this chapter is based on that presented in Chapters 1 to 3.

DISPERSION

The term "dispersion" refers to the variation of any quantity as a function of wavelength; more specifically, the variation of refractive index with wavelength (i.e., $dn/d\lambda$) is implied. Materials are characterized by *dispersion curves* (i.e., the plot of refractive index as a function of wavelength), which must be obtained experimentally for each substance. While these plots have certain common features, their form cannot be predicted in detail theoretically. Characteristic features of dispersion curves of transparent dielectric materials may be summarized as follows.

1. The values of refractive index for all substances approach 1.0 at very *short* wavelengths. At long wavelengths the value of n^2 approaches ϵ, the dielectric constant of the medium.

2. Discontinuities in the dispersion curve occur in the vicinity of absorption maxima.

3. Refractive index increases as absorption maxima are approached from longer wavelengths. Since all optical materials absorb at some wavelength in the ultraviolet, this means that larger refractive indices are observed at shorter visible wavelengths.

4. The level of values of refractive index is generally higher at wavelengths close to an absorption maximum.

The forms of so-called "anomalous" and "normal" dispersion curves are shown in Fig. 23-1a. The normal curve (dotted line) shows a slow decrease of refractive index at longer wavelengths. The "anomalous" curve (solid line) experiences, successively, a minimum and maximum at wavelengths shorter and longer, respectively, than some value λ_0. λ_0 is a wavelength of maximum absorption. Anomalous dispersion is, in fact, the normal behavior of the dispersion curve *in the vicinity of an absorption maximum*. Since most optical materials are transparent at all wavelengths in the visible and near ultraviolet, this type of dispersive behavior is not commonly observed. Anomalous dispersion is not considered at greater length here; however, comparable anomalous dispersion of optical rotatory power, which is known as the Cotton effect, is described on p. 593.

(Normal) dispersion curves within the visible and near ultraviolet range are given in Fig. 23-1b for three typical optical materials. "High dispersion flint glass" is characterized by a generally high refractive index, as well as by large values of dn/dx. This material is particularly useful for the construction of achromatic lens doublets. Crown glass, a category of material from which most standard optical parts are constructed, is characterized by lower values of both refractive index and dispersion. Fluorite (calcium fluoride), a material used for optical parts which must transmit ultraviolet radiation, has a still lower refractive index. Its dispersion, which is very low at the longer visible wavelengths, becomes appreciable in the ultraviolet. Note that the three curves shown in Fig. 23-1b differ somewhat in shape; although they are generally similar, they cannot be superposed.

The quantity of *dispersive power* is sometimes specified as a convenient numerical measure of dispersion. It is defined as

$$(23\text{-}a) \qquad \text{dispersive power} \equiv \frac{n_F - n_C}{n_D - 1},$$

where n_F, n_C, and n_D are the refractive indices with respect to the 'F', 'C', and 'D' lines of the solar spectrum; that is, for wavelengths of

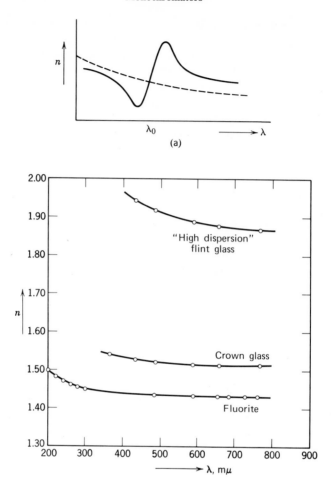

Fig. 23-1 (a) "Normal" and "anomalous" dispersion curves (Dotted line – normal; solid line – anomalous); (b) dispersion curves of optical glasses.

4861, 6563, and 5893 Å, respectively. In view of the variable shapes of dispersion curves, it is clear that the dispersive power is only a semi-quantitative index of dispersion in the visible range of the spectrum.

Note that, while Snell's law (p. 5) implies that refractive index is a constant, this is with respect to the angle of incidence of light, but at a fixed wavelength and at constant temperature and pressure of the medium. For many practical purposes, however, n may be very nearly constant over a considerable range of wavelengths.

PRISM MONOCHROMATORS

When a polychromatic light beam undergoes refraction, the dispersion of optical materials produces a physical separation of light of different wavelengths. This effect is exploited in prism monochromators, which are cut and illuminated at angles so chosen as generally to produce a maximum separation of the refracted components. Thus, in principle, the material used for a prism monochromator should have the largest possible dispersive power; unfortunately, since dispersive powers tend to be high only in the vicinity of absorption maxima, absorptive losses of light may limit the usefulness of highly dispersive materials.

Prisms used for the separation of wavelengths in the visible spectrum are made principally of optical glasses. In the ultraviolet range, quartz is used, while, in the infrared region, prisms are made of rock salt (NaCl), sylvite (KCl), potassium bromide, or cesium iodide. The latter materials provide satisfactory dispersion and transmission up to wavelengths of 15 or 20μ; other materials are available for the study of longer wavelength radiations.

Note that dispersive effects occur in other types of optical instrument, where the separation of refracted wavelengths is not desirable. Such systems are designed to minimize dispersion. The correction of chromatic aberration in optical lenses is a notable example (cf. p. 217).

A diagram of a typical prism monochromator is shown in Fig. 23-2. Light from a source is isolated by a slit and is incident upon a collimating lens. The aperture of the system is sufficient for illumination of the *entire* prism face by the collimated beam, a condition which ensures optimum resolution of component wavelengths (cf. p. 518). In Fig. 23-2, refraction of light of two wavelengths is shown. Light of a longer wavelength λ_1, is represented by solid lines; light of a shorter wavelength λ_2 is shown by

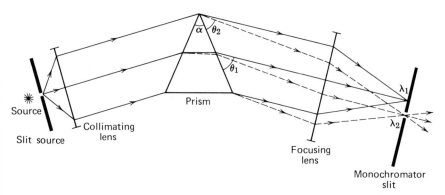

Fig. 23-2 Prism monochromator.

dotted lines. Note that the refractive index of the prism is larger with respect to λ_2 than with respect to λ_1. Rays leaving the prism are focused by a lens, forming images of the slit source in the plane of the second slit (the monochromator slit). The widths of the two slits, which are variable, and commonly equal, determine the *spectral purity* of the transmitted light. Thus the monochromator slit shown in Fig. 23-2 transmits rays of some wavelength range from $(\lambda_2 - \Delta\lambda)$ to $(\lambda_2 + \Delta\lambda)$. Reduction of the slit width eliminates part of this distribution. Spectral purity is defined quantitatively in terms of the *bandwidth*, which is the range of wavelengths for which the transmitted intensity is equal to or greater than half the intensity of the nominal wavelength λ_2.

The relative position of the monochromator slit in the focal plane determines the wavelength which is transmitted. (In practice, the slit is usually stationary, while the dispersing element is rotated.)

Many types of prism have been designed for use in various types of optical instruments; no attempt is made here to catalog these. Light beams may be inverted or redirected by the prism through any chosen angle or may be deviated to a variable degree. Silvered surfaces and/or the phenomenon of total internal reflection (p. 6) may be used for the redirection of light. For symmetrical prisms, a *prism angle* (α) may be specified; for example, the prism shown in Fig. 23-2 is a 45° prism. Note that the term "prism" is not applied to monochromators exclusively but is used for *any* device which consists primarily of a block of refractive material. An example is the Nicol prism (p. 98) which is used to produce plane-polarized light beams.

Important properties of a prism monochromator are its *deviation*, *angular dispersion*, and *resolving power*. Deviation specifies the angle through which incident light of any given wavelength is deviated by the prism. Illumination of prisms at an angle of incidence corresponding to minimum deviation is desirable. Angular dispersion (p. 514) is the quantity $(d\theta/d\lambda)$; that is, the difference, with wavelength, in the *angle* at which beams of different wavelength are refracted from the final face of a prism. The resolving power (p. 516) specifies that separation of wavelengths which can just be distinguished in the light refracted by a prism monochromator.

DEVIATION; CONDITION OF MINIMUM DEVIATION

The deviation by a prism is usually expressed as that of the sodium D line, which occupies a more or less central position in the visible spectrum.

Figure 23-3 shows the incidence of a ray of light upon a prism which is

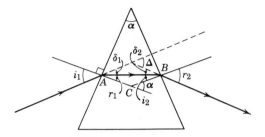

Fig. 23-3 Deviation by a prism.

characterized by the interfacial angle (apex angle) α. Note that, as shown, the angle between the normals to the sides of the prism is $(180-\alpha)°$. The deviation produced by the prism is, as shown, the angle Δ, between the direction of the incident ray and the direction of the ray leaving the prism.

In Fig. 23-3 a ray is shown incident upon the first face of the prism at angle i_1 which is refracted there at angle r_1, thus undergoing a deviation of δ_1. This ray is incident upon the second face of the prism at angle i_2, and undergoes a deviation of δ_2 there, thus leaving the prism at angle r_2. The algebraic sum of δ_1 and δ_2 constitutes the deviation by the prism; that is,

$$(23\text{-b}) \qquad\qquad \delta_1 + \delta_2 = \Delta.$$

It is readily shown by experiment that the deviation by a prism varies with the angle of incidence, and that there is a single minimum, occurring at an angle of incidence which is determined by the refractive index of the prism. That is, invariably, there is only one angle of incidence which effects minimum deviation in a given prism. From this observation it may be deduced that the angle of incidence at the first face must equal the angle of refraction at the second face if minimum deviation is to occur. In terms of the notation just cited, this condition is: $i_1 = r_2$. Note that, if this were not the case, the principle of reversibility (p. 3) would demand that *two* different angles of incidence produce minimum deviation.

Since, at minimum deviation, $i_1 = r_2$, then also the angle of refraction at the first face must equal the angle of incidence at the second; that is,

$$(23\text{-c}) \qquad\qquad r_1 = i_2.$$

It may be seen from Fig. 23-3 that α is the external angle of triangle ABC, of which r_1 and i_2 are the opposite interior angles. Therefore,

$$(23\text{-d}) \qquad\qquad \alpha = r_1 + i_2 = 2r_1 \quad \text{or} \quad r_1 = \tfrac{1}{2}\alpha.$$

Substitution of (23-d) into Snell's law then gives an expression for the angle of incidence corresponding to minimum deviation:

(23-e)
$$\sin i_{\text{min dev}} = \frac{n' \sin (\frac{1}{2}\alpha)}{n} ,$$

where n' is the refractive index of the prism, and n that of the surrounding medium.

In practice, the condition of minimum deviation is determined experimentally; its measurement is a sensitive method for obtaining the refractive index of the prism material with respect to the wavelength of light employed (cf. p. 386). In so doing, the prism angle (α) and the observed angle of minimum deviation are measured. n', the refractive index of the prism material, is then obtained from (23-g) below.

From Fig. 23-3 it is seen that

(23-f)
$$i_1 = r_1 + \delta_1 ,$$

while, at minimum deviation, $\delta_1 = \delta_2 = \frac{1}{2}\Delta$. Substitution of (23-d) and (23-f) in Snell's law thus gives

(23-g)
$$\frac{n'}{n} = \frac{\sin i_1}{\sin r_1} \quad \text{or} \quad \boxed{\frac{n'}{n} = \frac{\sin \left[\frac{1}{2}(\alpha + \Delta_{\text{min}})\right]}{\sin (\frac{1}{2}\alpha)}} .$$

Prisms used in optical instruments are set at an angle, determined experimentally or in accordance with (23-e), which gives minimum deviation of a representative wavelength. This has the effect of minimizing deterioration of the image (spectrum) which may result from imperfect collimation of the illuminating beam.

ANGULAR DISPERSION

Angular dispersion specifies the rate of change, with wavelength, of the angle of emergence θ of light rays from a prism (cf. Fig. 23-2). While θ varies according to the angle of incidence, the common use of prisms at a condition of minimum deviation provides a standard for expressing angular dispersion. Thus,

(23-h)
$$\text{dispersion} \equiv \left(\frac{d\theta}{d\lambda}\right)_{\Delta_{\text{min}}} .$$

Angular dispersion may be analyzed as the product of two factors: $d\theta/dn$, the rate of change of θ with refractive index, which is a function of prism geometry, and $dn/d\lambda$, which is the dispersion of the prism material, as discussed above on p. 508.

Some numerical values of $dn/d\lambda$ for a typical optical glass are given in Table 23-1.†

Table 23-1   Refractive Index and Dispersion of Borosilicate Crown Glass

Wavelength	n	$(-dn/d\lambda) \times 10^5$
399mμ	1.525	1.26
434	1.521	1.00
486	1.517	0.72
509	1.515	0.63
534	1.514	0.55
589	1.511	0.41
644	1.509	0.32

The factor $d\theta/dn$ may be determined from the geometry of the prism. In terms of the angles designated in Fig. 23-3, Snell's law specifies, for refraction at the second face of a prism of index n, located in air,

$$(23\text{-}i) \qquad \frac{\sin i_2}{\sin r_2} = \frac{1}{n}.$$

The variation of r_2 with n at any (constant) angle of incidence is, as obtained by differentiation of (23-i):

$$(23\text{-}j) \qquad \frac{dr_2}{dn} = \frac{\sin i_2}{\cos r_2}.$$

For a fixed value of i_2, dr_2 must be identical with $d\theta$. Thus, substituting for i_2 as obtained from (23-c) and (23-d),

$$(23\text{-}k) \qquad \left| \frac{d\theta}{dn} = \frac{\sin (\frac{1}{2}\alpha)}{\cos r_2} \right|,$$

where α is the prism angle and n the refractive index of the prism.

While (23-k) specifies $d\theta/dn$ for the prism, it is generally more convenient to measure lengths rather than angles. If, as shown in Fig. 23-4, B is the length of the prism base, s the length of the (illuminated) side, and b is the cross-sectional width of the emerging beam, then

$$(23\text{-}l) \qquad \sin (\tfrac{1}{2}\alpha) = \frac{B}{2s},$$

$$(23\text{-}m) \qquad \cos r_2 = \frac{b}{s}.$$

†Data from Jenkins and White, *Fundamentals of Optics*, 3rd Ed., p. 465.

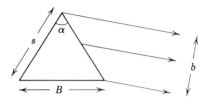

Fig. 23-4.

Substitution of (23-l) and (23-m) in (23-k), with cancellation of the common factor s gives

(23-n)
$$\frac{d\theta}{dn} = \frac{B}{b}.$$

The complete expression for angular dispersion by a prism is thus

(23-o)
$$\boxed{\frac{d\theta}{d\lambda} = \frac{B}{b}\frac{dn}{d\lambda}}$$

Note that, since $dn/d\lambda$ is not linear, *angular dispersion by a prism is not linear*; angular dispersion varies in no regular fashion with wavelength. This property is made obvious by observing the wavelength scale of any optical instrument which uses a prism monochromator. Settings are "crowded together" at the long wavelength end of the scale, but widely spaced at short wavelengths.

THE RESOLVING POWER OF A PRISM

In a prism monochromator, light of each wavelength forms an image of the slit source in the plane of the monochromator slit. The distribution of intensities in each of these images is that of a single slit diffraction pattern, as discussed on p. 56 and illustrated in Fig. 3-18. Each image of the slit source is thus of *finite* width, and adjacent images tend to be confused by the overlapping of intensities. Individual wavelengths are just resolved when the corresponding slit images are just distinguished as separate. In prism spectroscopy, as in microscopy, the Rayleigh criterion (p. 243) states that the two (line) images are just resolved when the central maximum of the diffraction pattern of one coincides with the first minimum of the diffraction pattern of the other. The criterion is illustrated by Fig. 23-5, in which light of wavelength $(\lambda + \Delta\lambda)$ (dotted lines) forms a central maximum which coincides with the first minimum of the diffraction pattern formed by light of wavelength λ (solid line).

The resolving power of a monochromator is expressed mathematically by the quantity $(\lambda/\Delta\lambda)$, that is, as the approximate wavelength divided

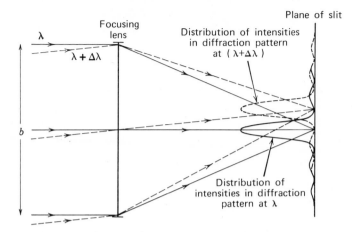

Plane of slit

λ

Focusing lens

λ + Δλ

Distribution of intensities in diffraction pattern at (λ+Δλ)

b

Distribution of intensities in diffraction pattern at λ

Fig. 23-5 Application of the Rayleigh criterion to monochromator output.

by the smallest resolvable difference in wavelength. For example, a prism which is just capable of resolving the sodium D line doublet (5890 and 5896 Å) has a resolving power of $5893/6 \cong 982$. Note that a *high* value of $(\lambda/\Delta\lambda)$ corresponds to a *high* resolving power. This convention is different from that of microscopy, where *small* values of d_{min} correspond to high resolving powers.

The derivation of an expression for the resolving power of a prism is considered in Fig. 23-6. As shown, CA is a wavefront of light of wave-

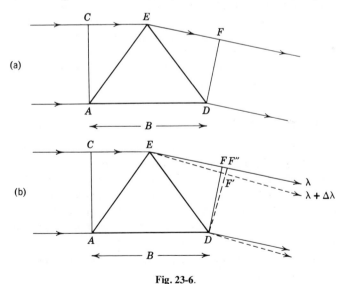

(a)

C   E   F

A   D

B

(b)

C   E   F F″   F′   λ

λ + Δλ

A   D

B

Fig. 23-6.

length λ incident upon a prism, while FD is a wavefront leaving it. $AD = B$, the prism base (cf. Fig. 23-4). The optical paths DA and CEF must be equal. Thus,

(23-p) $nB = 1.0 \times (CE + EF)$.

In Fig. 23-6b, the refraction of beams of wavelengths λ and $(\lambda + \Delta\lambda)$ are compared. CD is, as before, the incident wavefront, while the refracted wavefront for light of wavelength $(\lambda + \Delta\lambda)$ is now DF' or, approximately, DF''. Equation of the optical paths for this wavelength then gives

(23-q) $(n + \Delta n)B = 1.0 \times (CE + EF'')$.

Subtraction of (23-p) from (23-q) gives

(23-r) $\Delta nB = FF''$.

If the image formed by light of wavelength $(\lambda + \Delta\lambda)$ is just resolved, the distance FF'' must be just one wavelength since, as shown on p. 58, this is the condition for extreme rays which contribute to the first minimum of a single slit diffraction pattern. Thus,

(23-s) $\Delta nB = \lambda$,

or, dividing each side of (23-s) by $\Delta\lambda$ and equating $\Delta n/\Delta\lambda$ to $dn/d\lambda$,

(23-t) $$\boxed{\frac{\lambda}{\Delta\lambda} = \frac{dn}{d\lambda}B}.$$

Thus the resolving power of a prism is the product of the width of its base and the dispersion of the prism material. Since the value of $dn/d\lambda$ varies with wavelength, the resolving power of prisms also is different for different spectral regions. Note that, if the prism is incompletely illuminated, the effective width of the base decreases, and the resolving power deteriorates correspondingly.

GRATING MONOCHROMATORS

The description of the diffraction grating given on p. 59, may be summarized. The grating is a closely spaced array of parallel equidistant slits or rulings. Path and phase differences between rays which originate at the successive rulings vary according to the angle at which the light leaves the grating. When the beams diffracted by the grating are reunited by a lens, maxima and minima of intensity are thus formed as a consequence of interference between rays from successive slits which have traveled in any one direction. Angular positions of maxima and minima are given (for normal illumination) by

(3-tt) $$\text{maxima at } m\lambda = d \sin \theta,$$

(3-uu) $$\text{minima at } (m - \tfrac{1}{2}\lambda) = d \sin \theta,$$

where θ is the angle between the diffracted wavefronts and the grating, d is the spacing between rulings, and m is any integer. Successive *orders* of diffraction maxima, corresponding to the series of integral values of m, are formed. Because a large number of beams originate at the successive rulings, these diffracted maxima are extremely sharp.

The central or zero order maximum coincides for all wavelengths incident on the grating, but the subsequent maxima vary in position according to wavelength. Thus any maximum of a given wavelength can be isolated by a suitably positioned slit. Figure 23-7 is a diagram of a grating monochromator showing one each of the two first order beams produced by light of two different wavelengths. (It is convenient to illustrate a transmission grating instrument; in practice, however, reflection gratings are normally used.)

Grating monochromators, like prism monochromators, serve as spectrographs when the intensity passing through the monochromator slit is measured as a function of wavelength.

The deviation of beams produced by a grating is specified by (3-tt). The angular dispersion and the resolving power of grating monochromators are considered next.

ANGULAR DISPERSION OF THE GRATING MONOCHROMATOR

An expression for the angular dispersion $(d\theta/d\lambda)$ produced by a diffraction grating may be found by differentiating the grating equation (3-tt) to obtain

(23-u) $$\frac{d\theta}{d\lambda} = \frac{m}{d \cos \theta}.$$

Equation 23-u shows that angular dispersion increases with increasing

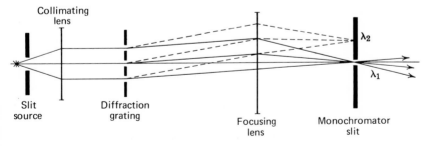

Fig. 23-7 Grating monochromator.

spectral order *m* and with diminishing grating spacing *d*. An important feature of (23-u) is that (cos θ) and thus ($d\theta/d\lambda$) are relatively insensitive to changes in θ when θ is not large. Thus the angular dispersion produced by a grating monochromator is very nearly *independent* of wavelength. This property of gratings contrasts markedly with the extreme and irregular variation of ($d\theta/d\lambda$) with wavelength as produced by prism monochromators. The diffraction thus is said to form a "normal spectrum" (rational spectrum).

RESOLVING POWER OF THE DIFFRACTION GRATING

The resolving power of a diffraction grating, like that of a prism, is expressed by the quantity ($\lambda/\Delta\lambda$) and can be evaluated by applying the Rayleigh criterion to the consideration of the overlapping images of the slit source which are formed by light of adjacent wavelengths.

If the Rayleigh criterion is just satisfied, the first minimum of the diffraction pattern formed by light of wavelength λ must coincide with the central maximum of the pattern formed by light of wavelength ($\lambda + \Delta\lambda$). The path differences between rays from the extreme top and bottom of the diffraction grating must be equal at the point of coincidence; equation of the path differences for light of the two wavelengths leads to an expression for the resolving power of the grating.

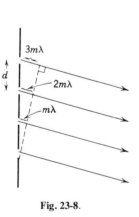

Fig. 23-8.

Figure 23-8 shows parallel beams of light of wavelength λ leaving the spacings of a grating in such a direction as to form the central maximum of the diffraction pattern of the respective order. As shown, the difference in paths between rays from two adjacent spacings is $m\lambda$, the cumulative difference between rays from three spacings is $2m\lambda$, and so forth. Thus the cumulative difference in path between rays from the top and bottom of the grating is $(N-1)m\lambda$, where N is the total number of spacings in the grating (i.e., the total number which is illuminated). Since N is always a very *large* number, this path difference can be expressed, to a good approximation, as $Nm\lambda$.

For rays of light of wavelength λ diffracted through a slightly different angle toward the *first minimum* of the diffraction pattern, the cumulative difference in path must be increased by *one wavelength* (as explained on p. 58). Thus Δ, the path difference between the extreme rays to the first minimum of the diffraction pattern for wavelength λ is

(23-v) $$\Delta = Nm\lambda + \lambda.$$

Δ', the cumulative difference in path to the first maximum for light of wavelength $(\lambda + \Delta\lambda)$ must be, in accordance with the same considerations,

(23-w) $$\Delta' = Nm(\lambda + \Delta\lambda).$$

The values of Δ and Δ' are identical when the diffraction patterns formed by light of the two wavelengths can just be resolved. Thus,

(23-x) $$Nm\lambda + \lambda = Nm(\lambda + \Delta\lambda).$$

Equation 23-x reduces directly to an expression, (23-y), which specifies the resolving power of the grating:

(23-y) $$\boxed{\frac{\lambda}{\Delta\lambda} = Nm}.$$

Resolving power is thus *independent* of the grating spacing d but is determined by the *total number* of lines N which are illuminated, as well as by the spectral order.

OVERLAPPING ORDERS

The formation by diffraction gratings of several orders of spectra may be advantageous in that improved resolution is achieved in the higher orders, as indicated by (23-y). However, the *overlapping* of light of different orders and wavelengths may interfere with attempts to isolate monochromatic light beams. For example, evaluation of (3-tt) shows that the second order of the Fraunhoffer C line ($\lambda = 6563$ Å) coincides with the third order of blue light of wavelength 4375 Å. Isolation of either wavelength may then be achieved with the use of an appropriate filter.

Note that the overlapping of orders is not a problem when only the first order spectra of visible light are observed. Thus the second order spectra of the shortest visible wavelengths (about 4000 Å) are diffracted through larger angles than are the first order spectra of the longest visible wavelengths (about 7000 Å).

When gratings are used as *spectrometers* (i.e., for high resolution analysis of the spectral composition of light beams), interference by over-lapping orders may be a serious problem. When the use of absorption filters is not feasible, the overlapping beams may be separated as first order reflections from a second instrument. Since *first* order spectra are normally the most intense (for an exception, see the discussion of blazed gratings below), overlapping orders are rarely a source of difficulty in the operation of *monochromators*.

BLAZED DIFFRACTION GRATINGS

At the surface of a diffraction grating, light is scattered in all directions so that the intensity of illumination produced in any one order may be low. This difficulty may be overcome by appropriate *shaping* of the individual grooves of the grating, a process called "blazing". As shown in Fig. 23-9, the blazed grooves are so sloped that reflection (which occurs, as at any surface, at an angle equal to that of incidence) produces a beam which leaves the grating in an essentially single direction. The *order* of the reflection so produced is determined by the angle of incidence of the illuminating beam.

Although the blazing of typical, very small, grating spacings is difficult, the same principle can be applied to larger spacings. *Echelon gratings* and similar devices which are used in high resolution spectroscopy have relatively large but precisely shaped spacings. Light may thereby be diffracted into very high orders, so that extremely small differences in wavelength can be resolved.

APPROPRIATE SPACINGS: PRODUCTION OF DIFFRACTION GRATINGS

In diffraction gratings which are used as monochromators, the grating spacing d should be such as to provide a good separation of wavelengths in the first diffracted order. In this way, interference from overlapping orders is eliminated. From (3-tt),

$$(23\text{-}z) \qquad\qquad d = \frac{m\lambda}{\sin\theta}.$$

Thus, for $m = 1$ and $\sin\theta \cong 1$, the appropriate grating spacing is approximately equal to the wavelength of the light to be isolated. Useful gratings commonly are ruled with about 10,000 to 50,000 lines per inch and thus have spacings in the range between about five and one wavelengths of visible light. Since $\sin\theta$ cannot be greater than one, the highest spectral order which can be observed for any wavelength is given by the next integer less than the quantity (d/λ).

The production of accurately ruled gratings with appropriate spacings

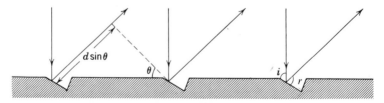

Fig. 23-9 Reflection from a blazed diffraction grating.

is technically formidable but can be achieved routinely under ideal conditions. Rulings are made by a diamond point, and the ruling engine is meticulously shielded from mechanical vibrations and from variations in temperature or humidity. As (23-y) shows, it is the total number of (evenly ruled) lines, rather than the value of d, that is of greatest importance in the production of gratings of high quality. Irregularities in ruling and uneven spacings must be avoided, since they may produce false maxima (called "ghosts") at positions other than those predicted by (3-tt). Many of the diffraction gratings now used in optical instruments are not produced directly by ruling engines but are copies (replicas) of such rulings.

COMPARISON OF PRISM AND GRATING MONOCHROMATORS

Both prism and grating monochromators are widely used in optical instruments at present. As the technology of the production of gratings has improved, the proportion of this type of instrument has tended to increase. Practical considerations related to availability favor the design of grating monochromators. The use of synthetic materials has extended the range of prism materials and the feasible size of individual prisms, but the high cost of polishing to the necessary tolerances tends to be prohibitive.

Diffraction gratings are advantageous in that they produce a rational spectrum (cf. p. 520), whereas prisms produce an irrational spectrum. This is important when electronic recording systems are to be coupled to the monochromator. Recording systems used with prisms require the use of a special cam which compensates for the nonlinear dispersion curve of the prism material. Light losses from absorption are substantial in prism monochromators, especially in spectral regions where dispersion by the prism is high. While some light loss occurs at grating surfaces, both by absorption and by scattering into unused orders, loss of intensity is usually a less serious problem in grating instruments. The resolving power of grating monochromators is generally greater than that of prisms of reasonable size.

Prism monochromators are free of the interference by higher order spectra or ghost lines or both which may, in extreme cases, lead to misinterpretation of data obtained from grating instruments. Diffraction gratings, replica gratings especially, must be handled with considerable care. While the prisms used for infrared spectroscopy are made of hygroscopic materials, and must be maintained in an atmosphere of controlled humidity, prisms used in the visible and ultraviolet regions of the spectrum are rugged and long lasting.

EMISSION SPECTROMETERS AND EMISSION SPECTROSCOPY

Thus far in this chapter, prisms and diffraction gratings have been discussed almost exclusively from the point of view of their roles as sources of monochromatic light beams. However, the same dispersing elements are also important as means of analyzing the wavelength composition of light beams, that is, as spectrometers. Thus the prism and grating monochromators illustrated in Figs. 23-2 and 23-7, respectively, may be converted into spectroscopic systems by replacing the monochromator exit slit by a viewing lens, or by a photographic plate or other recording system. Simple instruments in which the dispersed spectrum is viewed by eye are called *spectroscopes*, whereas high resolution instruments are generally termed *spectrometers*. Instruments which record spectra photographically may also be known as *spectrographs*.

The function of spectrometers is to measure the wavelengths of components of light beams. Since atoms of every type emit a characteristic spectrum upon excitation (as discussed in Chapter 7), spectrometers can be used (a) to study the fine structure of atoms in terms of their energy level transitions, and (b) in analysis, for the determination of the atomic composition of specimens. The first class of application, which in general demands the use of instruments of the highest possible resolving power, is beyond the scope of this book and is not discussed further. Analytical applications, however, are of considerable direct importance in biology. Spectroscopy in general provides a far more convenient and accurate method for determining the presence or concentration of small amounts of any element than do standard "wet chemistry" methods. Furthermore, emission spectroscopy has the particular advantage that many elements may be analyzed simultaneously in a small portion of a sample.

In brief, the spectroscopic method consists of the *excitation* of emission by the sample, the *dispersion* of the emitted wavelengths by a system of suitable geometry, the *recording* of the spectrum on a photographic plate (or by a photoelectric system), and the *analysis* of this record by comparison with the spectrum of a standard. Each of these aspects of emission spectroscopy is discussed briefly.

The nature of the *excitation* process determines the type of spectrum obtained. In general, the greater the energy supplied to the sample, the greater is the number and intensity of the emitted wavelengths. Excitation potentials for alkali and alkaline earth metals (i.e., the sodium and calcium series of elements) are relatively low, and spectra may be excited by many types of flame. The *flame photometer* is thus a spectrograph which excites emission from sodium, potassium, or other such elements in a flame of relatively low but controlled temperature. Classical types of spectrograph make use of more energetic arc or spark discharges for the excitation of the emission lines from a much larger group of elements; in these instruments the sample is usually supplied in the form of a liquid or powder. At the present time, however, *laser probe* spectrographs are being developed in which a very small spot (diameter of the order of microns) of intense laser light is focused on a specific area of a continuous solid specimen. With the use of such an instrument, it should, for example, become feasible to study the elemental composition of selected single cells, or even of cellular organelles. A similar principle is applied in the *electron microprobe*, in which a focused electron beam is used to excite characteristic *x-ray* spectra from selected small areas of specimens.

The mounting of the dispersing element and other optical components of spectrometers critically affects the sharpness and thus the resolution of the emitted lines. Prism instruments of the type diagramed in Fig. 23-2 are unsuitable for high resolution spectroscopy, since they require the use of refracting lenses for collimation and focusing, so that, even when highly corrected lenses are employed, chromatic aberration tends to limit spectroscopic resolving power. This difficulty is eliminated in spectroscopes in which a concave reflection grating combines the functions of dispersing element and focusing lenses. As noted on p. 218, reflection lenses are entirely free of chromatic defects, since the angle of reflection is independent of the refractive indices of media, and thus of the wavelength of the reflected light. A less desirable feature of reflection lenses is that they tend to form astigmatic images except when illuminated close to normal incidence. However, astigmatism is minimized in the *Wadsworth mounting* which, for this reason, is probably the most widely used type of spectroscopic mount. Nevertheless, high resolution prism spectrometers can also be built using mirror optics, exclusively, to effect focusing.

As shown in Fig. 23-10, in the Wadsworth mount, light from the specimen passes through a slit (the slit source) placed at the focal point of a concave mirror. The mirror thus serves to illuminate the concave grating with a collimated light beam. Light dispersed by the grating is then imaged by the focusing action of the curved grating, forming a series of images of the slit source on the photographic plate. The series of spectral lines formed in this type of mount are found to lie along a parabola, so that an entirely sharp spectrum cannot be formed on a flat plate. Usually, the lines are recorded on a film which lies in a suitably curved cassette. It may be noted that the radius of curvature of the grating and concave mirror which are used in Wadsworth spectrographs is of the order of one to ten meters; the curvature is so slight as to be almost undetectable upon direct inspection.

Other common types of spectroscopic mount make use of the *Rowland*

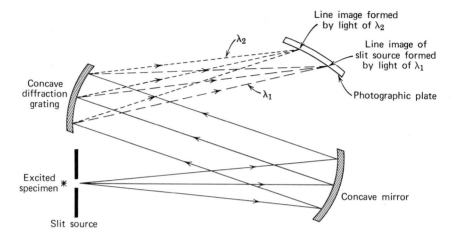

Fig. 23-10 The Wadsworth mount.

circle, which is illustrated by Fig. 23-11. As shown, the Rowland circle is *tangent* to the center of the concave diffraction grating, while its *diameter* is equal to R_g, the *radius* of curvature of the grating (i.e., the radius of the Rowland circle is half the radius of curvature of the grating). It is found that if the effective source of light (i.e., the slit source) lies anywhere on the Rowland circle, the dispersed spectrum is also brought to focus at positions lying along this circle.

The photographic record formed in a spectrograph consists of a row of lines (images of the slit source) of varying intensities. For qualitative analysis the wavelengths of the light forming these lines are determined, and sets of lines are identified as being associated with emission by a particular element. In quantitative analysis the photographic densities of the lines are also measured and are related to a standard calibration curve for each element. For accurate work, use of an internal standard is essential in order to account for variations in photographic processing or of the

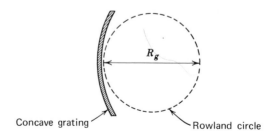

Fig. 23-11 The Rowland circle.

amount of specimen excited. A compound of iron is generally used as the standard, since this element provides a rather regularly spaced series of readily identifiable lines over a wide range of wavelengths in both the ultraviolet and visible portions of the spectrum. Thus identification of the iron emission lines provides a wavelength scale for the entire image. Availability of a series of calibration lines at rather short wavelength intervals is important, since it is found that the accuracy of grating spectrographs is greatest when these instruments are used to measure small differences between a known and an unknown wavelength, rather than for determinations of the absolute wavelength of isolated emission lines.

While the analysis of emission spectra is thus simple in principle, it must be understood that high resolution spectra contain very many lines, so that considerable skill is required in order to distinguish the set of lines emitted by any one element. Furthermore, the densitometry of very many emission lines may be an extremely tedious process. For large scale operations these difficulties can be overcome by the use of automated systems, in which photoelectric recordings are made of the intensities of a relatively small number of the characteristic emission lines of the element to be analyzed. These data may then be processed in such a way as directly to provide numerical values for the concentrations of the corresponding elements.

REFERENCES

F. A. Jenkins and H. E. White, *Fundamentals of Optics*, 3rd ed., McGraw-Hill, New York, 1963, Chapter 23 (Dispersion), Chapter 2 (Plane Surfaces) (this chapter discusses many of the properties of prisms), pp. 301–302 (a section on "Chromatic Resolving Power of a Prism"), and Chapter 17 (The Diffraction Grating).

R. P. Madden and J. Strong, in J. Strong, *Concepts of Classical Optics*, Freeman, San Francisco, 1958, pp. 596–615. This article, "Diffraction Gratings," develops equations for predicting the efficiencies of plane diffraction gratings in their various orders of spectra and evaluates the aberrations of the concave grating.

Strong, *op. cit.*, pp. 212–215. A section, "Spectroscopic Resolving Power," which includes a diagrammatic comparison of spectrographs, spectroscopes, and monochromators.

R. W. R. Baker and R. H. S. Thompson, in H. J. B. Atkins, Ed., *Tools of Biological Research*, Blackwell, Oxford, 1959, pp. 11–25. An article, "Flame Photometry," which describes equipment, techniques, and applications.

P. Connes, *Sci. Am.* **219**, 72 (1968). This article, "How Light Is Analyzed," emphasizes current techniques in spectroscopy, although with particular reference to astronomy, rather than chemistry or biology.

CHAPTER 24

Absorption and Spectrophotometry

The term "spectrophotometry" is used to refer to the measurement of light absorption as a function of wavelength. Three categories of spectrophotometric measurement may be described: (a) quantitative analysis of substances of known spectral characteristics, by means of the measurement of absorption at a suitable wavelength; (b) qualitative analysis, during which absorption spectra are determined and identified; (c) deduction of chemical structure by the recording and analysis of absorption spectra.

In this chapter quantitative relationships between absorption and the concentration of absorbing substances are derived and discussed. Absorption measurements of biological interest are usually those of solutions; while spectrophotometry may be carried out with gases or solids also, the study of solutions is principally considered here. The design, elements, and characteristics of spectrophotometers, as used for biological studies, are described, and the nature of spectrophotometric measurements is considered. The determination of molecular structure by analysis of infrared absorption spectra is a topic which is beyond the scope of this book; high resolution spectrophotometry of this type is not considered specifically here.

The physical nature of the process of light absorption has been explained in Chapter 7; familiarity with the concepts discussed there is assumed. Monochromators, which are an essential component of spectrophotometers, have been described separately in Chapter 23.

BEER'S LAW

Beer's law (more correctly called the Beer-Bougier or Beer-Lambert law) is the basic quantitative description of absorption. Its derivation is given here.

Experimentally, it can be observed that the passage of a light beam through any given thickness of a given medium results in the absorption of a constant fraction of the incident light, regardless of the initial level of light intensity. Thus, if a light beam is successively transmitted by identical solutions of constant thickness (i.e., constant light path), the greatest absolute amount of light is absorbed in the first solution, while successively less intensity is absorbed by subsequent portions; an example is illustrated in Fig. 24-1. Thinner portions of the absorbing solution than those shown in Fig. 24-1 might be considered; again, a constant though smaller fraction of the intensity incident on each would be absorbed. It is clear that, for solutions of finite thickness, total absorption is not directly proportional to the length of the light path through the solution. In the limit, however, an infinitely thin layer of absorber may be considered. Then, and only then, the quantity of light absorbed is directly proportional to the light path, as well as to the concentration of absorber. The fraction of light, $-dI/I$, which is absorbed by the infinitely thin layer is thus

(24-a) $$-\frac{dI}{I} = KC\,dl,$$

where K is a constant characteristic of the absorbing species, C is its concentration, and dl is the thickness of the layer.

An expression for the fraction of light absorbed by a layer of finite thickness, l, can then be obtained by *integration* of (24-a). In performing this integration, C, the absorber concentration, is held constant. Thus,

(24-b) $$-\int_0^l \frac{dI}{I} = \int_0^l KC\,dl = KC \int_0^l dl$$

Or, on integration between 0 and l, that is, over the complete length of the light path through the solution,

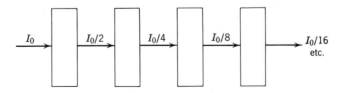

Fig. 24-1 Absorption by successive lengths of a solution.

(24-c)
$$\ln \frac{I_0}{I} = KCl \quad \text{or} \quad \frac{I}{I_0} = e^{-KCl},$$

where I_0 is the intensity incident upon the solution, and I is the intensity transmitted.

Equivalent forms of (24-c), expressed in terms of logarithms to the base *ten*, are

(24-d)
$$\log \frac{I_0}{I} = K'Cl \quad \text{and} \quad \frac{I}{I_0} = 10^{-K'Cl},$$

where $K' = K/2.303$.

Any of the expressions given in (24-c) and (24-d) can be designated *Beer's law*.

QUANTITATIVE EXPRESSION FOR THE ABSORPTION OF X-RAYS

Beer's law describes the absorption of any form of electromagnetic radiation and thus may be applied to that of x-rays. A slightly different form of the expression is then appropriate.

The *linear absorption coefficient* of materials with respect to x-rays, designated μ, may be defined such that

(24-e)
$$\ln \frac{I_0}{I} = \mu l;$$

μ, however, varies with the density of the absorber. Thus a more useful expression is the *mass absorption coefficient* μ_m, which is the linear absorption coefficient divided by the density of the absorber. Beer's law then becomes

(24-f)
$$\ln \frac{I_0}{I} = \mu_m \rho l,$$

where ρ is the density of the absorber. An equivalent useful form is

(24-g)
$$\ln \frac{I_0}{I} = \mu_m \frac{m}{A},$$

where m is the mass of the specimen, and A its cross-sectional area. In (24-f) and (24-g), μ_m is analogous to K of (24-c), while $\rho l = m/A$ are measures of the amount of absorbing substance encountered by the x-ray beam and are thus analogous to the product Cl.

μ_m as given in (24-g) and (24-f) is an average value for the specimen as a whole and is determined by the contributions to absorption of all atoms present in the specimen. Thus,

(24-h)
$$\mu_m = \sum_{\text{all atoms}} w_1\mu_{m_1} + w_2\mu_{m_2} + \dots,$$

where $w_1 w_2, \dots$ and $\mu_{m_1} \mu_{m_2}, \dots$ are the *weight fractions* and mass absorption coefficients, respectively, for each type of atom present in the specimen. Other aspects of the absorption of x-rays are discussed on p. 604.

ABSORPTION AND ABSORPTION COEFFICIENTS

A number of different quantities based on (24-c) and (24-d) are used to describe the *extent of absorption by solutions* and the *capacity of solutes to absorb light*. They are defined and explained here.

The *transmittance* of a solution, designated T, is the fraction (I/I_0); percent transmittance is $(I/I_0) \times 100$. The *optical density* (O.D.) of a solution is defined as the quantity $(-\log T)$. Thus,

(24-i)
$$\text{O.D.} \equiv -\log T = \log \frac{I_0}{I} = K'Cl.$$

Optical density is sometimes also called absorbance (A), extinction, or absorbancy. All of these terms are equivalent.

Optical density and transmittance are formally identical with photographic density and transparency, respectively. The latter quantities have been discussed in Chapter 21.

The combined optical density of two or more solutions placed in a light path is the sum of individual optical densities, whereas the combined transmittance is the product of individual transmittances (cf. Fig. 24-1). Consider, for example, two solutions which transmit 1/10 and 1/100, respectively, of incident intensity, that is, solutions characterized by 10% and 1% transmittance. The O.D. of the first is $\log (10/1) = 1.0$; that of the second is $\log (100/1) = 2.0$. The combined optical density is $1.0 + 2.0 = 3.0$. Thus,

(24-j)
$$\log \frac{I_0}{I} = 3.0 \quad \text{and} \quad \frac{I}{I_0} = 10^{-3},$$

corresponding to 0.1% transmittance. The same result is obtained by multiplying the individual transmittances:

(24-k)
$$0.10 \times 0.01 = 0.001 \quad \text{or} \quad 0.1\%.$$

The particular usefulness of specifying absorption in terms of optical densities is made evident by inspection of (24-i): *Optical density (unlike transmittance) is directly proportional to concentration.* A plot of optical density as a function of concentration (called a Beer's law plot) is, ideally, a straight line, as shown by Fig. 24-2. Exceptions to ideal behavior are discussed below.

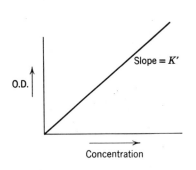

Fig. 24-2 A Beer's law plot.

It must be noted that I/I_0, $\log (I_0/I)$, and thus, also, transmittance and optical density are all unitless quantities. This means that the units used in specifying the factors which contribute to the product (KCl) or $(K'Cl)$ must cancel also. For example, if l is expressed in cm and C in g/cm³, then K (or K') must be expressed as cm²/g. Should C be given instead as millimoles per cm³, an equivalent, but numerically different, value for K (or K') expressed as cm² per millimole must be used.

Both optical density and transmittance are determined by the amount of absorber (specifically by solute concentration and the length of light path), while K and related quantities specify a characteristic property of the absorber (i.e., of the *solute*). This quantity is variously called the *absorption coefficient, extinction coefficient, absorptivity,* or *absorbancy index* of the material. (Only the first two of these terms are commonly encountered in biological literature, however.) Absorption coefficients may be expressed on either a weight or a *molar* basis. The *molar extinction coefficient*, designated by the symbol ϵ, specifies the total absorption by *one mole* of solute. Although textbooks often state that the molar extinction coefficient is the standard method of expressing absorption coefficients, the specific extinction is far more often cited, in the biological literature, in terms of *optical density units per mg per ml*, that is, on a *weight* basis. The reason for this convention is that accurate molecular weights of biological absorbers are often unknown.

According to one convention, Beer's law is written in a form equivalent to (24-d) as

$$(24\text{-}l) \qquad\qquad\qquad A = 10^{-abc}.$$

Here $A =$ absorbance (optical density), a is the absorption coefficient, $b = l$, the light path through the solution, and c specifies concentration. This notation, which is commonly used by chemists, usually implies the specification of a and c on a weight, rather than a molar, basis.

DEVIATIONS FROM BEER'S LAW

A Beer's law plot, as illustrated by Fig. 24-2, shows optical density as a function of concentration for illumination at any specified wavelength. In the absence of certain instrumental defects (discussed on p. 549), plots of this type are *linear* (i.e., Beer's law is obeyed) if (a) *the absorbing entity remains the same over the concentration range studied,* (b) the solute (absorbing entity) is *uniformly distributed,* and (c) the orientation of *dichroic* absorbers is *random.* Deviations which arise from failure to meet these conditions are considered here.

Changes in the nature of the solute, such as ionization, oxidation-reduction, or aggregation-disaggregation, alter the absorption spectrum. To cite an extreme example, absorption by acid-base indicator molecules changes totally upon the ionization which occurs as a function of pH. Other reactions are accompanied by significant, if less drastic, spectral changes. The extent of reactions is always influenced to some degree by the concentration of reactants. Thus ionization may be repressed, or aggregation favored, at high concentrations. The effects of such changes upon Beer's law plots is indicated in Fig. 24-3. Figure 24-3a represents a case in which an interaction reduces the concentration of an absorbing species, while Fig. 24-3b can be accounted for by the formation of a more highly absorbent species at high concentrations.

Interactions between two or more species may also result in failures of Beer's law. Interaction of solutes with their solvent may cause absorption to differ from that observed in an inert solvent. Similarly, the optical density of a mixture of two solutes may differ from the sum of the optical densities of the individuals if the two interact. Either of these effects may vary with concentration.

For some substances, absorption varies markedly as a function of temperature; this behavior may be correlated with structural transitions

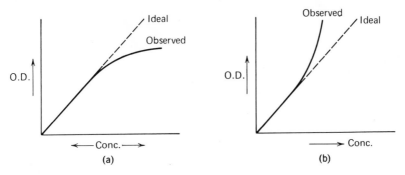

Fig. 24-3 Deviations from Beer's law.

in the absorbing molecule. A well known example of biological interest is the *hyperchromic effect* which is observed upon heating solutions of deoxyribonucleic acid (DNA). At relatively high temperatures specifically characteristic of nucleotide composition, these double-stranded molecules dissociate (under ideal conditions, reversibly) into the more highly absorbent single chains.

The effect of nonuniform distributions of solute is considered in Fig. 24-4. In an extreme case, as shown in Fig. 24-4a, 100% of the absorber is confined to one half of the cross section of the illuminating beam. Regardless of the concentration of the absorber, at least 50% of the incident illumination must thus be transmitted. The highest optical density which can be observed is

$$(24\text{-m}) \qquad\qquad \log\left(\frac{I_0}{I}\right)_{\text{lim}} = \log 2 = 0.301.$$

The corresponding Beer's law plot is shown in Fig. 24-4b. When the segregation of solute is less extreme than that considered in Fig. 24-4, higher values of the limiting optical density are obtained, but the general form of plot is similar — a limiting optical density value is approached exponentially at high concentrations.

These considerations suggests that, since substances in the solid state are rarely distributed uniformly, the spectrophotometry of solid objects is subject to much greater uncertainty than are measurements of the optical densities of liquid solutions or gases. While microspectrophoto-

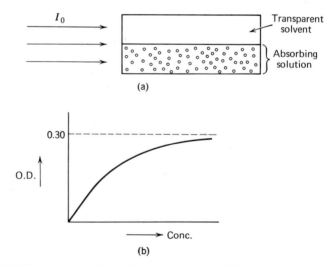

(a)

(b)

Fig. 24-4 Effect of a nonuniformly distributed solute: (a) light paths; (b) Beer's law plot.

metric measurements of tissue sections are in fact made with the aim of determining the intracellular distributions of substances (notably DNA or RNA) which exhibit characteristic absorptions, these measurements must always be corrected in some way in order to account for the *distribution* of absorbing structures within the cell.

Dichroic substance (cf. p. 96) absorb light preferentially according to its azimuth of polarization. The illuminating beam of a spectrophotometer is normally *unpolarized* and is thus equivalent from the point of view of absorption to the combination of a plane-polarized beam vibrating in the plane of preferential absorption by the dichroic specimen with a second plane-polarized beam (of equal intensity) polarized at right angles thereto. (For a discussion of the properties of polarized light beams, see Chapter 4.) The light which vibrates in an azimuth perpendicular to that which is preferentially absorbed by the specimen may, in the limit, be transmitted without attenuation. Thus up to 50% transmittance may result from illumination of a highly concentrated but dichroic specimen; the resulting Beer's law plot is then of the same form as given in Fig. 24-4b.

SPECTROPHOTOMETRIC MEASUREMENT OF CONCENTRATIONS

When the value of the absorption coefficient is known at a suitable wavelength, measurements of optical density provide a convenient and often highly specific method for determining the concentrations of substances in solution. The complete procedure for quantitative spectrophotometric analysis is now summarized:

1. The absorption spectrum, that is, a plot of optical density as a function of wavelength, is recorded.

2. A suitable wavelength for absorption measurements is selected. Considerations in the choice of a wavelength for measurement are illustrated by the spectrum shown in Fig. 24-5. The absorption maximum at λ_1 is more suitable that that at λ_2 since it is *broader*, so that errors in setting the wavelength scale are relatively unimportant. The maximum at λ_1 also corresponds to a higher value of the absorption coefficient than that at λ_3; thus measurements made at λ_1 are more sensitive. Interfering absorption by other components of the solution may of course influence the choice of a wavelength for measurement.

3. The absorption coefficient of the substance is determined at λ_1; that is, optical density is determined for a solution of known concentration. The concentration is obtained gravimetrically, or by means of other standard analytical procedures. Determination of the extinction coefficient must be carried out with extreme care, since the accuracy of future determinations of concentration depends upon the accuracy with which

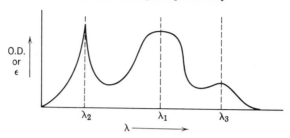

Fig. 24-5 Absorption spectrum.

K' is measured initially. Once this value is obtained, no further direct measurements of K' need be made.

4. A Beer's law plot of absorption at λ_1 is prepared; that is, the optical density at λ_1 is measured for a series of dilutions. Absorption need not obey Beer's law at all concentrations; it is simply necessary that a range of concentrations exist within which the plot is linear. If linearity is observed only at low concentrations, routine dilutions of solutions to be measured may be required.

5. Subsequently concentrations are obtained directly by making a single reading of optical density at λ_1. Then,

(24-n)
$$C = \frac{\text{O.D.}}{K'l}.$$

As noted above, K', l, and C must be expressed in a consistent set of units.

The analysis of proteins provides an example of the application of quantitative spectrophotometry in biology. Apart from a few proteins which contain prosthetic groups that absorb visible light, characteristic absorption by proteins occurs in the near ultraviolet region of the spectrum. The absorption is accounted for as that of the aromatic amino acids tryptophan and tyrosine and, to a much more limited extent, of phenylalanine. The variations of the molar absorption coefficients of these substances as a function of wavelength is shown in Fig. 24-6. Both tryptophan and tyrosine absorb maximally near 280 mμ; maximum absorption by proteins is also observed at or near this wavelength. The total absorption by a protein macromolecule is dependent upon its tryptophan and tyrosine composition; thus values of the absorption coefficients for complete protein molecules may vary considerably (on both weight and molar bases).

Interactions between solute and solvent do not interfere with quantitative analysis if concentrations in the linear range of the Beer's law plot are studied. (The absolute value of the measured absorption coefficient is,

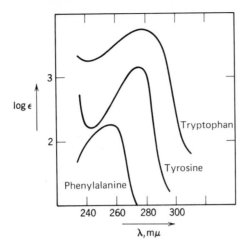

Fig. 24-6 Absorption spectra of the aromatic amino acids.

however, that of the solute-solvent complex, rather than that of the pure solute.)

Acidic and basic forms of an absorbing molecule may exist in equilibrium but may absorb maximally at different wavelengths, as in the case illustrated by Fig. 24-7. While it may not be feasible to maintain the pH of solutions of the molecules at an exactly constant value, the absorption spectra of the acidic and basic forms intersect at some wavelength. At this wavelength, which is termed the *isosbestic point*, spectrophotometric measurements may be made without regard to the condition of the acid-base equilibrium.

QUALITATIVE SPECTROPHOTOMETRIC ANALYSIS

A *chromophore* is a chemical grouping which absorbs at a specific wavelength. Examples of chromophores are: in the ultraviolet range, the

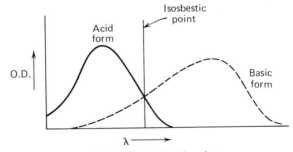

Fig. 24-7 The isosbestic point.

phenolic group, as in the tyrosine molecule; in the visible range, extended conjugated double bond sequences, as in the carotene molecule; in the infrared range, individual functional groups of organic molecules, such as the carbonyl or carboxyl groups, or specific types of chemical bond, such as $C—C$, $C=C$, $C—H$, etc.

Absorption at a wavelength characteristic of a known chromophore suggests the presence of that grouping and thus the presence in the solution of a particular compound or class of compounds. For example, observation of an absorption maximum near 280 mμ suggests the presence of tyrosine or tryptophan and thus of protein in the solution. While the absorption spectra of individual proteins are in general too similar to permit distinction between different types of protein by spectrophotometric means, many organic compounds can be identified precisely from their highly characteristic absorption spectra. In biological applications, however, qualitative spectrophotometric analysis is more often used to assign an unknown substance to a general class of compounds. A particular advantage of the method, as of quantitative spectrophotometric analysis, is that analysis of mixtures is often possible, provided only that absorption by interfering components is negligible at some wavelength of significant absorption by the unknown.

The precise wavelength of maximum absorption by a chromophore is influenced by its chemical environment. The study of such variations is particularly important in infrared spectroscopy, where the fine structure of absorption spectra may be analyzed as an aid to deducing the conformations of chemical bonds and the locations of functional groups. The hypochromic effect in DNA (discussed above on p. 534) is an example of the effect of (thermally induced) variations in the environment of chromophores upon absorption of light of shorter wavelengths.

Observations of changes in absorption spectra may be used to study changes which occur in the absorbing solution. The course of purifications may be followed spectrophotometrically. For example, separation of cellular nucleic acid from associated proteins is accompanied by an increase in the ratio

$$\frac{\text{O.D. at } 260 \text{ m}\mu}{\text{O.D. at } 280 \text{ m}\mu}.$$

The value of this ratio is about 0.6 for pure proteins and in excess of 2.0 for pure nucleic acids. (The specific nucleic acid value is rather highly dependent upon base composition.) Since the extinction coefficients, at either wavelength, of nucleic acid are higher by an order of magnitude than those of proteins, the value of the ratio is largely determined by nucleic acid content. For example, purified ribosomes, which contain about 60% RNA, have a ratio of about 2.0.

The *kinetics* of biochemical reactions may be studied in terms of the variation of absorption with time. A familiar example is that of oxidation-reduction reactions accompanied by conversions between DPN and DPNH (oxidized and reduced diphosphopyridine nucleotides, respectively). While both forms of this coenzyme exhibit characteristic absorption near 260 mμ, the reduced form, DPNH, posesses an additional absorption maximum in the near ultraviolet at 333 mμ. The rate of appearance (disappearance) of absorption of the latter wavelength is thus a measure of the rate of oxidation (reduction) of a substrate by the accompanying enzyme. The *stoichiometry* of such reactions may be studied simply by comparing initial and final absorptions.

SPECTROPHOTOMETRY AND RELATED METHODS

The term *spectrophotometer* describes instruments in which measurements of absorption are made, in which incident light is substantially monochromatic, and in which the incident wavelength can be continuously varied at will over a wide range. Two related types of instrument are the *colorimeter* and the *turbidimeter*. "Colorimeter" implies a device for the measurement of absorption in which a rather broad band of wavelengths is isolated by means of an absorption filter. Although colorimeters cannot be used to obtain absorption spectra, they are extremely useful as a simple and inexpensive type of instrument for performing routine quantitative analyses. However, since the wavelength distributions of output by the light source, and also of absorption by filters, are variable, results obtained on different colorimeters may not be directly comparable. Colorimetric analyses are often compared with a standard curve determined simultaneously.

Turbidimetry implies measurement of the attenuation of a light beam by means of scattering rather than by means of absorption. Instrumentally, turbidimetry differs from colorimetry or spectrophotometry only in that an incident wavelength is selected which is *not* absorbed by the solution. Thus turbidimetric measurements may be carried out in the same instrument used for absorption measurements. For example, concentrations of bacterial suspensions are often measured in the spectrophotometer with irradiation at about 600 mμ. Scattering may be distinguished from absorption in that it varies less rapidly with wavelength. A high degree of scattering is an indication of the presence, in suspension, of particles of colloidal size. Thus changes in the state of aggregation of components in solution can be followed by scattering measurements.

Nephelometry is an alternative means for performing quantitative analyses by means of light scattering. In this method scattered light is measured at right angles to the on-going beam. The instrumentation is then that of the fluorimeter or spectrofluorophotometer (see Chapter 25) rather than that of the colorimeter or spectrophotometer.

THE SPECTROPHOTOMETER

The basic block diagram for spectrophotometers (and for colorimeters or turbidimeters) is shown in Fig. 24-8. The source represents the lamp and monochromator, the modifier consists of the absorbing solution and its housing, and the detector consists of a light sensitive surface and read-out, including, if used, the recording system. Semischematic diagrams of single beam and double beam spectrophotometers are shown in Figs. 24-9 and 24-10, respectively.

The lamp must produce a continuous spectrum throughout the wavelength range of interest. An incandescent tungsten filament is normally used in the near infrared, and throughout the visible range, down to wavelengths of about 320 mμ. For measurements of ultraviolet absorption (250–350 mμ), hydrogen arc lamps are suitable. For measurements at wavelengths shorter than about 250 mμ, the very intense xenon lamp is required. Since fluctuations in the intensity of the source cause the precision of optical density readings to deteriorate, the power activating the lamp must be highly stabilized. Electronic equipment associated with spectrophotometers is of prime importance in high quality instruments; this aspect of instrumentation will not be discussed here, however.

Light from the lamp is collimated by a lens, which often is a curved mirror, rather than a refraction lens. The collimated beam is then incident upon a monochromator. Both grating monochromators (as shown in Figs. 24-9 and 24-10) and prism monochromators are used in spectrophotometers at the present time. The relative merits of the two have been discussed on p. 523. Usually the prism or grating is rotated in order to vary the wavelength incident upon the specimen; alternatively, the exit slit, sample, and recording system could be moved instead.

The width of the monochromator exit slit is variable; the width used should be the smallest possible, since this condition favors optimal spectrophotometric resolution (see p. 548). The minimum feasible slit

Fig. 24-8 Block diagram of instruments for measuring light absorption.

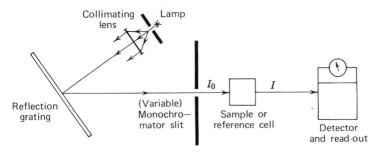

Fig. 24-9 Single beam spectrophotometer.

width in each specific case is determined by the resolving power of the monochromator, the wavelength distribution of intensities in the lamp output, the wavelength response characteristic of the detector, the extent of stray light present in the instrument, and the extent of absorption by the specimen. Errors which result from the use of excessive slit widths are discussed on p. 550.

The specimen is placed in a transparent cell of standard light path, usually one cm. The collimated light beam enters and leaves the cell normal to the faces (windows); thus variations in the light path through the solution due to refraction are avoided. Absorption and scattering of the light beam by the cell windows and by the solvent are compensated by making blank readings on a cell which contains only solvent.

A standard 1 cm path spectrophotometer cell is shown in Fig. 24-11. Cells with shorter light paths may be used when only limited quantities of solution are available; cells with longer light paths may be used to study absorption by very dilute solutions. If measurements are to be confined to the visible region of the spectrum, the cell may be constructed of optical glass; at wavelengths shorter than about 3500 Å, quartz (silica)

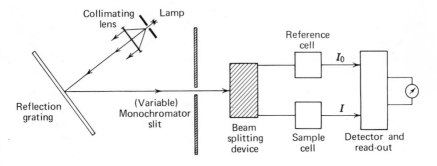

Fig. 24-10 Double beam spectrophotometer.

cells are required. Spectrophotometric measurements may be extended to wavelengths as short as 1850 Å by the use of quartz cells, of xenon lamps, and by passing a continuous flow of nitrogen through the light path of the instrument. Nitrogen outgassing reduces light losses due to absorption by oxygen, which is appreciable at wavelengths shorter than about 250 mμ.

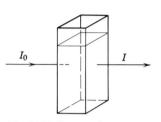

Fig. 24-22 Spectrophotometer cell.

Various types of specimen mount are used for infrared measurements. Some samples may be examined as solids, while rock salt cell windows are commonly used for the examination of nonaqueous specimens. The path lengths used in infrared spectrophotometry are usually shorter than those used for measurements at shorter wavelengths.

Light sensitive devices are described in the next section. The devices used in spectrophotometers are commonly photoemissive cells (phototubes) of suitable spectral sensitivity. Spectrophotometers are usually equipped with a meter from which either optical density or percent transmission can be read directly; often they are equipped for rapid automatic recording of spectra. A *linear* wavelength scale is normally desired for records; in the case of prism instruments, precise mechanical coupling of read-out and monochromator drive is required in order to compensate for the irrational spectrum produced by the prism. For this reason the use of grating monochromators is preferable in automatically recording instruments.

In the single beam design shown in Fig. 24-9, sample and reference cells are placed in the light beam alternately, at each wavelength. The recorder is set to read 100% transmission (O.D. = 0) when the reference (blank) cell is placed in the beam. Losses of intensity due to reflection at the cell surface or to absorption by the solvent are thereby compensated for.

In the double beam instrument (Fig. 24-10) the incident beam is split, usually by a chopper which deflects the beam to illuminate specimen and blank in rapid alternation. (The function of the chopper is to provide two beams of identical intensity, but there is no coherence requirement in this type of instrument.)

The detector is designed, electronically, to give a direct reading of either transmission or optical density; 100% transmission is set initially at one wavelength, by substituting a second blank for the sample. Double beam spectrophotometers are the most suitable type for use with automatic recording systems, since no physical translation of sample and reference cells is required during the recording of a spectrum.

PROPERTIES OF LIGHT SENSITIVE DEVICES

Light sensitive devices are characterized by their *spectral response, sensitivity, stability,* and *speed.*

The *spectral response* of a detector may be expressed in output per unit of incident energy or in output per incident photon; the energy basis of measurement is by far the most common. Calibration of detectors on an energy basis is appropriate when relative intensities are to be measured, as in the determination of percent transmission and optical density. If absolute intensities must be obtained, as in the determination of the quantum yield of fluorescence or of photochemical reactions (see Chapter 25), the detector response must be known in terms of the rate of photon arrival.

For light of any single wavelength, the energy response calibration can be converted to a per photon basis by numerical manipulation, as shown by (24-o) to (24-r):

$$(24\text{-o}) \qquad \text{energy per photon} = \frac{hc}{\lambda},$$

where h is Planck's constant, c is the velocity of light, and λ is the wavelength. This expression gives energy in *ergs* when c is expressed in cm per second and λ is expressed in centimeters (not in Å, $m\mu$, or μ!):

$$(24\text{-p}) \qquad 1 \text{ watt} = 1 \text{ joule/sec} = 10^7 \text{ ergs/sec.}$$

Thus, for a one watt beam,

$$(24\text{-q}) \qquad \text{photons/second} = \frac{10^7 \text{ ergs/second} \times \lambda \text{ cm}}{h \text{ erg/second} \times c \text{ cm/second}}$$

$$= \frac{10^7 \lambda}{6.63 \times 10^{-27} \times 3 \times 10^{10}}$$

$$= 6.34 \times 10^{-12} \lambda \text{ photons/sec/watt, at any}$$
$$\text{wavelength } \lambda;$$

or

$$(24\text{-r}) \qquad \text{response/photon/second} = \frac{1}{6.34 \times 10^{-12} \lambda} \times \text{response/watt}$$

$$= \frac{1}{\lambda} \times 1.99 \times 10^{11} \times \text{response/watt.}$$

The sensitivity of a detector is affected by the slope and linearity of the response curve and by the degree of freedom from *noise*. The noise level is established by the random response of the detector; that is, response which is unrelated to the incident signal. Freedom from noise is assessed in terms of the *signal-to-noise ratio* discussed on p. 467.

In Fig. 24-12 the response curves of two detectors are compared. One of them is an idealized device which, in the absence of noise, produces a linear response at all intensity levels. The other provides a linear response only *below* a certain intensity level, I_2; it is, however, a more sensitive detector over that limited range. The effect of the noise level N on the useful range of the detectors is considered in Fig. 24-12b (where N is taken to be the same for both of the detectors). As shown, the ideal detector can be used only at intensities greater than I_3, while the other produces a linear response of the range from I_1 to I_2.

In fact, the useful range of a detector need not be limited to that in which a strictly linear characteristic applies. Other response forms which are some simple function of incident intensity can be converted by suitable electronic instrumentation to give a final output which is directly proportional to the incident intensity. For example, if the response were proportional to the square of the incident intensity, electronic manipulation could be used to provide a final reading which would be proportional to the square root of the initial output from the detector.

The term *stability* implies considerations of both the noise level, as just discussed, and of the tendency of a device to exhibit *fatigue* effects. Exposure of light sensitive surfaces to very high light intensities, or to lower intensities for prolonged periods, tends to reduce their response per unit input. Recovery is generally achieved during a period of darkness, although the characteracteristic of the detector may gradually change irreversibly. For optical density and percent transmission determinations, in which only relative intensities are measured, detector fatigue is a problem only if it is so extreme as to impair substantially the overall sensitivity. However, if absolute intensity measurements are required (as in most fluorescence studies; see Chapter 25), the response of the detector to a standard illumination must be reestablished frequently.

The *speed* of response of a detector is defined quantitatively as the

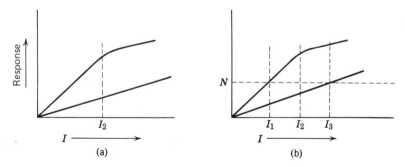

Fig. 24-12 Response curves of light sensitive devices.

time required for the device to yield $1/e = 37\%$ of the total response achieved at infinite time. When only single measurements are to be made, as in the determination of optical density at a single wavelength, a detector of high speed is not required. When spectra are to be scanned, or changes in spectra are to be recorded as a function of time, detector speed may be a property of critical importance.

TYPES OF LIGHT SENSITIVE DEVICES

Since electrical outputs are suited to the quantitation of data, the detectors used in modern spectrophotometers and many other types of instrument are almost invariably some form of photoelectric device. They include the *photovoltaic* or barrier layer cell, the *photoconductive* cell, and *photoemissive* tubes. The latter are most widely used in biological instrumentation. Other types of device which produce an electrical response are the *bolometer* and the *thermopile*, both of which are *heat* sensitive (total energy) elements. The response of heat sensitive devices, when expressed on an energy basis, is not a function of the incident wavelength; that is, the same total incident intensity produces the same response regardless of wavelength. These devices are used for source calibration and for measurements of infrared intensities. Nonelectrical light sensitive devices include the *eye*, the *photographic emulsion*, and the *chemical actinometer*.

The *photovoltaic cell* consists of a thin layer of a semiconductor (e.g., selenium) deposited on a metal surface (e.g., copper). Upon illumination, electrons travel from the surface layer into the metal, establishing a measurable voltage gradient. The device is most sensitive within the visible portion of the spectrum but is of generally low sensitivity. It is commonly used without amplification, at relatively high light intensities. A disadvantage of photovoltaic cells is that they are particularly subject to fatigue.

The *photoconductive cell* consists of a thin layer of lead sulfide (or of certain other lead salts) supported by a dielectric and protected from air oxidation either by lacquering or by evacuation. An external voltage is applied across this film. Upon illumination, electrons within the film are freed from their parent nuclei, so that electrical resistance falls and current flow is induced by the applied voltage in proportion to the incident intensity. While a high level of applied voltage is desirable, the level of current produced must not be so great as to produce significant heating, since the sensitivity of the photoconductive cell is extremely temperature dependent, being greatest at very low temperatures.

In *photoemissive* devices a stream of electrons is released into free space from an illuminated surface. (This is the photoelectric effect described on p. 20.) When an isolated photoemissive surface is exposed to light, some photoelectrons leave the material, creating a net positive charge on the surface. The charge rapidly builds up to a level which precludes further electron emission. The photoemissive surface must thus be used as an element in a circuit which continuously supplies electrons to the photoemissive region.

The basic schematic diagram of the phototube is given in Fig. 24-13. The photoemissive material acts as the cathode, while the emitted electrons are attracted to the positively charged anode. Alkali metals or alkaline earth oxides serve as the photosensitive surface. The tube must be evacuated, since the electron beam would be scattered at atmospheric pressure.

The sensitivity of phototubes is very high in the visible range, and their response is extremely rapid. Phototubes can also be made which are sensitive to ultraviolet wavelengths. Amplification of the output signal is readily achieved. The *photomultiplier* (electron multiplier tube) achieves amplification directly by using the emitted photoelectrons to stimulate secondary electron emission. Phototubes are subject to fatigue, but less so than is the photovoltaic type of detector. In summary, the properties which make photoemissive devices the detector of choice for spectrophotometry in the visible and ultraviolet are: high sensitivity, high speed, range of spectral sensitivity (approximately from 2000 Å to 1μ), and the convenience of amplification.

Photocells produce a certain level of *dark current* in the absence of illumination. This is accounted for by the thermal emission of electrons, and it can be minimized by lowering the temperature of the photocathode. (Note that this is the same process which is exploited in the electron source of the electron microscope, as described on p. 350.)

The *threshold* of a photocell is defined as the intensity of incident light required to just double the mean value of the dark current. Sensitivity is also expressed quantitatively in terms of the incident intensity required to produce a signal one hundred times as great as the dark current level.

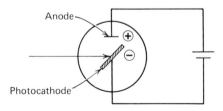

Fig. 24-13 Phototube.

Bolometers are constructed of materials in which electrical resistance changes markedly with temperature (i.e., with a high temperature coefficient of resistance). A sensitive bolometer is of *low* heat capacity; that is, the energy required to raise the temperature of the device by a fixed amount is relatively small. When a fixed voltage is applied across the bolometer, a current flows which is proportional to the incident intensity, as measured in energy units.

The *thermocouple* consists of a pair of bimetallic or semiconductor junctions, one of which is exposed to the radiation to be measured, while the second is kept at constant temperature. The voltage developed between the two junctions is proportional to the difference in their temperatures, as produced by the absorption of radiation. The *thermopile* consists of a number of thermocouples in series, one junction of each of which receives the heat absorbed by a blackened vane. Such a device may be of extreme sensitivity and must be carefully isolated from extraneous sources of heat. Its use in the calibration of other photoelements is described on p. 563.

The *eye* is a sensitive detector of small differences in intensity, especially at low intensity levels. Maximum sensitivity is in the central region of the visible spectrum, at wavelengths in the yellow and green. Since memory of intensities is poor, visual detection is useful only for direct comparison of intensities within the same field of view. Colorimeters of considerable sensitivity have been developed in which the color of a standard solution is matched by eye to that of an experimental solution. However, these instruments suffer from the disadvantage that precise matching is possible only when the wavelength distributions of the light leaving the specimen and of that leaving reference areas are very nearly the same.

Photographic recording of spectrophotometer intensities has been reported occasionally. Although relatively cumbersome, the method can be one of very high accuracy, and it has thus been used for exact determinations of extinction coefficients. Photographic densities of exposures are measured by a densitometer, as described on p. 471.

The chemical *actinometer* is discussed on p. 560.

RESOLUTION BY SPECTROPHOTOMETERS

Spectrophotometric resolution, while not defined numerically, is that quality which permits distinction of fine structure in absorption spectra. Thus, in Figs. 24-14a and 24-14b tracings of a portion of an absorption spectrum which are recorded at low and high resolution, respectively, are compared.

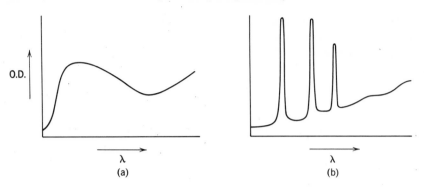

Fig. 24-14 Spectrophotometric resolution: (a) low; (b) high.

Resolution is determined by the effective monochromaticity of the illuminating beam. As Fig. 24-14b shows, the compound studied absorbs very strongly at three sharply defined wavelengths, but much less strongly at intermediate wavelengths. In the instrument used to record the tracing shown in Fig. 24-14a, however, the range of wavelengths contained in the illuminating beam at each setting is sufficient that only the *average* of the strong and weaker absorption can be observed. Thus peaks in the absorption spectrum are "smoothed out."

Spectrophotometric resolution is dependent upon use of a monochromator of high resolving power (cf. pp. 518 and 521), but a number of other factors may also be limiting. They are the same as those which determine the feasible slit width: the desired wavelength must be present in light from the lamp at adequate intensity, absorption by the specimen must not be excessive, and the transmitted beam must produce a signal which is distinguishable from the noise level of the detector. Resolution may be destroyed if an appreciable intensity of stray light (i.e., of wavelengths other than that isolated by the monochromator) is present in the instrument. In high resolution instruments, stray light is often minimized by the use of a double monochromator system. Since most of these factors vary with the wavelength of illumination, it is clear that spectrophotometric resolution is also wavelength dependent.

INSTRUMENTAL ERRORS IN SPECTROPHOTOMETRY

Properties of the solution which result in deviations from Beer's law have been discussed above on p. 533. Here instrumental factors which lead to spectrophotometric errors are summarized.

Any extraneous effect which removes intensity from or contributes it to the established light path of the spectrophotometer is a source of error. The transmission characteristics of specimen and reference cells should be matched at all wavelengths. Cells must be free of any scratches or other irregularities which might scatter light selectively. Solutions should be free of extraneous absorbing or scattering particles; interfering materials need not be visible to the eye.

The presence of light of unwanted wavelength (stray light) is a source of error. Consider, for example, the spectrum and Beer's law plot of a substance which absorbs maximally at some wavelength λ_1, as shown in Fig. 24-15. If, when set at λ_1, a faulty spectrophotometer passes an appreciable intensity of light of some other wavelength λ_2 which is not absorbed by the substance of interest, then, no matter how concentrated the solution, at least the fraction $[I\lambda_2/(I\lambda_1 + I\lambda_2)]$ of the incident light passes through the specimen and illuminates the detector. While, quantitatively, the response of the detector to λ_2 may not be identical with that to λ_1, qualitatively, the read-out produced by the two wavelengths is indistinguishable. Consequently, as shown in Fig. 24-15b, some limiting value of optical density is approached at high solution concentrations.

There exists a rather wide range of optical densities over which the sensitivity of spectrophotometers is approximately constant. This is approximately between O.D. values of 0.1 and 0.7 (20% to 80% transmission). At high optical densities, very low transmitted intensities must be measured which may not be distinguished accurately from detector noise. At very low optical densities, very small differences in the intensities transmitted by blank and sample are difficult to detect.

Irreversible changes in the sample may occur during spectrophotometric measurement. Ultraviolet illumination, especially, may catalyze photochemical alteration of the specimen, with accompanying spectral changes. Also, denatured substances may adsorb to the walls of the sample cell. Serious errors may then result if the same cell is used without

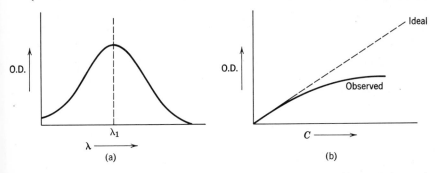

Fig. 24-15 Effect of stray light: (a) absorption spectrum of specimen; (b) ideal and observed Beer's law plots.

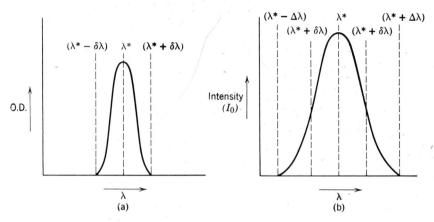

Fig. 24-16 Effect of finite slit width: (a) absorption spectrum of specimen; (b) spectral distribution of monochromator output.

thorough cleaning for measurements of absorption by different dilutions of the same solution.

The finite width of the monochromator slit results in an illuminating beam which is, to some extent, not strictly monochromatic. Thus optical density readings are always at least slightly smaller than that which would be obtained for a pure monochromatic beam of the maximally absorbed wavelength. Consider, for example, a substance which absorbs maximally at λ^*, and for which essentially all absorption is confined to some wavelength range from $(\lambda^* - \delta\lambda)$ to $(\lambda^* + \delta\lambda)$, as shown in Fig. 24-16a. This substance might be illuminated by a beam of nominal wavelength λ^* which, owing to finite slit width, in fact contains a distribution of wavelengths in the range from $(\lambda^* - \Delta\lambda)$ to $(\lambda^* + \Delta\lambda)$, as shown in Fig. 24-16b. If (to cite an extreme example) only 75% of the incident intensity thus falls within the wavelength range from $(\lambda^* - \delta\lambda)$ to $(\lambda^* + \delta\lambda)$, the transmission by the solution can never be less than 25%. In fact, it is always somewhat greater than this, since light of the extreme wavelengths $(\lambda^* - \delta\lambda)$ and $(\lambda^* + \delta\lambda)$ is less strongly absorbed than light of wavelength λ^*. Consequently, the measured optical density is erroneously low. The precise quantitative effects of slit width are determined by the shapes of both the absorption and spectral distribution curves.

REFERENCES

H. A. Strobel, *Chemical Instrumentation*, Addison-Wesley, Reading, Mass., 1960, Chapter 6 (Absorption Photometry) and Chapter 7 (Photometric Instrumentation). An excellent discussion of spectrophotometric methods and instrumentation.

R. B. Setlow and E. C. Pollard, *Molecular Biophysics*, Addison-Wesley, Reading, Mass., 1962, Chapter 7 (Absorption Spectroscopy and Molecular Structure).

Three chapters in G. W. Oster and A. W. Pollister, *Physical Techniques in Biological Research*, Vol. I, *Optical Methods*, Academic Press, New York, 1956, deal with spectrophotometry, with special reference to biological studies. They are: "Absorption Spectroscopy," by C. F. Hiskey; "Ultraviolet Absorption Spectrophotometry," by J. F. Scott; and "Infrared Spectrophotometry," by C. Clark.

E. J. Meehan, *Optical Methods of Analysis*, Wiley, New York, 1964. (The book is a reprint from I. M. Kolthoff and P. J. Elving, Eds., *Treatise on Analytical Chemistry*, Part I, Vol. 5, pp. 2707–2838), Chapter 54 (Fundamentals of Spectrophotometry) and Chapter 55 (Spectroscopic Apparatus and Measurements).

The problems and techniques of making spectrophotometric measurements of tissue sections are discussed in two chapters of Oster and Pollister, *op. cit.*, Vol. 3, *Cells and Tissues*. They are: "Microspectrophotometry with Visible Light," by H. Swift and E. Rasch, and "Ultraviolet Absorption Techniques," by P. M. Walker.

B. D. Cullity, *Elements of X-ray Diffraction*, Addison-Wesley, Reading, Mass., 1956, pp. 10–16. A discussion of x-ray absorption.

CHAPTER **25**

Measurements
of Luminescence

Luminescent processes consist of the *reradiation* of the light energy absorbed by a specimen. The measurement of fluorescence and phosphorescence, like that of light absorption, provides a means for specific quantitative and qualitative analysis of molecules of biological interest. The nature of molecular structures and interactions may also be studied by means of observations of luminescent phenomena.

In this chapter the nature of the processes of fluorescence and phosphorescence is briefly reviewed. The concept of *quantum yield* is discussed, and the determination of this quantity is described. Instrumentation for measurements of luminescence is described. In general, such instruments are characterized by a rectangular light path, as indicated in block diagram form by Fig. 25-1. Figure 25-1 may be compared with Fig. 24-8, which characterizes instruments used to measure light absorption, as described by the text on p. 540. *Scattered* intensities can also be measured in instruments of the basic design shown in Fig. 25-1; therefore instrumentation for *nephelometry* and for *Raman spectroscopy* is discussed here. Measurements of the *polarization* of fluoresced radiation are also considered briefly.

The theory of luminescent pro-

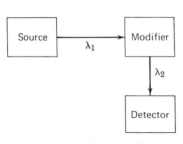

Fig. 25-1 Block diagram of instruments for fluorescence and related measurements.

cesses has been presented at some length in Chapter 7, which should be consulted for a fuller description of these phenomena than that given below. Luminescence instrumentation is similar in many respects to that used for the measurement of absorption; thus, much of the material presented in Chapters 23 and 24 is equally relevant to the discussions given here.

NATURE OF FLUORESCENCE AND PHOSPHORESCENCE

Fluorescence and phosphorescence are processes of quantum losses of energy from molecules which exist in excited electronic states. Electronic excitation of molecules is most commonly effected by the absorption of light but may also be produced by other mechanisms, such as chemical reactions or electric discharges.

Operationally speaking, fluorescence is the reradiation of quanta after delays of the order of only 10^{-9} to 10^{-8} seconds; that is, this is essentially immediate reradiation from most points of view. Phosphorescence is the reradiation of quanta after delays which may be as long as many seconds.

In quantum mechanical terms, fluorescence is radiation emitted in association with allowed transitions in electronic energy level (i.e., transitions between electronic states of identical multiplicity). Since the ground state of organic molecules in solution is normally a singlet state, fluorescence transitions, in general, are singlet-singlet transitions. Conversely, phosphorescence is radiation emitted in association with forbidden transitions in electronic energy level (i.e., transitions between electronic states of different multiplicity; these transitions are of low probability). Observed phosphorescence is usually associated with transitions between the lowest triplet state and the (singlet) ground state of a molecule.

QUANTUM YIELD

The quantum yield of fluorescence, Φ_{FL} (sometimes also called the quantum efficiency of fluorescence) has been defined on p. 143 as

(7-j)
$$\Phi_{FL} = \frac{\text{number of quanta fluoresced}}{\text{number of quanta absorbed}}.$$

Quantum yields for other luminescent processes are defined analogously.

While the intensity of light fluoresced usually varies very markedly with wavelength, the quantum yield of fluorescence tends to be invariant with wavelength. Thus variations in emission originate from differences in absorption rather than from differences in the efficiency of the re-emission. Quantum yield may, however, be a function of wavelength in molecules which contain more than one chromophoric group. This is the case if the chromophores differ in fluorescent yield, and if no transfer (or incomplete transfer) of energy occurs between them. In that case the net quantum yield should vary with wavelength as a function of the percentage of absorption by each group. An example is considered in Table 25-1 and Fig. 25.2, in which fluorescence by a molecule containing two independent chromophores is considered. The absorption spectra of the individual chromophores and of the complete molecule are shown in Fig. 25-2a, and the percentage of total absorption by group B and the net quantum yield of the molecule are plotted as a function of wavelength in Fig. 25-2b. Note that identical plots could be obtained for a mixture of two separate, noninteracting molecules.

While the chromophores within a complex molecule are sometimes found to be independent, in the way shown by Fig. 25-2b, in other cases, it must be inferred that transfer of absorbed energy occurs within or between molecules. If this transfer is efficient, invariance of quantum yield with wavelength is observed. An example is the photodissociation of carbonmonoxy myoglobin, in which CO is split off the protein and replaced by oxygen. The quantum yield for this dissociation is found to be the same at all wavelengths between 280 and 546 mμ. All absorption at 546 mμ is accounted for as that of the heme moeity of the myoglobin, to which CO is attached, whereas, at 280 mμ, about half of the total absorption is accounted for as that of the aromatic amino acid residues of the protein portion of the molecule. Thus quanta absorbed by these amino acids must be transferred to the heme group, where they are fully effective in energizing the dissociation of the carbon monoxide.

Values for the quantum yield of fluorescence vary according to the extent to which competing processes of de-excitation can occur. Losses of excitation energy by nonluminescent routes are collectively termed

Table 25-1 Fluorescence Yields for a Molecule Containing Two Independent Absorbers ($\Phi_A = 0.5$; $\Phi_B = 1.0$; all absorption and fluorescence expressed in terms of *relative numbers of photons*)

| Wavelength | Absorption | | Fluorescence | | Total Absorption | Total Fluorescence | Φ_{net} | % absorption by Group B |
	Group A	Group B	Group A	Group B				
500 mμ	0	1	0	1	1	1.0	1.00	100
525	0	2	0	2	2	2.0	1.00	100
550	1	4	0.5	4	5	4.5	0.90	80
575	2	2	1.0	2	4	3.0	0.75	50
600	4	1	2.0	1	5	3.0	0.60	20
625	3	0	1.5	0	3	1.5	0.50	0
650	2	0	1.0	0	2	1.0	0.50	0

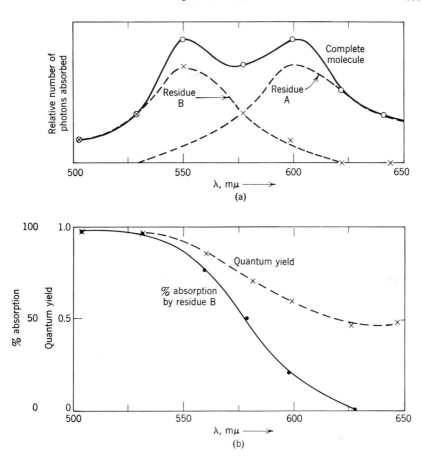

Fig. 25-2 Fluorescence of a molecule containing two independent chromophores: (a) fluorescence spectra; (b) quantum yield.

quenching processes. Collisional routes of energy loss are important in quenching and, accordingly, fluorescent yields are generally quite highly temperature dependent. Phosphorescent emissions, as a group, constitute an extreme example of temperature dependent interactions; quantum yields for virtually all phosphorescent emissions are negligible at ordinary temperatures. *Concentration* quenching, which results from the reabsorption of emitted fluorescence, as well as from other factors, becomes important at concentrations no greater than 10^{-2} molar; often concentration effects are observed at considerably higher dilutions. The mechanisms of quenching processes have been discussed in Chapter 7; the experimental measurement of quantum yields is described on p. 568.

QUANTITATIVE FLUORIMETRY

The quantitative features of light absorption are described by

$$(24\text{-c}) \qquad\qquad\qquad I = I_0 e^{-KCl},$$

where I_0 and I are incident and transmitted intensities, respectively, C is the concentration of the absorber, K is a constant characteristic of the absorber, and l is the path length of light through the solution. From this expression the *amount* of light absorbed by a solution is obtained as

$$(25\text{-a}) \qquad\qquad\qquad (I_0 - I) = I_0(1 - e^{-KCl}).$$

The corresponding intensity F fluoresced by a substance of quantum yield Φ_{FL} is thus

$$(25\text{-b}) \qquad\qquad\qquad F = \Phi_{FL} I_0(1 - e^{-KCl}).$$

An approximation to (25-b) may be obtained by considering the series expansion of the quantity e^{-KCl}:

$$(25\text{-c}) \qquad\qquad e^{-KCl} = 1 - KCl + \tfrac{1}{2}(KCl)^2 - \dots.$$

Substitution of this expression in (25-b) gives

$$(25\text{-d}) \qquad\qquad F = \Phi_{FL} I_0[KCl - \tfrac{1}{2}(KCl)^2 + \dots.$$

When a solution of *low* optical density is illuminated, the quantity (KCl) is *small*, so that $(KCl)^2$ and higher terms are negligible. Thus at low optical densities,

$$(25\text{-e}) \qquad\qquad \boxed{F = \Phi_{FL} I_0 KCl}.$$

That is, for a given substance, path length, and incident intensity, and at low concentrations, fluorescence is directly proportional to concentration. Measurement of fluoresced intensity can thus be used to determine concentration within a concentration range for which it can be established, experimentally, that (25-e) applies.

Note that, since the intensity of fluoresced light is measured directly, very low fluoresced intensities can be measured in an instrument which is free of stray light or other sources of disturbance. This may be compared with the difficulty of detecting very small differences between I_0 and I when dilute solutions are studied colorimetrically. Depending upon the quantum yield of the particular fluorescing substance, fluorimetric sensitivities may be as much as 10^3 times as great as those of absorption analyses. Furthermore, the emission of characteristic fluorescence is a somewhat more distinctive property than is absorption at a characteristic wavelength. Thus fluorimetry tends to be a highly *selective* as well as a highly *sensitive* method of analysis.

An important difference between (25-e) and expressions such as (24-c), which describe light absorption, is that F and I_0 represent intensities at different wavelengths, whereas I and I_0 represent intensity of the same wavelength. Since the wavelength response of detectors is variable, F and I_0 cannot be compared directly on an absolute basis. In quantitative fluorimetry, however, no absolute comparison of these two intensities is required. The fluoresced intensity is simply determined in a given instrument and compared with that emitted by a reference solution.

A simple fluorimeter is diagramed in Fig. 25-3. The light beam passes through a collimating lens and a filter before reaching the specimen cell. In general, illumination passed by the filter represents some distribution of wavelengths, $(\Sigma\lambda)_1$. The light passes through the cell to an absorbent surface, exciting fluorescence of wavelength distribution $(\Sigma\lambda)_2$ in the solution. The fluoresced intensity, which leaves the cell in all directions, is measured in a direction in which interference from the illuminating beam is avoided. Fluorescence emitted in all other directions is absorbed by the cell housing, as is the transmitted beam. A filter placed beyond the cell absorbs any scattered radiation in the wavelength range $(\Sigma\lambda)_1$. The filtered beam is then incident upon a detector.

The rectangular light path shown in Fig. 25-3 is suitable for measuring fluorescence from dilute solutions. However, if the concentration of the solute is such that more than, at most, 5% of incident intensity is absorbed, *inner filter* effects interfere with measurements of fluorescence. The origin of inner filter effects is twofold: (a) *Exciting* light may be absorbed before it reaches the region of the solution from which the observed fluorescence is emitted, so that the effective exciting intensity is less than that incident upon the sample cell. Thus, if only emission from molecules located near the center of the cell is observed, as in the arrangement shown in Fig. 25-4, the effective incident intensity is reduced

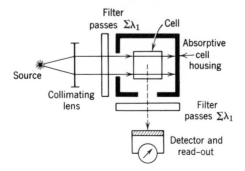

Fig. 25-3 Fluorimeter.

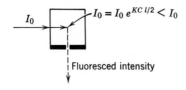

$$I_0 = I_0\, e^{KC\,l/2} < I_0$$

Fluoresced intensity

Fig. 25-4 Inner filter effect.

from I_0 by an amount determined by the optical density of the solution, and by the light path from the edge of the cell to the region of effective emission. (b) *Fluoresced* light may be reabsorbed before it escapes from the cell. When the fluorescence of concentrated solutions must be observed, use of a rectangular light path is unsuitable, and fluorescence from the cell surface is measured instead. The experimental arrangement is shown in Fig. 25-6 on p. 562 and is discussed in the text accompanying the figure.

THE SPECTROFLUOROPHOTOMETER

The spectrofluorophotometer is generally similar in basic design to the fluorimeter illustrated in Fig. 25-3, except that the absorption filters of the latter instrument are replaced by monochromators. Thus both the incident wavelength and the emitted wavelength to be measured may be selected at will. A diagram of a typical spectrofluorophotometer is given in Fig. 25-5.

The choice of components used in the spectrofluorophotometer is influenced by the fact that the level of intensity supplied to the detector tends to be limiting in that instrument. The intensity of the monochromatic excitation beam, and thus the intensity of fluorescence by the specimen, tends to be lower than in the fluorimeter, where the filtered illumination is usually much less highly monochromatic. Also, only a small portion of the total fluoresced intensity is transmitted by the emission monochroma-

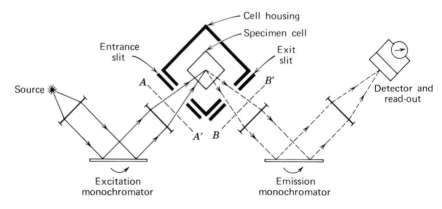

Fig. 25-5 Spectrofluorophotometer.

tor. (This situation may be compared with that of spectrophotometry, in which total absorption is measured.) Thus the light source used for spectrofluorophotometry should provide very high intensities at all wavelengths; a xenon vapor lamp is most suitable for the purpose. Diffraction gratings are almost invariably used as monochromators in order to avoid excessive light losses due to absorption by two prisms.

The standard specimen cells used for spectrofluorophotometry resemble the typical rectangular spectrophotometer cell shown in Fig. 24-11, but they have four transparent windows. Other designs include small cylinders (for measurements of very small solution volumes) and disc-shaped cells (for surface measurements from concentrated solutions). Since excitation is very often by ultraviolet wavelengths, these cells are normally made of quartz.

Stray light can be a serious problem in spectrofluorophotometry. The level of scattered radiation is reduced by using a series of entrance and exit slits (rather than just the single slits shown in Fig. 25-5). The sensitivity of the instrument can then be varied by changing the widths of these slits, as well as by adjustments of the recording system. Double monochromator systems have also been used.

Detection systems for spectrofluorophotometry are similar to those used in spectrophotometry, as described in Chapter 24.

TYPES OF SPECTROFLUOROPHOTOMETRIC MEASUREMENT

The spectrofluorophotometer can be used for *quantitative fluorimetry*. For this purpose the instrument has the advantage that a maximally absorbed wavelength can be selected for excitation and a wavelength of maximum emission for read-out. Nevertheless, the simplicity of the fluorimeter generally makes this instrument preferable for use in routine analyses.

The major application of the spectrofluorophotometer is in the recording of *fluorescence spectra* and the determination of *quantum yields of fluorescence*. These methods, which require the application of extensive corrections to data obtained instrumentally, are described on p. 565 and p. 568, respectively.

The spectrofluorophotometer may function as a *spectrophosphorimeter* when equipped for sequential interruption of the incident and emitted light beams. Phosphorescence measurements are discussed on p. 570. Note that, in the absence of sequenced shutters, spectrofluorophotometers and fluorimeters record both fluoresced and phosphoresced intensities without distinction.

The *polarization* of fluoresced radiation may be studied by placing polarizing devices (in practice, some form of polarizing prism) at the positions AA' and BB' which are indicated by dotted lines in Fig. 25-5. Measurements of depolarization of fluorescence are discussed on p. 571.

The spectrofluorophotometer may be used as a *nephelometer* for measurements of the intensity of light scattering, as described on p. 572.

MEASUREMENT OF THE PHOTON OUTPUT OF LIGHT SOURCES

In making quantitative spectrofluorometric or photochemical measurements it is often necessary to know the intensity of light beams in terms of number of quanta emitted per unit time. The most commonly used light-sensitive devices are either responsive to the total energy of a light beam (bolometers, thermopiles) or are normally calibrated in terms of response per watt or other energy unit (photocells). The photon output may, however, be measured directly by means of a *chemical actinometer* or by a *fluorescent screen*, for which the fluorescence yield is independent of wavelength. The photon intensity of *monochromatic* beams may also be determined from the readings of a calibrated *thermopile* or *phototube*. These methods for determining photon intensities are discussed in the following text. Here it may be noted that the *photon output at any wavelength is proportional to the product (energy output × wavelength)*. This follows from the fact that, as given in Chapter 1,

$$(25\text{-}f) \qquad\qquad E = h\nu = \frac{hc}{\lambda}$$

where E is the energy per photon, c is the velocity of light, and h is Planck's constant (6.6×10^{-27} erg sec). Thus;

$$(25\text{-}g) \quad \text{photon output} = \frac{\text{energy output}}{\text{energy per photon}} = \frac{\text{energy output} \times \lambda}{hc}.$$

Note that, whereas the total energy output of any polychromatic source can be measured directly by means of a thermopile, the total photon output cannot be so measured, since the distribution of wavelengths emitted by the source is in general unknown. Equation 25-g serves for interconversion of energy and photon readings *at a single wavelength* only.

THE CHEMICAL ACTINOMETER

The chemical actinometer is a solution in which a photochemical reaction of high, precisely known quantum yield occurs upon irradiation. The initial concentration of the reactants is known, and the final concentration of a product can be measured after irradiation of measured duration. Then,

(25-h)   source output in photons per unit time

$$= \frac{\text{number of incident quanta}}{\text{duration of irradiation}}$$

$$= \frac{\text{number of moles of product formed}}{\text{quantum yield of reaction} \times \text{duration of irradiation}}.$$

The type of actinometer which is used virtually exclusively measures the photodissociation of ferrioxalate ion to form ferrous oxalate, according to the reaction

(25-i) $2\text{Fe}(\text{C}_2\text{O}_4)_3^{3-} \xrightarrow{h\nu} 2\text{Fe}(\text{C}_2\text{O}_4) + 3\text{C}_2\text{O}_4^{2-} + 2\text{CO}_2.$

After irradiation, a solution of phenanthroline is added to the actinometer, forming a colored complex with the reduced (ferrous) iron. The concentration of the iron phenanthroline complex can then be determined by measurement of its absorption at 510 mμ.

The quantum yield of the reaction is not entirely invariant with wavelength, but it has been accurately determined at a number of wavelengths. These values are given in Table 25-2. The yield has also been shown to be independent of the level of incident intensity.

At wavelengths longer than 450 mμ, the quantum yield of the reaction falls rather rapidly. The reaction is thus suitable mainly for measurement of the quantum intensities of ultraviolet radiations.

Table 25-2   Quantum Yields for the Photodissociation of Ferrioxalate Ion (From Hatchard and Parker; see General References)

Wavelength	Φ
456 mμ	1.11
405	1.14
361,365	1.21
334	1.23
313	1.24
297,302	1.24
254	1.25

THE FLUORESCENT SCREEN

The fluorescent screen, or proportional photon counter, is a concentrated solution of a substance for which the fluorescence yield is high and independent of the incident wavelength. The relative intensities of light of different wavelengths emitted by such a solution is always the same, regardless of the wavelength of the incident light. Thus, if the screen absorbs the illuminating beam completely, the response of a phototube to the emitted fluorescence must always be proportional to the total number of quanta in the original beam. The spectral sensitivity curve of the photocell need not be known; it is necessary only that its sensitivity to the distribution of wavelengths emitted by the screen be moderately high. The light to be measured must be of sufficiently high quantum energy (short wavelength) to excite the complete fluorescence spectrum of the screen.

The fluorescent screen arrangement is diagramed in Fig. 25-6. Monochromatic light of wavelength λ_1 is incident upon the screen and is, essentially, completely absorbed close to the surface of the solution. Since total emission is to be measured, "inner filter" effects at the screen must be avoided (cf. p. 557); a surface light path, as shown, is therefore used. (For accurate measurements, however, corrections for reabsorption of fluorescence within the solution must nevertheless be made.) Light of the incident wavelength, which is scattered from the surface of the screen, is removed by a suitable filter, while the fluorescence emission is transmitted and is incident upon the photocell.

When the photon intensity of polychromatic beams is measured by the same technique, it may not be feasible to separate the incident and emitted wavelengths by means of a filter. Scattering of the incident beam can then be accounted for by a blank reading, for which the screen solution is replaced by a nonfluorescent solvent.

Substances which have been used for fluorescent screens include

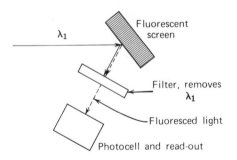

Fig. 25-6 The fluorescent screen.

1-dimethylamino-napthalene-5-sodium sulfonate (for measurement of intensities in the ultraviolet) and mixtures of acriflavin and rhodamin B (for measurement of intensities in the visible).

THERMOPILE CALIBRATION

The response of a thermopile is proportional to the total energy content of incident light, regardless of wavelength. The energy content of a monochromatic beam, as measured by this device, may be converted into intensity in photons by the application of (25-g). The calculated photon output is then accurate to the extent that the nominal wavelength setting of the monochromator correctly represents the wavelength of the emergent beam. If the monochromator slit width is appreciable, the calculated photon output must be corrected, with the aid of dispersion data for the monochromator, to account for the variation in thermopile response to the distribution of longer and shorter wavelengths which are present in the incident beam.

Calibration of the thermopile is obtained by measurement of its response to irradiation by a standard lamp of *known energy output*. As shown in Fig. 25-7, the standard lamp is placed at the center of an *integrating sphere*, which is a globe coated on its inner surface by a substance (usually barium sulfate) of high and diffuse reflectance. The sensitive surface of the thermopile is incorporated in the inner surface. A coated screen is placed between the lamp and the thermopile, so that the latter receives no direct light, but is illuminated entirely by reflections from the inner surface of the sphere. Light emitted by the lamp in any direction undergoes a series of reflections from the sphere and ultimately is absorbed by the thermopile. In this way, fluctuations in intensity which originate from angular inhomogeneities in the output of the lamp are of no consequence.

Ideally, light is absorbed within the sphere only by the surface of the

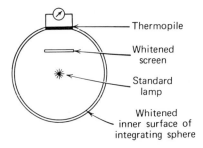

Fig. 25-7 Calibration of thermopile by means of an integrating sphere.

thermopile, so that the total output of the lamp is measured directly. In practice, a small correction must be made for absorption by the sphere, the reflectivity of which is of the order of 98%. Supports used to position the lamp at the center of the sphere, and any other equipment included therein, must also be whitened to avoid absorption.

PHOTOTUBE CALIBRATION

Phototube calibrations as supplied by manufacturers are usually expressed as response per watt as a function of wavelength, but may be converted into response per photon by the use of (25-g). Since response curves vary between individual phototubes, and since sensitivity changes with age, calibration should be repeated for accurate fluorescence measurements. Calibration data may be obtained by (a) comparing the phototube response to a series of monochromatic wavelengths with that of a calibrated thermopile or (b) determining the phototube response to monochromatic radiation isolated from the output of a lamp of known *color temperature*.

Discharge lamps, such as the mercury arc, produce a line emission spectrum which is characteristic of the energy levels of the specific excited element(s). Heated filaments, however, tend to radiate a continuous spectrum in which the distribution of intensities as a function of wavelength is characteristic of the temperature, rather than of the composition, of the emitter. To the extent that an incandescent filament is a perfect radiator (black body), the distribution of emitted wavelengths may be *calculated* from Planck's law of radiation:

$$(25\text{-j}) \qquad I_\lambda = \frac{8\pi ch}{\lambda^5}\left(\frac{1}{e^{ch/\lambda kT}-1}\right) \cong 8\pi ch\lambda^{-5}e^{(-ch/\lambda kT)},$$

where I_λ is the energy density of radiation at wavelength λ, c is the velocity of light, h is Planck's constant, k is the Boltzmann constant $(1.38 \times 10^{-16}$ erg/degree), and T is the absolute temperature. The approximate form of (25-j) is known as Wien's law. The derivation of these equations cannot be considered here. (See, however, [1].)

For the expression of spectral output in *photons*, combination of (25-j) with (25-g) yields

$$(25\text{-k}) \qquad\qquad I'_\lambda \cong 8\pi\lambda^{-4}e^{-(ch/\lambda kT)},$$

where I'_λ is the intensity output at wavelength λ expressed in numbers of photons.

The *color temperature* of any lamp which emits a continuous spectrum is defined as the temperature of a perfect black body which would emit the same distribution of wavelengths. An incandescent tungsten filament acts, to a good approximation, as an ideal black body, so that its precise spectral output can be calculated if the operating temperature of the filament is known.

Equations (25-j) and (25-k) show that emission at any temperature is maximal at a wavelength which becomes shorter as temperature increases. Thus an incandescent source which emits principally in the blue is characterized by a higher color temperature than one which emits principally in the red.

FLUORESCENCE SPECTRA AND THEIR CORRECTION

The spectrofluorophotometer may be used for measurement of both the excitation and the emission spectra of a specimen. The *excitation* spectrum is obtained by recording the detector output for successive wavelength settings of the excitation monochromator (at constant slit width), while the emission monochromator remains at a setting of maximum intensity. Ideally, if the quantum efficiency of fluorescence by the specimen is independent of wavelength, the excitation spectrum is identical with the *absorption* spectrum of the same compound. Use of the spectrofluorophotometer for such measurements can be advantageous in that absorption spectra can thus be obtained at much lower concentrations than is possible spectrophotometrically. Also, this method can be used to eliminate interference from other components of the solution which absorb but do not fluoresce, or which fluoresce at different wavelengths than does the substance of interest. However, as discussed below, corrections are required in order to obtain undistorted fluorescence spectra; spectrophotometric measurements of absorption are more reliable unless particular precautions are taken to avoid errors.

The *emission* spectrum is obtained by setting the excitation monochromator at a wavelength of maximum absorption and recording detector output as a function of the emission monochromator setting (at constant slit width). Ideally, the emission spectrum is specifically characteristic of the fluorescence of the specimen.

Both excitation and emission spectra, as recorded by standard, commercially available spectrofluorophotometers, are severely *distorted* as a consequence of: (a) variations with wavelength in the output of the light source-excitation monochromator combination; (b) variable light losses at the emission monochromator; (c) variable spectral bandwidth of radiation transmitted by the emission monochromator, and (d) variations with wavelength in the response of the light sensitive detector. Because of these effects, absorption maxima at wavelengths of low source output and emission maxima at wavelengths of low detector sensitivity are reduced and may even be overlooked. Thus fluorescence spectra recorded in the literature are (or should be) corrected from the apparent form obtained directly from the instrument to a standard form. The standard form chosen for the reporting of spectra is a plot of the *relative number of photons emitted* as a function of *frequency*. A scale linear with respect to frequency, rather than to the reciprocally related quantity, wavelength, is chosen in order to avoid compression of details at the short wavelength end of the spectra, and to represent more accurately the relation between the form of absorption and emission spectra (cf. p. 135). Reporting of correct data in a standard form makes it possible to compare results obtained with the use of different instruments. Since the fluorescence quantum yield may be temperature dependent, the temperature at which measurements are made should also be controlled and recorded.

The correction of excitation spectra must account for differences with wavelength in the photon intensity of the output of the excitation monochromator. Methods for obtaining source outputs on a number-of-photons basis have been described on p. 560. Variations in the output of the emission monochromator or in the response of the detector need not be considered in correcting excitation spectra, since emitted radiation is measured only at a single wavelength.

The correction of emission spectra need not account for differences in the spectral output of the source (since excitation is at one wavelength only), but the other three factors mentioned above must be considered. Corrections for bandwidth, light loss at the monochromator, and detector response effects may be made simultaneously in the visible, for which sources of known spectral output are available. Thus the specimen can be replaced by a lamp of specified color temperature (cf. p. 564), and the detector response at each wavelength compared with that obtained from the specimen. Since standard lamps are not available for ultraviolet wavelengths, readings in that portion of the spectrum must be corrected separately for the spectral response characteristic of the detector (calibrated with respect to the response of a thermopile), for the bandwidth (as calculated from the dispersive properties of the monochromator; cf. pp. 514 and 519), and for light losses at the monochromator. The latter are determined experimentally at a wavelength, such as $254\,m\mu$, for which monochromatic illumination is available; losses of intensity at adjacent wavelengths are then assumed to be identical.

More convenient alternatives to these fundamental methods for correcting emission spectra exist. Thus spectra may be corrected by comparison with the spectrum of a standard fluorescent substance (quinine sulfate) for which fully corrected spectra have been published. The fluorescent screen described on p. 562 is also useful.

Apart from corrections for instrumental factors, fluorescence spectra must be free of distortions introduced by interaction occurring at the sample. If appreciable *inner filter* effects occur (cf. p. 557), variations in the effective intensity of the exciting beam must be accounted for. *Concentration quenching* effects and *photodissociation* of the sample by the exciting radiation must be avoided. Dissolved oxygen substantially reduces the fluorescence efficiency of some substances, while others are unaffected; susceptibility of a system to oxygen quenching should always be tested.

Variations with time in source output or detector response often occur, and may be corrected by adjusting the sensitivity of the spectrofluorophotometer to give a constant reading for the fluorescence of a standard solution.

Although none has as yet come into general use, a number of instruments have been devised which record corrected fluorescence spectra more or less directly. A diagram of one such instrument is shown in Fig. 25-8. In this spectrofluorophotometer a *reference source* is provided in addition to the usual xenon lamp. The spectral output of the reference source must not vary with intensity, but need not be precisely known. Light transmitted by the sample (beam *S* in Fig. 25-8) and from the reference source (beam *R*) alternately illuminate a bolometer. The bolometer signal actuates an electronic *comparing element* which adjusts the power supplied to the reference source in such a way that its energy output is held in constant proportion to the energy transmitted by the sample. Beam *R'* from the reference lamp then passes through a partially

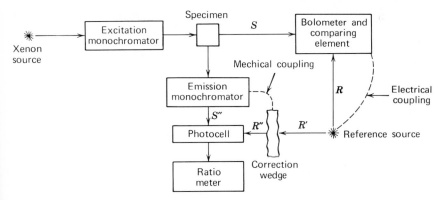

Fig. 25-8 Block diagram of a correcting spectrofluorophotometer.

absorbing *correction wedge*, the function of which is to account for differences with wavelength (i.e., with emission monochromator setting) in the response of the *photocell* to the output of the emission monochromator. Beam R'' from the correction wedge and beam S'' from the emission monochromator alternately illuminate the photocell, the response of which is recorded by a ratio meter. Fully corrected emission spectra are thus recorded directly. Because the bolometer which actuates the comparing element is an energy-sensitive device, excitation spectra as recorded by the instrument are plots of output per constant energy input. Equation 25-g must thus be applied in order to correct these spectra to a basis of constant photon input.

MEASUREMENT OF QUANTUM YIELD OF FLUORESCENCE

Measurements of the absolute quantum yield of fluorescence require the comparison of the number of quanta absorbed by a substance with the total number fluoresced. As the foregoing discussions indicate, accurate measurements of the absolute yield are not simple to make. Apart from application of all the corrections which have been discussed in the preceding section, it is necessary to measure all the quanta emitted within the complete solid angle of 4π rad or to devise some valid method for computing this total from measurements of the fluorescence within an aperture of limited angular extent. Standard instruments for the measurement of fluorescence, as shown in Figs. 25-3 and 25-5, are not suitable for measurement of total fluorescence, but simply detect those quanta which are emitted approximately at right angles to the exciting beam. These instruments are therefore primarily suited only to measurements of the *relative* intensity of fluorescence.

Measurements of the total spatial distribution of fluoresced quanta have been made in a number of ways: (a) The complete fluorescence is measured directly in an integrating sphere, as illustrated in Fig. 25-9. (For a description of the integrating sphere, see also p. 563.) Exciting

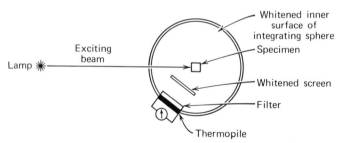

Fig. 25-9 Measurement of quantum yield of fluorescence by means of an integrating sphere.

radiation enters the sphere through a small opening and is incident upon the specimen cell, which is placed at the center of the sphere. Both transmitted and fluoresced radiation undergo multiple reflections from the surface of the sphere and are incident upon the surface of a calibrated thermopile after passing through a filter. Thermopile readings are recorded with a filter which absorbs the fluoresced radiation but transmits (a known fraction of) the exciting light, and with a filter which absorbs the exciting light, but transmits at other wavelengths. After correction for partial absorption by these filters, and by the surface of the integrating sphere, the ratio of the thermopile readings gives the quantum yield of fluorescence.

Other methods of measuring total emission include: (b) direct measurement of the solid angle subtended by the detector. If spherical symmetry of emission is assumed, the total fluorescence can then be calculated. (c) The comparison of fluoresced intensity with that scattered by a non-absorbing solution of equal apparent optical density. Measurements of the extent of polarization of the scattered and fluoresced intensities then permit calculation of total fluoresced intensity. (For a complete discussion of the latter type of correction see [2].)

The uncertainties in measuring absolute quantum yields of fluorescence are such that reliable values have been difficult to obtain. In practice, quantum yields are obtained by comparison with that of a standard substance, usually quinine sulfate, for which a reliable value of the quantum yield has been established by absolute methods. The relative measurements, which have been considered more reliable than any single absolute measurement, can be made in a standard spectrofluorophotometer in the following manner: The complete corrected emission spectra of the unknown and of the standard are obtained, and the integrated areas under these curves, F_x and $F_{quinine}$, respectively, are measured. (For a discussion of graphical integration methods, see Appendix G.) In each case the F value is proportional to the product $(I_0 \times \Phi \times \text{O.D.})$, where I_0 is the intensity of the exciting radiation, Φ is the quantum yield of fluorescence, and O.D. is the optical density of the solution. Thus Φ for the unknown substance may be obtained from

(25-b)
$$\boxed{\frac{F_x}{F_{quinine}} = \frac{\Phi_x}{\Phi_{quinine}} \frac{(\text{O.D.})_x}{(\text{O.D.})_{quinine}}}.$$

It must be emphasized that the F values used in (25-b) are *not* readings of maximum emission obtained directly from the spectrofluorophotometer, but are the integrated areas under the complete fluorescence curves. This is essential, since the ratio of maximum ordinate to total area under the curve varies according to the shape of the spectra of different substances.

PHOSPHORIMETRY

Phosphorimetry is the measurement of radiation which is re-emitted from an illuminated specimen after a short delay. Instrumentation for phosphorimetry differs from that used in fluorimetry or spectrofluorophotometry in the inclusion of provision for intermittent illumination of the specimen with subsequent observation of re-emission; often a rotating shutter is used. As shown in Fig. 25-10a, the specimen is illuminated at one position of the shutter, while fluorescent (i.e., immediate) reradiation is absorbed. After a time interval Δt, as shown in Fig. 25-10b, the shutter has moved to a position at which the exciting emission is absorbed and the re-emitted radiation (phosphorescence) can reach the detector. Maximum detector response is obtained when the speed of rotation of the shutter is such that Δt is equal to the average lifetime of the excited stage from which phosphorescent emission occurs.

Phosphorescent measurements usually must be carried out at or near liquid nitrogen temperatures ($-196°C = +77°K$), since collisional quenching usually precludes the observation of phosphorescence at ordinary temperatures. Suitable solvents have the property of forming clear glasses (rather than crystals) upon cooling; in general, nonpolar solvents have been used.

Where applicable, phosphorimetry is a technique of even higher sensitivity and selectivity than is fluorimetry. In addition to quantitation by the measurement of phosphorescent intensity, and identification by means of recording phosphorescence spectra, a substance may be characterized by determination of the lifetime of the excited (triplet) stage. Although applications of phosphorimetry in biology have not been extensive to date, a number of compounds of biological interest are known to emit characteristic phosphorescence spectra.

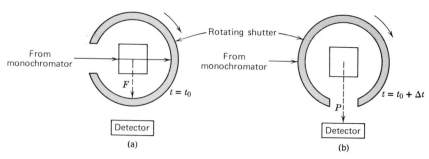

Fig. 25-10 Phosphorimetry.

DEPOLARIZATION OF FLUORESCENCE

When a solution absorbs polarized light, the subsequent fluorescence is depolarized to a degree which is determined intrinsically by the *nature of the electronic transitions* associated with absorption and emission, and extrinsically by the extent to which processes of *energy transfer* or molecular *rotation* of the emitter occur during the brief (10^{-9} to 10^{-8} sec) lifetime of the excited state. Although the intrinsic and extrinsic effects tend to superimpose, it is possible to deduce their contributions to the depolarization by adjusting the *concentration* and the *viscosity* of the fluorescent solution. Intermolecular energy transfer is negligible at low concentrations, while rotation of molecules is inhibited at high viscosity. Thus only intrinsic depolarization should be observed in dilute, viscous solutions. (Propylene glycol is commonly used as a solvent for such measurements.) A combination of intrinsic and energy transfer effects should be observed in concentrated but viscous solutions, and a combination of intrinsic and rotational effects contributes to depolarization in dilute solutions of low viscosity.

Polarization (p) has been defined quantitatively in Chapter 7 as

(7-m)
$$p = \frac{I_{\parallel} - I_{\perp}}{I_{\parallel} + I_{\perp}},$$

where $I_{\parallel}$ is the intensity polarized in the same plane as is the exciting beam, while $I_{\perp}$ is the intensity polarized at right angles thereto. As explained on p. 147, the value of p may in general range from -1 to $+1$, but it can be shown that for fluoresced radiation it may vary only within the range from $-\frac{1}{3}$ to $+\frac{1}{2}$. As noted above, p can be measured in a standard spectrofluorophotometer equipped with a pair of polarizers at positions AA' and BB', as shown in Fig. 25-5.

A *fluorescence polarization spectrum* is a plot of p as a function of *excitation wavelength*. Basically, such spectra tend to consist of maxima or minima located at wavelengths of maximum absorption. They may, however, be more complex in form when more than one electronic transition contributes to a given absorption peak. Comparison of polarization and absorption spectra is useful in the interpretation of electronic energy level transitions within the fluorescent molecule.

Studies of depolarization at concentrations of the order of 0.1 to 1.0 molar have contributed to the development of a theoretical description of intermolecular energy transfer processes. Intramolecular energy transfer also contributes to depolarization but is more difficult to analyze, since the effect cannot easily be distinguished from that of intrinsic polarization related to the nature of energy level transitions. However, anomalously low apparent intrinsic polarizations have been interpreted in terms of intramolecular energy transfer; an example of biological interest is the low polarization of fluorescence from chloroplasts.

Measurements of depolarization originating from rotational diffusion are used (in combination with other physical techniques) to deduce the *molecular volumes* and shapes of both large and small molecules. The technique has been applied particularly to studies of the shapes and interactions of protein molecules.

USE OF THE SPECTROFLUOROPHOTOMETER FOR MEASUREMENTS OF LIGHT SCATTERING

The intensity of light scattered by suspensions of molecules or other small particles may be measured in order to determine concentrations. The use of colorimeters and spectrophotometers as turbidimeters, for the measurement of the intensity transmitted by scattering solutions, has already been mentioned on p. 539. Comparably, a fluorimeter or spectro-fluorophotometer may be used as a nephelometer, to measure the intensity of radiation scattered at right angles to the incident beam. For this purpose both excitation and emission monochromators are set at the same wavelength, which must be one for which absorption by the specimen is negligible. For quantitative analysis the intensity of the scattered radiation is compared with a standard curve.

Scattered intensity tends to be a function of the square of particle size and is therefore highly sensitive to changes in molecular size or state of aggregation. Thus the effects of chemical agents or of enzymes upon particles of colloidal size may often be deduced from the observation of changes in scattered intensity. For example, the addition of detergent solutions to bacterial suspensions may sharply reduce scattering by lysing the organisms. Measurements of scattering may also be used to deduce the kinetics of such reactions.

The *angular* distribution of scattered intensities, which is critically related to the size and shape of scattering particles, can *not* be measured in a standard spectrofluorophotometer. This type of measurement is not considered in this book, but some references to discussions of the theory and instrumentation for light scattering measurements are included in the General References for this chapter.

RAMAN SPECTROSCOPY

Raman emission, which has been discussed on p. 149, consists of an alteration of the wavelength of scattered radiation by amounts which correspond to rotational or vibrational energy level changes in the scatterer. The principal instrumental requirement for the recording of Raman spectra is the availability of a source of *very high intensity*, since the intensities of Raman lines are only about 10^{-4} as great as that of the exciting radiation. The illumination must also be *monochromatic* in order to avoid the formation of an overlapping series of Raman spectra. In the past, one of the mercury arc lines has been used, but lasers now provide a more intense and highly monochromatic source of exciting radiation. Particular attention must be paid to the elimination of stray radiation of the exciting wavelength. While Raman spectra can be observed in sensitive spectrofluorophotometers, Raman spectrophotometers, which are specifically designed for the observation of these spectra, are available commercially.

The principal application of Raman spectroscopy has been in studies of the structure of organic molecules. Improvements in technique should favor extensions of the method to studies of compounds of biological interest. Observation of the Raman spectra of solvents has been proposed as a means of obtaining a standard for the correction of fluorescence spectra.

REFERENCES

General References

D. M. Hercules, ed., *Fluorescence and Phosphorescence Analysis*, Interscience, New York, 1966. The following chapters are of particular interest: Chapter 1 (Theory of Luminescence Processes, by D. M. Hercules), Chapter 2 (Luminescence Instrumentation and Experimental Details, by D. W. Ellis), Chapter 5 (Analytical Uses of Phosphorescence, by J. D. Wineforder), Chapter 6 (Fluorescence in Biomedical Research, by B. L. Van Duren), and Chapter 8 (Polarization of the Fluorescence of Solutions, by G. Weber).

H. A. Strobel, *Chemical Instrumentation*, Addison-Wesley, Reading, Mass., 1960, Chapter 8 (Light Scattering Photometry).

C. G. Hatchard and C. A. Parker, *Proc. Roy. Soc. (London)* **A235**, 518 (1956). This article describes "A New Sensitive Chemical Actinometer II. Potassium Ferrioxalate as a Standard Chemical Actinometer."

C. A. Parker and W. T. Rees, *Analyst* **85**, 587 (1960). A particularly lucid and informative article on "Correction of Fluorescence Spectra and Measurement of Fluorescence Quantum Efficiency."

L. S. Forster and R. Livingston, *J. Chem. Phys.* **20**, 1315, 1952, "The Absolute Quantum Yields of Fluorescence of Chlorophyll Solutions." This article describes the use of an integrating sphere for obtaining quantum yields.

W. H. Melhuish, *J. Phys. Chem.* **64**, 762 (1960), "A Standard Spectrum for Calibrating Spectrophotometers."

G. K. Turner, *Science* **154**, 183 (1965), "An Absolute Spectrofluorophotometer."

C. Tanford, *Physical Chemistry of Macromolecules*, Wiley, New York, 1961, Chapter 5 (Light Scattering). A general discussion of light scattering with reference to studies on macromolecules of biological interest.

Specific References

[1] F. K. Richtmeyer, E. H. Kennard, and T. Lauritsen, *Introduction to Modern Physics*, McGraw-Hill, New York, 1955, Chapter 4 (The Origin of the Quantum Theory).
[2] G. Weber and F. W. J. Teale, *Trans. Faraday Soc.* **53**, 646, 1957.

CHAPTER 26

The Determination
of Refractive Index

Determination of the value of refractive index is required in many applications of optical instrumentation. Thus, for example, determination of refractive index may aid in the identification of substances. Knowledge of the index of glasses and cements is essential in the design of optical components. Interpretation of quantitative measurements made with many different types of instrument, for example, the phase contrast microscope, the polarizing microscope, or the ultracentrifuge, requires measurement of the refractive index of the specimen.

Refractive index can be measured by a number of different techniques, the method chosen being determined by the level of accuracy required and by the physical state of the specimen, which may be a gas, pure liquid, solution, crystal, amorphous particulate solid, or a block of glass or other extended solid. A variety of optical phenomena is exploited in making these measurements. Thus geometric *refractometers* may determine the *critical angle for total internal reflection* or may directly measure the *deviation* of the refracted rays. Interferometric refractometers (*interferometers*) are used to measure the *optical path* of a specimen of known length. The refractive index of small solid particles can be determined by observation of *phase contrast effects* in either an ordinary or phase contrast light microscope (the *immersion method*). The index of glasses can be found by determination of the *angle of minimum deviation* by a prism made of the substance. The principles upon which these methods are based have been discussed at various places in this book, including

Chapters 1, 3, and 23. Each of the methods is described or reviewed in this chapter.

APPLICATION OF CORRECTIONS IN REFRACTOMETRY:

All substances exhibit some degree of dispersion of refractive index (i.e., variation of refractive index with wavelength; cf. p. 508 and Fig. 23-1); exact values of n are thus always associated with a specified wavelength. Tables of indices usually cite n_D, the refractive index for sodium D line light. For some substances, dispersion is so slight that, when extreme accuracy is not required, the value of n_D applies adequately over the entire visible spectrum. For refractometric measurements of high precision, monochromatic illumination is essential. Nevertheless, the effects of dispersion can largely be avoided in geometrical refractometers by the use of compensators of the type illustrated in Fig. 26-3.

Refractive index is a function of temperature to an extent which varies substantially between different types of specimen. For liquids the temperature coefficient of variation in refractive index is usually of the order of 5×10^{-4} refractive index units per degree centigrade, but it may vary within an order of magnitude. For the majority of solids the variation is much less, being of the order of 10^{-5} or 10^{-6} unit per degree. However, a few solids, notably certain plastics, possess much larger temperature coefficients. The values of n decrease as the temperature is raised; these effects can be related directly to the thermal expansion of the specimen. Control of temperature or correction for temperature effects may be required in refractometry, depending upon the degree of accuracy sought and the nature of the specimen. Commercially available refractometers are often equipped for operation at a controlled temperature.

CRITICAL ANGLE REFRACTOMETERS

The most widely used types of refractometer measure the critical angle for total internal reflection. As explained on p. 5, rays which are incident upon an interface between media of different refractive indices from the side of higher index are totally reflected, rather than refracted, *if* they are incident at an angle (i.e., with respect to the normal) which is greater than the critical angle α_c. The value of α_c at any interface was shown in Chapter 1 to be given by the expression

$$(26\text{-a}) \qquad \sin \alpha_c = \frac{n_2}{n_1} ,$$

where n_2 is the refractive index of the medium (of higher index) in which reflection occurs. Correspondingly, light rays which enter a medium of higher index at a grazing angle (i.e., at an angle very close to 90° with respect to the normal to the interface) are refracted at angle α_c with respect to the normal.

In critical angle refractometers the specimen (of refractive index n_x) is in contact with a prism of known refractive index, n_p ($n_p > n_x$). Light refracted into the prism from the prism-specimen interface is contained within the critical angle, so that the boundary between illuminated and unilluminated regions can be observed by a movable telescope system.

In the *Pulfrich* refractometer, diagramed in Fig. 26-1a, light is focused by a lens onto the specimen-prism interface. Ray ①, which is incident at a grazing angle, is refracted through the critical angle α_c, while other rays, such as ② and ③, make larger angles with respect to the surface and are refracted closer to the normal. Ray ① thus forms a boundary between light and dark areas at a position which is determined by the relative values of the refractive indices of prism and specimen. A telescope (which may include a system for the compensation of dispersion, as described below) is then so positioned that its field of view is equally divided between bright and dark areas, as shown in Fig. 26-1b. The position of the telescope is noted on a ruled scale. (Other arrangements which are geometrically equivalent may also be used.)

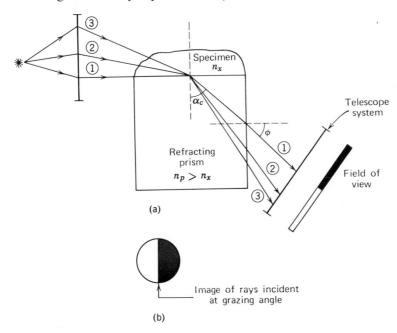

(a)

(b)

Image of rays incident at grazing angle

Fig. 26-1 Pulfrich refractometer.

In principle, n_x, the refractive index of the specimen, may then be calculated from the known refractive index of the prism, and from the angle ϕ measured by the scale of the telescope arm. This may be done by noting, first, that at the specimen-prism interface, the law of refraction gives

(26-b)
$$\frac{\sin 90°}{\sin \alpha_c} = \frac{n_p}{n_x} \quad \text{or} \quad \sin \alpha_c = \frac{n_x}{n_p}.$$

For refraction at the prism-air interface, the corresponding expression is

(26-c)
$$\frac{\sin (90 - \alpha_c)°}{\sin \phi} = \frac{1}{n_p},$$

or, since $\sin (90 - \alpha_c)° \equiv \cos \alpha_c$,

(26-d)
$$\cos \alpha_c = \frac{\sin \phi}{n_p}.$$

Equations 26-b and 26-d may be squared and added, giving, with the use of the trigonometric identity ($\sin^2 \alpha_c + \cos^2 \alpha_c \equiv 1$),

(26-e)
$$1 = \frac{n_x^2}{n_p^2} + \frac{\sin^2 \phi}{n_p^2},$$

or, upon rearranging and taking square roots,

(26-f)
$$\boxed{n_x = \sqrt{n_p^2 - \sin^2 \phi}} \ .$$

In practice, computations with the use of (26-f) can be avoided, since the telescope scale may be ruled directly in refractive index units. Calibration of the scale may be checked by measurement of specimens of known refractive index; glass blocks are sometimes supplied for this purpose.

The *Abbe* refractometer differs from the Pulfrich in that the specimen is, typically, used to form a thin layer between a pair of prisms, as shown in Fig. 26-2. The first prism directs the light into the layer of liquid specimen, and the second (refracting) prism performs the same function as that of the Pulfrich instrument. Both types of refractometer are very widely used, although the Abbe instrument requires somewhat more care in alignment than does the Pulfrich, and it may form a less sharply defined boundary. The *dipping refractometer* is a modification of the Abbe design in which the refracting prism can be lowered into the surface of the liquid to be studied.

When thermostated and used with monochromatic illumination, critical angle refractometers can measure refractive indices to within

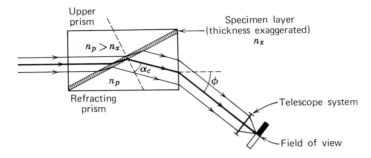

Fig. 26-2 Abbe refractometer.

10^{-5} units. Use of non-monochromatic light in an uncompensated system reduces precision to the level of about 5×10^{-3}. However, compensation of dispersion can be achieved with the use of a pair of *Amici prisms*. As shown in Fig. 26-3, the Amici prism is a cemented set of three prisms which are constructed of crown glass (low index and dispersion), flint glass (high index and dispersion), and crown glass, respectively. The components are so cut as to transmit D line (yellow) light without deviation, while wavelengths in the blue and red are deviated on opposite sides of the D line. A pair of these prisms, when mutually rotated, tend to reunite the dispersed wavelengths (i.e., by producing equal deviations in opposite directions). While the redirection is exact at two wavelengths only, the use of such a device can improve the precision of white light refractive index measurements to one part in 10^4.

The usefulness of critical angle instruments is limited to some extent by the requirement that the index of the refracting prism be higher than that of the specimen. If the reverse is true, rays incident upon the sample-prism interface are totally reflected within the sample, so that the field of view of the telescope remains dark. The upper limit of measurable indices is thus of the order of 1.80.

Critical angle refractometers are most commonly used for studies of

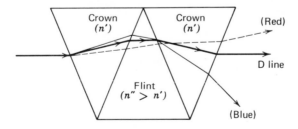

Fig. 26-3 Amici prism.

liquids but may also be used to measure solids, provided that at least one face of the solid specimen is polished. The polished face is placed in contact with the refracting prism, optical contact being established by means of a thin layer of liquid of index higher than that of the specimen. When solids are studied in an Abbe refractometer, the upper prism is removed. A disadvantage of critical angle measurements is the fact that only the surface layer of the specimen influences the reading obtained. If the properties of the surface of a solid differ from those of the bulk of the material, refractive indices obtained from critical angle refractometers may be quite misleading.

DIRECT MEASUREMENT (IMAGE DISPLACEMENT) REFRACTOMETERS

Several types of refractometer effectively measure the angle of refraction directly. Figure 26-4 shows an instrument in which a specimen of index n_x is included in a double prism of index n_p. Light from a slit is collimated by a lens and is incident normally on the face of the double prism. If $n_x = n_p$, no refraction occurs but, if $n_x \neq n_p$, the transmitted rays are deflected. A virtual image of the slit is then observed which is displaced through a distance determined by the difference in refractive indices. Because critical angle reflection is not involved, n_p may be either smaller or larger than n_x. Also, the bulk refractive properties of the specimen are measured. A number of commercial instruments designed on this principle vary with respect to the precision which can be obtained.

Another type of direct measurement refractometer uses the specimen as the dispersing element of a prism spectroscope. In this type of instrument, either two faces of a solid specimen must be polished to form a prism, or a liquid specimen is contained in a hollow prism-shaped cell. As shown in Fig. 26-5, monochromatic light from a slit source illuminates the specimen as a collimated beam. The light which leaves the prism is focused to form an image of the slit source *at a position which is deter-*

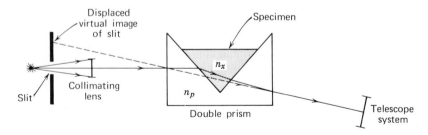

Fig. 26-4 Image displacement refractometer.

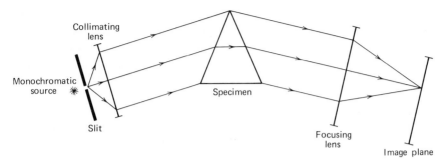

Collimating lens

Monochromatic source

Slit

Specimen

Focusing lens

Image plane

Fig. 26-5 Prism refractometer.

mined by the refractive index of the specimen. (Note that the instrument shown in Fig. 26-5 is essentially the same as the prism spectroscope illustrated in Fig. 23-2.) Under optimum conditions this type of refractometer can provide measurements of refractive index to one part in 10^6. The requirement for forming two polished surfaces on solid specimens is inconvenient, however.

The prism spectroscope principle is also exploited in one type of differential refractometer in which two prism-shaped cells contain the sample and a reference liquid, respectively. When the cells are mutually rotated by 180°, small resultant deviations (corresponding to small *differences* in refractive index) which can be measured very accurately are obtained.

INTERFEROMETRIC REFRACTOMETERS

The general principle of the interferometer has been discussed on p. 71. To summarize: Interference between pairs of coherent beams forms a system of fringes which is shifted if the optical paths of the beams differ. Optical path Δ is, as defined in Chapter 3 by

(3-dd) $$\Delta \equiv \sum nl.$$

Thus, if a pair of coherent beams travel an equal distance l before they superpose, the observed fringe shift which results from their interference can be related to differences in the refractive index of the media traversed. In interferometric refractometry the optical path through a standard cell, containing a medium of known refractive index, is compared with that of an unknown. The number of fringes, Δm, displaced when substances of indices n_0 and n_x are compared interferometrically is thus

(26-g) $$\Delta m = \frac{(n_x - n_0)l}{\lambda},$$

where l is the physical length along which the refractive indices differ, and λ is the wavelength of (monochromatic) light.

In the *Jamin interferometer*, illustrated in Fig. 26-6, two coherent beams are formed by reflection at the upper partially silvered and lower fully silvered surfaces, respectively, of a "Jamin plate." For a plate of appreciable thickness, well-separated beams are obtained. After passage through the interferometer cells, these beams are recombined by reflection at a second, identical plate which is in antiparallel orientation with respect to the first. (Note that, since each beam undergoes one reflection at an upper surface and one at a lower surface, the two make equal contributions to the amplitude of the recombined beam.)

The *Rayleigh interferometer* is shown in Fig. 26-7. In this instrument, light from a narrow slit is collimated by a lens, and it forms a pair of parallel coherent beams after incidence upon a pair of narrow slits of equal widths. The beams pass through the matched interferometer cells and are focused by a condensing lens, forming a diffraction image of the slit source in the focal plane of that lens.

In either type of interferometer a magnifying lens may be used to form a conveniently enlarged image of the diffraction pattern. As in any interferometer, the zero order fringe is centrally located when the optical paths through the two cells are identical. If the optical paths differ, a similar pattern is obtained, but its center is displaced in the direction of the solution of greater optical path. (For example, if the solution in cell B

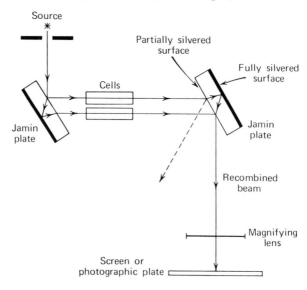

Fig. 26-6 Jamin interferometer.

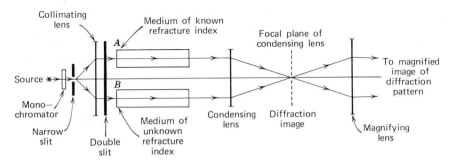

Fig. 26-7 Rayleigh interferometer.

of Fig. 26-7 is of higher refractive index than that in cell A, the zero order fringe formed in the focal plane of the magnifying lens is displaced downwards.)

Differences in refractive index between a standard liquid in one interferometer cell and the specimen in the other can be quantitated by observation of fringe shifts. The fringe shift may be measured directly by comparison with a reference pattern formed by beams which have undergone equal optical paths (i.e., after location of zero order fringes by observations under illumination by white light). Automatic selection systems have also been devised.

Under optimum conditions the precision of interferometric measurements of refractive indices approaches one part in 10^8 for gases. The precision of measurements on liquids tends to be limited by the difficulty of obtaining adequately homogeneous samples, but may approach one part in 10^7. Although the adjustment of interferometer systems is far more critical and time-consuming than that of other types of refractometer, interferometric measurements are the method of choice when refractive index values of the highest possible accuracy are sought.

IMMERSION METHODS

The refractive indices of small particles, whether amorphous or crystalline, may be determined by immersing these objects in a series of media of known index. The media and suspended particles are then observed in a microscope.

Immersion measurements of refractive index with the use of the phase contrast microscope have been described on p. 300. In that instrument the specimen (usually a cell or subcellular particle) is suspended in a series of inert solutions (usually consisting of different dilutions of serum albumin in water) of known index. The refractive index of the immersion media may be calculated from the known specific refractive increment of the solute or may be measured directly in a refractometer. When the indices of medium and particle are identical, the optical paths through either must also be identical; thus, for an appropriate choice of immersion medium, phase contrast is eliminated, and (if amplitude contrast due to different absorption of light by the particle is absent) the particle disappears from the image.

While the use of the phase contrast microscope permits rather sensitive measurements of refractive index to be made, a similar principle may be applied to observations made in an ordinary light microscope. Although phase contrast is then absent or negligible at in-focus settings, a series of phase effects can be observed as the microscope objective is raised or lowered from its in-focus position. When the indices of specimen and immersion medium match, these effects disappear; if the match is exact, particles lacking amplitude contrast disappear entirely from the image.

The liquids used as immersion media must not dissolve or otherwise react with the particles to be studied. If possible, they should be of high stability and low volatility. Series of suitable fluids, which are available commercially, include such substances as benzene ($n_D = 1.504$), anise oil (1.55), nitrobenzene (1.553), carbon disulfide (1.632), α-monobromo-napthalene (1.658), and methylene iodide (1.74).

Mismatches of refractive index are determined by means of observation of the *Becke line*. This is a fringe or halo which at a certain focal level appears outside the geometrical edges of particles suspended in a medium of different refractive index. As exact focus is approached, the fringe moves inward to coincide with the particle edge. Beyond this point the bright region moves into the image of the particle, tending to form a bright spot at its center.

When particles are of higher refractive index than their surround, the Becke line moves inward as the objective lens is raised, whereas, for particles which are of lower index than the surround, the halo moves inward as the objective lens is lowered. The formation of the Becke line can be explained, in geometrical terms, as a consequence of the (imperfect) lenslike action of spherical or approximately spherical particles. This is illustrated for particles of lower and higher index than their surround in Figs. 26-8a and 26-8b, respectively. The Becke line has also been discussed on p. 259.

By observations of the Becke line, not only may mismatches of index be detected, but also the direction of mismatch may be determined. With the use of an adequate series of immersion media, indices may thus be determined to 0.01 refractive index units or better.

The refractive index of the majority of crystals is a function of their *orientation* and of the *azimuth of polarization* of incident light (i.e., this

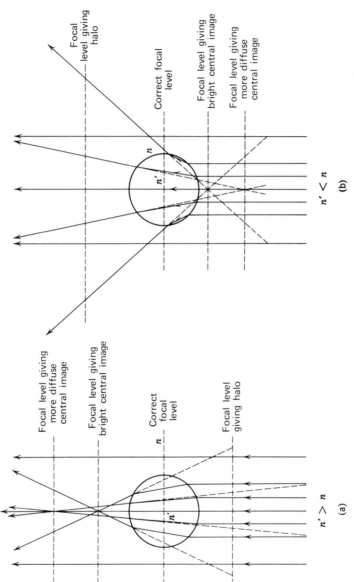

Focal
level giving
halo

Correct focal
level

Focal level giving
bright central image

Focal level giving
more diffuse
central image

n'

n

$n' < n$

(b)

Focal level giving
more diffuse
central image

Focal level giving
bright central image

Correct
focal
level

Focal level
giving halo

n

n'

$n' > n$

(a)

Fig. 26-8 Formation of the Becke line.

585

is the case for optically anisotropic crystals; cf. Chapter 6). Microscopes which are used for crystallographic refractometry must thus be equipped with polarizing devices, and with some means for controlling the orientation of the specimen.

MEASUREMENT OF THE ANGLE OF MINIMUM DEVIATION BY A PRISM

The refractive properties of prisms are described in part by the *angle of deviation* experienced by rays of any given wavelength upon transmission by the prism. As explained on p. 512, there exists, for any prism and wavelength, a *unique* angle of incidence for *minimum* deviation. This angle, Δ_{min}, may be determined experimentally. The relation of Δ_{min} to the properties of the prism is then given in Chapter 23 by

$$(23\text{-g}) \qquad \frac{n'}{n} = \frac{\sin \frac{1}{2}(\alpha + \Delta_{min})}{\sin (\frac{1}{2}\alpha)},$$

where n' is the index of the prism material, n the index of the surrounding medium (usually air), and α is the prism angle (i.e., the angle between its refracting faces).

Accurate values for the refractive indices of glasses and other solids may be obtained with the help of (23-g) by using the material to construct a prism of known geometry. Note that, since no restriction is placed on the value of α in (23-g), a "prism" is simply any homogeneous solid with *two* polished plane faces. Δ_{min} (which is the angle of incidence corresponding to minimum deviation) is then determined experimentally with the use of monochromatic light of any desired wavelength.

REFERENCES

H. A. Strobel, *Chemical Instrumentation*, Addison-Wesley, Reading, Mass., 1960, Chapter 9 (Refractometry). A good discussion of all types of refractometers.

F. A. Jenkins and H. E. White, *Fundamentals of Optics*, 3rd ed., McGraw-Hill, New York, 1957, pp. 256–258. A section on "Index of Refraction by Interference Methods."

R. S. Longhurst, *Geometrical and Physical Optics*, Longmans, London, 1957, pp. 87–88, 147–149, 223–225. Descriptions of various types of refractometer.

H. S. Bennett, in Jones, Ed., *McClung's Handbook of Microscopical Technique*, 3rd ed., Hafner, New York, 1950, pp. 663–668. A section on "The Measurement of the Index of Refraction of an Object" which discusses the immersion method. A table of immersion fluids is included.

R. Barer, in G. Oster and A. W. Pollister, Eds., *Physical Techniques in Biological Research*, Vol. III, *Cells and Tissues*, Academic Press, New York, 1955, pp. 55–68. A section on "Immersion Refractometry of Living Cells."

Handbook of Chemistry and Physics, Chemical Rubber Co., Cleveland, Ohio, p. E103, gives a table of "Liquids for Index by Immersion Method."

CHAPTER 27

Optical Rotatory Dispersion (ORD) and Circular Dichroism (CD) by Bernard Talbot

Optical rotatory dispersion (ORD) and circular dichroism (CD) provide a means of studying the three-dimensional configurational and conformational asymmetry of molecules in solution. The origin, measurement, and interpretation of these phenomena are discussed in this chapter.

ORD and CD are phenomena associated with interactions between polarized light beams and molecules, as briefly described in Chapter 6. Familiarity with the properties and interactions of polarized light beams, as described in Chapters 4, 5, and 6, is assumed here.

OPTICAL ACTIVITY

When light impinges upon a molecule, its electrons are forced to undergo oscillations at the frequency of the incident electromagnetic waves. The existence of optical activity in a molecule requires the simultaneous displacement of electron charge along and around a given axis in the molecule leading to simultaneous electric and magnetic dipole moments. General expressions for the optical activity of molecules have been derived from quantum theory. To date, however, it has been impossible to use these general expressions to predict theoretically the optical activity of molecules because the electronic structure of complicated molecules is not adequately known.

Optically active molecules have the property of rotating the plane of polarization of incident, plane-polarized light. In order to be optically active a molecule must possess a *chromophore* (a chromophore is a chemical group which absorbs light at a particular wavelength; cf. p. 537), and this chromophore must be in an asymmetric environment. The asymmetry can be *configurational* (e.g., an asymmetric carbon atom) or *conformational* (e.g., a right- or left-handed α-helix). Pasteur first formulated the requirement of asymmetry for optical activity by stating that to be optically active a molecule must be such that it cannot be brought into coincidence with its mirror image; that is, the molecule has neither a plane nor a center of symmetry.

Optically active molecules may be characterized by measurements of the rotation of the plane of polarization of light of a single wavelength, usually the sodium D line. A *polarimeter*, as used for such measurements, is diagramed in Fig. 27-1. In this instrument the light from the sodium lamp is incident upon a filter which permits only the sodium D line to pass, is converted by the polarizer into a plane-polarized beam, passes through the sample and a rotatable analyzer, and is incident upon the detector; both photoelectric detectors and visual intensity matching are used.

When an optically inactive specimen is placed in the sample tube, no intensity is transmitted by the analyzer if the latter is crossed with respect to the polarizer. Rotation of the analyzer through 90° (i.e., to a position "parallel" to the polarizer) results in the passage of maximum intensity. If an optically active specimen is substituted for the optically inactive sample, the analyzer must be rotated through some angle α in order again to transmit maximum intensity. This angle α is termed the *observed rotation*. As discussed on p. 116, this rotation by the optically active sample is due to a different index of refraction for right-hand circularly polarized light (RHCP) and left-hand circularly polarized light (LHCP). Thus,

(27-a) $$\alpha = \frac{\pi d}{\lambda}(n_l - n_r),$$

where $\alpha =$ the observed rotation, in radians, $d =$ the thickness of the sample, in cm, and n_l and n_r are, respectively, the index of refraction for left-hand and right-hand circularly polarized light.

For d in decimeters and α in degrees,

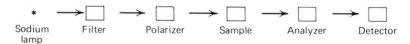

Sodium lamp   Filter   Polarizer   Sample   Analyzer   Detector

Fig. 27-1 Block diagram of a polarimeter.

(27-a')
$$\alpha = \frac{1800d}{\lambda}(n_l - n_r).$$

It is seen that α is positive when $n_l > n_r$. A substance with a positive α is called *dextrorotatory*; a substance with negative α is called *levorotatory*. The *specific rotation* $[\alpha]$ of the solute is defined as

(27-b)
$$[\alpha] = \frac{\alpha}{dC},$$

where α is the observed rotation, in degrees, $d =$ the length of the light path through the sample, in decimeters, $C =$ the concentration of the solute, in g/cm³. The temperature and wavelength at which the specific rotation is measured are specified by superscript and subscript, respectively. Thus $[\alpha]_\lambda^{20}$ specifies the value of specific rotation at 20°C with illumination by plane-polarized light of wavelength λ. In particular, for standard measurements using sodium D line light, the quantity of $[\alpha]_D^{20}$ is specified. Typical values of $[\alpha]_D^{20}$ in aqueous solution are $+112.2$ for α-D-glucose and -6.8 for L-serine.

The *molecular rotation*, $[M]_\lambda$ or $[\phi]_\lambda$, is defined as

(27-c)
$$[\phi]_\lambda = [M]_\lambda = \frac{M[\alpha]_\lambda}{100},$$

where M is the molecular weight of the optically active compound. For macromolecules the *mean residue rotation* $[m]_\lambda$ may also be specified. To obtain this quantity the mean residue weight m is substituted in (27-c) for the total molecular weight M. For proteins, m, the mean amino acid residue weight, is about 115, and varies from one protein to another according to the protein's amino acid composition.

The *effective residue rotation* is defined as

(27-d)
$$[m']_\lambda = \frac{3[m]_\lambda}{n^2+2} = \frac{3m[\alpha]_\lambda}{100(n^2+2)},$$

where n is the refractive index of the solution.

THE MEASUREMENT OF OPTICAL ROTATORY DISPERSION

Measurement of α at a single wavelength serves principally to provide a numerical value for the identification of compounds of low molecular weight. Studies of the variation of α as a function of wavelength, that is, of the optical rotatory *dispersion*, are of much greater intrinsic interest, since optical rotatory dispersion is related to the structure of molecules. Until the mid-1950s optical rotatory dispersion measurements were performed by recording photographically and usually required several days

Source   Monochrometer   Polarizer   Sample   Analyzer   Detector
                                                                                            (photomultiplier)

Fig. 27-2 Block diagram of a spectropolarimeter.

to record the spectrum of a single substance. The introduction of commercial photoelectric recording spectropolarimeters in 1955 has greatly simplified measurements. Such a spectropolarimeter is shown schematically in Fig. 27-2.

In addition to the elements shown in Fig. 27-2, a Faraday modulator (see p. 597) may be present between the polarizer and analyzer. The source is commonly a high pressure xenon arc. This arc is constructed of two tungsten electrodes surrounded by xenon filled at about 8 atm pressure at room temperature, rising to about 25 atm pressure at operating temperature, and housed in a quartz container. The monochrometer may be a diffraction grating, but it is more commonly a quartz prism; quartz optical parts are essential throughout the instrument because often measurements are carried well into the ultraviolet. Current instruments permit measurements to wavelengths as short as 185 mμ.

CIRCULAR DICHROISM

As discussed on p. 116 and 588, rotation of the plane of polarization by an optically active medium is due to a difference in the indices of refraction in the medium for left-hand circularly polarized light (LHCP) and right-hand circularly polarized light (RHCP). A difference of refractive indices for LHCP and RHCP is always accompanied by a difference in extinction coefficient for LHCP and RHCP. Such a difference in extinction coefficient for LHCP and RHCP is called *circular dichroism*.

Circular dichroism is measured in terms of the *differential circular dichroic absorption* $\Delta \in$, defined as

$$(27\text{-e}) \qquad\qquad \Delta \in = \in_l - \in_r,$$

where $\in_l$ and $\in_r$ are the molar extinction coefficients for LHCP and RHCP, respectively. Circular dichroism is thus positive if LHCP is absorbed more strongly than RHCP. $\Delta \in$ is the quantity measured experimentally, but the circular dichroism of a medium may also be expressed in terms of *ellipticity* of the emergent beam by using (27-i).

When plane-polarized light passes through a circularly dichroic medium, the emerging light is elliptically polarized (cf. p. 86). Experimentally the ellipse is always extremely elongate, and under such circumstances it can be shown that incident plane-polarized light after passing through unit length of the sample, emerges with *ellipticity*,

(27-f) $$\theta = \frac{k_l - k_r}{4},$$

where k_l and k_r are, respectively, the absorption coefficients for LHCP and RHCP.

The *molar ellipticity* [θ] is defined as

(27-g) $$[\theta] = \theta \frac{18M}{\pi C},$$

where C is the concentration, in g/cm³, of optically active material, and M is its molecular weight.

The absorption coefficient k is related to the more common molar extinction coefficient $\in$ by

(27-h) $$k = \frac{\in C'}{\log e} \cong 2.303 \in C',$$

where C' is the concentration of absorbing material, in moles per liter.

Combination of (27-f), (27-g), and (27-h) leads to

(27-i) $$[\theta] \cong 3300 (\in_l - \in_r).$$

In principle, $\Delta \in$ may be measured in a spectrophotometer in which a *polarizer* and *quarter wave plate* are placed between the monochromator and the specimen as shown in Fig. 27-3. As described on p. 114, the quarter wave plate can produce a circularly polarized light beam of either sense when appropriately oriented with respect to the plane of polarization of the beam transmitted by the polarizer. Thus, by rotation of the quarter wave plate, either LHCP or RHCP can be produced, and the absorption by the medium of LHCP and RHCP can be measured separately.

In fact, the conventional quarter wave plate is a device of *fixed* birefringence which produces a circularly polarized beam only for incident plane-polarized light of a single wavelength. The practical *dichrograph* makes use of an equivalent device in which a variable birefringence is induced electrically. As the voltage applied across a crystal of ammonium dihydrogen phosphate is varied, the induced birefringence of the crystal

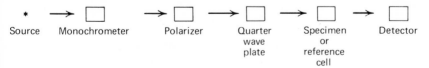

| Source | Monochrometer | Polarizer | Quarter wave plate | Specimen or reference cell | Detector |

Fig. 27-3 Block diagram of a dichrograph.

varies. This is the *Pockels effect* which is also exploited in the phase modulated interference microscope discussed on p. 317. By using the Pockels effect, LHCP and RHCP can be produced alternately at each wavelength and, with the use of an appropriate recording system, the dichrograph can thus produce directly a plot of $\Delta \in$ as a function of wavelength. Dichrographs have been available commercially only since 1960.

THE COTTON EFFECT

Consider wavelengths close to λ_i, where λ_i is the wavelength of maximum absorption of an isolated absorption band of an optically active chromophore. Plots of the molar extinction coefficient $\in$, the differential circular dichroic absorption $\Delta \in$, and the specific rotation $[\alpha]$, all as a function of wavelength λ, are shown in Fig. 27-4. The two parts of Fig. 27-4 represent identical idealized optically active chromophores, but the environment of the chromophore considered in Fig. 27-4b is the *mirror*

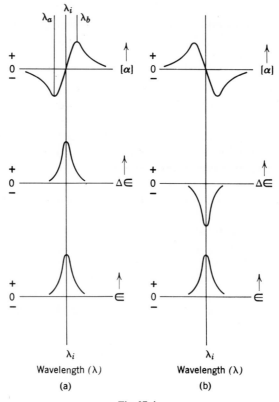

Fig. 27-4.

image of that in 27-4a. λ_i is known as the wavelength of the absorption band. It is seen that, if $\in$ is maximal at λ_i, $\Delta\in$ has a maximum or minimum at λ_i, while $[\alpha]$ passes through zero and has an inflection point at λ_i with a maximum (peak) and minimum (trough) at wavelengths near λ_i. The combined behavior of $\Delta\in$ and $[\alpha]$ in the vicinity of an absorption band of an optically active chromophore is called a *Cotton effect*, named after A. Cotton, who first described it in 1896. When the maximum value of $[\alpha]$ occurs at the long wavelength side of λ_i, as in Fig. 27-4a, the Cotton effect is called *positive*. When the peak value of $[\alpha]$ occurs at shorter wavelength than λ_i, as in Fig. 27-4b, the Cotton effect is negative. A positive Cotton effect is always associated with a positive $\Delta\in$, and a negative Cotton effect with a negative $\Delta\in$. As shown in Fig. 27-4, at wavelengths progressively more distant from λ_i, both $\Delta\in$ and $[\alpha]$ tend to zero. $\Delta\in$, however, tends to zero much more quickly than does $[\alpha]$. Circular dichroism always accompanies optical rotation at wavelengths near an optically active absorption band. At wavelengths far from an optically active absorption band, optical rotation may be found without circular dichroism, since $\Delta\in$ reaches zero before $[\alpha]$.

Consider a positive Cotton effect as shown in Fig. 27-4a. When measuring $[\alpha]$ as a function of wavelength beginning at high λ and progressively decreasing λ toward λ_i, the observed values of $[\alpha]$ are found to increase monotonically, reach a maximum at λ_b, and then fall to zero. If one restricts measurement to wavelengths above λ_b, the entire measured curve is a monotonic increase in $[\alpha]$ as λ is decreased. Such a monotonic variation of $[\alpha]$ with λ is called *simple dispersion* or *normal dispersion*. If the measurements, however, include a Cotton effect peak or trough, the variation of $[\alpha]$ with λ is not monotonic; this is called *anomalous dispersion* (cf. p. 509).

In a molecule containing many optically active absorption bands the *total* specific rotation of the molecule as a whole is just the *sum* of the individual contributions by each of the optically active absorption bands [see (27-k)].

Note that in Fig. 27-4a the chromophore is dextrorotatory at wavelengths above λ_i and levorotatory at wavelengths below λ_i. This change from dextrorotatory to levorotatory at some wavelength(s) is a property of all optically active molecules and is a result of the shape of the Cotton effect curve.

INTERPRETATION OF ORD SPECTRA

When ORD measurements are performed experimentally over a range which includes wavelengths at and near an optically active absorption band, the sign of the observed Cotton effect provides information concerning the asymmetry of the environment of the chromophore. For example, the configuration at position C_5 of a steroid molecules depends on whether the steroid A and B rings are *cis*-fused or *trans*-fused. The carbonyl chromophore by itself is symmetric. When the chromophore is placed in a proper asymmetric environment, however, asymmetry is induced in the electrical distribution of the carbonyl chromophore, and Cotton effects may be observed. Thus the sign of the carbonyl Cotton effect in a steroid molecule with a lone carbonyl group at position C_3 can provide information about the configuration at position C_5. Thus compound I of Fig. 27-5, which has the A and B rings *trans*-fused, shows a positive Cotton effect with peak at 307 mμ, while compound II of Fig. 27-5, which has the A and B rings *cis*-fused, shows a negative Cotton effect with trough at 307 mμ. In addition to steriochemical information, ORD spectra of steroids and other compounds can give further information about chemical structure. Thus, for example, a carbonyl group in a steroid at position C_1 or C_7 gives an ORD spectrum different from that given by a carbonyl group at position C_3.

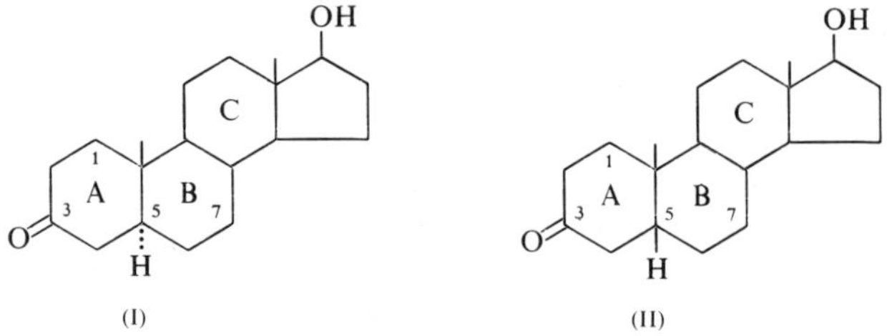

Fig. 27-5.

In the case of proteins and also polypeptides and nucleic acids, ORD measurements have been used to show conformational asymmetries of molecules (e.g., the presence of the right-handed α-helix). Before 1960, all optical rotatory dispersion measurements on proteins were limited to the region above 250 mμ because of the great experimental difficulty encountered at lower wavelengths, which is the result of decreased output of light sources, decreased sensitivity of light detectors, and the rapidly increasing absorption of proteins. In the region above 250 mμ no Cotton effects due to the peptide group were observed, and the experimental results showed generally a "simple" rather than "anomalous" dispersion. Using this type of experimental data and certain approximate or phenomenological equations such as the one-term Drude equation (27-l) or the Moffitt equation (27-m), estimates can be made of the fraction of the polypeptide chain in right-handed α-helical conformation.

Consider the isolated optically active absorption band shown in Fig. 27-4a. For $\lambda < \lambda_a$ or $\lambda > \lambda_b$, it can be derived from quantum mechanical theory that

(27-j)
$$[\alpha]_\lambda = \frac{K_i}{\lambda^2 - \lambda_i^2},$$

where $[\alpha]_\lambda$ is the specific rotation as defined in (27-b), λ_i is the wavelength of the absorption band, and K_i is a constant proportional to the "rotatory strength" of the absorption band. For a molecule consisting of many optically active absorption bands, the total specific rotation of the molecule is the sum of the individual contributions of each of the optically active absorption bands, as expressed in the *Drude equation*,

(27-k)
$$[\alpha]_\lambda = \sum_i \frac{K_i}{\lambda^2 - \lambda_i^2}.$$

If experimental measurements are all performed at wavelengths well above the wavelengths of all optically active absorption bands, then (27-k) can be well approximated by the *one-term Drude equation*,

(27-l)
$$[\alpha]_\lambda = \frac{K}{\lambda^2 - \lambda_c^2},$$

or the *Moffitt equation*,

(27-m)
$$[m']_\lambda = \frac{a_0 \lambda_0^2}{\lambda^2 - \lambda_0^2} + \frac{b_0 \lambda_0^4}{(\lambda^2 - \lambda_0^2)^2}.$$

In the Drude equation, K and λ_c are constants. In the Moffitt equation, $[m']$ is the effective residue rotation as defined in (27-d), and a_0, b_0, and λ_0 are constants. In using the Moffitt equation for proteins and polypeptides, the experimental data of α versus λ are converted using (27-d) to $[m']_\lambda$ versus λ, and this is used in (27-m), assuming $\lambda_0 = 212$ mμ, to calculate a_0 and b_0. Or, alternatively, the data of $[m']_\lambda$ versus λ are used to calculate λ_0, a_0, and b_0. It has been found for a number of synthetic polypeptides that, when they are in a conformation which contains no α-helix, their ORD data fit the Moffitt equation with $b_0 = 0$, and that, when they are in a conformation which contains 100% right-handed α-helix, their ORD data fits the Moffitt equation with $b_0 = -630$. For proteins the Moffitt equation b_0 parameter divided by -630 is taken to indicate the fraction of the molecule in right-handed α-helical conformation.

Beginning in 1960, the extension of experimental measurements to shorter wavelengths has made direct observation of peptide Cotton effects in proteins possible. For example, in right-handed α-helical polypeptides and proteins there is a negative Cotton effect with a trough at 233 mμ and an inflection at 225 mμ. In nonhelical molecules this Cotton effect is absent. In left-handed α-helical polypeptides there is a positive Cotton effect at the same wavelength, that is, peak at 233 mμ, inflection at 225 mμ. For most proteins which contain some areas of right-handed α-helix and some areas of no helix, one can estimate from the depth of the trough at 233 mμ what fraction of the protein's polypeptide chain is helical. Synthetic polypeptides which contain 100% right-handed α-helix are used as a calibration standard.

The Cotton effect discussed above with inflection at 225 mμ and another with inflection at about 190 mμ are associated with an absorption band of the helical polypeptide backbone. Other weaker protein Cotton effects are associated with absorption by amino acid side chains such as tyrosine. These Cotton effects due to the polypeptide backbone and amino acid side chains are called *intrinsic Cotton effects*. Cotton effects arising from prosthetic groups bound to the protein are called *extrinsic Cotton effects*. An example of an extrinsic Cotton effect is the Cotton effect associated with the visible absorption bands of the heme group in the heme proteins such as hemoglobin and myoglobin. Also, extrinsic Cotton effects have been observed which are associated with coenzymes bound to enzymes, such as DPNH bound to liver alcohol dehydrogenase, with metals bound to metalloproteins, and with dyes bound to proteins and synthetic polypeptides. Consider a dye which has strong absorption bands at its characteristic wavelengths, but which is not optically active since it is symmetric. If this dye is now bound to a polypeptide such that it is in an asymmetric environment, Cotton effects now appear at the dye absorption bands. If the Cotton effect is positive when the dye is in an asymmetric environment a, the Cotton effect is negative when the dye is in an asymmetric environment which is the mirror image of a. For example, acridine orange, when bound to poly-L-glutamic acid (which forms a right-handed α-helix) shows a *negative* Cotton effect, with an inflection at 465 mμ, while the same dye, when bound to poly-D-glutamic acid (which forms a left-handed α-helix) shows a *positive* Cotton effect of the same magnitude, also with inflection at 465 mμ. The Cotton effect disappears when the polypeptide is converted into a nonhelical structure.

THE RELATION OF ORD AND CD MEASUREMENTS

The same type of information which is revealed by measurements of optical rotatory dispersion (i.e., information concerning the configurational and conformational asymmetry of molecules) can also be obtained from studies of circular dichroism. In fact, ORD and CD curves can be interconverted by calculation. The transformation from one curve to the other involves a well known mathematical form called the Kronig-Kramer transform. Moscowitz has given formulas for ORD to CD and CD to ORD transformation for the case in which the circular dichroism is a Gaussian function (see Djerassi).

When measurements must be confined to a spectral range far from the absorption maxima of the optically active chromophores of a molecule, only ORD can be studied, since CD falls rapidly to zero at wavelengths of negligible absorption. Thus, in the past, instrumental limitations have dictated the study of ORD rather than CD. CD measurements are to be preferred, however, when an adequate spectral range can be studied. $\Delta \in$ contributions from different chromophores tend to be *separate*, whereas $[\alpha]$ contributions from different chromophores tend to *superimpose* and ORD measurements at any wavelength include contributions from many chromophores. Accordingly, continued extension of the spectral range of instruments tends to favor studies of the "purer" circular dichroism data.

THE FARADAY EFFECT. MAGNETO-OPTICAL ROTATION (MOR)

Faraday discovered in 1846 that optical activity can be produced artificially in any isotropic medium if the medium is placed in a magnetic field whose field lines run parallel to the direction of propagation of polarized light. This is called the *Faraday effect* or *magneto-optical rotation* (cf. p. 118). The angle through which the plane of polarization is rotated, α, is given by

$$(27\text{-n}) \qquad\qquad \alpha = VHL,$$

where L is the sample path length, H is the magnetic field strength, and V is a constant, known as the *Verdet constant*, which varies with the material in the field and the wavelength of the incident light.

The Faraday effect is used as an instrumental aid in some recording spectropolarimeters (used to measure ORD, not MOR). In these recording spectropolarimeters a Faraday modulator consisting of a quartz or fused silica rod centered within a coil is placed between the polarizer and the analyzer. When alternating current is passed through the coil, the Faraday modulator causes the plane of polarization of light to be shifted with the sinusoidal alternating current. Initially polarizer and analyzer are crossed and, if the sample is not optically active and no current is flowing through the Faraday modulator coil, no light reaches the detector. If current is now passed through the Faraday modulator coil, a light intensity is received at the detector in a symmetric sinusoidal ac curve with equal intensity coming in each half of the ac cycle. If now the wavelength of incident light is changed to a wavelength at which the sample is optically active, the sample optical activity superimposed on the Faraday modulator activity results in unequal intensity of light striking the detector during the two halves of the ac cycle. This inequality of detected intensity is converted into an electrical signal which drives a motor to rotate the polarizer until the detected intensities during each half of the ac cycle are equal again. The polarizer rotation is a measure of the optical rotation of the sample.

More interesting than this use of the Faraday effect as an instrumental aid in measuring ORD is its use in measurement of the MOR spectra of materials. In measuring MOR spectra the instrument used is the same as that shown in Fig. 27-2, with the addition of a magnetic field across the sample. This technique has just begun to be extensively investigated, and it holds much promise. The study of ORD is limited to naturally optically active molecules and cannot be applied to the vast majority of compounds, which are optically inactive. MOR, however, is observed in all molecules. The Faraday effect has been observed not only in the visible and ultraviolet regions of the electromagnetic spectrum but also in the microwave, infrared, and x-ray regions. When measuring MOR, all substances become optically active in the magnetic field, including the air between the magnet pole pieces, the cell windows, and the solvent. The quantity of interest, the rotation of the solute, is very small compared to the rotation of solvent and of the cell. Three independent measurements are thus made, the MOR spectra of

$$1. \text{ the empty cell} = K,$$
$$2. \text{ the cell with solvent} = R_0,$$
$$3. \text{ the cell with solution} = R_s.$$

The rotation α of the solute is then calculated point for point at each wavelength by the formula

(27-o) $$\alpha = R_s - K - \left(\frac{\rho_s - C}{\rho_0}\right)(R_0 - K),$$

where ρ_s = solution density, ρ_0 = solvent density, C = concentration (g/cm^3) of the solute. The term $(\rho_s - C)/\rho_0$ corrects for solvent molecules replaced by solute when the solution is measured.

Another method involves the use of two cells through which the polarized light beam passes sequentially, one filled with solvent and one with solution. The two cells are placed in magnetic fields of equal strength and opposite direction. If the path length of the cell with solvent, l_2, and the path length of the cell with solution, l_1, are adjusted so that

$$l_2 = \frac{\rho_s - C}{\rho_0} l_1,$$

then a direct reading can be made of α versus λ, since the total rotation of the system equals the rotation of the solute. Just as optical rotation close to an optically active absorption band is accompanied by circular dichroism, the Faraday effect is accompanied by magnetic circular dichroism (the *inverse Zeeman effect*).

REFERENCES

C. Djerassi, *Optical Rotatory Dispersion*, McGraw-Hill, New York, 1960.

L. Velluz, M. Legrand, and M. Grosjean, *Optical Circular Dichroism*, Academic Press, New York, 1965.

H. A. Strobel, *Chemical Instrumentation*, Addison-Wesley, Reading, Mass., 1960, Chapter 10 (Polarimetry).

J. A. Schellman and C. Schellman, in H. Neurath, Ed., *The Proteins*, Vol. II, 2nd ed., Academic Press, New York, 1964, pp. 69–96.

D. Ridgeway, in J. H. Lawrence and J. W. Gofman, Eds., *Advances in Biological and Medical Physics*, Vol. 9, Academic Press, New York, 1963, pp. 271–353.

S. Beychok, *Science* **154**, 1288–1299 (1966), "Circular Dichroism of Biological Macromolecules."

The first two references are comprehensive books with chapters on theory, instrumentation, and results, with emphasis on small organic compounds and also discussion of macromolecules.

The last three references are review articles concentrating on theory, and on results with polypeptides and proteins.

CHAPTER 28

X-rays and
X-ray Microscopy

The x-ray region of the electromagnetic spectrum has been described briefly in Chapter 2. Here the nature of x-ray radiation and the general features of instrumentation for the production and detection of x-rays are discussed. Methods for x-ray microscopy (microradiography) are also described.

ORIGIN OF X-RAYS

X-rays may be defined as electromagnetic radiation in the wavelength range from about 0.1 to about 1000 Å. X-ray radiation is excited by the impinging of energetic electrons upon atoms. Two types of interaction lead to the emission of x-ray photons. One type of emission, described as *white radiation*, results from interactions of incident electrons with the electric field in the vicinity of atomic nuclei. The interaction decelerates the impinging electron, causing a decrease in its energy which appears as an x-ray quantum of corresponding size. At most, the *entire* kinetic energy of an incident energy can be lost upon collision. Thus a lower limit for the x-ray wavelength (i.e., an upper limit for x-ray quantum energy) is determined by the energy of the incident electrons. Electronic energy, in turn, is determined by the potential difference through which the electrons are accelerated. Therefore,

(28-a) $$E = eV = h\nu_{max} = \frac{hc}{\lambda_{min}} \quad \text{or} \quad \lambda_{min} = \frac{hc}{eV},$$

where h is Planck's constant, V the accelerating voltage, and e the electronic charge. Numerically this gives

(28-b)
$$\lambda_{min} = \frac{12{,}400\,\text{Å}}{V},$$

when V is expressed in volts. 'White' x-ray radiation may be of any wavelength longer than that determined by (28-a) and (28-b), corresponding to partial losses of energy by incident electrons upon interaction with nuclei. Wavelengths of maximum *emission* are found to vary as a function of the accelerating voltage; in general, the wavelength intensity curve for white radiation is of the form shown in Fig. 28-1. Such spectra vary somewhat with the nature of the target but are primarily determined by the level of accelerating voltage.

When the accelerating voltage is sufficiently high, the so-called *characteristic spectrum* is superimposed upon the white spectrum shown in Fig. 28-1. The characteristic emission consists of approximately monochromatic radiation at a number of wavelengths which are specifically determined by the elemental composition of the emitter. The general form of the complete x-ray emission spectrum is thus as shown in Fig. 28-2.

The characteristic x-ray spectrum arises from electronic transitions within the atoms of the target. Whereas the electronic transitions associated with the emission of visible and ultraviolet radiation are transitions in *valence* electron energy levels (as explained in Chapter 7), x-rays are emitted as a consequence of transitions involving the *inner electronic shells* of atoms.

For example, the magnesium atom, which is diagramed in Fig. 28-3, contains electrons in the $K, L,$ and M shells. An impinging electron might dislodge an electron from the innermost (K) shell of the atom. The missing inner electron could then be replaced from either the L shell or the M shell. Either of the replacement transitions is accompanied by a loss of energy in the form of an x-ray photon. The process of photon emission is

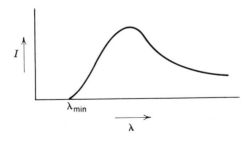

Fig. 28-1 Continuous x-ray emission spectrum.

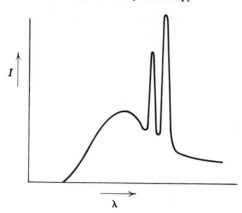

Fig. 28-2 Complete x-ray emission spectrum.

illustrated in Fig. 28-4. Figure 28-4a shows the excited atom produced by electron bombardment; one electron is missing from the (innermost) *K* shell. Figure 28-4b shows the replacement of the missing electron by one from the *L* shell of the atom, a process known as K_α *emission*.

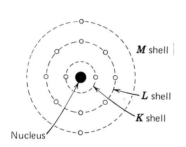

Fig. 28-3 The magnesium atom.

Similarly, Fig. 28-4 shows the replacement of the missing electron by one from the *M* shell, and the associated K_β *emission*. Since the $M \rightarrow K$ transition represents a larger decrease in the total energy of the atom than does the $L \rightarrow K$ transition, K_β radiation is of shorter wavelength than K_α. In an analogous manner, the removal of an *L* electron by an impinging electron can lead to x-ray emission associated with the $M \rightarrow L$ transition, or (in larger atoms than those shown in Figs. 28-3 and 28-4) with $N \rightarrow L$ or $O \rightarrow L$ transitions. The *L* emission lines correspond to smaller changes in the total energy of the atom than do the *K* emission lines; thus they are observed at longer wavelengths.

None of the characteristic x-ray radiations can be observed unless the energy of the incident electron beam is sufficient to remove, initially, an electron from the target atom. Thus only the white spectrum shown in Fig. 28-1 is observed at low accelerating voltages. At a voltage characteristic of the target element, the emission spectrum changes to the mixed type illustrated by Fig. 28-2.

The process of emission of characteristic x-ray spectra is analogous to

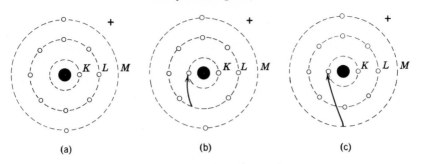

Fig. 28-4 X-ray emission by the magnesium atom.

fluorescent emission of visible light, as described on p. 135. However, the term *x-ray fluorescence* is applied specifically to the *secondary* emission of x-rays which may occur when primary x-ray quanta (excited by electron bombardment) are sufficiently energetic to cause the ejection of electrons from other atoms, leading to further electronic transitions in the x-ray quantum energy range.

X-RAY SCATTERING EFFECTS

The nature of x-ray scattering, like that of excitation, is twofold. One type of scattering occurs when x-ray photons incident upon an atom of a target encounter free or loosely bound electrons. These collisions may result in a slight recoil of the target electron, corresponding to a partial loss of energy by the incident quantum. Since the total energy and the total momentum of the system must both be conserved, the scattering, in an altered direction, of an x-ray quantum of slightly *lower* energy results. That is, the wavelength of the scattered x-rays is slightly *longer* than that of the incident x-rays. This process, the *Compton effect*, has been described on p. 149. At the time of its discovery the Compton effect constituted a striking proof of the particulate nature of electromagnetic radiation. Since Compton scattering results in the loss of definite phase relationships, it is also referred to as *incoherent scattering*.

Alternatively, the incident x-rays may encounter rigidly bound electrons within the target atoms. *Coherent* scattering (also known as *Bragg scattering*) then occurs, with no transfer of energy to the target. Both the phase relationships and the wavelength of the incident beam are preserved. In the study of the substructure of matter by x-ray diffraction methods, it is necessary to minimize the effects of white radiation, x-ray fluorescence and/or Compton scattering in order to observe *coherent scattering of monochromatic x-ray beams*.

X-RAY ABSORPTION

X-ray absorption spectra, like optical absorption spectra, exhibit a general correspondence with the features of the respective emission spectra. The general form of a typical x-ray absorption spectrum is shown in Fig. 28-5. There is a continuum of absorption at all wavelengths, with certain abrupt discontinuities known as *absorption edges*. For long wavelengths ("soft x-rays," for which the wavelength approaches 1000 Å), absorption (by all elements) is substantial, while all elements are relatively transparent to "hard," short wavelength x-rays.

The absorption edges occur at wavelengths for which the quantum energy is just sufficient to ionize the absorber. The K absorption edge shown in Fig. 28-5 corresponds to the energy just required to remove an electron from the K shell of the atom, while the L edges correspond to the less energetic process of removal of an electron from the L shell. (Three rather than a single L absorption edge are in fact observed as a consequence of the distribution of energy levels with the atom. Similarly, five M absorption edges and seven N absorption edges exist.) The approximately twenty-fold difference in absorption coefficients at the K and L edges, as shown in Fig. 28-5, is typical.

The energy required for removal of an electron from the K shell of an atom is somewhat greater than that released when the K shell is filled by an electron from an outer electronic shell. Therefore absorption edges occur at slightly shorter wavelengths (higher quantum energies) than the corresponding characteristic emissions. The decreasing absorption coefficients to the left of each of the absorption edges corresponds to an increasing probability that energetic quanta may pass through the atom without effecting ionization.

Note that the absorption of x-rays of a wavelength just shorter than, for example, the K absorption edge removes a K electron from the atom. The K shell may subsequently be filled by an electron from *any* of the outer

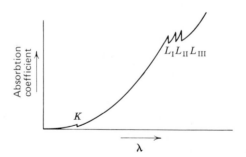

Fig. 28-5 X-ray absorption spectrum.

shells. Thus the absorption of a single wavelength (of adequate quantum energy) can excite emission of the complete corresponding series of characteristic emission lines (K_α, K_β, etc.).

Figure 28-5 illustrates the x-ray absorption spectrum of a single element. The corresponding absorptions (and emissions) occur at successively *shorter* wavelengths for elements of higher atomic number; that is, as would be expected, removal of electrons from the inner shells of larger atoms is a more energetic process. When a mixture of different elements is irradiated by x-rays, the absorbtion spectra of the individual elements are superimposed.

The *quantitation* of absorption of x-rays of any given wavelength is expressed by equations of the same form as Beer's law, as given on p. 530.

$$\text{(24-g)} \qquad \ln\frac{I_0}{I} = \mu_m \frac{m}{A},$$

where μ_m is the mass absorption coefficient and m and A are the total mass and cross-sectional area of the specimen, respectively.

The *atomic absorption coefficient* μ_a of an element may also be defined as

$$\text{(28-c)} \qquad \mu_a = \frac{\mu_m A'}{N_0},$$

where A' is the atomic number of the element, and N_0 is Avogadro's number. Over the complete range of x-ray wavelengths, μ_a is found to be approximately proportional to the fourth power of the atomic number Z. At wavelengths used in biological instrumentation (ca. 1 Å), however, μ_a is more nearly proportional to Z^3.

X-RAY SOURCES

Practical x-ray sources utilize the beams emitted at characteristic wavelengths when a target of appropriate composition is bombarded by an electron beam. A copper target is very often used, producing a beam of Cu K_α radiation at a wavelength of 1.54 Å. The general principles of tube design apply, however, to the use of other elements also; silver, molybdenum, nickel, cobalt, iron, and chromium are among those used for the purpose.

A schematic diagram of an x-ray tube is given in Fig. 28-6. Electrons leave a source, which may be a heated tungsten filament (cf. p. 350) and are focused by an electron lens (cf. Chapter 17). The high voltage which accelerates the electron beam is applied between the electron source (cathode) and the target (anode); accelerating voltages are in the

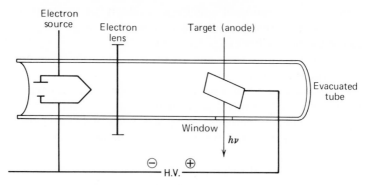

Fig. 28-6 X-ray tube.

neighborhood of 50 kV. Electrons which impinge upon the target excite x-rays; the latter leave the tube through a window region of appropriate properties.

The properties of the x-ray generator must be such as to provide, with operational safety, an intense, highly monochromatic beam. Often, the beam must be of limited cross section. In general, tubes must be suitable for prolonged operation (i.e., over periods of several days).

Safety features include shielding and the construction of tubes, except for the window region, of leaded glass or other x-ray opaque material. Because of the imperceptibility of the beam and the cumulative pathogenic effects of x-ray dosage, precautions must be observed routinely by operators.

The window region of the x-ray tube is made of a material which is, insofar as possible, transparent to x-rays of the desired wavelength. Since a vacuum must be maintained within the tube (in order to prevent scattering and absorption of the exciting electron beam), requirements for mechanical stability limit the thinness of window which may be used. Beryllium metal, mica, or a boron-lithium-beryllium mixture known as *Lindeman glass* are the materials commonly used. The window may also contain a suitable x-ray filter, as described below.

The excitation of x-rays by electron bombardment is, at best, an inefficient process from the point of view of energy consumption. Heating of the anode (target) is a serious problem. The target may become so hot during prolonged operation that other elements of the assembly are damaged, or may itself develop pitting, etc., which severely limits the lifetime of the tube. The problem is combated by continuous water cooling of the target. Modern tubes may also make use of *rotating anodes*, which continuously present a newly cooled surface for bombardment. The efficient conduction of heat by copper makes this element particularly suitable for use as the target.

Monochromatization of the x-ray beam is effected by *filters* (see below) and by the use of *crystal monochromators* (see below). White radiation from the target is found to be minimal when the excitation voltage level is about four times the minimum required for excitation of the characteristic emission.

X-RAY FILTERS

As explained above, the positions of characteristic emission lines and absorption edges vary continuously with atomic number. A consequence is that the characteristic emission by an element of atomic number Z tends to overlap, in part, the corresponding absorption edge for the element of atomic number $(Z-1)$. Thus a slab of material of the $(Z-1)$ element may transmit one of the characteristic emission lines of an element of atomic number Z, while absorbing much of the remaining x-radiation from that element. The effect is illustrated in Fig. 28-7. As shown, white and K_β are very substantially absorbed by the filter, while K_α radiation is only slightly absorbed. For a *copper* target $(Z=29)$ the appropriate filter material is *nickel* $(Z=28)$. A convenient thickness of nickel serves to eliminate about 98% of copper K_β radiation, resulting in a moderately monochromatic beam of radiation at the copper K_α wavelength, 1.542 Å.

X-RAY MONOCHROMATORS

Highly monochromatic beams can be produced with the use of crystal monochromators. The crystal monochromator is, in principle, a *diffraction grating* for short wavelength radiations, in which the spacing between

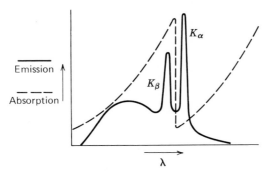

Fig. 28-7 Action of x-ray filters.

atoms of the crystal corresponds to the spacing between the ruled lines of an optical diffraction grating. Focusing of the beam on the object to be diffracted may be achieved by crystals which are cut and bent, as illustrated in Fig. 28-8. For incidence on the curved crystal at a given angle, only radiation of a single wavelength produces reflected intensity (i.e., only radiation of that wavelength which satisfies the Bragg law is reflected; cf. p. 627). In general, therefore, a monochromatic beam is obtained directly from the bent crystal.

X-RAY DETECTORS

Photographic emulsions and *counters* and related devices are the means used for the detection and quantitation of x-radiation. The principles which apply to the use of photographic techniques in x-ray work are generally similar to those which apply at optical wavelengths (discussed in Chapter 21). In the past, photography has been the method of choice for the recording of x-ray intensities, since a permanent and directly visible record of quite high accuracy is produced. More recently, however, the scope of x-ray diffraction studies has expanded to a point at which the time consumed in measurement of photographic records has become a limiting factor. The trend in x-ray instrumentation is thus to the use of automatic counting techniques. Systems have been developed in which intensities are measured by means of proportional counters or scintillation counters. The electrical output of these detectors can be supplied directly to a computer. However, since crystals may decompose during prolonged x-irradiation, high speed recording is also an important consideration. From this point of view, photographic methods are superior to counting techniques, since, in the latter, each maximum of intensity must be recorded individually. Therefore effort is currently being directed to the development of adequate methods for automatic scanning of the intensities of photographic records.

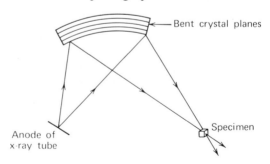

Fig. 28-8 Crystal monochromator for x-rays.

The *proportional counter* consists of a tube in which the central filament (anode) is maintained at a potential of the order of 1 kV with respect to the grounded metallized inner surface (cathode). A continuous gas flow is maintained through the tube at or slightly above atmospheric pressure. An inert gas (usually helium or argon) is used, together with about 10% of a polyatomic gas which serves as a quencher.

An entering x-ray photon *ionizes* the material within the tube, producing electrons and positive ions. The electrons move toward the anode, experiencing an acceleration as they approach this central, positively charged wire. At appropriate levels of tube voltage, these photoelectrons are sufficiently energetic to produce an avalanche of secondary ionizations of the gas atoms *within a cylindrical volume surrounding the original x-ray track*. The avalanche of ionizations produces a "pulse" of current which is proportional in magnitude to the energy of the incident photon (or of other ionizing particles, such as α particles or accelerated electrons). Thus, with the use of appropriate electronic instrumentation, only those x-rays with energies within some precisely specified range (i.e., x-rays which produce pulses of some specified size) need supply a signal to the computer. In this way the proportional counter effectively monochromatizes the diffraction output. In addition, the counter is a device of relatively high sensitivity to radiations of low intensity.

Scintillation counters are substances (liquids or crystals) with which incident x-rays interact to produce fluorescent emission, usually in the blue or ultraviolet regions of the spectrum. Sodium iodide, activated by small amounts of thallium, is the substance most commonly used as a scintillator in biological diffraction studies. The fluoresced radiation is received and amplified by a photomultiplier tube (cf. p. 546). Scintillation counters are characterized by high sensitivity and rapid response, although background noise levels may be a problem.

REFRACTION OF X-RAYS

To a first approximation, x-ray beams do not show refraction effects comparable to those observed with electromagnetic radiations of longer wavelengths. Careful measurements, however, show that refractive indices of all materials for x-rays are very slightly *less* than one, the difference from unity being of the order of a few parts per million. For example, slight refraction effects become evident when attempts are made to measure x-ray wavelengths very accurately by study of Bragg reflections from crystals (cf. p. 627). Values of the quantity $(2d \sin \theta)/m$ are found to decrease slightly as the order of reflection, m, increases.

A refractive index of less than one implies a radiation velocity greater

than c. This conclusion is apparently in contradiction with the fundamental premises of the theory of relativity. It must be remembered, however, that the velocity $v = c/n$ is, in fact, a phase velocity, which specifies
the velocity of an idealized sine wave of infinite extent. The transfer of
energy, however, is always associated with finite pulses of radiation,
which are equivalent to the superposition of a group of sine waves of
varying frequency and amplitude. The observable velocity is always that
of the group of waves as a whole. The group velocity of x-rays, as of other
radiations, is never greater than c. (A discussion of phase and group
velocities has also been given on p. 73.)

X-RAY MICROSCOPY

The wavelengths of x-rays are of atomic dimensions; beams in the
wavelength range from 1 to 10 Å are readily obtained. Furthermore,
x-rays penetrate specimens to a much greater depth than do either electrons or visible light. Specimens 20 to $50\,\mu$ thick may be highly transparent, while much thicker specimens can be observed if the diffuseness
of the image which results from scattering can be tolerated. X-ray beams
travel through air for appreciable distances without attenuation. Over
the wavelength range from 1 to 10 Å, coefficients for the absorption of
x-rays are approximately proportional to the *cube* of the atomic number
of the absorbing element, so that the extent of absorption is very sensitive
to the atomic composition of specimens.

In principle, therefore, x-ray radiation should provide a method for
obtaining high resolution, high contrast microscope images of specimens
which could be hydrated and/or of considerable thickness. Microscopy
would not need to be carried out in a vacuum; thus hydrated and even
living specimens could be observed. It would seem that x-ray microscopy offers an ideal method for overcoming the limitations of both
optical and electron microscopy. The fatal snag is, of course, that *x-rays
cannot be effectively focused.* The very limited refraction of x-rays by
matter precludes the development of refractive lenses of the type used for
visible light, while the absence of charge on x-ray photons precludes any
possibility of focusing by electrostatic or magnetic fields, as is done with
electron beams. The application of holographic methods of focusing
(p. 249) has thus far been precluded by the unavailability of any coherent
x-ray source.

Some approaches have nonetheless been made to the use of x-rays in
microscopy; they are *contact microradiography* and the *point projection*
and *reflection focusing* systems. Resolving powers approaching that of
the light microscope have been achieved, while, in principle, resolution

down to about 100 Å seems possible. Prospects for the application of *holographic techniques* in x-ray microscopy are being actively considered at present, although practical systems have not been developed.

Even at the relatively limited available resolving powers, x-ray systems afford some advantages not provided by other types of microscope. Thick, opaque and/or living specimens may be observed, and quantitative elemental microanalyses may be performed. While, on account of the Z^3 dependence of x-ray absorption coefficients, analysis is most sensitive for elements of high atomic number, analysis of elements of biological importance is also possible with the use of relatively long wavelength radiation.

CONTACT MICRORADIOGRAPHY

In the method of contact microradiography, x-rays from a small source illuminate a specimen placed in contact with a photographic emulsion, as shown in Fig. 28-9a. The emulsion is subsequently developed and is observed in the light microscope. As shown by Fig. 28-9b, a source of finite size produces an image which is blurred to a degree determined by the effective distance of the specimen from the emulsion. Use of the smallest possible source is thus essential; sources about 0.1 mm in diameter are available. With the use of fine grained photographic emulsions, the resolution limit of contact microradiography is that $(0.1–0.2\,\mu)$ of the optical microscope used as an enlarging system.

POINT PROJECTION MICROSCOPY

In the point projection or *shadow microscope* the specimen is, ideally, coincident with a point source of radiation. An enlarged image is then formed on an emulsion placed beyond the source and specimen.

The factor which limits the resolution of any point projection system is the effective size of the "point" source. This size is ultimately limited by

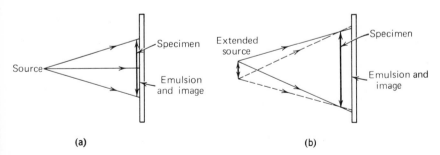

(a)   **(b)**

Fig. 28-9 Contact microradiography: (a) ideal; (b) blurring of image by an extended source.

diffraction effects. In point projection x-ray microscopes a source of about $1\,\mu$ effective diameter is formed by focusing an electron beam on the target; the latter then emits from a limited area only. Since the specimen is placed immediately in front of the source, smaller sources are practicable in this technique than in that of contact microradiography; i.e., from the point of view of adequate illumination. The point projection x-ray microscope is diagramed in Fig. 28-10.

REFLECTION FOCUSING X-RAY MICROSCOPE

Since the refractive indices of media with respect to visible light are greater than one, total *internal* reflection occurs when a medium-to-air interface is illuminated at angles greater than the critical angle (cf. p. 5). Similarly, since the refractive indices of media with respect to x-rays are less than one, total *external* reflection may occur when surfaces are illuminated by x-rays at appropriate angles. Since the refractive indices are very close to unity, the critical angle for total external x-ray reflection is always very nearly 90°; beams must be incident at angles of about 0°20′ with respect to a surface in order to undergo total reflection. Pairs of curved surfaces so illuminated may be used to form a point focus, as shown in Fig. 28-11. Such a system does, in fact, constitute a type of x-ray lens. Unfortunately, the images formed by these lenses, as by all mirror lenses, suffer rather seriously from geometrical aberrations, especially third order astigmatism (cf. p. 218). For this reason, only quite small apertures of illumination can be used, and the resolving power is correspondingly limited. Resolutions of the order of $0.5\,\mu$ have been obtained; a theoretical limit of about 100 Å is predicted.

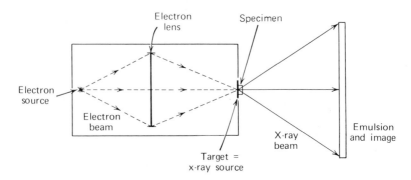

Fig. 28-10 Point projection x-ray microscopy.

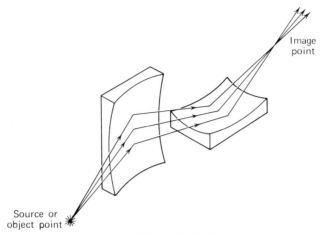

Fig. 28-11 X-ray reflection lens.

X-RAY HOLOGRAPHY

In the method of holographic image formation (p. 249), the diffraction pattern formed by an object is recorded directly, together with that of a reference beam. This pattern, the *hologram*, is subsequently illuminated, forming the reconstructed image of the object. The technique is potentially a most important one since it would offer a method for obtaining images in which the inherent resolution of x-ray wavelengths could be preserved, while using visible light to focus the reconstructed image. The difficulty at present is that a coherent source of x-ray illumination is required to form the hologram itself. While no x-ray lasers have yet been constructed, the development of such devices now seems feasible (see [1]).

Possible holographic methods which use partially coherent x-rays have also been considered. The proposed methods would employ the radiation passed by an aperture of very limited dimensions. In one of them, the *mirror* method illustrated in Fig. 28-12, light from the (incoherent)

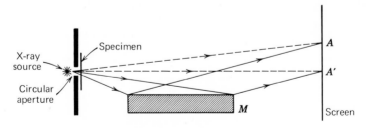

Fig. 28-12 Mirror method for x-ray holography.

source passes through a circular aperture of about $50\,\mu$ diameter, and then through the object O. Part of the transmitted beam is then incident at a grazing angle on a plane x-ray mirror, while the remainder of the beam travels to the screen without interruption. Rays incident upon the mirror undergo total external reflection and thus superimpose upon the direct illumination of the screen in the region AA'. It is calculated that a pair of holograms recorded at AA' should then serve to form a reconstructed image of the object, which would be of the general form shown in Fig. 28-13. The minimum distance resolved in this image should be of the order of

$$(28\text{-d}) \qquad\qquad d_{\min} \cong \frac{\lambda}{\alpha_c}$$

(where α_c is the critical angle for total external reflection at the x-ray mirror), or about 150 Å for $\lambda \cong 1$ Å and $\alpha_c \cong 0°20'$ or 0.03 rad.

In the *pinhole* method a reference pinhole and a specimen of unknown structure are illuminated simultaneously by an apertured x-ray source. The hologram so formed consists of their superimposed diffraction patterns, while the reconstructed image is, again, of the general form shown in Fig. 28-13. (This type of image may in fact be observed in optical analogs of the pinhole system.)

It is estimated that the resolution limit in this method is approximately equal to the diameter of the pinhole (reference object) and that the pinhole and specimen can be separated by only a few microns. Since apertures even as small as $10\,\mu$ in diameter are extremely difficult to construct, these conditions seem impossibly restrictive. However, if it were possible to replace the reference pinhole by an object of molecular dimensions and known structure, application of this form of holography might become practicable.

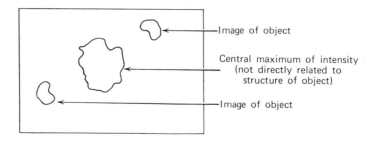

Fig. 28-13 Form of x-ray hologram.

REFERENCES

General References

H. P. Klug and L. E. Alexander, *X-ray Diffraction Procedures*, Wiley, New York, 1954, Chapter 2 (The Production and Properties of X-rays).

F. K. Richtmyer, E. H. Kennard and T. Lauritsen, *Introduction to Modern Physics*, McGraw-Hill, New York, 1955, Chapter 8 (X-rays).

A. Engstrom, in G. Oster and A. W. Pollister, Eds., *Physical Techniques in Biological Research*, Vol. 3, *Cells and Tissues*, Academic Press, New York, 1956, pp. 489–583. This chapter on "Historadiography" includes a short discussion of x-ray scattering and absorption, a table of "Mass Absorption Coefficients for Elements of Biological Interest," and descriptions of x-ray microscope methods and the properties of photographic emulsions.

H. A. Strobel, *Chemical Instrumentation*, Addison-Wesley, Reading, Mass., 1960, pp. 574–578. Sections on "Gas Ionization Detectors" and "Geiger-Mueller Counters." Table 18-2 summarizes "Characteristics of Some Nuclear Radiation Detectors."

B. D. Cullity, *Elements of X-ray Diffraction*, Addison-Wesley, Reading, Mass., 1956, Chapter 1 (Properties of X-rays).

J. T. Winthrop and C. R. Worthington, *Physics Letters* **15**, 124 (1965); **21**, 413 (1966). These articles on "X-ray Microscopy by Successive Fourier Transformation" discuss the theory of two possible methods for holographic x-ray microscopy and describe the results obtained in the optical analog of one type of system.

Specific Reference

[1] A. L. Schawlow *Scientific American* **219**, 136 (1968).

CHAPTER 29

Crystalline Properties of Matter as Related to X-ray Diffraction

Observation of x-ray diffraction patterns is a powerful method of studying the structure of matter which is organized in a regular or semiregular manner (i.e., of studying crystalline or paracrystalline specimens). A complete description of the methods used by crystallographers to obtain and to interpret diffraction patterns would be beyond the scope of this book, partly because of the voluminous nature of the subject, and partly because of the highly mathematical character of diffraction theory. Chapters 29 to 31 are thus intended only to provide the biologist or biophysicist with a sound general knowledge of the *nature* of diffraction studies of biological interest. In this chapter the methods used to categorize organized structures are described, and basic relationships between crystal structure and diffraction pattern spacings are derived. Methods for obtaining diffraction patterns and for deducing structures from these patterns are described in Chapters 30 and 31. A knowledge of the effects of diffraction of electromagnetic radiation, in general, as presented in Chapter 3, is essential for the study of this and the subsequent chapters.

CRYSTALS

Macroscopically, a crystal is a regular polyhedral solid. The sides of the polyhedron are called the crystal *faces*. In terms of atomic structure, the superficial regularity of crystals is found to be derived from a regularly repeated arrangement of the constituent atoms. The regular arrangement, or *lattice*, of atoms extends through space in three dimensions.

Depending upon the conditions under which crystallization takes place, crystals of a given substance may exhibit various *habits*; that is, the various crystal faces which are permitted by the nature of the atomic lattice may be developed to varying degrees. The resulting crystals may be superficially quite different in appearance. It is found, however, that, regardless of habit, the *interfacial angles are constant* in all crystals of a given substance.

Paracrystalline materials are those which possess a regularly repeating structure in only one or two dimensions. For example, the group of rodlets shown in Fig. 29-1 is a paracrystalline structure. The rodlets are arranged in a regular two-dimensional *array*, but their rotational orientation is random.

A regular arrangement of objects in one dimension is termed a *row*; in two dimensions an *array*, and in three dimensions a *lattice*. A row, an array, and a lattice are illustrated in Figs. 29-2a, 29-2b, and 29-2c respectively. Strictly speaking, the term "lattice" is correctly applied only to a three-dimensional arrangement of idealized *points*; the term is often used, however, to specify the three-dimensional collection of atoms in a crystal.

All crystals may be classified as belonging to one of six *crystal systems*. Three directions are selected as the *crystallographic axes* of the crystal, designated *a*, *b*, and *c*, respectively. The relative lengths of these axes,

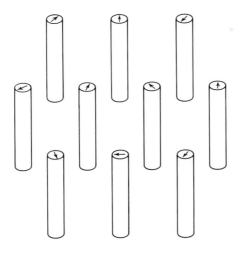

Fig. 29-1 A paracrystalline array.

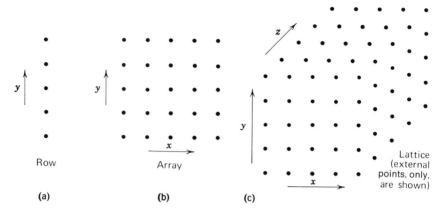

Fig. 29-2 Regular arrangements of points.

and the angles between them, determine to which system the crystal belongs. Table 29-1 summarizes the defining characteristics of the six crystal systems. (Seven crystal systems have sometimes been described. However, the seventh one — the rhombohedral system — may be shown to be a transformation of the hexagonal system.) Note that the interaxial angles are conventionally designated as α (between b and c axes), β (between a and c), and γ (between a and b). It is often possible to assign a crystal to more than one system; in general, the system of highest symmetry is chosen.

SYMMETRY OPERATIONS

The *movement* of a body may be defined as an operation by which the distance between any pair of points within the body remains unchanged. *Translations*, *rotations* about axes, and *reflections* across planes are examples of movements, as are combinations of these operations. Note,

Table 29-1   The Crystal Systems

System	*Axial Ratios*	*Interfacial Angles*
Cubic	$a = b = c$	$\alpha = \beta = \gamma = 90°$
Tetragonal	$a = b; c \neq a, b$	$\alpha = \beta = \gamma = 90°$
Orthorhombic	$a \neq b \neq c$	$\alpha = \beta = \gamma = 90°$
Monoclinic	$a \neq b \neq c$	$\alpha = \gamma = 90°; \beta \neq 90°$
Hexagonal	$a = b; c \neq a, b$	$\alpha = \beta = 90°; \gamma = 120°$
Triclinic	$a \neq b \neq c$	$\alpha \neq \beta \neq \gamma$

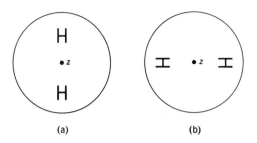

(a)   (b)

Fig. 29-3 Objects with axes of rotational symmetry.

for example, that any displacement of an object may be effected by combining an appropriate translation and rotation.

The symmetry properties of objects or of lattices may be characterized by specifying those movements which bring the object or lattice into coincidence with itself, that is, into an orientation indistinguishable from an initial one. Consider, for example, the disc illustrated in Fig. 29-3a. Rotation of the disc through $(360/4)° = 90°$ about the z axis (the axis through the center of the disc and perpendicular to the plane of the paper) produces the orientation illustrated by Fig. 29-3b, which is readily distinguished from that shown in part (a) of the figure. Rotation of the disc through $(360/2)° = 180°$, however, produces a structure which is indistinguishable from that shown in (a). The z axis of the disc is thus said to be a *two-fold axis of rotational symmetry*, but not a four-fold axis. Note that, crystallographically, the term n-fold refers to the possibility of performing an operation $1/n$ times in order to obtain an identical structure. The term does *not*, as in general usage, imply the existence of n copies.

Like *all* axes, the z axis of the disc shown in Fig. 29-3 is also a *one-fold* axis of rotational symmetry; any object may be brought into coincidence with itself by rotation through 360°. This rotation through 360° is termed the *identity operation.*

The disc of Fig. 29-3 lacks three- or higher-fold axes of rotational symmetry. Note that the somewhat similar disc shown in Fig. 29-4 possesses *no* rotational axes except the one-fold axis.

It may be shown that lattices can possess 1-, 2-, 3-, 4- or 6-fold axes of rotational symmetry, but that lattices of 5- or 7- or higher-fold rotational symmetry cannot exist.

As shown in Figs. 29-5a and 29-5b, respectively, each of the discs of Figs. 29-3 and 29-4 has *reflection planes* in the y direction, and the

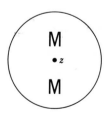

Fig. 29-4 An object which lacks axes of rotational symmetry.

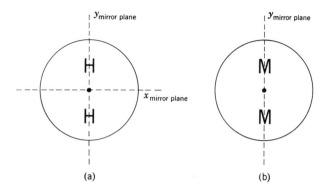

Fig. 29-5 Reflection planes.

disc of Fig. 29-3 has a reflection plane in the x direction also. Reflection of one half of the object across these reflection planes generates a structure which is indistinguishable from the original.

Further symmetry operations may be defined with consideration of three-dimensional structures. Figure 29-6 depicts a disc from which points can be imagined to project, to equal distances, above and below the plane of the page, as designated by (+) and (−), respectively. This object may be described as having an x axis which is a two-fold axis of rotational symmetry. Alternatively, however, the original orientation may be reproduced by combining *rotation* through 360° about the z axis with inversion across the origin. The structure thus possesses a *one-fold inversion axis*. More generally, an *n-fold inversion axis* specifies rotation through $(360/n)°$ coupled with inversion through a point on the axis of rotation. (Note that the order in which these operations are considered to be performed is optional.) Note also that the direction in which the inversion is performed must be correctly chosen if the original structure is to be reproduced. An example of a structure possessing a four-fold axis of inversion is shown in Fig. 29-7. A *center of inversion* or *center of*

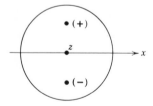

Fig. 29-6 An object possessing a one-fold inversion axis.

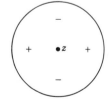

Fig. 29-7 An object possessing a four-fold inversion axis.

symmetry is defined as the intersection of a two-fold rotational axis with a reflection plane perpendicular to the rotational axis. Thus, in Fig. 29-8, *AA'* is an axis of two-fold rotational symmetry, and *BB'* a mirror plane (or vice versa), so that *C* is a center of inversion (center of symmetry).

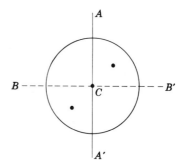

Fig. 29-8 An object possessing a center of symmetry.

Other symmetry operations may be defined which apply to lattices of infinite extent. When translated through a suitable repeat distance, an infinite lattice is indistinguishable from that lattice in its original position. The symmetry operations already discussed may thus be combined with translations. A *glide plane* is present, for example, in the repeating structure shown in Fig. 29-9. The glide plane combines reflection about the *x* axis with translation through (in this case) one half of the repeat distance in the *x* direction.

Similarly, a *screw axis* combines rotation about an axis with translation along that axis. Again using (+) and (−) to denote equal projections above and below the plane of the page, Fig. 29-10 depicts a structure in which a *two-fold screw axis* is present. Rotation about the *x* axis of $(360/2)°$ coupled with a translation of one half the repeat distance in the *x* direction regenerates the original structure.

For screw axes which are higher than two-fold, more than one possibility exists for the extent of translation. Figures 29-11a and 29-11b illustrate, in two-dimensional projection, two types of four-fold screw axes. The structure shown in Fig. 29-11a must be rotated through 90° and translated through one half the repeat distance to achieve coincidence, while that of Fig. 29-11b requires translation through one fourth of the repeat distance.

Other types of screw symmetries combine rotation about one axis with translation along a different axis.

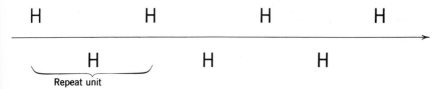

Fig. 29-9 Glide plane.

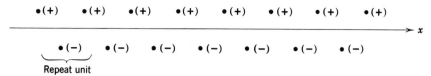

Fig. 29-10 Two-fold screw axis.

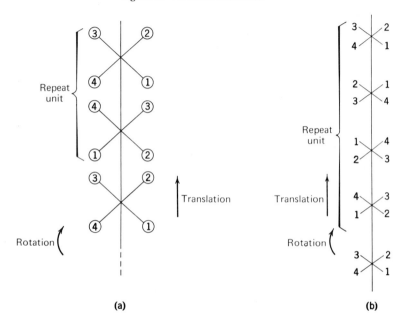

Fig. 29-11 Four-fold screw axes.

POINT GROUPS, SPACE LATTICES, AND SPACE GROUPS

A *point group* is the set of symmetry operations which may be used to bring an object, or a finite set of points, into coincidence with itself. Point groups are used to characterize the symmetry of objects and lattices. Note that, whereas point groups may be illustrated by an arrangement of points to which the appropriate symmetry operations apply, the point groups are in fact *not* sets of points but sets of operations.

Point groups include the rotations, reflections, and combinations of rotation and reflection discussed in the preceding section. Operations which include translations do *not* belong to point groups, since finite objects cannot be brought into coincidence by operations of translation. If *n* different symmetry operations apply to an object, the object is said to belong to a point group of the *n*th order. As an example, the operations which describe the symmetry of a hexagon (approximated by the benzene molecule) are illustrated in Fig. 29-12. A total of 24 operations or combinations of operations exist which produce identical orientations. (Note that the vertices of the hexagon are all identical, although designated differently as 1 . . . 6 in Fig. 29-12 in order to make clear the nature of each movement.) The hexagon thus belongs to a point group of the 24th order. Whereas the operations described by Figs. 29-12h and 29-12i are trivial for a plane figure, analogous operations would constitute important specifications of symmetry for a solid figure.

Altogether, thirty-two different point groups exist. The simplest of them is the completely "asymmetric" condition of a one-fold rotational axis (identity operation) which applies to single points. Arrangements of higher symmetry may require as many as 48 points for representation. The details of point group specifications are described in a number of crystallography texts and need not be considered here.

The *space lattice*, or *Bravais lattice*, is an arrangement of points in space such that the environment of each point is equivalent; i.e., it may be stated that the structure of a space lattice is "homogeneous". In these lattices a line joining any pair of points can be projected to pass through a continuing, similarly spaced sequence of points. All lines parallel to each other and passing through points of the lattice encounter the same spacing of points. There are 14 possible space lattices, one or more of which correspond to each of the crystal systems summarized in Table 29-1. For example, the cubic crystal system includes the simple cubic, body-centered cubic, and face-centered cubic space lattices.

A *space group* is the set of all symmetry operations which may be used to bring an infinite lattice into coincidence with itself, that is, including operations of translation. Space groups thus consist of combinations of rotations, reflections, and translations.

Certain of the space groups may be derived by placing a set of points representing a point group at each position of a space lattice. This implies the existence of at least $32 \times 14 = 448$ space groups; many of them, however, are identical. Additional space groups result from the operation of glide planes and screw axes with respect to such structures. Altogether 230 nonidentical space groups may be specified.

THE PRIMITIVE CELL AND THE UNIT CELL

In a lattice (or a two-dimensional array) the *smallest complete unit of pattern* is known as a *primitive cell*. The *unit cell* of a crystal is a unit of pattern which may be repeated in three dimensions to build up the complete structure of the crystal. While the primitive cell may always

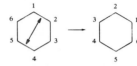

(a) Identity (1-fold rotation) (1 operation)

(b) 6-fold rotation in either sense about center (2 operations)

(c) 3-fold rotation in either sense about center (2 operations)

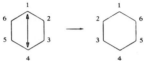
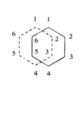

(d) 2-fold rotation about center (1 operation)

(e) 2-fold rotations about lines joining vertices 1–4, 2–5, or 3–6 (3 operations)

(f) 2-fold rotations about lines joining midpoints of sides (3 operations)

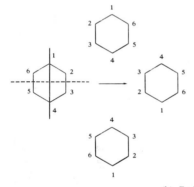

(g) Center of inversion at center = 2-fold rotation + reflection (1 operation)

(h) Reflection in plane of hexagon (as for any two-dimensional figure) (1 operation)

(i) 6-fold or 3-fold rotation + reflection in plane of hexagon (4 operations)

(j) Reflections in any of the planes perpendicular to the plane of the hexagon and passing through opposite vertices (3 operations)

(k) Reflections in any of the 3 planes perpendicular to the plane of the hexagon and passing through the midpoints of opposite sides (3 operations)

Fig. 29-12.

be chosen as a unit cell, the unit cell actually designated may be the equivalent of several primitive cells. This is illustrated by Fig. 29-13 in which a body-centered cubic unit cell is compared with the corresponding primitive cell. In this case the volume of the unit cell is three times that of the primitive cell. Clearly, there exist various possibilities for the choice of the unit cell to be used in specifying the structure of a crystal. This choice is dictated by considerations of convenience in handling the crystallographic data. As Fig. 29-13 suggests, the choice of a rectangular unit cell makes the visualization of the complete crystal structure relatively convenient.

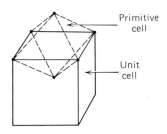

Fig. 29-13 Primitive and unit cells of body centered cubic system.

The size of the unit cell is determined by the lengths of its axes, which may be designated a_0, b_0, and c_0, and by the angles between these axes. a_0, b_0, and c_0 are absolute lengths, proportional to the relative values specified in Table 29-1; a, b, and c may be used to denote distances along the respective axes. For some purposes it is convenient to specify the unit cell axes and their directions as vectors **a**, **b**, and **c**.

CRYSTAL PLANES AND MILLER INDICES

Any regular lattice may be thought of as constituted by a set of parallel planes on which the lattice points constitute a regular array. This concept is illustrated, in two dimensions, by Fig. 29-14, in which the array illustrated can be considered to be composed of a set of parallel, densely populated rows separated by a distance d_a. Alternatively, the array can be considered to consist of a set of less densely populated rows, separated by the smaller distance d_b or of any of an infinite number (in principle) of sets of rows. The direction and spacing of planes is of course not arbitrary but is determined by the nature of the array, that is, by the structure of a crystal.

A convenient method for specifying the orientation of such planes, with respect to the crystal axes, is by means of the *Miller indices* characteristic of each plane. A *unit* or *parametral* plane is defined which has the property of making *equal intercepts* with each of the crystal axes. The choice of a unit plane is thus equivalent to the choice of a set of axes, the designation of all crystal planes being determined by the choice. The axes, in turn, are selected in a manner which gives a convenient shape of unit cell, as discussed above.

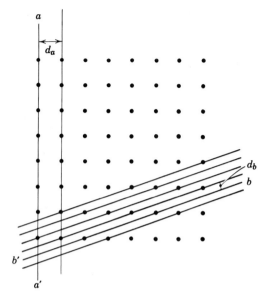

Fig. 29-14 Crystal planes.

Any plane which contains lattice points of the crystal intercepts the axes at positions which may be designated ma, nb, and pc. For the unit plane, m, n, and p are each equal to one. Since the ratios of the intercepts m, n, and p remain constant if a plane is moved parallel to itself, these numbers can always be expressed as integers or, for planes parallel to one of the axes, as infinity. Because in practice it is convenient to avoid values of infinity, the intercepts of planes with the crystal axes are specified by a second set of integers which are proportional to the *reciprocals* of the intercepts m, n, and p. The reciprocal intercepts are designated h, k, and l, and are the Miller indices of the plane. That is,

$$(29\text{-}a) \qquad h \propto 1/m; \qquad k \propto 1/n; \qquad l \propto 1/p.$$

Consider, for example, a plane which is parallel to the c axis of a crystal but makes intercepts in the ratio $2:1$ with the a and b axes, respectively, as shown in Fig. 29-15. m, n, and p are thus 2, 1, and infinity, respectively, while h, k, and l are proportional to $1/2$, $1/1$, and $1/\infty = 0$. Appropriate integral values of h, k, and l are thus 1, 2, and 0, and the plane shown in Fig. 29-15 is accordingly termed the $(1, 2, 0)$ plane. The smallest integral combination of h, k, and l is chosen in preference to other combinations such as 2, 4, 0. The Miller indices of the unit plane are, of course $(1, 1, 1)$.

Negative as well as positive values of the Miller indices are possible. For example, the plane for which $h=+1$, $k=-1$, and $l=+1$ is designated the $(1, \bar{1}, 1)$ plane. Figure 29-16a and 29-16b, respectively, illustrate segments of the $(1, 1, 1)$ and $(1, \bar{1}, 1)$ planes. Note that, by reversing the positive direction of the b axis, the plane shown in Fig. 29-16b could have been *defined* as the unit plane of the crystal.

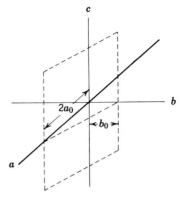

Fig. 29-15 The $(1, 2, 0)$ plane.

THE BRAGG LAW

The Bragg law is the fundamental relationship which specifies the relation between spacings in a crystal and the diffraction pattern formed by the crystal. As discussed in the previous section, a crystal may be thought of as constituted by sets of equidistant parallel planes. Bragg's law may then be derived by considering that incident x-rays (or other radiation) are reflected from each member of any set of planes in a manner consistent with the law of reflection. Thus, as illustrated by Fig. 29-17, light incident upon the upper surface of a crystal at angle θ to the surface is reflected in part by the uppermost plane of atoms, also at angle θ to the surface. The reflected beam thus suffers a total angular deviation, with respect to the incident beam, of 2θ. At the same time, part of the beam is transmitted through the crystal and is partially reflected, again at angle θ, from each of the subsequent planes of atoms. These planes are spaced at the characteristic distance d_{hkl}. All the reflections through angle θ superimpose to form the beam reflected (diffracted) by the (h, k, l) planes of the crystal.

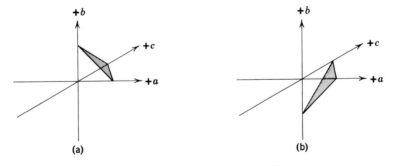

Fig. 29-16 (a) The $(1, 1, 1)$ plane; (b) the $(1, \bar{1}, 1)$ plane.

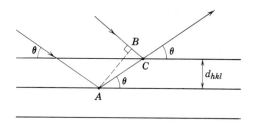

Fig. 29-17 Reflection from crystal planes.

The components of the reflected beam which originate from different planes differ successively in phase to a degree which determines the resultant intensity of the diffracted beam. In order to derive the condition for a maximum of diffracted intensity, with reference to Fig. 29-17 note that beams from successive crystal planes differ in path by $(AC - BC)$. If this distance is an integral number of wavelengths, reinforcement occurs. The condition for a diffraction maximum is thus

(29-b) $$AC - BC = m\lambda,$$

where m in any integer. From Fig. 29-17,

(29-c) $$AC = \frac{d_{hkl}}{\sin\theta},$$

(29-d) $$\angle ACB = 2\theta,$$

(29-e) $$\cos 2\theta = \frac{BC}{AC}.$$

Combination of (29-c), (29-d), and (29-e) then yields

(29-f) $$BC = d_{hkl}\frac{\cos 2\theta}{\sin\theta}.$$

Substitution of (29-c) and (29-f) in (29-b) then gives

(29-g) $$\frac{d_{hkl}(1 - \cos 2\theta)}{\sin\theta} = m\lambda.$$

Making use of the trigonometric identity

(29-h) $$(1 - \cos 2\theta) \equiv 2\sin^2\theta$$

yields

(29-i) $$\boxed{2d_{hkl}\sin\theta = m\lambda,}$$

which is the Bragg law.

The Bragg law thus states that, for a set of reflecting planes separated by a distance d_{hkl}, light of wavelength λ, incident upon the planes at angle θ, is diffracted with maximum intensity through an angle 2θ. As there are many reflecting planes in a crystal of reasonable size, a sharply defined maximum of intensity tends to be diffracted at the Bragg angle 2θ, while intensity scattered in all other directions is essentially zero. However, to the extent that Compton scattering effects are appreciable, loss of phase relationships can result in a less sharply defined distribution of intensities.

It is important to understand that the Bragg law specifies that a given set of planes produces a diffracted intensity *only* when the planes are illuminated at precisely the correct angle θ_{hkl}. Within any crystal there are many sets of planes, each of which (except in certain specific cases considered below) diffracts at an appropriate value of θ. In general, however, a *single, stationary* crystal illuminated by *monochromatic* x-rays produces few or no diffracted intensities, since conditions for Bragg scattering are *not* met precisely for most planes or for any planes. Useful diffraction patterns, which include intensities corresponding to many different values of (h, k, l) may be obtained by effectively varying the orientation of the diffracting object in some way. Techniques for obtaining diffraction patterns are surveyed in Chapter 30.

Any given set of planes, characterized by a value d_{hkl}, produces only a finite number of diffracted orders of x-rays of a single wavelength, since $(\sin \theta)$ cannot be greater than one. The larger the value of d_{hkl}, the more different values may *m* of (29-i) assume.

The Bragg law is similar in form to the diffraction grating equation $(3\text{-}c^*)$. The former applies, however, to sets of uniformly spaced planes in *three* dimensions, rather than to uniform grating spacings in *two* dimensions.

THE LAUE EQUATIONS AND THE BRAGG LAW

The Bragg law, just derived in terms of reflections, may also be obtained from consideration of interference between beams which are scattered at crystal planes. This somewhat more sophisticated approach may now be considered.

Scattering from any point or plane of a crystal may be described in terms of the directions of incident and diffracted rays, as indicated in Fig. 29-18. The incident ray can be represented by a vector of unit length, s_0, and the diffracted ray by a second unit vector s_1. (For an elementary discussion of vectors see appendix B.) The relationship between s_0 and s_1 is conveniently described by a single vector S such that

(29-j) $$S = (s_1 - s_0).$$

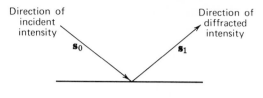

Fig. 29-18.

The geometric representation of **S** is given by Fig. 29-19.

The separation of two identical scattering points P and P' within a crystal may specify a vector $\mathbf{r}$ as shown in Fig. 29-20a. When a beam is incident in a direction given by $\mathbf{s_0}$, as shown in Fig. 29-20b, the phase of the beam at some point M is equal to that at P. At the second scattering point P', the phase of the beam differs by an amount determined by the distance MP'; that is, by $(2\pi/\lambda)(MP')$. As shown in Fig. 29-20c, the radiation scattered by P and P' may be observed in a direction $\mathbf{s_1}$. The optical paths of the rays from the two scattering points are equal from $P'N$ onward, but the rays differ in phase by an amount determined by the distance PN, that is, by $(2\pi/\lambda)(PN)$. Thus the total difference in phase between the rays diffracted in the direction $\mathbf{s_1}$ is $(2\pi/\lambda)(PN - MP')$. If the length $(PN - MP')$ is an integral multiple of λ, a *maximum* of diffracted intensity occurs in the direction $\mathbf{s_1}$. Thus the condition for a diffraction maximum is

(29-k) $$(PN - MP') = m\lambda.$$

MP' is in the direction $\mathbf{s_0}$ of the incident ray and thus, as is evident from Fig. 29-20b, this distance is given by the dot product of $\mathbf{r}$ and $\mathbf{s_0}$; that is,

(29-l) $$MP' = \mathbf{r} \cdot \mathbf{s_0}.$$

Likewise, from Fig. 29-20c,

(29-m) $$PN = \mathbf{r} \cdot \mathbf{s_1}.$$

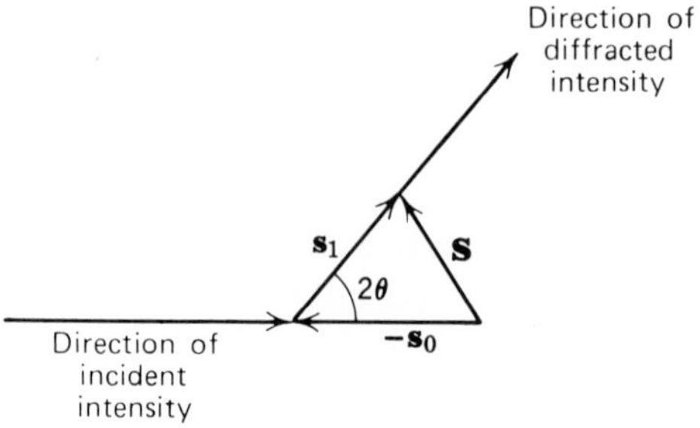

Fig. 29-19.

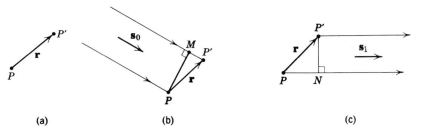

Fig. 29-20.

Substitution of (29-1) and (29-m) into (29-k) gives the condition for a maximum of diffracted intensity as

(29-n) $$m\lambda = \mathbf{r} \cdot (\mathbf{s}_1 - \mathbf{s}_0),$$

or in terms of $\mathbf{S}$ as defined above,

(29-o) $$\boxed{m\lambda = \mathbf{r} \cdot \mathbf{S}}.$$

Equation 29-o is known as *Von Laue's condition*.

It may be noted from Fig. 29-21 that diffraction is equivalent to reflection from a mirror plane to which both $\mathbf{s}_0$ and $\mathbf{s}_1$ are inclined at angle θ. As shown in Fig. 29-21, ABC is an isosceles triangle (since both $\mathbf{s}_0$ and $\mathbf{s}_1$ are vectors of unit length), and the mirror plane bisects the angle 2θ. It is convenient to specify a third unit vector $\mathbf{n}$ in the direction of $\mathbf{S}$. Note that the magnitude of $\mathbf{S}$ is obtained from the geometry of Fig. 29-21 as

(29-p) $$|\mathbf{S}| = 2\sin\theta.$$

Then since, by definition, $\mathbf{S} \equiv |\mathbf{S}|\mathbf{n}$,

(29-q) $$\mathbf{S} = (2\sin\theta)\mathbf{n},$$

Von Laue's conditions may thus also be written as

(29-r) $$\boxed{(2\sin\theta)(\mathbf{r}\cdot\mathbf{n}) = m\lambda}.$$

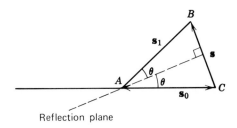

Reflection plane

Fig. 29-21.

r, like any vector, may be expressed as the sum of its projections along three axes specified by the unit vectors **i**, **j**, and **k**. In considering diffraction by a crystal, it is convenient to substitute for **i**, **j**, and **k** the unit vectors **a**, **b**, and **c**, which lie in the direction of the axes of the unit cell. For cubic unit cells, **a**, **b**, and **c** are mutually perpendicular, and thus directly equivalent to **i**, **j**, and **k**; for nonrectangular unit cells these axes are mutually inclined at angles which must be accounted for in making the transition. Clearly, however, the position of any atom in the unit cell, relative to an arbitrarily selected position at which **r** = 0, may be specified by a vector **r** = m**a** + n**b** + p**c**. For diffraction maxima to occur, the projections of **a**, **b**, and **c**, on **S** must *all* satisfy (29-o); that is, they must all be integral multiples of one wavelength. This requirement is expressed by the equations known as the *Laue equations*:

(29-s)
$$\boxed{\mathbf{a}\cdot\mathbf{S}=h;\quad \mathbf{b}\cdot\mathbf{S}=k;\quad \mathbf{c}\cdot\mathbf{S}=l}\,,$$

where h, k, and l may be considered *any* combination of integers, and they are not restricted to the lowest integral multiples as in the case of the Miller indices discussed above.

In a manner analogous to (29-r), the Laue equations may also be expressed in the form

(29-t)
$$\boxed{\begin{aligned}(2\sin\theta)(\mathbf{a}\cdot\mathbf{n}) &= h \quad\text{and}\\ (2\sin\theta)(\mathbf{b}\cdot\mathbf{n}) &= k \quad\text{and}\\ (2\sin\theta)(\mathbf{c}\cdot\mathbf{n}) &= l.\end{aligned}}$$

From (29-t),

(29-u)
$$\frac{2\sin\theta}{\lambda}=\frac{\mathbf{a}\cdot\mathbf{n}}{h}=\frac{\mathbf{b}\cdot\mathbf{n}}{k}=\frac{\mathbf{c}\cdot\mathbf{n}}{l}.$$

Thus, for the set of crystal planes to which **n** is normal,

(29-v)
$$\frac{\mathbf{a}}{h}\cdot\mathbf{n}=\frac{\mathbf{b}}{k}\cdot\mathbf{n}=\frac{\mathbf{c}}{l}\cdot\mathbf{n}=d_{hkl},$$

which gives

(29-w)
$$d_{hkl}=\frac{2\sin\theta}{\lambda}.$$

Equation 29-w is the Bragg law; this expression is identical with (29-i) except for omission of the integral factor m. The absence of m derives from the fact that h, k, and l have been considered here as *any* combination of integers which may, therefore, have a common divisor m.

SYSTEMATIC ABSENCES

Diffraction maxima tend to appear at angle 2θ to the incident beam when values of d_{hkl}, θ, and λ are such as to satisfy the Bragg law (29-i) or (29-w). In certain cases, however, the symmetry of the diffracting crystal may be such as to cause the intensity at these positions to be zero or, in the case of crystals containing more than one type of atom, to be small. Such systematic absences of diffraction spots provide important clues to the symmetry properties of the crystal.

An example of a structure which produces systematic absences is the body-centered cubic lattice, the unit cell of which has been illustrated in Fig. 29-13. Since the cell is cubic, the unit vectors $\mathbf{a}$, $\mathbf{b}$, and $\mathbf{c}$ may be considered to be equal in length to the corresponding cell axes. The position of the central atom of the unit cell is then given by the vector $\mathbf{r}$, such that

$$(29\text{-x}) \qquad \mathbf{r} = \tfrac{1}{2}\mathbf{a} + \tfrac{1}{2}\mathbf{b} + \tfrac{1}{2}\mathbf{c}.$$

Diffraction maxima occur when (29-o) and (29-p) are satisfied simultaneously. That is, when

$$(29\text{-y}) \qquad m\lambda = \tfrac{1}{2}(\mathbf{a} + \mathbf{b} + \mathbf{c}) \cdot \mathbf{S} = \tfrac{1}{2}h\lambda + \tfrac{1}{2}k\lambda + \tfrac{1}{2}l\lambda,$$

which gives

$$(29\text{-z}) \qquad 2m\lambda = h + k + l.$$

That is, the sum $(h + k + l)$ must be even. Reflections from the $(1,1,0)$ plane, for example, are observed, whereas those from, for example, the $(1,1,1)$ plane are not. The observation of such absences is thus diagnostic of the symmetry of the diffracting structure. More specifically, systematic absences may be correlated with the space group to which the crystal belongs.

Systematic absences are complete only if the points of the crystal lattice are of *identical scattering power*. As discussed in Chapter 31, x-ray scattering by atoms is roughly a function of the square of their atomic number; thus, if a crystal is made up of atoms of different atomic numbers, destructive interference is incomplete at positions corresponding to the systematic absences, and a reduced but nonzero intensity results.

THE CONCEPT OF RECIPROCAL SPACE

The *reciprocal lattice*, existing in reciprocal space, is a conceptual method for considering the three-dimensional diffraction patterns of crystals. The reciprocal lattice has been an extremely fruitful approach in the formulation of diffraction problems. Although consideration of the somewhat sophisticated concept of reciprocal space is not essential to an elementary discussion of x-ray diffraction methods, a short description of the nature of the reciprocal lattice, and of its geometrical interpretation in terms of diffraction, is given here. The application of these ideas in diffraction experiments is indicated in succeeding chapters.

The reciprocal lattice provides a simple means for expressing both the *spacings* between the planes of a crystal and the slopes of those planes. The reciprocal lattice can be considered to be built up in the following way: vectors are drawn, from a single origin, *normal* to each set of planes in the real crystal and of a length proportional to $1/d_{hkl}$, the reciprocal of the interplanar spacing. Thus

$$(29\text{-aa}) \qquad\qquad \sigma_{hkl} = C\frac{1}{d_{hkl}}$$

where C is an arbitrary constant and σ_{hkl} is the length of the reciprocal lattice vector. The reciprocal lattice then consists of the collection of points lying at the ends of the lines so constructed. It can be shown that this collection of points is, in fact, a regular lattice; this general proof is assumed here.

The construction of the reciprocal lattice can conveniently be illustrated by a two-dimensional example. Figure 29-22a illustrates a portion of a two-dimensional rectangular array of points, characterized by the spacings a and b. Sets of parallel lines drawn through the array may be specified by two Miller indices, h and k. Figure 29-22b shows, for example, a set of such parallel lines which makes intercepts of 2 and 1 with the a and b axes, respectively, and which thus are characterized by the indices $(1, 2)$. A single unit cell of the array is shown in Fig. 29-22c; its edges are of the absolute lengths a and b. One of the lines shown in Fig. 29-22b is included; it intercepts the cell edges at intercepts $a/h = a/l = a$ and $b/k = b/2$. The length MN of the segment passing through the unit cell can then be obtained from the Pythagorean theorem:

$$(29\text{-bb}) \qquad\qquad MN = \sqrt{\left(\frac{a}{h}\right)^2 + \left(\frac{b}{k}\right)^2}.$$

A line adjacent to MN passes through the origin of the unit cell, as shown in Fig. 29-22d. The two line segments are separated by the perpendicular distance $d = OP$. OPN and MON form similar triangles. Thus,

$$(29\text{-cc}) \qquad\qquad \frac{OP}{ON} = \frac{MO}{MN} = \frac{d_{hk}}{(a/h)} = \frac{(b/k)}{MN},$$

and, substituting MN from (29-bb),

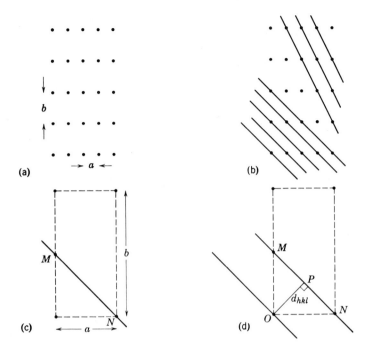

Fig. 29-22 (a) Two dimensional array; (b) sets of parallel lines within array; (c) single unit cell; (d) formation of reciprocal vector.

$$(29\text{-dd}) \qquad d_{hk} = \frac{(a/h)\,(b/k)}{\sqrt{(a/h)^2 + (b/k)^2}}.$$

d_{hk} is the separation of rows in the real array; d_{hk} as drawn in Fig. 29-22d also represents a direction perpendicular to these rows.

The reciprocal array vector may now be obtained in the manner defined above by (29-aa), taking the constant C equal to unity. From (29-dd),

$$(29\text{-ee}) \qquad \sigma_{hk} = \frac{hk}{ab}\sqrt{\left(\frac{a}{h}\right)^2 + \left(\frac{b}{k}\right)^2} = \sqrt{\left(\frac{k}{b}\right)^2 + \left(\frac{h}{a}\right)^2}.$$

Equation 29-ee may be applied to the definition of the reciprocal lattice to generate the reciprocal array from any real, two-dimensional array. An array of the form illustrated by Fig. 29-22, and in which $a = 1$ cm and $b = 2$ cm, may be considered as a specific example, as shown in Fig. 29-23a. The expression for the reciprocal vector, σ_{hk}, is then

$$(29\text{-ff}) \qquad \sigma_{hk} = \sqrt{\left(\frac{k}{2}\right)^2 + \left(\frac{h}{1}\right)^2}.$$

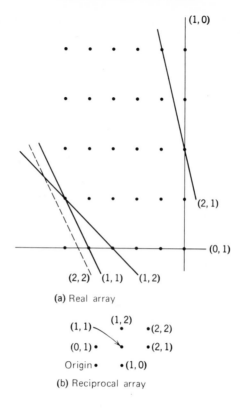

(a) Real array

(b) Reciprocal array

Fig. 29-23 Real and reciprocal arrays.

The computation of the distances σ_{hk}, which correspond to a number of planes of the real array, is summarized by Table 29-2. The points of the reciprocal array lie along the directions normal to the lines indicated in

Table 29-2 Distances σ_{hk} to Points of a Reciprocal Array

h	l	$(\sigma_{hk})^2$	σ_{hk}
0	1	$(\frac{1}{2})^2$	0.5 cm
1	0	1	1.0
1	1	$(\frac{1}{2})^2 + 1 = \frac{5}{4}$	$\sqrt{5}/2 \cong 1.12$
1	2	$1 + 1 = 2$	$\sqrt{2} \cong 1.41$
2	1	$4^2 + (\frac{1}{2})^2 = \frac{17}{4}$	$\sqrt{17}/2 \cong 2.06$
2	2	$1 + 4 = 5$	$\sqrt{5} \cong 2.24$

Fig. 29-23a. The reciprocal array so obtained is given by Fig. 29-23b. Note that the points corresponding to the largest values of d_{hk} lie closest to the origin of the reciprocal array. Additional points of the reciprocal array could be generated by applying (29-ff) to planes of the real array which are more closely spaced than the $(2, 2)$ set. Clearly, each point of a reciprocal lattice or array represents a spacing present in the *entire* real structure.

It is evident from Fig. 29-23 that, for a real lattice in which the b (vertical) dimension is greater than the a (horizontal) dimension, a reciprocal lattice is obtained in which the larger spacing lies in the a direction. More generally, the reciprocal of a rectangular lattice is one oriented at 90° to the original. The axes of three-dimensional reciprocal lattices are conventionally designated a^*, b^*, and c^*, such that

$$(29\text{-gg}) \qquad a^* = \frac{1}{d_{1,0,0}}; \qquad b^* = \frac{1}{d_{0,1,0}}; \qquad c^* = \frac{1}{d_{0,0,1}}.$$

Figure 29-24a shows an array in which the unit cell is a *parallelogram*; the corresponding reciprocal array is illustrated in Fig. 29-24b. In the reciprocal array the ratio of axial spacings is, as before, reciprocal to that of the real array, while the interaxial angle is the supplement of that found in the real array. That is,

(29-hh)
interaxial angle of real array + interaxial angle of reciprocal array = 180°.

In this more general case also, similar considerations apply to three-dimensional lattices.

The properties of reciprocal lattices are such that *the real lattice may be generated from the reciprocal lattice by the same set of operations as used to effect the original transformation.*

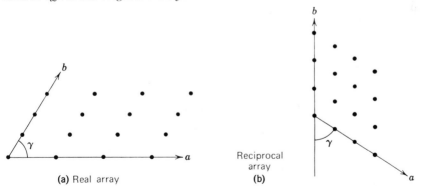

(a) Real array   Reciprocal array   (b)

Fig. 29-24 Angular relationship between real and reciprocal arrays.

GEOMETRICAL INTERPRETATION OF THE RECIPROCAL LATTICE

Bragg's law, written in the form

(29-ii)
$$\sin\theta_{hk} = \frac{\lambda\,(1/d_{hk})}{2},$$

may be represented geometrically in two dimensions by Fig. 29-25; the arc QA is proportional to the quantity λ/d_{hk} for a circle of diameter two. From (29-aa). $\sigma_{hk} = C(1/d_{hk})$. Thus, if the constant C is chosen to be λ rather than unity as in the cases considered in the preceding section, the length QA represents the reciprocal lattice spacing.

If OA represents the direction of an *incident* beam, OQ represents the orientation of a set of crystal planes which satisfies Bragg's law. QA is normal to OQ, and thus to the diffracting planes. QA corresponds in direction, as well as in length, to the reciprocal lattice vector σ_{hk}. Accordingly, Q is the reciprocal lattice point which corresponds to real lattice planes in the direction OQ when the origin of reciprocal space is chosen to lie at point A.

In accordance with Bragg's law, the beam *diffracted* from the OQ planes must make an angle of 2θ with the incident beam. It will be recollected from elementary geometry that the angle QCA in Fig. 29-26 is twice angle QOA. Thus the direction of the diffracted beam is represented by CQ. Geometrically, then, the condition for observation of diffraction maxima is that crystal planes be so oriented that Q, the corresponding reciprocal lattice point (i.e., the tip of the reciprocal lattice vector QA), lies on the circle of unit radius shown in Figs. 29-25 and 29-26. Reciprocal lattice points which do *not* lie on this unit circle correspond to planes which are incorrectly oriented for the production of diffraction maxima; that is, they correspond to directions in which there

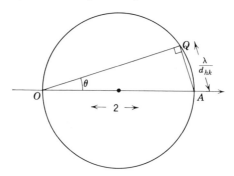

Fig. 29-25 Geometrical representation of Bragg's law.

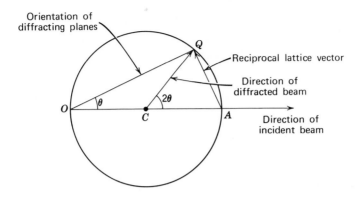

Fig. 29-26 The sphere of reflection and the reciprocal lattice vector.

is destructive interference between the amplitudes reflected from successive planes of the crystal. The same considerations apply to diffraction from a three-dimensional lattice, the condition for observation of reciprocal lattice points then being that they lie on the surface of a sphere of unit radius termed the *sphere of reflection.*

A crystal may be irradiated in any direction. Experimentally observable reciprocal lattice points thus have values of σ_{hkl} which may be, at most, two. This consideration defines the *limiting sphere*, illustrated by Fig. 29-27.

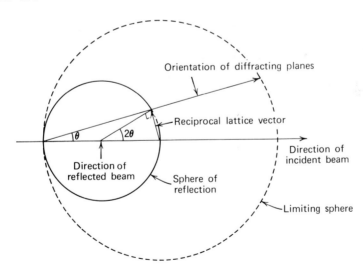

Fig. 29-27 The sphere of reflection and the limiting sphere.

REFERENCES

H. P. Klug and L. E. Alexander, *X-ray Diffraction Procedures*, Wiley, New York, 1954, Chapter 1 (Elementary Crystallography).

C. Tanford, *Physical Chemistry of Macromolecules*, Wiley, New York, 1961, pp. 16–27.

N. H. Hartshorne and A. Stuart, *Practical Optical Crystallography*, American Elsevier New York, 1964. Chapter 1 (The Morphology of Crystals).

W. L. Bragg, *The Crystalline State*, G. Bell, London, 1933.

C. W. Bunn, *Chemical Crystallography*, 2nd ed., University Press, Oxford, 1961.

CHAPTER 30

X-ray Diffraction Patterns and Their use in the Determination of Spacings

As a first approximation, crystals may be considered lattices of idealized geometrical points; the spacings between these points may, in general, be determined by measurements of the related spacings which appear in the corresponding x-ray diffraction patterns. In Chapter 29 the nature of crystal lattices and the relation of crystal spacings to diffraction spacings have been discussed. Here the forms of the diffraction patterns which arise in certain types of experiment are described. The manner in which information can be obtained from these patterns by the *measurement of spacings* is then discussed. It may be noted here, however, that real crystals are in fact lattices of atoms which are of finite dimensions. Different types of atom differ in their x-ray scattering powers. Hence real x-ray diffraction patterns differ from the hypothetical patterns diffracted by ideal lattices in that the diffracted intensities and also the sharpness of the spots or lines are variable. Generally, therefore, "complete" determination of the structure of a crystal depends upon measurement not only of the spacings, but also of the *intensities* (ideally, of the amplitudes and phases) of the diffraction maxima. For this approach, to be discussed in Chapter 31, considerations based upon the measurement of spacings are quite inadequate. In some relatively simple cases, however, the detailed atomic structures of crystals may be deduced by measurements of spacings alone; in all cases the separations of crystal planes, the dimensions of the unit cell of the crystal, and related data are required for the detailed determination of structure.

641

In obtaining and processing measurements of diffraction pattern spacings and intensities, a number of important technological factors must be taken into account, including the geometry of the diffraction apparatus, and such perturbing effects as the polarization or absorption of x-rays or the thermal motions of molecules. Despite their importance, these factors cannot be discussed in this book, which endeavors only to explain the approaches and potentialities of x-ray diffraction methods.

DIFFRACTION FROM A ROW OF ATOMS

Insight into the physical form of diffraction patterns may be gained by considering the diffraction from a hypothetical uniformly spaced *row* of atoms illuminated by x-rays along the normal, as indicated in Fig. 30-1a. The separation of the atoms may be expressed vectorially as $\mathbf{d}$, where $|\mathbf{d}|$ is the interatomic spacing. The conditions limiting the positions of maxima of scattered intensity are less stringent than in the case of a three-dimensional crystal, since only a single Laue equation (29-s or 29-t) need be satisfied. The appropriate form of Laue equation is

$$(30\text{-a}) \qquad \mathbf{d} \cdot \mathbf{s} = \mathbf{d} \cdot (\mathbf{s}_1 - \mathbf{s}_0) = m\lambda,$$

where $\mathbf{s}_0$ and $\mathbf{s}_1$ are unit vectors in the directions of incidence and diffraction, as defined in Chapter 29, and m is any integer. As shown in Fig. 30-1b, x-rays incident perpendicular to the row in the direction of $\mathbf{s}_0$ may be scattered through an angle α in the direction of $\mathbf{s}_1$. By definition of the dot product (cf. Appendix B),

$$(30\text{-b}) \qquad \mathbf{d} \cdot \mathbf{s}_0 = d \cos 90° = 0,$$

$$(30\text{-c}) \qquad \mathbf{d} \cdot \mathbf{s}_1 = d \cos (90° - \alpha) = d \sin \theta.$$

Fig. 30-1 Diffraction from a row of atoms.

Equation 30-a, the condition for maxima, thus becomes

(30-d)
$$\boxed{\sin \alpha = \frac{m\lambda}{d}}.$$

All values of α in three dimensions which satisfy (30-d) are directions in which maxima of diffracted intensity may be observed. Thus there results the formation of hollow cones of illumination above and below the row of atoms, corresponding, respectively, to values of $m = 0 \pm 1, \pm 2, \ldots$ The effect is illustrated in two-dimensional section through the row of atoms, and in perspective, by Figs. 30-2a and 30-2b, respectively. For sharply defined cones of illumination to be formed, the diffracting row must contain many atoms.

If now a cylindrical strip of photographic emulsion surrounds the row of diffracting atoms at any given distance, the diffracted cones, or orders, intercept the emulsion at distances from the position of the incident beam which increase with m. When the emulsion is subsequently unrolled, these interceptions appear as straight lines, as shown in Fig. 30-3a. The lines of maximum intensity are known as *layer lines*.

If the emulsion is not rolled into a cylinder around the diffracting row, but is exposed as a plane at some distance from the atoms, the cones of

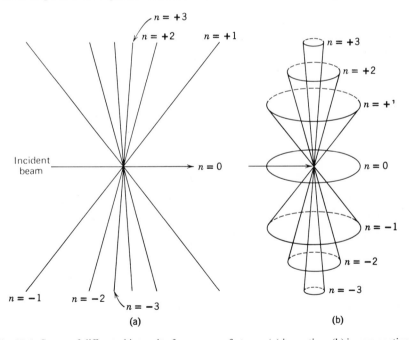

Fig. 30-2 Cones of diffracted intensity from a row of atoms: (a) in section; (b) in perspective.

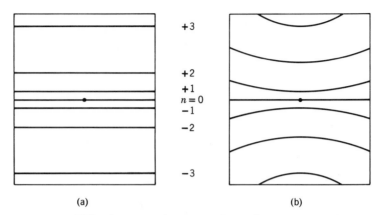

Fig. 30-3 Diffraction pattern from a row of atoms, as recorded on a film.

diffracted intensity intersect the film along an arc. Curved layer lines result, as shown in Fig. 30-3b.

Equation 30-d is a fundamental relation pertaining to diffraction spacings. It is important to understand, however, that it is not merely a statement of the Bragg law, the apparent similarity to (29-i) not withstanding. The Bragg law,

(29-i) $m\lambda = 2d_{hkl} \sin \theta$,

describes the angular distribution of beams that are reflected from *mirror planes* in a crystal, and which are recorded as points on a diffraction photograph. Equation 30-d, on the other hand, describes scattering from series of points. The angle of scattering, α, is fixed with respect to one dimension, but the direction of scattering in planes perpendicular to the row of scatterers is arbitrary.

THE THREE-DIMENSIONAL DIFFRACTION PATTERNS OF CRYSTALS

When crystals are irradiated by x-rays in a direction perpendicular to one of the crystal axes, for example, the *a* axis, the resulting diffraction pattern may be considered to be the resultant of diffraction from a parallel series of *rows* of atoms. It is evident therefore that, whatever the effects of interference between diffracted intensities from successive rows of atoms, *diffracted intensities can only lie along the layer lines which correspond to diffraction by individual rows.*

Figure 30-4a shows a portion of a crystal, the *a* axis of which is perpendicular to the plane of the page, and which is irradiated in a direction perpendicular to the *a* axis. Rows of atoms can be considered to ex-

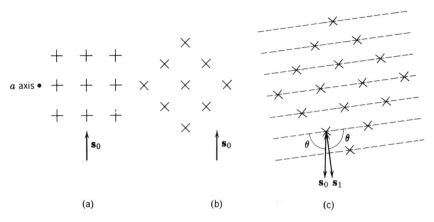

Fig. 30-4 (a) Irradiation of a crystal in a direction perpendicular to its a axis; (b) rotation of the crystal by 90°; (c) reflection from $(1, 1, 1)$ planes of the crystal.

tend above and below the plane of the page. Figure 30-4b shows the same crystal after rotation through 90° about a direction parallel to the a axis. The relative orientations of the incident beam and the b and c axes have changed, but the incident beam remains perpendicular to the a axis and to the original rows of atoms. Thus the condition that diffracted intensities must lie on the corresponding layer lines persists.

Crystal planes may be defined in Fig. 30-4 in the manner described in Chapter 29. Figure 30-4c shows, for example, the intersection of the $(1,1,1)$ planes of the lattice with the plane of the page. Note that the $(1,1,1)$ planes include the atoms which lie in the plane of the page, but are tilted at 45° with respect to the a axis. In Fig. 30-4c the orientation of the crystal is chosen as that for which the first order Bragg reflection is formed by the $(1,1,1)$ planes, that is, that for which

$$(30\text{-e}) \qquad\qquad 2 \sin \theta = \frac{\lambda}{d_{111}}.$$

For this relative orientation of crystal and beam, therefore, a spot of diffracted intensity appears on the first layer line of the diffraction pattern. Figure 30-5 shows the physical arrangement corresponding to the formation of this part of the diffraction pattern. Incident x-rays which are not diffracted form the central, or zero order, spot of the pattern. X-rays diffracted through angle θ of (30-e) form the $(1,1,1)$ spot on the first layer line. Other parts of the diffraction pattern are not formed while the crystal is in the orientation shown.

The deviation of the diffracted beam illustrated in Fig. 30-5 may be described in terms of the two angles θ and α as specified by the Bragg

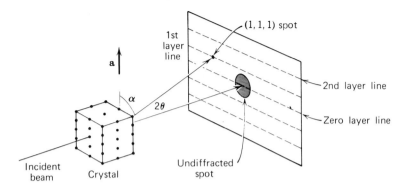

Fig. 30-5 Formation of layer lines by diffraction from a single crystal.

law and by (30-d), respectively. The total angular deviation of the $(1,1,1)$ reflection is 2θ. Note that 2θ is an angle lying in the plane defined by the $(0,0,0)$ beam (direct beam) and the $(1,1,1)$ beam. The angle α which satisfies (30-d) lies in the plane defined by the diffracted $(1,1,1)$ beam and by the a axis of the crystal (i.e., more generally, by the axis of the crystal which is perpendicular to the incident beam).

It is evident from this discussion, as well as that in Chapter 29, that *only certain specific relative orientations of beam and crystal can satisfy the Bragg law* and thus give rise to a spot of diffracted intensity. In general, a fixed crystal illuminated with monochromatic x-rays produces no diffracted intensity. Crystals of proteins and other large macromolecules may constitute an effective exception to this generality, however. Since the unit cells of these crystals are relatively very large, there exist a large number of different sets of reflecting planes. Correspondingly, there exist very many diffraction maxima (i.e., the reciprocal lattice of protein crystals is densely populated). Thus the probability of observing maxima at any single orientation of the crystal is high.

In terms of the geometry of reciprocal space (cf. p. 638) it may be said that the condition for observation of diffracted intensities is that reciprocal lattice points must lie on the surface of the unit sphere. In order to obtain a useful diffraction pattern, the factor θ or λ of the Bragg law must be varied continuously in some way.

THE VON LAUE DIAGRAM

One way to observe diffraction maxima from many sets of crystal planes is to irradiate the crystal with polychromatic x-rays. This was done initially by Von Laue and co-workers in establishing the electromagnetic nature of x-rays and their approximate wavelengths. For each set of planes in a crystal there may be some wavelength in the incident beam for which the Bragg law is just satisfied. This is illustrated (in two dimensions) by Fig. 30-6a.

The diffraction pattern of a crystal illuminated by polychromatic radiation is of the general form shown in Fig. 30-6b. A central maximum of intensity represents the undiffracted beam, while each of the other spots is formed by x-rays of a different wavelength, diffracted from a different set of crystal planes. In terms of the reciprocal lattice, it is evident that C (29-aa) is not a constant in Fig. 30-6b, but is proportional to the variable λ. A continuous series of reciprocal lattices thus corresponds to the single real lattice; portions of some of these lattices are recorded in the Von Laue diagram.

It is immediately obvious that there are too many variables in this type of diffraction pattern. Both wavelength and d_{hkl} are unknown for each

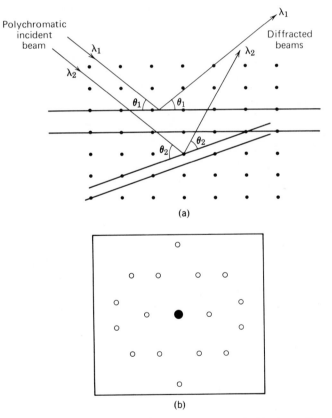

(a)

(b)

Fig. 30-6 (a) Diffraction of polychromatic X-rays; (b) Von Laue diagram.

spot and can be deduced only in simple cases. Although methods exist for determining the symmetry of crystals from their Von Laue patterns, the method is mainly one of historical interest.

POWDER PATTERNS

Many substances are readily available as microcrystalline powders, that is, in the form of a very large number of small *randomly oriented* crystals. If the number of crystals in such a specimen is sufficiently large, some are always so oriented that the Bragg law is satisfied for various values of d_{hkl}. Note that, since the orientations of the crystallites are *random*, there is no question of irradiating the sample parallel to any fixed axis; thus (30-d) does not apply. The Bragg law may, however, be specified in *any* direction with respect to that of the incident radiation. Accordingly, *rings* of diffracted intensity tend to be formed, each of which corresponds to a particular value of d_{hkl}, by combinations of spots diffracted by different crystallites in different directions through angles of $2\theta_{hkl}$. The intensities corresponding to equally spaced sets of planes, such as the $(1, 1, 1)$, $(\bar{1}, 1, 1) \ldots , (\bar{1}, \bar{1}, \bar{1})$ planes, coincide in the powder pattern. As in the case of diffraction from a row of atoms, the powder pattern may be thought of as consisting of a series of concentric cones. The cones are made up of many reflections, each originating from a different crystal. The pattern is recorded in a direction perpendicular to the incident beam, as shown in Figs. 30-7a and 30-7b. Figure 30-7a shows a complete powder pattern, while Fig. 30-7b represents the pattern formed by a sample containing a relatively small number of crystallites. The latter pattern consists of discreet spots, all of which lie along the positions of the rings shown in Fig. 30-7a. A sufficiently large number of such spots combine in forming the complete pattern. Figure 30-8 shows, in projection, the cones of illumination which form the powder diffraction pattern.

Powder pattern diffraction has been used to deduce complete structures of crystals in which only a limited number of interatomic spacings are of importance. More often, powder patterns are recorded for the purpose of making qualitative analyses, since the ring spacings, which are quite highly characteristic, are well known for many substances.

Observations of the *width* of the powder pattern rings (the Scherrer breadth) provides an indication of the *size* of the crystals comprising the powder. Expressions which relate ring breadth to particle size are of the form

(30-f)
$$\bar{D} = \frac{K\lambda}{\beta \cos \theta},$$

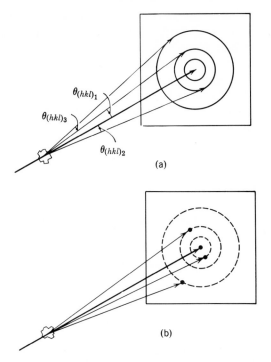

(a)

(b)

Fig. 30-7 Formation of powder pattern: (a) complete pattern; (b) contributions of intensities from individual crystals.

where D is the average dimension of the crystals, θ is the angle of diffraction (i.e., the Bragg angle), λ the x-ray wavelength, K is a constant related to the *shape* of the crystals, and β is the width of the ring. Absolute

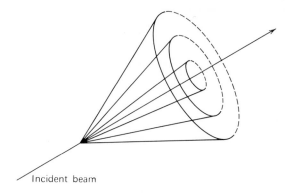

Incident beam

Fig. 30-8 Cones of diffraction from a crystalline powder.

evaluations of $\bar{D}$ from the Scherrer breadth are not straightforward, because of uncertainties in both K and β. The *observed* line broadening may originate in part from experimental conditions (imperfect mono-chromaticity, etc.) as well as from the size of the diffracting crystals. K is of the order of unity (e.g., $K = 1.0747$ for a spherical particle), but its exact value may be uncertain if the shape of the particles is either unknown or irregular. In general, however, (30-f) shows that ring breadth is inversely proportional to crystallite size, and thus provides at least a qualitative index of dimensions.

In practice, x-ray powder patterns are usually obtained with a rotation or agitation of the specimen during irradiation, thus assuring that many orientations of the crystallites are exploited.

TECHNIQUES FOR OBTAINING SINGLE CRYSTAL PATTERNS

Single crystals give rise to diffracted intensities only, as explained above, when irradiated in those directions which satisfy the conditions

$$(30\text{-}d) \qquad \sin\alpha = \frac{m\lambda}{d},$$

$$(29\text{-}i) \qquad \sin\theta = \frac{m\lambda}{2d_{hkl}},$$

where d is the distance between scattering atoms in the direction perpendicular to the direction of incidence, and d_{hkl} is the separation between any set of reflecting planes in the crystal. *Rotation* of the crystal relative to the incident beam is thus a requirement for obtaining a useful number of diffracted intensities. The experimental arrangement for obtaining a *rotation photograph* from a single crystal is indicated in Fig. 30-9. If,

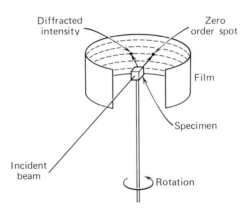

Fig. 30-9 Recording of a single crystal rotation photograph.

as shown in Figs. 30-3 and 30-4, the single crystal is illuminated in a direction perpendicular to its a axis and is rotated around that axis, the diffraction pattern is of the form shown in Fig. 30-10. All spots in the pattern lie along the layer lines which satisfy the equation

(30-g) $$\sin \alpha = \frac{m\lambda}{a}.$$

In (30-g), m is equivalent to h, the Miller index which specifies the a intercept of the diffracting planes. Thus this equation may be written

(30-h) $$\boxed{\sin \alpha = \frac{h\lambda}{a}}.$$

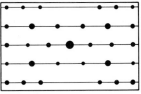

Fig. 30-10 Single crystal rotation photograph.

Each layer line of the rotation photograph thus corresponds to a specific value of h. For example, the 0th layer line (equator) of the pattern consists of spots with Miller indices of the form $(0,k,l)$, the first layer line of spots $(1,k,l)$, and so forth. The dimension a of the unit cell can be obtained from (30-h) simply by measuring the angular separation between successive layer lines.

Similar considerations apply when crystals are irradiated in directions perpendicular to their b or c axes. Thus irradiation perpendicular to the b axis produces a rotation pattern with $(h,0,l)$ spots on the equator, $(h,1,l)$ spots on the first layer line, etc., and yields a value for the b dimension of the unit cell. If irradiation is *not* in a direction exactly perpendicular to one of the axes, the rectilinear form of pattern shown in Fig. 30-10 is in general not obtained. Precise orientation of diffracting crystals is thus an important experimental consideration. While single crystal rotation diagrams of the type shown by Fig. 30-5 may in fact contain all possible diffraction spots, they can be inadequate from the point of view of obtaining all possible information from the diffraction record. The exploitation of diffraction data requires that a set (h,k,l) of Miller indices be assigned unequivocally to each spot on the diffraction pattern. In particular, when diffracted intensities are to be measured, it is essential to be able to assign a diffracted intensity to each set of crystal planes individually.

The distribution of diffracted intensities is three-dimensional; that is, as explained in Chapter 29, the reciprocal lattice of a three-dimensional crystal is itself a three-dimensional structure. The photographic recording of a diffraction pattern, however, can only be a two-dimensional projection of the complete pattern. The resulting difficulty in assigning sets of indices unequivocally to spots may be illustrated by a specific example.

Consider the rotation about its axis of a crystal in which two very similarly spaced sets of planes M and N are present. M and N share one Miller index, h_0. The two sets of planes are, however, inclined to each other with respect to the b and c directions, as indicated by the respective Miller indices $k_1 \neq k_2$ and $l \neq l_2$. These planes are illustrated in Fig. 30-11, in which the a axis is perpendicular to the plane of the page. As the crystal is rotated about the a axis through angle ω, the M and N planes each come to lie in an orientation which satisfies the Bragg law. The values of ω at which this occurs are quite different for the two sets of planes. Nevertheless, since $d_{h_0 k_1 l_1}$ is equal, or nearly equal, to $d_{h_0 k_2 l_2}$, the values of $\theta_{h_0 k l}$ are likewise identical or nearly so. Thus, *for a constant direction of the incident beam, and for a stationary film, the diffraction spots formed by the two sets of planes superimpose.* This is particularly true of the diffraction patterns formed by macromolecules, which contain very large numbers of maxima. The indexing of spots is accordingly uncertain to the extent that similar spacings occur in the crystal.

These difficulties can be surmounted by separating the diffraction pattern into a series of plane layers, so that single layers of the patterns (i.e., plane sections of the reciprocal lattice) can be photographed separately. The positions of spots in the plane can then be specified in two dimensions and, since the index corresponding to the plane as a whole may be known, all three indices are obtained unequivocally for each spot. Experimentally, this may be achieved if a means is devised for isolating a single layer of the pattern and spreading the spots out into two dimensions. One index is that which corresponds to the layer line chosen for isolation. θ (determined by d_{hkl}) may be read from the spacing along one axis of the pattern; ω is given by the spacing along the other axis. Note that ω is determined both by d_{hkl} and by the orientation of the diffracting plane in a direction perpendicular to the axis of rotation of the crystal. A number

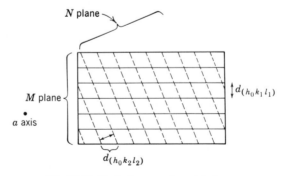

Fig. 30-11 Similarly spaced crystal planes.

of techniques which have been devised for the separation of these parameters, and which are known as *moving film methods*, are described in the
following sections.

THE WEISSENBERG DIFFRACTION CAMERA

The isolation of single layer lines from the diffraction pattern of a crystal
may be accomplished in the instrument shown in Fig. 30-12. The crystal
is surrounded by an opaque cylindrical shield. An entrance port admits
the x-ray beam, while a circular aperture allows one cone of diffracted
intensity (layer line), designated AA' in Fig. 30-12, to escape from the
apparatus. Other layer lines, such as the cone BB', are absorbed by the
shield. Vertical adjustments of the shield are used to select the layer
line to be studied.

During irradiation the crystal is rotated, and diffracted intensities are
produced at appropriate values of 2θ, as in the rotation photograph. At
the same time, however, the recording film is moved in some different
direction. One of several possible experimental arrangements is diagramed
in Fig. 30-13. Here rotation of the crystal is coupled with an upward
translation of the film. The undiffracted spot (direct beam) at the time
shown (corresponding to ω_2) is B, while an initial undiffracted spot at
zero time (or ω_0) is shown at A. At ω_2, a set of crystal planes separated
by spacing d diffracts through the corresponding angle 2θ forming a spot
on the film at C. For some other rotation ω_1, an identically spaced but
distinct set of planes diffracts through the identical angle 2θ, forming

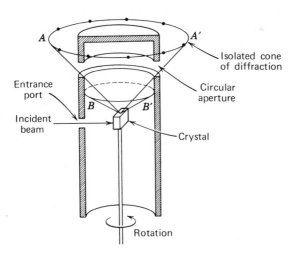

Fig. 30-12 Isolation of a single layer line by the Weissenberg diffraction camera.

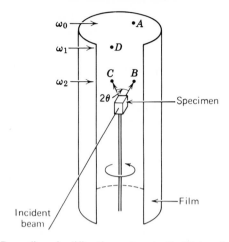

Fig. 30-13 Recording of a diffraction pattern by the Weissenberg camera.

the entirely separate diffraction spot D. The isolating shield shown in Fig. 30-12 is omitted from Fig. 30-13 for clarity but must, of course, be included in the experimental apparatus.

THE OSCILLATION CAMERA

The principle of the oscillation method of recording diffraction patterns is illustrated in Fig. 30-14. In Fig. 30-14a the x-ray beam is incident upon a crystal at some angle θ_1. If a set of crystal planes of corresponding dimensions is appropriately oriented, a reflection results. The reflected intensity is recorded on the photographic emulsion at x. Figure 30-14b shows the same system after rotation (oscillation) of both crystal and film through an identical angle. (In practice, the angle is a small one.)

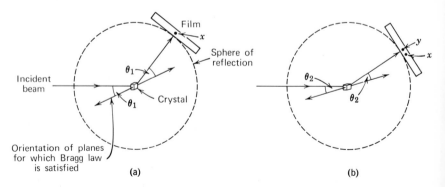

Fig. 30-14 Recording of a diffraction pattern by the oscillation camera.

Since the x-ray source is *not* rotated, the beam is incident upon the crystal at some different angle θ_2, which may produce a Bragg reflection from a different set of planes from that considered in Fig. 30-14a. (Note that neither of the reflected beams necessarily lies *in* the plane of Fig. 30-14.) The diffraction maximum is then recorded at some different position y. In this way the series of Bragg maxima are spread out over the surface of the film as the crystal and film oscillate. This method of recording diffraction patterns produces well-separated spots, but the pattern is distorted, so that the indexing of spots is not straightforward.

THE BUERGER PRECESSION CAMERA

The precession camera is the instrument of choice for recording diffraction patterns of proteins and other biological macromolecules. An undistorted record of isolated layer lines is produced. This record includes all maxima for which α is less than or equal to any specified (up to a practical limit of about 40°.) Thus the precession camera may be adjusted to record either low angle or wide angle diffraction from a given specimen.

Figure 30-15a illustrates the irradiation of a crystal in a direction which makes some angle α with respect to a selected crystal axis. Diffracted intensities may lie along the positions of the successive diffracted cones, which are shown by dotted lines. (As explained on p. 645, the intersections of these cones with the photographic emulsion form the layer lines of the recorded pattern.) The cones are centered with respect to the projection of the crystal axis. Therefore a film placed beyond the crystal and perpendicular to the crystal axis records the layer line intensities as circles. The positions of the layer lines on the photographic record are as indicated in Fig. 30-15b.

Any desired layer line can be isolated by interposition of an appropriately spaced annular aperture. A system in which the zero layer line can be recorded is shown in Fig. 30-16a. The form of the corresponding pattern formed on the emulsion is as indicated in 30-16b, which, like Fig. 30-15b, indicates the location of the layer line(s). In fact, as for any single crystal diffraction pattern, few if any maxima of intensity are recorded if the crystal remains stationary. (For protein crystals, however, it is quite probable that some maxima may be recorded at any given fixed orientation of the crystal; cf. p. 646). In order to record an appreciable number of maxima, the crystal, aperture, and film undergo a coupled rotation about an axis which is *coincident with the direction of the incident beam*. During this rotation the angle α (between the crystal axis and the incident beam) remains constant. Thus the position of the film traces out a *cone* the frustrum of which is at the position, x, of the zero order

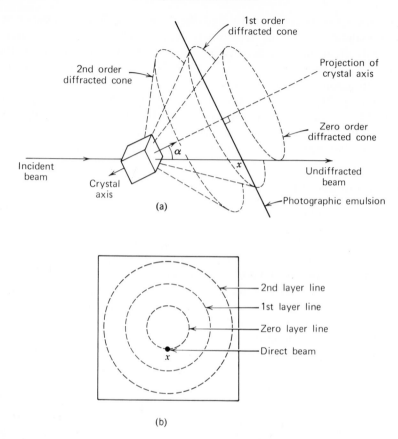

Fig. 30-15 (a) Cones of diffracted intensity formed by irradiation of a single crystal; (b) corresponding positions of layer lines.

spot (direct beam), as shown in Fig. 30-16. For example, after rotation through 180° the components shown in Fig. 30-16a are located as shown in Fig. 30-17a. The corresponding zero layer line position is illustrated by Fig. 30-17b. (Note that, since the film, as well as the crystal, has rotated through 180°, the images formed as shown in Figs. 30-16 and 30-17) are *coincident*. However, during the rotation (precession) the position of the isolated layer line rotates through 360° with respect to the central spot as the film moves. The positions of the zero layer image at successive times during the precession are as shown in Fig. 30-18. However, different sets of crystal planes produce the Bragg reflections at each orientation of the crystal, so that a separated set of spots is formed in the complete precession photograph.

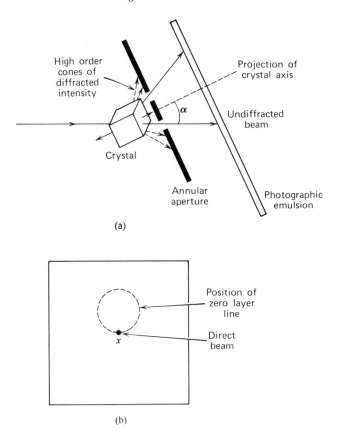

Fig. 30-16 (a) One orientation of precession camera; (b) corresponding diffraction pattern.

Whereas the angle α between incident beam and crystal axis remains constant during the recording of a precession photograph, the Bragg angle 2θ of the recorded intensities (i.e., the angle between the incident and reflected beams) may vary through all values up to and including 2α. This becomes clear upon considering the positions at which intensities are transmitted by the annular aperture. In Fig. 30-19 position 0 marks that of the direct beam, for which $2\theta = 0$. Position 1 marks that of a diffracted beam, for which there is a corresponding value of 2θ. Position 2, 180° removed from the direct beam, corresponds to a maximum possible value of 2θ; since the angle between the direct beam and the center of the aperture is α, this maximum value of 2θ must be identical with 2α. If it is desired to record diffraction spots which correspond to larger values of α (i.e., spots reflected from smaller spacings in the crystal), the value of

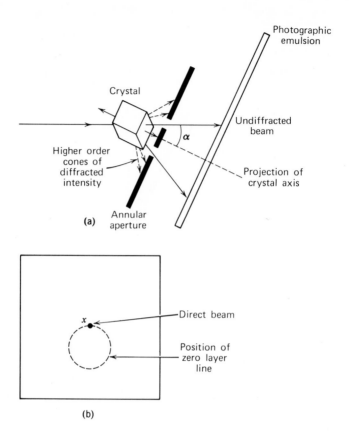

(a)

(b)

Fig. 30-17 (a) Subsequent orientation of precession camera; (b) corresponding diffraction pattern.

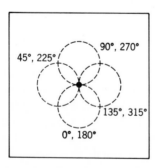

Fig. 30-18 Location of intensities recorded at successive orientations of precession camera.

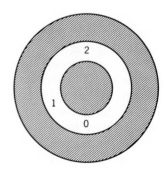

Fig. 30-19 Transmission of intensities by annular aperture of the precession camera.

θ must be increased. That is, the incident beam must make a larger angle with the crystal axis. At the same time the film and circular aperture must remain perpendicular to the crystal axis; otherwise, the recorded pattern is distorted, and indexing of spots is no longer straightforward.

SINGLE CRYSTAL PATTERNS AND THE RECIPROCAL LATTICE

In Chapter 29 it has been shown, in terms of a two-dimensional analogy, that the lattice of a crystal (the real lattice) corresponds to a three-dimensional reciprocal lattice. The points of the reciprocal lattice which correspond to the largest spacings between planes in the crystal are those which lie closest to the origin of the reciprocal lattice. The origin itself corresponds to an interplanar spacing of infinity. This origin may be considered to be coincident with the position of the crystal. Directions in the reciprocal lattice are specified in terms of normals to directions in the real crystal. *If the real crystal rotates through some angle ω, the reciprocal lattice rotates with it through the same angle.* (Note that this property contrasts with the inverse relationship of dimensions and interaxial angles in real and reciprocal lattices.) It may then be said that *diffraction maxima appear on a film placed beyond the crystal wherever a point on the reciprocal lattice coincides with the surface of the film.*

As a crystal rotates about an axis, the sphere of reflection rotates about the diameter of the limiting sphere, tracing out a toroidal path, as shown in Fig. 30-20. (Compare also Figs. 29-26 and 29-27.) The toroid contains all those diffraction maxima which form in space during the irradiation of the crystal at the given angle. Those reciprocal lattice points are re-

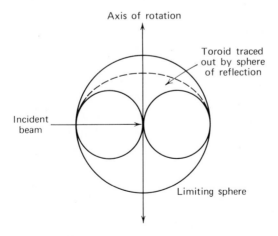

Fig. 30-20 Movement of the sphere of reflection during rotation of a crystal.

corded for which a film placed on the diameter of the limiting sphere intercepts the toroid.

FIBER DIAGRAMS

The diffraction patterns formed when groups of parallel fibers are irradiated by x-rays in a direction perpendicular to the fiber axes are called *fiber diagrams*. A group of parallel fibers constitutes a paracrystalline arrangement of matter, of the form illustrated in Fig. 29-1. The structure of such a specimen is (ideally) regular in two dimensions, but a random arrangement persists in the third dimension. Thus, in principle, irradiation of a group of oriented fibers in a direction perpendicular to their axes is equivalent to the formation of a single crystal rotation photograph in which the crystal is irradiated perpendicular to one of its axes and rotated about a direction parallel to that axis. In fact, fiber diagrams, although generally similar in form to single crystal rotation photographs, differ in respects which reveal the imperfect crystallinity of the specimen.

Fibers differ from true paracrystals in that they may contain amorphous regions and that crystalline regions are of limited extent and may be imperfectly aligned with respect to the fiber axis. The amorphous regions tend to scatter x-rays in all directions without producing well defined maxima of intensity. Thus fiber diagrams tend to suffer from high background intensity levels. At the same time the limited size of crystalline regions accounts for the fact that the maxima themselves are more diffuse than those formed from single crystals, while spots corresponding to *large* values of d_{hkl} are in general absent. The imperfect alignment of the crystalline regions results in smearing out of the diffraction spots into arcs. These characteristic features of fiber diagrams are illustrated in Fig. 30-21.

In principle, fibers are less favorable objects for study by x-ray diffraction techniques than are single crystals; however, fiber diagrams have in fact provided a fruitful means for determining the structures of biologically important macromolecules. Oriented groups of fibers are much more easily obtained than are single crystals of adequate size, while fiber diagram spacings are relatively readily correlated with known chemical structures of linear polymers. The "KMEF" proteins (keratin, myosin,

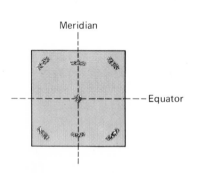

Fig. 30-21 Fiber diffraction diagram.

elastin, and fibrin) and the nucleic acids have been extensively studied by the technique.

Development of a theory of diffraction by *helical* structures provided an important advance in the interpretation of fiber diagrams. Although a discussion of helical diffraction theory is beyond the scope of this book, a few conclusions of the treatment may be noted. The diffraction patterns of helical structures are characterized by distinctive systematic absences (or variations of intensity between layer lines; cf. p. 633), so that a diffraction pattern may rather easily be recognized as that of a helical structure by the experienced diffractionist. From the nature of the systematic absences observed it is possible to determine the spacing between successive turns of the helix (termed the *pitch* of the helix) and also the spacings between atoms along a direction parallel to the axis of the helix.

CHARACTERIZATION OF CRYSTALS BY MEASUREMENTS OF SPACINGS: A SUMMARY

In characterizing crystalline materials the superficial appearance of the crystal is first observed. This gives a general (although possibly misleading) indication of the crystal system to which the specimen belongs, and thus of the type of unit cell to be expected. Subsequently, single crystal diffraction diagrams are obtained by irradiation during rotation about each of the three crystal axes. A process of trial and error is required for accurate location of these axes; if illumination is not precisely perpendicular to an axis, the layer lines of the resulting diffraction pattern are skewed.

From the spacings between the layer lines on the diffraction patterns, the dimensions, *a, b, and c of the unit cell* are obtained in the manner explained on p. 651. *Miller indices are then assigned* to each diffraction spot. In order to index all spots unequivocally it may be necessary to obtain precession photographs of the original layer lines, as described on p. 655. Note that the initial choice of a unit cell is to some degree arbitrary. The indices assigned to the spots are associated with the particular choice of unit cell.

Observations of *systematic absences* among the diffraction spots, together with a knowledge of the character of the unit cell, then makes possible the assignment of the crystal to one, or possibly only to any of a small number of space groups (cf. p. 633). The properties of the space group, in turn, not only define the symmetry of the unit cell, but also specify possibilities for the *number of molecules per unit cell*. Having collected these data, it may be possible to deduce an arrangement of atoms which is consistent with the properties of the unit cell, the observed values of d_{hkl}, and with any available chemical information. In general, however, the analysis of spacings cannot be so extended; study of the intensities of the diffraction spots is required as is discussed in Chapter 31).

OPTICAL TRANSFORMS

The correctness of structures deduced from x-ray measurements can be tested by the construction and diffraction of an *optical transform*. An arrangement of holes corresponding to the assumed structure are punched in a card, and the dimensions of the array of holes are reduced photographically. Ideally, the holes in the demagnified mask are comparable in size to the wavelength of visible light, just as the dimensions of interatomic spacings are comparable to the wavelength of diffracted x-rays. The demagnified mask is illuminated by a coherent collimated beam of monochromatic light (preferably a laser beam), and its diffraction pattern (optical transform) is observed. A correct structure produces a diffraction pattern which is generally similar to that obtained originally by the diffraction of x-rays.

Presently, optical transforms are of little practical use in testing the trial structures deduced from diffraction patterns. With the use of high speed computers it is now feasible to determine the correctness of the trial structures by the more accurate method of calculation. Nevertheless, the optical transform remains an interesting demonstration of the basic similarity of diffraction effects at greatly different wavelengths. The optical filtering method for studying electron microscope images of crystalline structures is an extension of the optical transform technique (cf. p. 447).

THE DETERMINATION OF MOLECULAR WEIGHTS FROM DIFFRACTION DATA

X-ray diffraction data, when available, can be used to obtain very accurate molecular weights of macromolecules. The *volume of the unit cell* can be calculated from the dimensions a, b, and c which are obtained, as explained above, from the layer line spacings of single crystal rotation diagrams. Then;

$$(30\text{-i}) \qquad \boxed{M = \frac{\rho \ \text{g/cm}^3 \cdot N/\text{mole} \cdot V \ \text{cm}^3}{n}},$$

where ρ is the density of the crystal, N is Avogadro's number, V the volume of the unit cell, and n is the number of molecules per unit cell. ρ can be measured directly, while n, the number of molecules per unit cell, can be deduced from more approximate values of M obtained otherwise. For example, Mn might be found from (30-i) to be 104,000 for some protein, while hydrodynamic methods might yield a value of 24,000 for the same substance. It could then be concluded that there are four molecules per unit cell in the crystal, and that the molecular weight of each is exactly 26,000. Data so obtained include the contribution of any solvent which is incorporated into the crystal, and correction for solvent content is essential. This is particularly true in the case of protein crystals which may incorporate substantial quantities of solvent.

REFERENCES

R. B. Setlow and E. C. Pollard, *Molecular Biophysics*, Addison-Wesley, Reading, Mass., 1962. Chapter 5 (X-ray Analysis and Molecular Structure), pp. 113–128.

H. P. Klug and L. E. Alexander, *X-ray Diffraction Procedures*, Wiley, New York, 1954, pp. 111–125, discuss "The Geometry of Diffraction." Chapters 4–10 provide detailed discussions of powder diffraction techniques. Chapter 12 discusses "Small Angle X-ray Scattering."

R. W. James, *The Crystalline State*, Vol. II, *The Optical Principles of the Diffraction of X-rays*, G. Bell, London, 1941.

W. Cochran, F. H. C. Crick, and V. Vand, *Acta Cryst.* **5**, 581 (1952). A classical paper describing the theory of diffraction by helical structures.

H. R. Wilson, *Diffraction of X-rays by Proteins, Nucleic Acids and Viruses*, St. Martins Press, New York, 1966. In addition to describing the results of x-ray diffraction studies on biological macromolecules, this short (131 pp.) text summarizes much of the material presented in chapters 29 and 30. A short discussion of "Diffraction by Helical Molecules" is included on pp. 38–49.

CHAPTER 31

The Determination of Electron Density Distribution by X-ray Diffraction

The ultimate goal of x-ray diffraction studies is the complete determination of molecular structure. The study of spacings, which has been discussed in the two preceding chapters, provides information related to molecular structure, including values of interatomic spacings, and the symmetry conditions which limit the possibilities for arrangement of the atoms within a crystal. In practice, this information is insufficient in all but the simplest cases to allow the deduction of the mutual orientation of the atoms within the molecules. In general, a function which directly describes the atomic positions must be sought from diffraction data; this function is the *electron density distribution*. In order to obtain the electron density distribution it is necessary first to know the symmetry properties of the crystal, the unit cell dimensions, and the h, k, l, indices of the diffraction pattern maxima. The procedures for obtaining these data have been described in Chapter 30. The manner in which measured *intensities* of indexed spots are combined with certain other data to obtain the electron density distribution is discussed in this chapter. Knowledge of the material covered in Chapters 29 and 30 is required, as is an appreciation of the exponential form of wave equations, of the basic properties of complex numbers summarized in Appendix C and of the properties of Fourier integrals as described in Chapter 10.

THE FOURIER TRANSFORM CONCEPT OF OBJECT AND IMAGE

An approach to microscopy and diffraction which has been presented in Chapter 10 may first be reviewed. Qualitatively, the idea may be expressed by saying that *the formation of a diffraction pattern by an object and the formation of an image from the diffraction pattern are processes which may be described by mathematical treatments of identical form.* More specifically, if the structure of an object may be represented by some function $f(x)$, the diffraction pattern formed when the object is irradiated may be represented by another function $g(\omega)$ which is, mathematically, the Fourier transform of $f(x)$. $g(\omega)$ is in fact the wave equation for the total wave scattered by the object. The image subsequently formed in a microscope is then represented by $f_1(x)$, where, ideally $f_1(x)$ is identical with $f(x)$ and thus represents the *inverse transform* of $g(\omega)$. In practice, $g(\omega)$ is always incompletely utilized in forming the microscope image, so that $f_1(x)$ is always an imperfect representation of $f(x)$; that is, the resolution of the image is limited.

In microscopy the physical action of a *lens* forms the inverse transform, or image, just as the interaction of light with the object physically forms the transform, or diffraction pattern. The information required for formation of the image is already present in the diffraction pattern, but interpretation of this information is required. In the case of x-irradiation, no adequate lenses are available to interpret the data by physical formation of an image. Instead, one attempts to resynthesize the image by *computation*, utilizing data obtained from the diffraction pattern.

RELATION OF ELECTRON DENSITY DISTRIBUTION TO THE IMAGE

As explained in the discussion of contrast in Chapter 11, the kind of image formed by any type of radiation is determined by the nature of the interactions which occur between the radiation and the specimen. X-rays, as explained in Chapter 28, interact with the electrons of the specimen, principally with the inner shell electrons of the atoms. The degree of interaction of x-rays with a specimen is thus determined by the density of electrons present in the specimen. Accordingly, an x-ray "image" may be viewed as a "map" of the *electron density distribution* in three dimensions or as its projection into two dimensions. Such a map contains sharply defined maxima at positions which correspond to the centers of atoms; the higher the atomic number, the more intense is the maximum. The distribution of electron density is, however, a continuous function throughout space. The function, which is designated $\varphi(x, y, z)$ is approximated by the set of coordinates of centers of atoms,

$$\sum_{n=1}^{N} (x_n, y_n, z_n).$$

The determination of $\rho(x, y, z)$ is the central problem in x-ray diffraction studies. The complete arrangement of atoms in space may be deduced from the electron density distribution with the aid of knowledge of the chemical composition of the diffracting specimen. Note that a plot of the electron density distribution consists of a series of contours representing electron density levels; for example, as shown for the (planar) benzene molecule in Fig. 31-1a. From contour maps of this type, conventional chemical formulas, as shown for benzene in Fig. 31-1b, must be deduced.

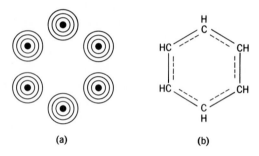

(a)   (b)

Fig. 31-1 The benzene molecule: (a) electron density contours; (b) conventional structural formula.

THE ATOMIC SCATTERING FACTOR

When electromagnetic radiation is incident upon an object which is small with respect to the wavelength, radiation is scattered equally in all directions. When objects of dimensions comparable to the wavelength are illuminated, however, radiations scattered from different parts of the object differ in phase and interfere to produce differences with direction in the level of scattered intensity. An electron may be considered a small object with respect to x-irradiation, while atoms are of sizes comparable to the wavelength. Scattering of x-rays by an atom containing Z electrons thus need not be exactly Z times as great as scattering by an individual electron. Scattered amplitudes are, however, of the order of Z times as great as the amplitude scattered by an isolated electron, but vary as a function of the angle 2θ between incident and scattered beams. To a first approximation, atoms may be considered spherical, so that scattered amplitudes do not vary as a function of atomic orientations. (In refining calculations, however, it is necessary to take account of ellipticities which result from thermal perturbations.) Scattered intensities are proportional to the squares of scattered amplitudes and thus vary approximately as a function of Z^2.

The atomic scattering factor, designated f, is defined as *the ratio of amplitude scattered by an atom to that which would be scattered by an isolated electron at the same position*. Numerical values of f, as a function of θ have been determined, and are available in the literature.

THE MOLECULAR TRANSFORM

When a molecule (or any other group of atoms regardless of the nature of their binding) is irradiated, scattered radiation from each atom of the group interferes, producing a resultant scattered wave which forms the diffraction pattern of the molecule. The resultant wave is analogous to that formed by the interference between the radiation scattered by the different electrons within a single atom. An expression for the resultant wave is conveniently written in the exponential form (see Appendix C). For waves of a single wavelength (i.e., of a single frequency) the wave equation may be written

(31-a)
$$A = A_0 e^{-i\varphi}$$

where A_0 is the amplitude of the wave. A its instantaneous amplitude, and φ is the phase of the wave. In the evaluation of x-ray scattering in terms of (31-a) the contribution of each atom of a molecule to the amplitude of scattering is simply the atomic scattering factor f as defined above. Thus,

(31-b)
$$A_0 = \sum_{n=1}^{N} f_n$$

for a molecule containing N atoms.

The phase of each contribution to the amplitude of the scattered wave is determined by the position of the atom relative to some common, arbitrarily selected origin. The position of the atom with respect to the origin may be specified by an appropriate vector **r**. As shown in Fig. 29-18, the incident and diffracted directions of the x-ray beam may be represented by s_0 and s_1, while **S** was defined in Chapter 29 as

(31-c)
$$S = s_0 - s_1.$$

Here it is convenient to consider s_0 and s_1 as vectors of length $1/\lambda$ rather than as unit vectors. The magnitude of **S** is correspondingly altered by the factor $1/\lambda$ also.

The phase of an amplitude arising at the origin of the unit cell may be designated, arbitrarily, as zero. The path difference between this amplitude at position r and one originating *at r* (i.e., from an atom located at the tip of the vector **r** is then

(31-d)   path difference $= \lambda (\mathbf{r} \cdot \mathbf{S})$

(in which the factor λ appears in order to compensate for the correspond-
ing change in $|\mathbf{S}|$, as explained above). The corresponding phase differ-
ence between the amplitudes is $2\pi/\lambda$ times this amount; that is,

(31-e)   phase difference $= 2\pi (\mathbf{r} \cdot \mathbf{S})$.

Combining (31-b) and (31-e) with (31-a) and specifying the scattering
factor and position of the nth atom of the molecule as f_n and $\mathbf{r}_n$, respectively, yields an equation for the wave scattered by that atom as

(31-f) $A_n = f_n e^{-2\pi i (\mathbf{r}_n \cdot \mathbf{S})}$

The wave scattered by all N atoms of the molecule is then simply the
sum of all the A_n for the individual atoms. Note the convenience of using
the exponential form of the wave equation in obtaining such a sum. The
summation is conventionally designated $G(S)$ and is termed the *molecular
scattering* factor or *molecular transform*. This quantity is given by

(31-g) $G(S) = f_1 e^{-2\pi i (\mathbf{r}_1 \cdot \mathbf{S})} + f_2 e^{-2\pi i (\mathbf{r}_2 \cdot \mathbf{S})} + \cdots$

$$= \sum_{n=1}^{N} f_n e^{-2\pi i (\mathbf{r}_n \cdot \mathbf{S})}.$$

Equation 31-g implies that there will be a *continuous* distribution of
scattered intensity throughout the space surrounding the molecule. This is
indeed true for an isolated molecule or for any randomly arranged group
of molecules as a whole. Therefore the molecular transform is sometimes
referred to as the *continuous transform*.

If, however, the molecule exists as a unit in an ordered structure, such
as a crystal, interference between the radiation scattered by sucessive
molecules limits the observation of scattered intensities to those which
satisfy the condition for diffraction maxima from a regular lattice. That
is, the directions in which the molecular transform may be observed
must be such as to satisfy the von Laue equations as expressed by (29-s)
or (29-t).

Note that, whereas the structure of a single molecule or a random array
might, in principle, be determined from its diffraction pattern (molecular
transform), in practice the intensity diffracted from a single molecule is
too weak for measurement, so that the discontinuous spot diffraction
pattern of crystalline or paracrystalline specimens must be observed
instead. Such patterns sample the continuous molecular transform at
positions which satisfy the von Laue conditions.

THE STRUCTURE FACTOR

In the preceding section it has been shown that the diffraction pattern of an isolated molecule is represented mathematically by its molecular transform, as defined by (31-g), but that for a molecule present in a crystal diffraction maxima are observed only when the Laue conditions,

(29-s) $$\mathbf{a} \cdot \mathbf{S} = h; \quad \mathbf{b} \cdot \mathbf{S} = k; \quad \mathbf{c} \cdot \mathbf{S} = l,$$

are satisfied. In other words, it may be stated that the molecular transform is sampled, in reciprocal space, at those points which are specified by integral sets of values of h, k, and l. Thus the molecular transform determines the intensities of the diffraction spots at the points specified by integral values of h, k, and l, but, as a consequence of destructive interference between amplitudes originating from different molecules, the molecular transform is unobservable at all other positions in space.

The *structure factors* of the diffracting molecule may then be defined as *the values of the molecular transform $G(S)$ which satisfy the conditions for the formation of diffraction maxima.* A structure factor F_{hkl}, may be specified for *each* (h, k, l) spot of the diffraction pattern. The structure factor, F_{hkl} may also be defined, in a manner analogous to the definition of the atomic scattering factor f, as *the ratio of the amplitude scattered by the (h, k, l) plane of a crystal to that scattered by a single electron.*

In characterizing the structure factors further, it may be noted that $\mathbf{r}_n$ of (31-f) and (31-g) may, like any vector, be analyzed as the sum of its projections in three dimensions. The three directions may be chosen here as those of the unit cell axes a, b, and c. Thus,

(31-h) $$\mathbf{r}_n = x_n \mathbf{a} + y_n \mathbf{b} + z_n \mathbf{c}.$$

Equation 31-g may then be written

(31-i) $$G(S) = \sum_{n=1}^{N} f_n \, e^{-2\pi i (x_n \mathbf{a} \cdot \mathbf{S} + y_n \mathbf{b} \cdot \mathbf{S} + z_n \mathbf{c} \cdot \mathbf{S})},$$

and, from (29-s) and the definition of F_{hkl},

(31-j) $$\boxed{F_{hkl} = \sum_{n=1}^{N} f_n \, e^{-2i(hx_n + ky_n + lz_n)}}.$$

Equation 31-j defines the structure factor of the (h, k, l) spot formed by diffraction from a lattice. Note that (31-i) is the wave equation describing the radiation which forms the (h, k, l) spot. Specifically, the structure factor (wave equation) for the $(0, 1, 2)$ spot would be

(31-k) $$F_{(0,1,2)} = \sum_{n=1}^{N} f_n \, e^{-2\pi i (y_n + 2z_n)}$$

for a molecule containing N atoms located at specified positions $y_1, z_1 \ldots,$ y_N, z_N in the unit cell and associated with values of the atomic scattering factor $f_1 \ldots f_N$.

Equation 31-j contains, in general, for a unit cell of N atoms, $3N$ unknowns specifying the x, y, and z coordinates for each of the N atoms, respectively. Thus, in principle, measurement of F_{hkl} values of $3N$ different spots from the diffraction pattern of a crystal would permit one to solve equations of the form of (31-i), yielding values of all three atomic coordinates for each atom of the unit cell. Knowledge of f_n for each atom and indexing of each spot would be required. The set of coordinates x_n, y_n, and z_n would specify the position of the center of each atom, thereby specifying the complete structure of the unit cell.

In practice, it is usually more informative to obtain the continuous electron density distribution function $\rho(x, y, z)$ rather than the set of atomic coordinates $\sum_n (x_n, y_n, z_n)$. In terms of the electron density distribution the structure factors may be expressed as the triple integral

$$(31\text{-}l) \qquad \boxed{F_{hkl} = \int_{x=0}^{x=a} \int_{y=c}^{y=b} \int_{z=0}^{z=c} \varphi(x, y, z)\, e^{-2\pi i(hx+ky+lz)}\, dx\, dy\, dz}.$$

The manner in which (31-l) is treated to obtain $\varphi(x, y, z)$ is outlined on p. 678. There exists, however, a serious difficulty: the values of F_{hkl} cannot in fact be obtained readily from the data provided by diffraction patterns. The problem of obtaining the F_{hkl} is discussed below.

THE PHASE PROBLEM

Experimentally, the intensities, but not the amplitudes, of the spots of a diffraction pattern may be measured. In complex form the equation of a wave function is

$$(31\text{-}a) \qquad\qquad\qquad A = A_0 e^{-i\varphi},$$

while the intensity is the product of the amplitude A and its complex conjugate A^* (cf. Appendix C); that is

$$(31\text{-}m) \qquad I = AA^* = (A_0 e^{-i\varphi})(A_0 e^{+i\varphi}) = A_0^2.$$

Thus $\sqrt{I}$ remains indeterminate within a factor $e^{-i\varphi}$. Physically, this is tantamount to a statement of the fact, noted in Chapter 1, that *the absolute phase of a wave cannot be measured*.

The structure factors, whether defined by (31-j) or (31-l), are complete wave equations of the form of (31-a), in which both amplitude and phase are expressed. Experimental determination of I, however, yields only the absolute value of the amplitude, while *information pertaining to the phase of the wave forming the* (h, k, l) *spot is lost.* It is therefore *not* possible to measure directly the values of the structure factors.

It might appear, therefore, that attempts to determine the complete structures of molecules are doomed. In fact, however, as described below, the phases of the amplitudes forming the diffraction spots may often be estimated or deduced, and the corresponding structure factors evaluated.

It may be noted first that, in some not unusual cases, the phases of the diffraction spots may be known except for their sign. Then

$$(31\text{-n}) \qquad F_{hkl} = +\sqrt{I_{hkl}} \quad or \quad F_{hkl} = -\sqrt{I_{hkl}}.$$

If the unit cell of the diffracting crystal has a *center of symmetry* (cf. Chapter 29), that is, if it is *centrosymmetric*, then, for each atom located at some position specified by (x_n, y_n, z_n), there must also exist an identical atom at a position specified by $(-x_n, -y_n, -z_n)$. Correspondingly, for each term $f_n \exp[-2\pi i(hx_n + ky_n + lz_n)]$ of (36-j), there must also be a term $f_n \exp[+2\pi i(hx_n + ky_n + lz_n)]$. The structure factor then consists of a set of terms of the form

$$(31\text{-o}) \qquad f_n(e^{-i\varphi} + e^{+i\varphi}).$$

Noting that

$$(\text{C-e}) \qquad \cos\theta = \tfrac{1}{2}(e^{i\theta} + e^{-i\theta}),$$

(31-o) is seen to be equivalent to

$$(31\text{-p}) \qquad \frac{\cos\varphi_n}{f_n}.$$

Recalling that $\cos\varphi_n$ is the real part of the expression $z = \cos\varphi_n + i \sin\varphi_n$, it is evident that, in this case of centrosymmetry, the structure factor F_{hkl} is entirely real. For real values of F_{hkl} the term $e^{i\varphi}$ can only be equal to unity, as is the case when $\varphi = 0$. Although the *sign* of F_{hkl} remains in doubt, as indicated by (31-n), all other numerical values of F_{hkl} are excluded. Similarly, in systems of less complete symmetry (e.g., in the presence of axes of rotational symmetry), *some* of the atoms present in the unit cell are paired, and some of the (h, k, l) reflections are corresponding ambiguous in phase only with respect to sign.

TRIAL AND ERROR METHODS

A procedure of trial and error may, at least in principle, be used to *guess* the phases of the diffraction spots and thus to determine the complete structure of the specimen. In so doing, the size and symmetry of the unit cell and the indices of the diffraction spots are determined by the methods described in Chapter 30. The intensities of the spots are measured. The atomic composition and chemical nature of the material are known from independent data. From this information, one or a few probable distributions of the atoms are deduced. Values of the F_{hkl} are then calculated on the basis of the atomic positions in the trial structure(s), that is, from (31-l) or (31-j) and from the values of the atomic scattering factors. The corresponding intensities $I_{hkl} = F_{hkl}F_{hkl}^{*}$ are then compared with the observed diffraction pattern. If the correspondence of the observed and calculated intensities is reasonably good, it is assumed that the trial structure is basically correct. The trial structure is then modified slightly in such a way as to cause the calculated intensities to correspond more closely with the observed pattern. Several rounds of successive calculation and refinement may be required.

Although the method cannot be considered to be generally satisfactory, the availability of high speed computers has made extensive use of trial and error techniques feasible. It is remarkable that the structure of so complex a molecule as DNA was deduced in considerable detail by this approach before the perfection of present computer techniques!

THE PATTERSON SYNTHESIS

The electron density distribution $\rho(x, y, z)$ is obtained from the structure factors F_{hkl} when they can be deduced. A related procedure yields a function $P(x, y, z)$, known as the *Patterson function*, which incorporates the (measurable) intensities rather than the amplitudes of the diffraction maxima. The Patterson function is defined as

$$(31\text{-}q) \qquad P(x, y, z) = \frac{1}{V} \sum_{h=-\infty}^{h=+\infty} \sum_{k=-\infty}^{k=+\infty} \sum_{l=-\infty}^{l=+\infty} I_{hkl}\, e^{-2\pi i(hx + ky + lz)},$$

in which V is the volume of the unit cell.

It is found that $P(x, y, z)$ represents the *superposition* of electron density distributions which would be obtained by placing each atom, successively, at the origin of the unit cell. The nature of the Patterson plot is diagramed in Fig. 31-2 for a cell which contains four distinguishable "atoms." Figure 31-2a shows the unit cell, while Fig. 31-2b is the corresponding plot obtained by placing each of the "atoms" in turn at position O (i.e., coupled with appropriate translations along the cell axes so that all projections are shown within the same unit cell). Even in this hypothetical case of a very few, easily distinguished "atoms" the Patterson plot is extremely confusing! The unit cell structure is not readily deduced by observation of Fig. 31-2b. The plot can, however, serve to rule out certain possible structures such as, for example, that shown in Fig. 31-2c.

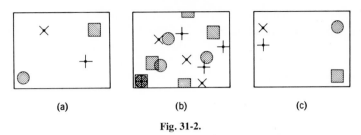

(a) (b) (c)

Fig. 31-2.

The Patterson plot may thus be used as an aid in eliminating structures which might otherwise serve as unprofitable models for trial and error calculations.

An important application of the Patterson function is the location of the sites of attachment of atoms of high atomic number within crystals composed primarily of atoms of low atomic numbers. This is required in the method of isomorphous replacement, as discussed below. When only one or two heavy atoms are present per unit cell, the contributions to scattered intensity by these atoms may greatly exceed that of all other atoms. The contributions of the heavy atoms to the Patterson plot may then be obvious, and thus the positions of these few heavily scattering atoms can be deduced relatively easily.

THE EFFECT OF HEAVY ATOMS; ISOMORPHOUS REPLACEMENT

The amplitudes of the structure factors, as given by (31-j) or (31-p), are determined by the values of the atomic scattering factors f_n. Values of f_n vary between atoms approximately in proportion to the atomic number Z. Thus the contribution of heavy atoms to the intensities in the diffraction pattern is disproportionately large in terms of the numbers of these atoms present in the structure. Diffraction patterns of crystals of heavy metal oxides, for example, in the main consist of the diffraction by the metal atoms, modified only to a limited degree by the presence of the oxygens.

It is found, experimentally, that heavy metal derivatives of proteins may be prepared which crystallize in a form identical with that of crystals of the unaltered molecule. That is, the dimensions of the unit cell, and the space group of the crystals are unchanged by the addition of the heavy atom. The process of forming such derivatives is called *isomorphous replacement*. The derivatives must have the property that *the heavy atoms occupy a defined site within the unit cell*; in general, the heavy atom must be covalently bonded to the protein.

The diffraction pattern of the isomorphous derivative is the resultant of the diffraction pattern of the unaltered protein with that of the regular arrangement of heavy atoms. That is, the Fourier transform of the sum (protein + heavy atom) is equal to the sum of the transforms of protein and heavy atoms separately. This relationship gives the "isomorphous replacement equation,"

$$(31\text{-r}) \qquad\qquad F_{M+H} = F_M + F_H,$$

where F_M is the structure factor of the native molecule, F_{M+H} is that of the isomorphous derivative, and F_H is that of the heavy atoms alone (i.e., as they exist in the derivative).

For large molecules it can be expected that scattering by heavy atoms, while of significant magnitude, is nonetheless less than the total scattering by the atoms of the protein. (Note, however, that scattering by the heavy atom, for which f is of the order of 100, is compared with the resultant of scattering by all other atoms; the latter is in general much smaller than the *sum* of the scattering factors of those atoms.) The diffraction pattern of the isomorphous crystal should thus be recognizably similar to that of the original macromolecule, although the distribution of intensities is significantly altered by the presence of the heavy atom.

STRUCTURE FACTORS OF A CENTROSYMMETRIC CRYSTAL

Given a centrosymmetric crystal and a single isomorphous heavy metal derivative, the phases, and thus the structure factors, of diffracted intensities may (at least in principle) be determined in the following way. The diffraction patterns of both native macromolecule and isomorphous derivative are recorded and indexed, and the intensities of all diffraction spots are measured. Thus a series of values $I_{(h,k,l)M}$ and $I_{(h,k,l)M+H}$ are obtained. Then,

$$(31\text{-s}) \qquad\qquad \sqrt{I_{(h,k,l)_M}} = |F_M|_{h,k,l},$$

$$(31\text{-t}) \qquad\qquad \sqrt{I_{(h,k,l)_{M+H}}} = |F_{M+H}|_{h,k,l}.$$

For a centrosymmetric crystal, as explained above, only the *sign*, $+$ or $-$, of these structure factors is unknown. Thus, from (31-r), the amplitudes $|F_H|$, which are contributed by the heavy atoms, are seen to be

$$(31\text{-u}) \qquad\qquad F_H = \pm(|F_{M+H}| - |F_M|);$$

that is, the *absolute values* $|F_H|$ of the contributions of heavy atoms to each diffraction spot can be computed. The *positions* of the heavy atoms can then be deduced. A *difference Patterson synthesis* is performed by substituting the series of values $(F_H)^2_{h,k,l}$ for $I_{h,k,l}$ in (31-q). Since only one or a few heavy atoms are present in each unit cell, it is feasible to determine their positions from the relatively simple difference Patterson plot. The signs of the corresponding heavy atom structure factors can then be computed from the known structure of heavy atoms. Finally, the values of F_M (and of F_{M+H}) can be obtained by substitution in (31-r).

A numerical example may be considered. The relative intensities of some diffraction spot, h_0, k_0, l_0, might be found to be 225 and 100 in the patterns of a native protein and its isomorphous derivative, respectively. Therefore, from (31-s) and (31-t), respectively;

(31-v) $$F_M = \pm 15 \quad \text{and} \quad F_{M+H} = \pm 10,$$

and, from (31-u),

(31-w) $$F_H = \pm (10 - 15) = \pm 5, \text{ so that } F_H^2 = 25.$$

From the difference Patterson synthesis it might be deduced that the value of F_H is *plus* 5. In general, this value can be consistent with only *one* possible set of signs for F_M and F_{M+H}. Since $-10 = -15 + 5 \neq +15 + 5$, and $+10 \neq +15 + 5 \neq -15 + 5$, it is deduced that $F_M = -15$ and $F_{M+H} = -10$. Similar evaluations of F_M can be carried out for *each* set of indices, h, k, l. (In practice, experimental errors may result in ambiguous determinations of some of the structure factor values.)

STRUCTURE FACTORS OF NONCENTROSYMMETRIC CRYSTALS

Phases of arbitrary value may be determined by methods which involve comparison of diffraction patterns formed by several different isomorphous derivatives. In principle, two derivatives suffice; in practice, study of several different derivatives may be required in order to yield a series of unambiguous values of the structure factors. The phases are obtained either graphically or analytically. The graphical method, which is generally the most satisfactory, is now outlined here.

For each diffraction spot characterized by a set of indices h, k, l, measurement of the intensities gives the values $|F_M|$, $|F_{M+H_1}|$, and $|F_{M+H_2}|$ for the molecule and its two heavy derivatives, respectively. These values may be considered to be the lengths of the vectors which represent the corresponding amplitudes.

The complete structure factors F_{H_1} and F_{H_2}, which give the amplitudes and phases of the heavy atom contributions, are determined by means of a difference Patterson synthesis, as described above.

In making the geometrical construction, vectors of lengths $-|F_{H_1}|$ and $-|F_{H_2}|$ are drawn from an arbitrarily located origin, and at the correct relative orientation, as shown in Fig. 31-3a.

From the same origin a circle of radius $|F_M|$ is drawn as shown in Fig.

31-3b. The tip of the vector which represents F_M must lie at some point on this circle. Determination of the phase of the radiation which forms the h, k, l spot thus consists of determining the angular orientation of that vector.

A second circle is then added to the construction, as shown in Fig. 31-3c. This circle originates at the tip of the vector F_{H_1} and is of radius

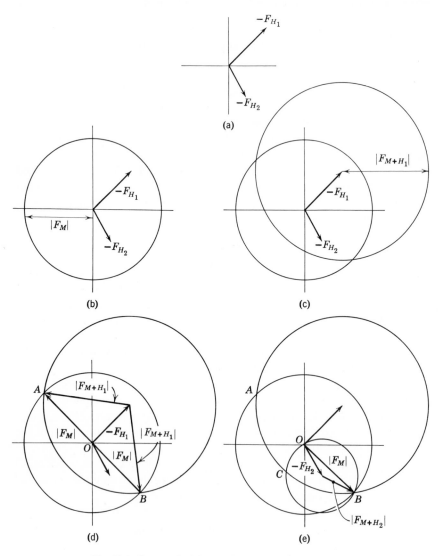

Fig. 31-3 Geometrical determination of arbitrary phases.

$|F_{M+H_1}|$. The rationale for construction of the second circle is made evident in Fig. 31-3d. The two circles intersect at two points, at each of which vector addition gives

(31-x) $$|F_{M+H_1}| - F_{H_1} = |F_M|.$$

Thus F_M is given either by OA or by OB of Fig. 31-3d. The ambiguity is resolved by constructing a third circle, of radius $|F_{M+H_2}|$, centered at the tip of the vector F_{H_2} as shown in Fig. 31-3e. At the points of intersection of the first and third circles, vector addition gives

(31-y) $$|F_{M+H_2}| - F_{H_2} = |F_M|.$$

Thus the complete structure factor F_M could be given by either OB or OC. OB must thus represent the structure factor F_M, since it is consistent with results from both isomorphous derivatives. That is, the tip of the vector representing the structure factor of the unaltered molecule must lie at the intersection of the three cricles shown in Fig. 31-3e.

Experimentally, the three circles may not precisely intersect at a point. For this reason, examination of more than two isomorphous derivatives may be required.

The complete series of structure factors for all combinations of the indices h, k, and l is determined by drawing a series of diagrams analogous to Fig. 31-3e.

APPLICABILITY OF ISOMORPHOUS REPLACEMENT STUDIES

The isomorphous replacement approach has made possible the determination, with resolutions of the order of 1 Å, of the structures of hemoglobin, myoglobin, and a number of other proteins. It is evident that these extremely high resolutions are achieved only at the expense of extremely tedious procedures involving the preparation of isomorphous derivatives (an accomplishment which may not be possible for every protein one might wish to study), the indexing and measurement of intensity of, literally, many thousands of diffraction maxima, and the computation of $\varphi(x, y, z)$ from these data. In recent years much effort has been directed to the development of automated systems for the recording of diffracted intensities and the subsequent performance of computations. While these developments make projects in high resolution x-ray diffraction more feasible, the task of determining molecular structures remains formidable.

It is clear that isomorphous replacement methods are not generally applicable but can be used only for substances for which appropriate derivatives can in fact be prepared. Protein crystals, which contain large amounts of included water, have been particularly suitable for the preparation of derivatives in which the protein unit cell is not appreciably distorted.

COMPUTATION OF ELECTRON DENSITY DISTRIBUTIONS FROM THE STRUCTURE FACTORS

Assuming that the structure factors can be evaluated by some means, it is necessary to use these quantities for evaluation of the function $\varphi(x, y, z)$. The structure factors, as defined by (31-l) or (31-j), collectively represent the wave motions which form the diffraction pattern of a structure. According to the views reviewed at the beginning of this chapter, the diffraction pattern is the Fourier *transform* of the structure of the object. The structure of the object, that is, the function $\varphi(x, y, z)$, can thus be obtained by computing *the inverse transform of the structure factors*.

The electron density distribution in a crystal is clearly a *periodic function* in terms of the unit cell dimensions. That is

$$(31\text{-z}) \qquad \varphi(x, y, z) = \varphi(x + m_1 a, y + m_2 b, z + m_3 c),$$

where m_1, m_2, m_3 are any set of integers and $a, b,$ and c are the dimensions of the unit cell. $\varphi(x, y, z)$ could likewise, with a change of scale, be expressed as $\varphi(2\pi x, 2\pi y, 2\pi z)$, that is, as a function repeating periodically at intervals of $2\pi x, 2\pi y,$ and $2\pi z$. Accordingly, it can be expressed as a Fourier series:

$$(31\text{-aa}) \qquad \varphi(x, y, z) = \sum_{h=-\infty}^{h=+\infty} \sum_{k=-\infty}^{k=+\infty} \sum_{l=-\infty}^{l=+\infty} C_{hkl} e^{-2\pi i(hx+ky+lz)}.$$

Values C_{hkl} are then given by

$$(31\text{-bb})$$

$$C_{hkl} = \frac{\displaystyle\int_{x=0}^{x=a} \int_{y=0}^{y=b} \int_{z=0}^{z=c} \varphi(x, y, z)\, e^{-2\pi i(hx+ky+lz)}}{dx\, dy\, dz}$$

$$= \frac{\displaystyle\int_{x=0}^{x=a} \int_{y=0}^{y=b} \int_{z=0}^{z=c} \varphi(x, y, z) e^{-2\pi i(nx+ky+lz)}}{V},$$

where V is the volume of the unit cell.

The numerator of (31-bb) is just the function F_{hkl} defined by (31-l). Thus,

$$(31\text{-cc}) \qquad\qquad\qquad C_{hkl} = \frac{F_{hkl}}{V}.$$

Substitution of (31-cc) in (31-aa) then gives

(31-dd)
$$\varphi(x, y, z) = \frac{1}{V} \sum_{n=-\infty}^{h=\infty} \sum_{k=-\infty}^{k=\infty} \sum_{l=+\infty}^{l=\infty} F_{hkl} e^{-2\pi i(hx+ky+lz)}.$$

Given the volume V of the unit cell and the values F_{hkl} of all structure factors, the electron density distribution at any point (x, y, z) can be determined directly from (31-dd).

A less complete but often more feasible method (i.e., from the point of view of the availability of values for F_{hkl}) is the computation of functions such as $\varphi(y, z)$, which is the *projection* of the electron density distribution on the plane normal to the x direction, or a axis, of the unit cell. $\varphi(y, z)$ is given by (31-ee), which is analogous to (31-dd).

(31-ee)
$$\varphi(y, z) = \frac{1}{A} \sum_{k=-\infty}^{k=+\infty} \sum_{l=-\infty}^{l=+\infty} F_{0kl} e^{-2\pi i(ky+lz)},$$

where A is the area of the unit cell projection in the plane of the b and c axes. Similarly, projections of the electron density on planes normal to the b or c axes, respectively, could be obtained from expressions of the form of (31-ee) in which only $(h, 0, l)$ or $(h, k, 0)$ structure factors were summed. Although difficult to interpret in the case of large unit cells, these projections of electron density are of considerable assistance in revealing the structures of simpler unit cells.

DETERMINATION OF PROTEIN STRUCTURES FROM X-RAY DIFFRACTION DATA; A SUMMARY

Assuming that crystals of a protein and of one or more isomorphous derivatives are available for study, the procedures followed in determining the molecular structure of the protein may be summarized as follows.

1. The space group and the unit cell dimensions of the protein and derivative(s) are determined in the manner described on p. 650 [see especially (30-h)]. These parameters must be essentially identical for the native protein and any useful derivative.

2. The single crystal diffraction patterns of each are obtained in such a way as to record all possible spots, h, k, l, separately. Since the diffraction patterns produced by proteins may be highly detailed, the recording of a number of precession photographs is required, as described on p. 655. The spots appearing on these diagrams are indexed; that is, a set of (h, k, l) values is assigned to each.

3. The intensity of each diffraction spot is measured. Whereas, in the past, intensities have been recorded photographically, these methods

become extremely tedious when the intensities of literally thousands of spots must be measured. Newer automated systems combine the recording and quantitative measurements of intensities, producing a direct electrical response to the diffracted radiation.

4. For each isomorphous derivative the Patterson function, given by (31-q), is plotted. From the Patterson plot the positions of the heavy atoms are deduced. The complete contributions of the heavy atoms to the structure factors F_{hkl} as given by (31-j) can then be calculated.

5. The structure factors F_{hkl} of the reflections from the protein itself can then be computed for each set of indices, h, k, l. In the case of centrosymmetric crystals this can be done in the manner indicated by (31-r) through (31-u). More generally, diagrams such as Fig. 36-3e, or equivalent methods, are employed.

6. Equation 31-dd [or, if necessary, (31-ee)] is then evaluated for a series of sets of values of x, y, and z. That is, a three- (or two-) dimensional evaluation of the electron density distribution function is made.

7. $\varphi(x, y, z)$, the electron density distribution, is then obtained as a three-dimensional plot after evaluation of the function at a large number of positions. The correctness of the electron density distribution so evaluated may be tested by comparison of the diffraction pattern computed therefrom with that observed.

8. The atomic structure of the protein is deduced by comparing the plot of electron density distribution with chemical data, which usually include the amino acid sequence.

REFERENCES

H. P. Klug and L. E. Alexander, *X-ray Diffraction Procedures*, Wiley, New York, 1954. Appendix VII (pp. 680–681) lists values of atomic and ionic scattering factors.

C. Tanford, *Physical Chemistry of Macromolecules*, Wiley, New York, 1961, pp. 27–73. A description of methods for determining the electron density distributions of centrosymmetric crystals and of the results of diffraction studies of macromolecules of biological interest.

D. Harker, *Acta Cryst.* **9**, 1 (1956). A classical paper which gives a lucid description of methods for obtaining arbitrary phases from isomorphous replacement diffraction patterns.

H. Lipson and C. A. Taylor, *Fourier Transforms and X-ray Diffraction*, G. Bell, London, 1958. A short and authoritative account of theories which relate diffraction pattern intensities to object structure.

K. C. Holmes and D. M. Blow, in D. Glick, *Ed.*, *Methods of Biochemical Analysis*, Vol. XIII, Interscience, New York, 1965, pp. 113–241. An article on "The Use of X-ray Diffraction in the Study of Protein and Nucleic Acid Structure," which describes both methods and results.

H. R. Wilson, *Diffraction of X-rays by Proteins, Nucleic Acids and Viruses*, St. Martins Press, New York, 1966.

CHAPTER 32

Ultracentrifuge
Optical Systems

In the ultracentrifuge, solutions are subjected to large gravitational fields, which effect the sedimentation of solutes, thereby establishing concentration gradients within the solutions. Two general types of measurement are commonly made in the ultracentrifuge; they are sedimentation *velocity* determinations, in which the position of a boundary formed between solvent and solution is measured as a function of *time*, and sedimentation *equilibrium* studies, in which equilibrium concentrations are measured as a function of *position* in the ultracentrifuge cell. Optical methods for measurement of concentration and concentration gradients are described in this chapter; other aspects of ultracentrifugation are beyond the scope of this book.

Three types of optical system have been developed for use in the ultracentrifuge. The *schlieren* optical system measures concentration gradient as a function of position and is thus particularly suited for use in studies of sedimentation velocity. For many years optical systems of the ultracentrifuge were based, almost exclusively, on the schlieren principle. More recently, interest in equilibrium measurements has created a demand for detection systems which provide data directly as concentrations, and which are of high selectivity, sensitivity and accuracy. *Interference* and *absorption* optical systems have, accordingly, been developed. Each type of detection system will be described here.

Familiarity with the properties of light waves and of lenses, as discussed in Chapters 1 and 8, is necessary for understanding of this material. The description of the interference system also requires an understanding of diffraction effects, as described in Chapter 3, while the quantitation of light absorption is discussed in Chapter 24.

GENERAL FEATURES OF ULTRACENTRIFUGE DATA

The ultracentrifuge cell is sector-shaped in one dimension but of constant thickness in the two directions at right angles, as shown in Fig. 32-1. Gravitational force is applied in the direction indicated by $+x$ in that figure, while in all types of optical system the illuminating beam passes through the cell in the direction shown. The depth of standard ultracentrifuge cells is 12 mm. Sedimentation causes solutes to accumulate toward the bottom of the cell. After a period of ultracentrifugation, therefore, the solution is more concentrated near the cell bottom, and the refractive index in this part of the cell is correspondingly increased. Refractive index is directly related to concentration according to

$$(32\text{-}a) \qquad\qquad n = n_0 + \alpha c,$$

where n is the refractive index of the solution, n_0 is the refractive index of the solvent, c is the concentration of solute, usually expressed in g/cm³, and α is the *refractive increment, dn/dc*. The value of the refractive increment is normally found to be a *constant* independent of concentration.

Because of the direct relationship between concentration and refractive index, plots of either concentration or of refractive index as a function of position in the ultracentrifuge cell are of the same form, as shown in Fig. 32-2.

A plot of refractive index or concentration *gradient* as a function of position is of the form shown in Fig. 32-3. The maximum value of *dn/dx* or *dc/dx* indicates the position of maximum gradient; however,

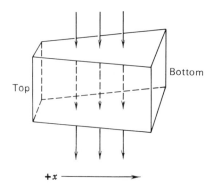

Fig. 32-1 Ultracentrifuge cell.

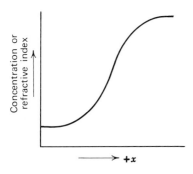

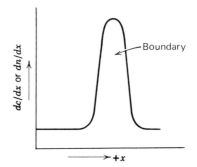

Fig. 32-2 Variation of concentration or refractive index in an ultracentrifuge cell after a period of sedimentation.

Fig. 32-3 Gradient of concentration or refractive index in an ultracentrifuge cell after a period of sedimentation.

note that, unless this peak is infinitely sharp, solute is present at positions to the left of this maximum. The region of maximum gradient is termed the *boundary*.

EFFECT OF A GRADIENT UPON A PLANE WAVEFRONT

A plane wavefront of light incident normally upon a solution of uniform refractive index is not deviated. The form of the wavefront, and the direction of energy transfer by the wavefront are unaffected by passage through a homogeneous solution of constant thickness, as indicated in Fig. 32-4a. If, however, a concentration gradient exists with respect to the x direction shown in Fig. 32-4, each portion of the wavefront is transmitted by the solution with a different velocity. Noting that a wavefront is a surface of uniform phase, it is evident that a uniform refractive index gradient produces a uniformly inclined wavefront, as shown in Fig. 32-4b. Since refractive index is inversely proportional to the velocity of transmission, the wavefront is most retarded by the region of *highest* refractive index. The direction of a light beam is that of the normal to the wavefront; thus the light transmitted by the gradient is deviated *toward the direction of highest refractive index*, as shown in Fig. 32-4b, even though the beam is incident normally upon the cell. (Analogous effects in the refraction of electron beams are illustrated in Fig. 19-5).

Normally an ultracentrifuge cell does not contain a uniform concentration gradient, but regions of uniform refractive index exist at the top and bottom of the cell, while a gradient is established between them. The gradient tends to be steepest at its center and to decrease symmetrically on either side of that position. A typical plot of refractive index as

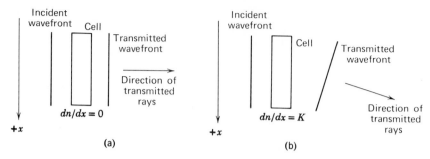

Fig. 32-4 Transmission of a plane wavefront: (a) by a uniform solution; (b) by a refractive index gradient.

a function of position is shown in Fig. 32-5a, and the deviation of a collimated light beam, incident normally, by a gradient of this form is illustrated in Fig. 32-5b. Portions of the incident beam which travel through regions of uniform refractive index remain undeviated, while other portions are deflected *in the direction of increasing refractive index.* The greatest deflection is experienced by light which passes through the center of the gradient. *The extent of deviation of the light is proportional to the magnitude of the refractive index gradient.*

IMAGING OF A CELL CONTAINING A REFRACTIVE INDEX GRADIENT

The schlieren optical system exploits the deviation of light by refractive index gradients in such a manner as to record, directly, the value of that gradient as a function of position in the cell. Note that a special

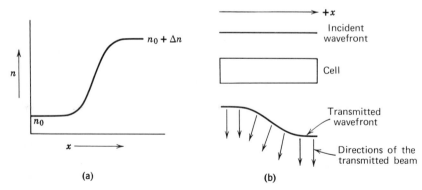

Fig. 32-5 (a) Refractive index as a function of position in ultracentrifuge cell; (b) corresponding distortion of a plane wavefront.

optical system is in fact required in order to accomplish this! Thus Fig. 32-6 shows a cell which contains solutions of uniform refractive index in regions ① and ③, while a gradient exists in region ②. A collimating lens illuminates the cell with a parallel beam of light, while a second lens, called the condensing lens, forms an image of the cell at some position beyond. Rays ① and ③ are focused by the condensing lens to form images of the respective portions of the cell at ①′ and ③′. Since these rays remain parallel until they reach the condensing lens, they intersect on the axis of the system at the second focal point of that lens. Rays from region ②, however, are deviated by the gradient, so that they travel at an angle to the axis and thus cross the focal plane of the condensing lens at some off-axis position. Nevertheless, ray ② reaches the corresponding point ②′ on the image of the cell, since, by definition (cf. p. 156), any lens causes rays leaving a point on the object in any direction to reunite at the corresponding image point. In short, no distinction is made, in the image formed by the condensing lens, between points from which light has or has not been deviated.

In the focal plane of the condensing lens, however, deviated and deviated rays are separated. Rays which are deviated through any given angle form separate images of the light source in this plane (i.e., of the "slit source" shown in Fig. 32-6). The image formed by undeviated rays is centered about the axis, while the images formed by deviated rays lie off-axis. The relative positions of the slit source, collimating lens, cell, condensing lens, and slit images are shown in perspective in Fig. 32-7.

The location of a refractive index gradient may be detected by exploiting the separation of the rays in the rear focal plane. A relatively simple type of system which accomplishes this is shown in Fig. 32-8. Rays passing through regions A and C, which are of uniform refractive index, are incident upon the condensing lens parallel to the axis, and thus, as before, they cross the axis at the second focal point. From this point they diverge, illuminating corresponding positions on a screen placed beyond, at the image plane of the condensing lens. Rays which pass

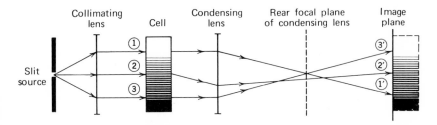

Fig. 32-6 Imaging of a cell containing a gradient.

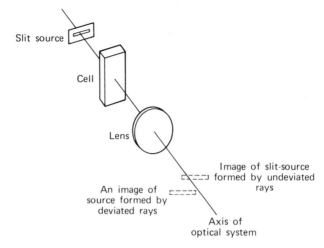

Fig. 32-7 Elements of an ultracentrifuge optical system.

through the edge of the gradient (represented by dotted lines in Fig. 32-8) are deviated slightly in the direction of higher refractive index, and thus intersect in the focal plane at a position slightly below the axis. As shown, the knife edge is so placed in the focal plane that those rays are just allowed to pass and thus, also, to illuminate the screen. A ray from the center of the gradient (represented by a solid line) is more strongly deviated and is thus intercepted by the knife edge. The corresponding position

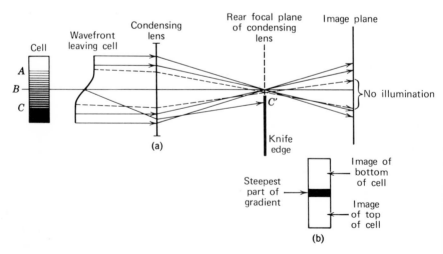

Fig. 32-8 (a) Imaging of a cell containing a gradient by a system which includes a knife edge in the focal plane; (b) corresponding image.

in the image of the cell remains unilluminated; that is, as shown in Fig. 32-8b, there is a shadow in the image at positions corresponding to the steepest refractive index gradient in the cell. The extent of the shadow is determined by the position of the knife edge; an edge so placed as just to miss the axis of the system would produce a shadow in all regions where any gradient exists, while an edge placed at position C' would result in the formation of a very narrow shadow in a position corresponding to that of the highest refractive index gradient.

Changes in position of the gradient during sedimentation could be recorded in the form of changes in position of the shadow of a knife edge located in the rear focal plane of the condensing lens. In fact, however, practical schlieren systems make use of a more elegant optical system which forms a continuous trace of gradient as a function of position in the cell.

SCHLIEREN OPTICAL SYSTEMS

It has been shown above that a lens may form an image of a cell containing a gradient, in which each point on the image corresponds to a point in the cell, but which does not reveal the presence of the gradient. At the second focal plane of the lens the distribution of illumination *is* determined by the position and extent of the refractive index gradient in the cell, but there is no point-to-point correspondence between the distribution of intensities in the focal plane and positions in the cell. The achievement of the schlieren optical system is to superpose an image of the focal plane onto the image of the cell, thus providing a direct indication of the nature of the gradient present in the cell.

The distinguishing element of the schlieren optical system is an *inclined slit* (or its equivalent). In addition, the system incorporates a slit source, collimating lens, cell, condensing lens, and screen, as already shown in Fig. 23-7, and also a *cylindrical lens* and a second focusing lens (known as the *camera lens*). The inclined slit is placed in the focal plane of the condensing lens, and the cylindrical lens is placed between the inclined slit and the screen with its long axis parallel to the gradient in the cell. The camera lens, which is placed between the inclined slit and the cylindrical lens, contributes to the focusing action which forms an image of the cell on the screen. The arrangement of the optical elements of the schlieren system is shown in Fig. 32-9.

Light which leaves the cell in any one direction forms an image of the slit source in the focal plane of the condensing lens (as shown by Fig. 32-7). Since the gradient deviates light through a continuous range of angles, a continuous series of such images is formed. The position of each

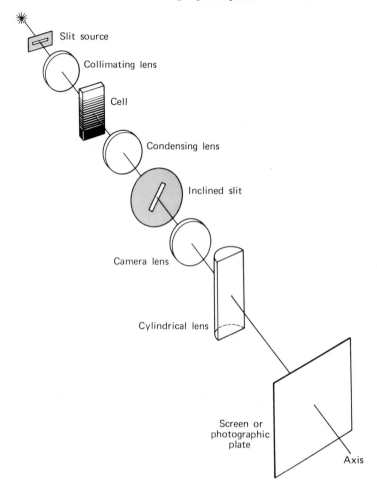

Fig. 32-9 Schlieren optical system.

image, individually, is determined by the angle at which light leaves the cell, and thus by the value of the gradient at that point. (For a general discussion of the nature of the distribution of intensities in the rear focal plane of a lens see p. 222 and in particular Fig. 10-2.) When the inclined slit is placed in the focal plane, only *one* portion of each image of the slit is transmitted, as shown in Fig. 32-10. Thus the position in the y direction of light transmitted by the inclined slit is determined by the value of the refractive index gradient at the point in the cell from which the beam originated. The position of the transmitted rays with respect to the

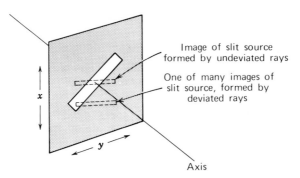

Fig. 32-10 The inclined slit of the schlieren optical system.

x direction corresponds, as in any lens system, to position along the x direction of the cell.

The portions of the slit source images which are passed by the inclined slit illuminate the cylindrical lens. The lens serves to magnify differences in position with respect to the y direction. As noted in Fig. 32-11, the cylindrical lens does not deflect rays which are incident along its mid-line, while a maximum deviation is experienced by rays which are incident along its sides. (For a discussion of the action of cylindrical lenses, see also p. 202.)

Undeviated rays, originating from regions of uniform refractive index within the cell, are transmitted by the axial portion of the inclined slit and thus are incident on the cylindrical lens near its center, where they experience no deflection. Rays which are deviated to various degrees by the refractive index gradient in the cell are transmitted by extra-axial portions of the inclined slit and thus are incident upon the cylindrical lens at a series of positions away from its center. The nearer the edge of the

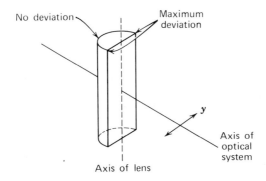

Fig. 32-11 Action of a cylindrical lens.

cylindrical lens is the point of incidence, the more strongly is the ray deflected in the y direction.

Rays from the gradient-free portions of the cell form portions of a line image of the cell in the plane of the screen. The line image, which is interrupted at positions corresponding to the parts of the cell in which a refractive index gradient is present, constitutes the *base line* of the schlieren pattern. The formation of the base line is diagramed in Fig. 32-12, where positions ① and ② in the schlieren image correspond to positions ① and ②, respectively, in the cell. (Focusing action by the camera lens is ignored in Figs. 32-12 and 32-13 for clarity.)

Figure 32-12 may be compared with Fig. 32-13, which illustrates the contribution of two levels of equal refractive index gradient to the schlieren image. Rays from the two levels of the cell are deviated through the same angle and are focused by the condensing lens to form an off-axis image of the slit source in the plane of the inclined slit. Rays from positions ③ and ④ in the cell contribute to that portion of the slit source image which is transmitted. These rays are incident upon the cylindrical lens part way to its edges, and they undergo a deviation in the y direction.

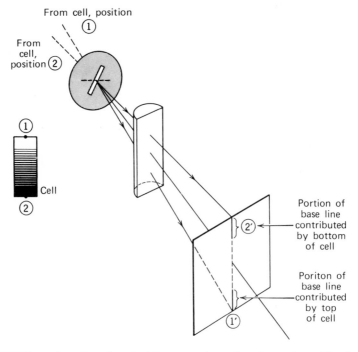

Fig. 32-12 Formation of baseline of schlieren trace by gradient-free regions of the ultra-centrifuge cell.

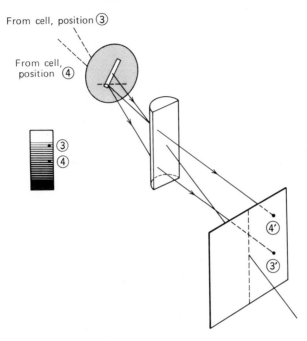

Fig. 32-13 Formation of portions of schlieren trace by regions of the ultracentrifuge cell which contain a gradient.

At the screen the rays contribute to a line image of the portions of the cell in which the specified value of the refractive index gradient exists. As shown in Fig. 32-13, this line image is confined to the two points ③ and ④. Similar partial line images of other portions of the cell, in which other values of the refractive index gradient exist, are formed in other positions parallel to the base line.

The complete schlieren tracing and its relation to positions in the cell are shown in Fig. 32-14.

Within limits, the scale of the schlieren pattern may be selected by adjusting the angle of the inclined slit. If the slit is inclined more sharply than shown in Fig. 32-9 and other figures, the schlieren pattern is enlarged in the $+y$ direction. The process is limited by the possibility of rotating the slit so far that *no* portion of the deviated line images is transmitted, as would be the case for a vertical slit.

Schlieren imaging may fail if (a) the refractive index gradient in the cell is so large as to deviate light entirely beyond the inclined slit, or (b) if absorption by the solution reduces the intensity of light transmitted by the cell.

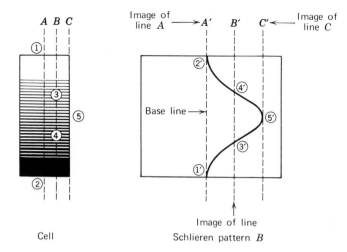

Fig. 32-14 (a) Positions in ultracentrifuge cell; (b) corresponding schlieren image.

The schlieren system just described differs from that actually used in ultracentrifuges in that the latter includes either an inclined *bar* or a *phase plate* rather than an inclined *slit* for the selection of the transmitted rays. The effect of using an inclined bar is simply to form a *dark* image on a bright field, rather than vice versa. Thus only those rays shown here as passing through the inclined slit are excluded from the schlieren image. The phase plate makes use of diffraction effects, rather than a physical obstacle, to select the rays which contribute to the schlieren image.

INTERFERENCE OPTICAL SYSTEMS

In the interference optical system of the ultracentrifuge, a fringe system is formed by the interference of light transmitted by the sedimenting solution with that transmitted by an identical length of solvent. This pattern is magnified and superimposed upon an image of the cells. Concentration at any point in the cell can then be determined by measurement of the shift of fringes, at that point, with respect to a reference fringe pattern.

The Rayleigh type of interferometer system has been most widely used in ultracentrifuges. A diagram of the Rayleigh interferometer has been given in Fig. 26-7, and its basic features are explained in the text accompanying that figure. A more general discussion of interferometers has also been given on p. 71.

The adaptation of the Rayleigh system for use in the ultracentrifuge is illustrated in Fig. 32-15. The monochromator shown in that figure is usually an interference filter. As in the standard interferometer, the double

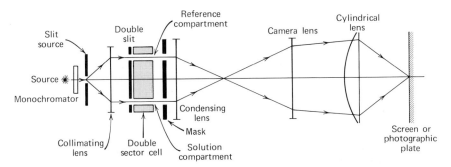

Fig. 32-15 Ultracentrifuge interference optical system.

slit serves as a beam splitter. A double sector cell, as shown in Fig. 32-16, is used in place of the standard ultracentrifuge cell illustrated in Fig. 32-1 and is the equivalent of the interferometer tubes shown in Fig. 26-7. Each compartment of this cell is a sector which is centered about a radius to the axis of rotation of the centrifuge. The cell must be transparent to the ultraviolet wavelengths usually employed for interference measurements. During high speed centrifugation, quartz cells undergo distortions which significantly alter the optical thickness of the cell; difficulties of this sort are minimized by the use of *sapphire* cell windows.

Another difficulty which is introduced by the conditions of ultracentrifugation is that oblique illumination of the double sector cell during rotation of the rotor tends to produce a background of misplaced interference fringes. This effect is minimized by placing a second double slit mask *beyond* the cell in order to limit the angle of rotation "seen" at the image. The position of the mask is shown in Fig. 32-15 and by dotted lines in Fig. 32-16.

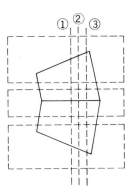

Fig. 32-16 Double sector ultracentrifuge cell.

Solution is placed in one compartment of the double sector cell and solvent in the other. Usually the depth of solution is only 1–3 mm (as compared with the total cell capacity of 12 mm), while that of solvent is slightly greater. In this way the time required for establishment of equilibrium concentration gradient can be minimized.

The condensing lens reunites rays which have traveled through these chambers, forming the interference pattern in the rear focal plane of that lens. As in the schlieren optical system, the camera lens serves primarily to form a magnified image of the cell, while the cylindrical lens effects unidirectional magnification of the focal plane of the condensing lens (i.e., in this case, of the interference pattern).

Reference interference fringes are formed by light which passes through two holes in an otherwise opaque counterbalance cell. The counterbalance is placed in the rotor at a separation of 180° from the double sector cell.

THE NATURE AND INTERPRETATION OF INTERFERENCE PATTERNS OBTAINED IN THE ULTRACENTRIFUGE

The precise form of ultracentrifuge interference patterns varies with the solution volume and with the speed of the rotor, according to the manner in which sedimentation data are to be processed. For the purpose of illustration it is convenient to consider the shift of fringes which takes place during a sedimentation *velocity* run in a cell that is completely filled by solution and solvent, respectively. In fact, however, interference optics have more often been used in *equilibrium* studies.

If each of the sectors *A* and *B* of the cell shown in Fig. 32-16 contains only solvent, optical paths through the cell are identical, and a zero order fringe is formed on the axis of the optical system, as illustrated by Fig. 32-17a. (Note that, in interferometer patterns formed by monochromatic light, the zero order fringe is not distinguishable from the higher order fringes; this fringe is represented by a heavy line in Fig. 32-17 only for the purpose of illustration. With polychromatic illumination, however, the white zero order fringe is easily distinguished from the higher order fringes formed by light of any given wavelength.)

If sector *A* of the cell contains solvent, and sector *B* a homogeneous solution (as at the start of an ultracentrifuge run), the zero order fringe is displaced to an extent proportional to the difference in refractive index (and thus to the difference in concentration) between solvent and solution. This displacement is expressed in terms of *number of fringes*, designated *J*. *J* need not be an integral number; for example, *J* is about 1.9 fringes in the case shown in Fig. 32-20b.

To relate the value of *J* to the difference in refractive index between solvent (reference) and solution, note that, for an ultracentrifuge cell of length *l*, the path length through the medium of higher index is increased by $l \Delta n$ wavelengths. Thus, for light of any wavelength λ,

(32-b) $$\boxed{J = \frac{l \Delta n}{\lambda}}.$$

The quantity Δn is equal to the product of the specific refractive increment

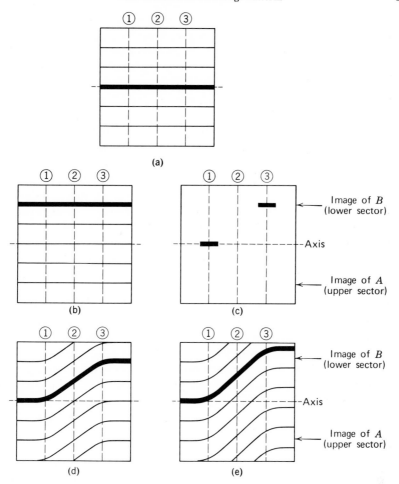

Fig. 32-17 Formation of ultracentrifuge interference trace.

α of the solute and the concentration of the solution. Thus, if α is known, the concentration C can be obtained from

(32-c)
$$C = \frac{J\lambda}{l\alpha}.$$

During centrifugation a refractive index gradient is established in the cell which contains the solution. Near the top of the solution (position ① as indicated in Fig. 32-16) pure solvent may exist in both solution and reference sectors. The optical paths through both sectors are then equal at this position, so that the zero order fringe lies on the axis of the corresponding part of the image. At the bottom of the cell (position ③ of Fig. 32-16) a region of high but constant refractive index is formed in sector B. The zero order fringe corresponding to that portion of the cell is accordingly displaced. The locations of the parts of the zero order fringe corresponding to positions ① and ③, only, are shown in Fig. 32-17c. In the intermediate region the value of the refractive index and thus the position of the zero order fringe change continuously; the complete pattern is as shown in Fig. 32-20d. As sedimentation continues, the solute-free region at the top of the cell extends (or, more generally, the concentration of solution at the top of the cell decreases), while the concentration at the bottom of the cell increases, so that the interference pattern shifts to that shown in Fig. 32-17e. The fringe positions continue to shift as sedimentation proceeds. Thus the appearance of the patterns actually recorded during a sedimentation velocity run is as shown by Fig. 32-18.

The form of interference pattern obtained during an equilibrium run is illustrated in Fig. 32-19a, and corresponding positions in the double sector cell are shown in Fig. 32-19b. The separate fringe blocks A and A' are images of the reference holes in the counterbalance cell. The function of these images is to provide an absolute measure of distance from the axis of rotation. Region B of the interference pattern corresponds to the air space in the upper parts of the ultracentrifuge cell; since the optical path through both sectors is identical in this region, a straight fringe system is formed. Part C of the pattern represents a region in which the light path through solvent (in the reference cell) is compared with air space (in the solution cell). In this zone the optical path difference between the sectors

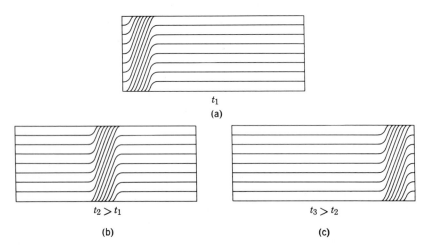

t_1

(a)

$t_2 > t_1$

(b)

$t_3 > t_2$

(c)

Fig. 32-18 Successive interference traces obtained during a sedimentation velocity study.

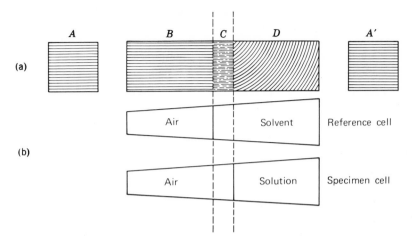

Fig. 32-19 (a) Ultracentrifuge pattern obtained in a sedimentation equilibrium study; (b) corresponding regions in double sector cell.

is so large that no well defined fringes are observed. Region D of the specimen cell contains the sedimenting solution; thus this part of the interference pattern consists of continuously shifting fringes. The *number* of fringes crossing the pattern can be counted at any position x, and gives the concentration of solute at the corresponding position in the cell. Thus the form of the refractive index gradient may be determined by making fringe counts at many points along region D. The precise form in which data are recorded depends upon the nature of the run being made, a topic which is not discussed here. Usually, the measurements are made after the establishment of an equilibrium gradient.

Apart from the convenience of obtaining direct measurements of concentration for equilibrium sedimentation studies, interference optical systems are advantageous in that they are of high accuracy and that lower solute concentrations may be studied. Accuracy is determined by the number of fringes which can be viewed, resolved, and counted. With care, counting to within ± 0.02 fringe is possible. Protein concentrations of the order of $10\ mg/ml$ may thus be determined to an accuracy of $\pm 0.05\%$. Interference systems are well suited to the measurement of molecular weights by equilibrium sedimentation and have been widely applied for this purpose. Interference imaging is also useful when sedimentation velocity runs must be made at low concentrations. However, the precise and time-consuming alignment procedures required to obtain interference diagrams are not justified if the same information can be obtained more easily from schlieren traces.

ABSORPTION OPTICAL SYSTEMS

The first analytical centrifuges to be developed made use of absorption optics for producing a record of sedimentation. In these systems, as shown in Fig. 32-20, a collimated beam of light of a wavelength specifically absorbed by the solute is incident upon the cell, while a focusing lens subsequently forms an image of the cell on a photographic plate. The plate (or series of plates exposed during the course of sedimentation) is analyzed by recording a densitometer tracing along the x' direction. (For a discussion of the densitometry of photographic exposures see p. 471.)

The absorption method, as first evolved, suffered from the disadvantage of requiring the development and densitometry of the photographic plates; no immediate check on the course of sedimentation could be made. Also, separate centrifugation of a cell containing solvent was required in order to provide a "base line" which would account for unevenness of illumination and for absorption by the solvent. These cumbersome techniques were virtually abandoned as soon as the more convenient schlieren systems became available.

Recent developments have obviated the difficulties just mentioned and have otherwise improved absorption optical systems. Efficient and essentially instantaneous methods for photoelectric scanning of the absorption image have been devised. Information can thus be supplied *directly* in the form of a trace of optical density or optical density gradient or both as a function of position in the cell. Electronic "holding" circuits have also been developed which make it possible not only to compare absorption by solution and solvent during a single ultracentrifuge run, but even to study several different sedimenting solutions simultaneously. More versatile systems have resulted from the use of monochromators, rather than filters, for the production of monochromatic illumination, while light sources which provide workable intensities at wavelengths as short as 225 mμ have increased the sensitivity of absorption techniques.

Very high sensitivity is in fact the most impressive advantage of the absorption system. Limiting concentrations are determined by the absorp-

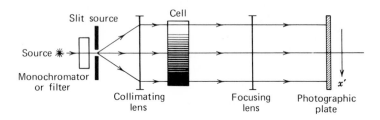

Fig. 32-20 Ultracentrifuge absorption optical system.

tion coefficient of the substance studied. For illumination at 280 mμ, protein concentrations of the order of 0.1 mg/ml may be used; for illumination at 230 mμ, ultracentrifugation of proteins at concentrations as low as 3 μg/ml has been reported.

The *selectivity* of absorption systems is also an important advantage. Since refractive index varies as a function of concentration, both schlieren and interference tracings are similar in form for all types of solute. Absorption spectra, however, are highly characteristic for each substance; thus the sedimentation of proteins, nucleic acids, or other compounds which absorb at specific wavelengths can be differentiated according to the wavelength dependence of the absorption pattern. However, because many substances absorb strongly at short wavelengths (below about 230 mμ), selectivity is reduced in the far ultraviolet region, where sensitivity is highest.

REFERENCES

E. C. Pollard and R. B. Setlow, *Molecular Biophysics*, Addison-Wesley, Reading, Mass., 1962. Chapter 4 (Physical Methods of Determining the Sizes and Shapes of Molecules) includes a general description of ultracentrifugal methods and a short account of schlieren and absorption optical systems.

Beckman Instrument Company Manual: Ultracentrifuge Model E. Section 4 (Optical Systems) includes a complete description and elegant diagrams of the schlieren optical system.

P. E. Hexner, R. D. Boyle, and J. W. Beams, *J. Phys. Chem.* **66**, 1948 (1962). This article describes an interference optical system which is based upon the Jamin interferometer.

H. K. Schachman, *Biochemistry* **2**, 887 (1963). A review of recent work on interference and absorption systems.

H. J. Schachman and S. J. Edelstein, *Biochemistry* **5**, 2681 (1962). A review of recent developments in and applications of absorption systems.

APPENDIX A

Angular Measure

By definition, there are 360 *degrees* or 2π *radians* in a complete circle.

The subdivisions of angular measure, as expressed in degrees, are *minutes* and *seconds* of arc. Thus, for example,

(A-a) $$1° = 0°60' = 0°59'60''.$$

Angular sizes, as expressed in degrees, may be converted into radian measure by means of

(A-b) $$\frac{\text{size of angle, in degrees}}{2\pi} = \text{size of angle, in radians}.$$

Thus,

(A-c) $$\frac{360°}{2\pi} = 1 \text{ rad} = 57°17'44'' \quad \text{or} \quad 360° = 6.28\ldots \text{ rad}.$$

Note that angular size is expressed by radians on a decimal basis. Mathematical formulas which include angular quantities imply expression of these angles in radians.

For small angles, values of the sine and tangent are approximately equal to the value of the angle itself, *as expressed in radians*. For example,

(A-d) $$10^{-2} \text{ rad} = 0°34'23'',$$
$$\sin (0°34'23'') \cong \tan (0°34'23'') = 0.00999 \cong 10^{-2}.$$

Solid angles similarly are expressed in *steradians*, the value of the total solid angle about a point being 4π steradians.

Vectors

Vector quantities are associated with both *magnitude* (length) and *direction* (azimuth), whereas *scalar* quantities are associated with magnitude only. Geometrically, vectors are represented by arrows, the lengths of which are proportional to magnitude. For any vector **r**, the magnitude, which is a scalar quantity is designated $|\mathbf{r}|$. $|\mathbf{r}|$ is also called the *modulus* of **r**.

In any consideration of vectors, a vector of arbitrary length and orientation may be defined as a *unit vector* **i** (e.g., as shown in Fig. B-1a). Additional unit vectors **j** and **k** are equal in magnitude (length) to **i** but are perpendicular to **i** and to each other, as shown in Fig. B-1b. Any vector **r** may then be specified in terms of its *projections* in the directions of these unit vectors. Thus, in general,

$$\text{(B-a)} \qquad \mathbf{r} = a\mathbf{i} + b\mathbf{j} + c\mathbf{k},$$

where a, b, and c are the magnitudes of the projections made by **r** on the respective unit vectors. For example, the vector **r′**, shown in Fig. B-1c is so specified as

$$\text{(B-b)} \qquad \mathbf{r}' = 2\mathbf{i} + 2\mathbf{j} + 0\mathbf{k}.$$

The magnitude of **r′** may be obtained (from the Pythagorean theorem) as

$$\text{(B-c)} \qquad |\mathbf{r}'| = \sqrt{2^2 + 2^2} = \sqrt{8} = 2\sqrt{2} \quad (\times \text{ arbitrary length of unit vector)}.$$

More generally, the magnitude of a vector is given by

$$\text{(B-d)} \qquad |\mathbf{r}| = \sqrt{a^2 + b^2 + c^2}.$$

The sum of two (or more) vectors, termed their *resultant*, is also a

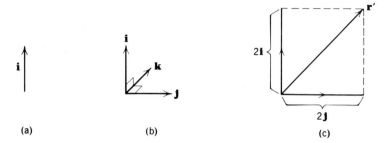

(a)   (b)   (c)

Fig. B-1 Vectors: (a) unit vector; (b) unit vectors; (c) specification of arbitrary vector in terms of unit vectors.

vector. The geometrical interpretation of the addition of vectors is shown in Fig. B-2a: the vectors to be added (**m** and **n**) are drawn head-to-tail in the correct relative orientation, while their resultant, (**m** + **n**), connects the "tail" of the first vector (**m**) with the "head" of the last (**n**). If **m** and **n** are expressed algebraically in terms of unit vectors, that is, if

(B-e) $\mathbf{m} = a_m\mathbf{i} + b_m\mathbf{j} + c_m\mathbf{k}$ and $\mathbf{n} = a_n\mathbf{i} + b_n\mathbf{j} + c_n\mathbf{k}$,

then their *sum* (resultant) is expressed algebraically by *separately adding the coefficients of each unit vector*. That is,

(B-f) $(\mathbf{m} + \mathbf{n}) = (a_m + a_n)\mathbf{i} + (b_m + b_n)\mathbf{j} + (c_m + c_n)\mathbf{k}$.

Similarly the *difference* of the two vectors, as represented geometrically in Fig. B-2b, is

(B-g) $(\mathbf{m} - \mathbf{n}) = (a_m - a_n)\mathbf{i} + (b_m - b_n)\mathbf{j} + (c_m - c_n)\mathbf{k}$.

The "multiplication" of vectors may be considered in two ways. One process gives the *scalar product* (*dot product*) (**a.b**). The dot product may simply be called "the projection of **a** on **b**" or, equivalently, "the projection of *b* on *a*"; these two equivalent geometrical interpretations are shown in Fig. B-3. Note that the dot product is simply a length (magni-

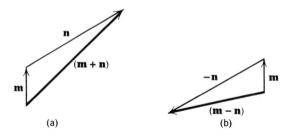

(a)   (b)

Fig. B-2 (a) Addition of two vectors; (b) subtraction of vectors.

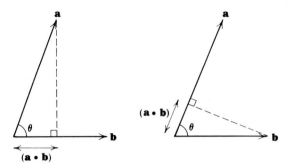

Fig. B-3 Two equivalent representations of the dot product of two vectors.

tude); that is, it is a scalar quantity. The numerical value of the dot product is

(B-h) $$(\mathbf{a} \cdot \mathbf{b}) = |\mathbf{a}||\mathbf{b}| \cos \theta,$$

where θ is the angle between $\mathbf{a}$ and $\mathbf{b}$, as shown in Fig. B-3, and $|\mathbf{a}|$ and $|\mathbf{b}|$ are obtained from (B-d).

These few considerations are sufficient to describe the vector notation used in the text. However, it may be noted that two vectors may also be "multiplied" to form the *vector product* or *cross product*, designated $(\mathbf{a} \times \mathbf{b})$. Geometrically, the cross product is represented by a vector oriented *perpendicular* to the plane containing $\mathbf{a}$ and $\mathbf{b}$. Its magnitude is

(B-i) $$|\mathbf{a} \times \mathbf{b}| = |\mathbf{a}| |\mathbf{b}| \sin \theta.$$

Derivations of (B-h) and (B-i) and further discussions of the properties of vectors may be found in many mathematical texts.*

*See, for example, I. S. Sokolnikoff and R. M. Redheffer, *Mathematics and Physics of Modern Engineering*; McGraw-Hill, 1958, p. 287 ff.

The Complex Representation of Wave Motions

In relatively simple cases the analysis of interfering wave motions by means of the algebraic and geometrical approaches developed in Chapter 3 leads readily to results of obvious physical significance. Thus these treatments are applied exclusively to the discussion of most of the cases of interference phenomena considered in this book. However, when the intensity and phase of the resultant of *many* superposing wave motions must be computed, use of the simple methods becomes unduly cumbersome. Application of the wave equation in a form which makes use of *complex numbers* then provides a much more convenient method of computing the properties of resultant motions.

In this appendix the properties of complex numbers are reviewed briefly, the complex form of the wave equation is derived, and the application of this equation to computation of the amplitude and phase of resultant motions is illustrated. An example of the use of the complex form of the wave equation for the computation of intensities transmitted by interference filters is given on p. 491. The complex equation is also applied to discussions of the interpretation of x-ray diffraction patterns in Chapter 31.

COMPLEX NUMBERS

Such algebraic quantities as x, y, etc., are examples of *real* numbers. The quantity $\sqrt{-1}$, here designated i (also designated as j in the literature) is an *imaginary* number; such products as ix, iy, etc., are *pure imaginary numbers*. Combinations of real and imaginary numbers, as, for example, $z = (x + iy)$, are called *complex numbers*. If $z = (x + iy)$, it is said that

(C-a)
$$z = \text{Re}(z) + i\,\text{Im}(z),$$

where Re $(z) = x$, the real part of z, and Im $(z) = y$, the imaginary part of z.

The *complex conjugate*, z^*, of a complex number z is that number in which the *sign* of i is reversed throughout. Thus, if $z = (x + iy)$, then $z^* = (x - iy)$, while, if $z = e^{-ix}$, then $z^* = e^{+ix}$.

Complex numbers may be added, subtracted, multiplied, or divided according to the rules of ordinary algebra, provided that the property $i^2 = -1$ is taken into account. The product of two imaginary numbers is thus neither imaginary nor complex, but real. For example;

(C-b)
$$(iA)(iB) = i^2 AB = -AB.$$

In general, however, the performance of algebraic operations on complex numbers produces other complex numbers. This is true, for example, of the process of addition. Thus,

(C-c) $\quad (A + iB) + (C + iD) = A + C + iB + iD = (A + C) + i(B + D).$

In (C-c) the real part of the sum is $(A + C)$, and the imaginary part is $(B + D)$, while the sum itself is of the form $(x + iy)$, as were its components singly.

Sine and cosine functions may be expressed in complex *exponential* form as

(C-d)
$$\sin \theta = \frac{1}{2i}(e^{i\theta} - e^{-i\theta}),$$

(C-e)
$$\cos \theta = \tfrac{1}{2}(e^{i\theta} + e^{-i\theta}).$$

Combination of (C-d) and (C-e) provides an expression, known as *Euler's formula*, for $e^{i\theta}$:

(C-f)
$$e^{i\theta} = \cos \theta + i \sin \theta.$$

Complex numbers may be represented either as *points* or *vectors* on the so-called *complex plane*, as shown in Figs. C-1a and C-1b, respectively. The x axis of the complex plane is the real axis, specifying Re (z), while the y axis is the imaginary axis, specifying Im (z). When complex numbers are represented as vectors, the *projection* of z on the real axis specifies Re (z) and that on the imaginary axis specifies Im (z). As for real vectors (cf. Appendix B), the length (modulus) of z is obtained from the projections Re (z) and Im (z) as

(C-g)
$$z = \sqrt{\text{Re}(z)^2 + \text{Im}(z)^2}.$$

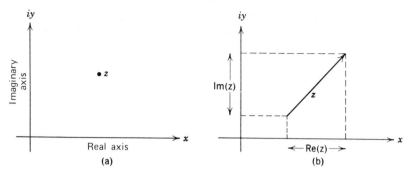

Fig. C-1 Graphical representation of a complex number: (a) as a point in the complex plane; (b) as a vector in the complex plane.

The angle θ between the vector and the x (real) axis is given by

(C-h)
$$\tan \theta = \frac{\mathrm{Im}\,(z)}{\mathrm{Re}\,(z)}.$$

These properties suggest the applicability of complex numbers to the description of wave motions since, as discussed in Chapter 3, a wave motion of any given frequency may be represented by a vector of a length proportional to the *amplitude* of the wave motion and of an orientation determined by the *relative phase* of the wave.

THE WAVE EQUATION

The cosine form of the fundamental wave equation, as developed in Chapter 1, is

(1-m)
$$A' = A_0 \cos\,(\omega t - \alpha')$$

or, equivalently,

(C-i)
$$A' = A_0 \cos\,(\omega t - \delta),$$

where A_0 is the amplitude of the wave motion, A' its instantaneous amplitude at time t, ω is the angular frequency of the motion, α' or δ the phase. The quantity given by (C-i) is in fact the real part of the complex expression

(C-j)
$$\begin{aligned} A' &= \mathrm{Re}\,(A') + i\,\mathrm{Im}\,(A') \\ &= A_0 \cos\,(\omega t - \delta) + iA_0 \sin\,(\omega t - \delta). \end{aligned}$$

Combination of (C-j) with (C-f) gives

(C-k)
$$A' = A_0 e^{i(\omega t - \delta)} = A_0 e^{i\omega t} e^{-i\delta}.$$

When interference of waves of a single frequency only is considered (as is often the case in optical applications), the factor $e^{i\omega t}$ in (C-k) determines the *instantaneous* amplitude but is independent of either the *amplitude* (A_0) or *phase* of the wave motion. Thus the wave equation may be written

(C-l) $$A = A_0 e^{-i\delta},$$

where $A = A'e^{-i\omega t}$.

The quantity A, which is defined by (C-l), specifies both amplitude and phase of the vibratory motion and is called the *complex amplitude* of the wave. (In some texts, the complex amplitude is defined as the quantity $A_0 e^{+i\delta}$. Either definition is "correct", provided that signs are consistently applied in making computations.)

The advantage of describing wave motions in complex notation is that *the complex amplitude of the resultant of two or more superposed wave motions* (of the same frequency) *can be obtained directly by addition of the complex amplitudes of the individual wave motions*. From this sum *the real amplitude and phase of the resultant follows directly*. An example is considered in the following section.

The *intensity* of a wave motion is given by the *product* of its complex amplitude A and A^*, the complex conjugate of the complex amplitude. Thus,

(C-m) $$I = AA^* = (A_0 e^{-i\delta})(A_0 e^{+i\delta}) = A_0^2.$$

The value of the intensity so obtained thus specifies the absolute value of the *amplitude*, but not the *phase* of the wave. This result is in accordance with the physical fact that the phase of a wave is unobservable.

While (C-l) has been derived here with respect to two dimensions only, the treatment may be extended to three dimensions, thus making it possible to consider the interaction of waves which are polarized in different planes.

THE COMPUTATION OF THE AMPLITUDE AND PHASE OF RESULTANT MOTIONS

Algebraic computations given on p. 36 lead to the derivation of expressions for the amplitude and phase of the resultant of two interfering wave motions. These expressions are

(C-n) $$(A')^2 = A_{0_1} + A_{0_2}^2 + 2A_{0_1}A_{0_2} \cos(\delta_1 - \delta_2)$$

and

(C-o) $$\tan\delta = \frac{A_{0_1} \sin\delta_1 + A_{0_2} \sin\delta_2}{A_{0_1} \cos\delta_1 + A_{0_2} \cos\delta_2},$$

in which A_{0_1}, δ_1, A_{0_2} and δ_2 refer to the amplitudes and phases of the respective interfering waves, while A and δ are the respective quantities for the resultant. Equivalent expressions follow directly from considerations of the complex amplitude; the case of *three* interfering wave motions (of the same frequency) is described here.

The individual complex amplitudes of the component motions may be written

(C-p) $$A_1 = A_{0_1} (\cos \delta_1 + i \sin \delta_i),$$
$$A_2 = A_{0_2} (\cos \delta_2 + i \sin \delta_2),$$
$$A_3 = A_{0_3} (\cos \delta_3 + i \sin \delta_3),$$

while A, the complex amplitude of the resultant, is

(C-q) $A = A_1 + A_2 + A_3 = (A_{0_1} \cos \delta_1 + A_{0_2} \cos \delta_2 + A_{0_3} \cos \delta_3)$
$$+ i (A_{0_1} \sin \delta_1 + A_{0_2} \sin \delta_2 + A_{0_3} \sin \delta_3).$$

Thus the expression for A is of the form $(x + iy)$; that is, the real part of the complex amplitude is the quantity $(A_{0_1} \cos \delta_1 + A_{0_2} \cos \delta_2 + A_{0_3} \cos \delta_3)$, while the imaginary part of the complex amplitude is the quantity $(A_{0_1} \sin \delta_1 + A_{0_2} \sin \delta_2 + A_{0_3} \sin \delta_3)$. The complex number A thus represents (as described above on p. 705) a vector in the complex plane, the *modulus* and *angular orientation* of which represent the amplitude and phase of the resultant wave. These quantities can be evaluated by substitution in (C-g) and (C-h), respectively, giving

(C-r) Amplitude of resultant $= [(A_{0_1} \cos \delta_1 + A_{0_2} \cos \delta_2 + A_{0_3} \cos \delta_2)^{2+}$
$$+ (A_{0_1} \sin \delta_1 + A_{0_2} \sin \delta_2 + A_{0_3} \sin \delta_3)^2]^{1/2}$$

and

Phase of resultant $= \tan^{-1}\left(\dfrac{A_{0_1} \sin \delta_1 + A_{0_2} \sin \delta_2 + A_{0_3} \sin \delta_3}{A_{0_1} \cos \delta_1 + A_{0_2} \cos \delta_2 + A_{0_3} \cos \delta_3}\right).$

It is evident that (C-r) and (C-s) are analogous to (C-n) and (C-o), respectively.

Distribution of Amplitudes in the Diffraction Pattern of a Single Slit

The distribution of amplitudes in the Fraunhofer diffraction pattern (i.e., the diffraction pattern formed upon illumination by a collimated beam) of a narrow slit is considered in this appendix. Figure D-1 illustrates the illumination of any point P on a screen placed at the focal plane of a lens which collects the light transmitted by a single slit of width s. P is thus illuminated by the light which leaves each portion of the slit at some angle, θ. (As explained in Chapter 10, mutually parallel rays leaving an object are brought to focus in the rear focal plane of any lens.) The width of the slit, as shown in Fig. D-1, is vastly exaggerated; in reality

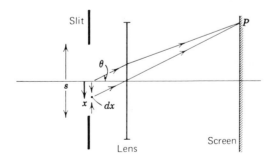

Fig. D-1 Formation of single slit diffraction pattern.

the slit is so narrow that the distance D from any point of the slit to point P on the screen is effectively constant. Illumination of the screen is by light of any single wavelength, λ.

A_0, the amplitude at the center of the diffraction pattern formed on the screen, is directly proportional to the slit width, δ and inversely proportional to the slit-screen distance D. Thus at the center of the screen the wave equation may be written

(D-a)
$$A_0 = \frac{a_0 s}{D} \sin (\omega t - \delta)$$

where $(a_0 s)$ is the amplitude in the plane of the slit.

The amplitude A at P is the resultant of infinitesimal contributions of amplitude, dA, which originate from each element dx of slit width. In general,

(D-b)
$$dA = \frac{a_0 s}{D} \sin (\omega t - \delta) \, dx,$$

where the specific value of δ is determined by the relative position of the contributing slit element.

As Fig. D-2 shows, rays from points in the slit which are separated by a distance x differ in *path* by $(s \sin \theta)$ and thus in *phase* by $2\pi x \sin (\theta/\lambda)$. Therefore, if $(\omega t - \delta)$ is the phase reaching P from the center of the slit, the phases of the waves from positions x units on either side of the slit must be $\omega t - \delta - 2\pi x \sin (\theta/\lambda)$ and $\omega t - \delta + 2\pi x \sin (\theta/\lambda)$, respectively. The resultant amplitude at P may then be computed by *summing* the pairs of amplitudes contributed by all portions of the slit. That is,

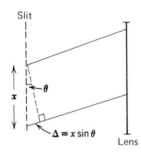

Slit

$\Delta = x \sin \theta$

Lens

Fig. D-2 Path difference between rays contributing to single slit pattern.

(D-c)
$$A = \int_{x=0}^{x=s/2} da_x + da_{-x}$$

$$= \frac{a_0 s}{D} \int_{x=0}^{x=s/2} \Bigg[\sin (\omega t - \delta)$$

$$- 2\pi x \sin \left(\frac{\theta}{\lambda}\right) + \sin (\omega t - \delta)$$

$$+ 2\pi x \sin \left(\frac{\theta}{\lambda}\right) \Bigg] dx.$$

Equation D-c may be transformed with the aid of the trigonometric identities:

(D-d) $\sin M + \sin N \equiv 2 \cos \tfrac{1}{2}(M - N) \sin \tfrac{1}{2}(M + N),$

(D-e) $\cos (-M) \equiv \cos M,$

to give

(D-f)
$$A = \frac{2a_0}{D} \sin{(\omega t - \delta)} \int_{x=0}^{x=s_2} \cos{\left(2\pi x \sin{\frac{\theta}{\lambda}}\right)} dx.$$

Thus, since θ and λ are constants, the integral reduces to the form

(D-g)
$$\int \cos{mx} \, dx = \frac{1}{m} \sin{mx},$$

where $m = 2\pi \sin{(\theta/\lambda)}$; therefore,

(D-h)
$$A = \frac{2a_0}{D} \sin{(\omega t - \delta)} \left(\frac{\lambda}{2\pi} \sin{\theta}\right) \left[\sin{\left(2\pi x \sin{\frac{\theta}{\lambda}}\right)}\right]_0^{s/2}$$

$$= \frac{a_0 s}{D} \sin{(\omega t - \delta)} \left(\frac{\lambda}{\pi s} \sin{\theta}\right) \sin{\left(\pi s \sin{\frac{\theta}{\lambda}}\right)}$$

or, making the substitution

(D-i)
$$\gamma = \pi s \sin{\frac{\theta}{\lambda}} \qquad \text{[as used also in (3-zz)]},$$

(D-j)
$$A = \frac{a_0 s}{D} \sin{(\omega t - \delta)} \frac{\sin{\gamma}}{\gamma}.$$

Substitution of (D-A) in (D-j) then gives

(D-k)
$$\boxed{A = A_0 \frac{(\sin{\gamma})}{\gamma}}.$$

Squaring of (D-k) gives

(D-l)
$$\boxed{I = I_0 \frac{(\sin^2{\gamma})}{\gamma^2}}.$$

Equation D-l relates I_0, the observed intensity at the center of the screen, to I, the intensity at any position on the screen as specified by the quantity γ, which has been defined by (D-i). (Note that, for any given wavelength and slit width, γ is a function only of θ, the angle at which light leaves the slit.)

Minima of the single slit diffraction pattern occur when $\sin{\gamma} = 0$, as is the case for $\gamma = m\pi$, where m is any integer (excluding zero). Thus the condition for minima is, as obtained from geometrical considerations in Chapter 3,

(3-yy)
$$\boxed{d \sin{\theta} = m\lambda}.$$

Maxima of the pattern occur at values of γ for which

(D-m)
$$\frac{dA}{d\gamma} = 0.$$

Differentiation of (D-k) with respect to γ yields

(D-n)
$$dA = \frac{A_0 d\,(\sin\gamma)}{\gamma} = \frac{A_0\left(\dfrac{\cos\gamma}{\gamma} - \sin\gamma\right)}{\gamma^2}.$$

so that the condition for maxima becomes

(D-o)
$$\cos\frac{\gamma}{\gamma} = \sin\frac{\gamma}{\gamma^2} \quad \text{or} \quad \boxed{\tan\gamma = \gamma}.$$

APPENDIX **E**

Image Location

This appendix is a summary of definitions, sign conventions, and formulas required for calculation of the positions and magnifications of images formed by lenses. Many of these expressions are discussed in Chapter 8.

The formulas included here are strictly applicable to paraxial rays; that is, their derivation is based on the first order approximation, $\sin \theta \cong \theta$. The effects of lens aberrations are ignored by these expressions.

Note that the use of formulas for location of images and determination of magnifications requires strict adherence to the appropriate definitions and conventions.

DEFINITIONS OF SYMBOLS

A_1, A_2 The positions of the vertices of the first and second surfaces, respectively, of a lens.

d The thickness of a thick lens at its axis.

f The focal length, which,

for a single surface, is the distance between the vertex and the focal point;

for a thin lens, is the distance between the lens and the focal point;

for a thick lens, is the distance between the (first or second) principal plane and the (first or second) focal point.

f' The first focal length of a single surface.

f'' The second focal length of a single surface.

f'_1 The first focal length of the first surface of a thick lens.

f''_1 The second focal length of the first surface of a thick lens.

f'_2 The first focal length of the second surface of a thick lens.

f_2'' The second focal length of the second surface of a thick lens.

F_1, F_2 The focal points, that is, the positions on the axis at which light entering a lens parallel to the axis from the right (F_1) or the left (F_2) is brought to a focus.

H_1, H_2 The principal points, that is, the positions at which the first and second principal planes, respectively, cross the axis.

m Magnification; more specifically, m_{lat} = image height/object height.

m_{ang} Angular magnification = angular subtense of image/angular subtense of object.

n Refractive index. In the equations given here, n' designates the refractive index of the lens. In equations describing refraction by lenses suspended in media other than air, the refractive index of the medium to the left of the lens is designated by n, and that to the right of the lens by n''.

R The radius of curvature of a surface.

s The object distance, which,
for a single surface or thin lens, is the distance between the object and the vertex;
for a thick lens, is the distance between the object and the first principal plane.

s' The image distance, which, as used here
for a single surface or thin lens, is the distance between the vertex of the lens and the image;
for a thick lens, is the distance between the second principal plane and the image.
Note, however, that in the main text the following convention is used: s_1 = object distance with respect to the first surface or principal plane of a lens; s_1' = image distance with respect to the first surface or principal plane of a lens; s_2 = object distance with respect to the second surface or principal plane of a lens; s_2'' = image distance with respect to the second surface or principal plane of a lens.

x The distance between the object and the first focal point of a lens.

x' The distance between the second focal point of a lens and the image.

Most of these conventions are illustrated in Fig. E-1.

SIGN CONVENTIONS

1. s (or, in general, the *object* distance) is positive when measured to the *left* of a vertex.

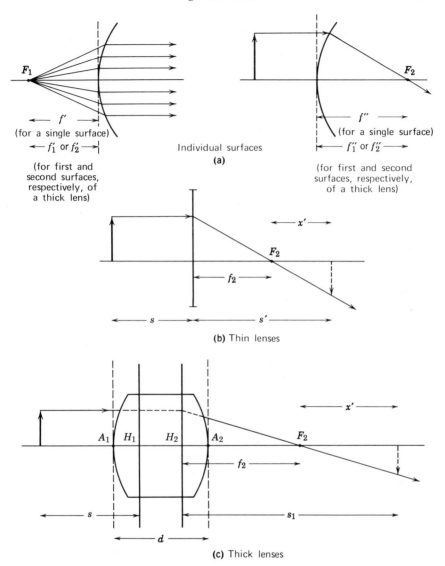

Individual surfaces

(a)

(b) Thin lenses

(c) Thick lenses

Fig. E-1.

2. s' (or, in general, the *image* distance) is positive when measured to the *right* of a vertex.

3. Surfaces which are *convex* with respect to a ray entering from the *left* of a figure are assigned a *positive* value of R; concave surfaces are assigned a negative value of R.

4. The value of f is *positive* for *converging* systems and *negative* for *diverging* systems.

5. *Erect* images are assigned a *positive* magnification; *inverted* images are assigned a *negative* magnification.

6. *Positive* signs signify distances measured to the *right* of a reference vertex; *negative* signs signify distances measured to the *left*.

EQUATIONS FOR SINGLE SPHERICAL SURFACES

(E-a) $\dfrac{n}{s} + \dfrac{n'}{s'} = \dfrac{n'-n}{R}.$ Location of object and image (8-m) p 162).

(E-b) $\dfrac{n}{f_1} = \dfrac{n'-n}{R} = \dfrac{n'}{f_2}.$ Location of focal points [(8-p) and (8-q) p. 164]

(E-c) $m = \dfrac{-(s'-R)}{(s+R)}.$ Determination of magnification [(8-kk) p. 183].

EQUATIONS FOR THIN LENSES

(E-d) $\dfrac{f_1}{f_2} = \dfrac{n}{n''}.$ Relation of first and second focal lengths.

(E-e) $\dfrac{1}{s} + \dfrac{1}{s'} = \dfrac{1}{f}.$ Image location in air [equivalent to (8-z) p. 167].

(E-f) $xx' = f^2 = f_1 f_2.$ Image location (Newtonian convention).

(E-g) $\dfrac{1}{f} = (n-1)\left(\dfrac{1}{R_1} - \dfrac{1}{R_2}\right).$ Determination of focal length (the "lens maker's equation").

(E-h) $m = -\dfrac{s'}{s}.$ Determination of mangification [equivalent of (8-mm) p. 183].

(E-i) $m = -\dfrac{f}{x} = -\dfrac{x'}{f}.$ Determination of magnification [equi-[Newtonian convention; (8-pp) p. 184].

EQUATIONS FOR THICK LENSES

I. Equation E-b is used to compute the focal lengths of the separate surfaces; that is,

(E-b$_1$) $\qquad\qquad\qquad \dfrac{n}{f_1'} = \dfrac{n'-n}{R_1} = \dfrac{n'}{f_1''},$

(E-b$_2$)
$$\frac{n'}{f'_2} = \frac{n' - n''}{R_2} = \frac{n''}{f''_2}.$$

II. The focal length of the lens as a whole is found with the use of

(E-j)
$$\frac{n}{f_1} = \frac{n'}{f''_1} + \frac{n''}{f''_2} - \frac{dn''}{f''_1 f''_2} = \frac{n''}{f_2}.$$

III. The positions of the principal and focal points with respect to the vertices are given by

(E-k)
$$A_1 F_1 = -f_1\left(1 - \frac{d}{f'_2}\right),$$

(E-l)
$$A_2 F_2 = +f_2\left(1 - \frac{d}{f''_1}\right),$$

(E-m)
$$A_1 H_1 = +\frac{f_1 d}{f''_2},$$

(E-n)
$$A_2 H_2 = -\frac{f_2 d}{f''_1}.$$

IV. Images are located with the use of (E-e), noting that s and s' (s''_2) are measured from H_1 and H_2, respectively.

Properties of Axially Symmetric Electrostatic and Electromagnetic Fields

In this appendix the properties of the electrostatic and electromagnetic fields which focus electron beams are discussed. That axially symmetric electrostatic fields act as electron lenses is proved in detail, while the nature of a comparable proof of the focusing action of electromagnetic fields is outlined. (The electrostatic rather than the electromagnetic case is chosen because of the convenience of considering focusing action in two, rather then three, dimensions. The treatment of electrostatic fields is also simpler mathematically.) It is shown also that the net action of either type of electron lens, when surrounded by field-free space, is inevitably convergent.

GENERAL NATURE OF THE PROOF THAT AN AXIALLY SYMMETRIC ELECTROSTATIC FIELD ACTS AS A LENS

The proof that a single glass surface of spherical curvature acts as a lens (given on p. 159) consists in demonstrating that, for the imaging of an axial object located at s, the image distance s' is *independent* of the distance h from the axis, at which rays reach the surface. Similarly, proof of electron lens action consists in showing that image distance is independent of the *slope r_0'* ($= dr/dx$ at $z = 0$) at which electrons enter the field.

Whereas the refractive properties of a glass surface are very simply characterized (by means of the equations summarized on p. 716) in terms of the radius of curvature R, and the refractive indices of the media on either side of the surface, characterization of the refractive properties of an electrostatic field is less straightforward, since these properties change continuously as a function of position. Thus proof of lens action must begin with the development of an equation which specifies the potential ϕ, at any distance z, along the axis of the field and at any radius r from the axis, as a function of the applied axial potential $\Phi(z)$. This equation is then combined with a general equation which describes the motions of electrons in electrostatic fields, thus obtaining an equation which specifically describes the trajectories of electrons in axially symmetric electrostatic fields. The last equation is known as the *paraxial ray equation* since its derivation, like that of (8-10) which describes the refraction of light at spherical glass surfaces, depends upon the use of approximations in which (in this case) terms in powers of r greater than the first are deleted. Thus the equation is applicable to electrons which travel only a very small distance r from the axis, that is, to paraxial rays. The paraxial ray equation is a differential equation in r; its solution must finally be shown to be such that image distances are the same for *all* rays entering the field from a given object point.

DERIVATION OF A DIFFERENTIAL EQUATION CHARACTERIZING THE AXIALLY SYMMETRIC ELECTROSTATIC FIELD

According to Coulomb's law, an electron located in a vacuum exerts a radial force F such that at a distance r from the electron

(F-a)
$$F = \frac{e^2}{r^2},$$

where e is the electronic charge. The electrostatic field strength in the space around the electron is then

(F-b)
$$E = \frac{F}{e} = \frac{E}{r^2},$$

while the *vector* field strength in any direction, s, is equal to the *gradient* of the scalar electrostatic potential ϕ. That is,

(F-c)
$$\mathbf{E} = \frac{d\phi}{ds}.$$

In the space surrounding an electron, the field strength at opposite faces of a cubic element of volume (for which $V = dx\,dy\,dz$) is E_x and $(E_x + \partial E_x / \partial x)$, respectively, in the x direction (and likewise for the other two pairs of faces in the y and x directions). The *difference* in electrostatic flux into and out of the opposing faces is thus $(\partial E_x/\partial_x)\, dx\, dy\, dz$ (and likewise for the other pairs). The *sum* of these fluxes must be zero if the volume element lacks any sources or sinks of electrostatic potential, so that

(F-d)
$$\left(\frac{\partial E_x}{\partial x}+\frac{\partial E_y}{\partial y}+\frac{\partial E_z}{\partial z}\right) dx\, dy\, dz = 0.$$

Substituting, from (F-c),

(F-e)
$$\partial E_x = \frac{\partial^2 \phi}{\partial x^2};\qquad \partial E_y = \frac{\partial^2 \phi}{\partial y^2};\qquad \partial E_z = \frac{\partial^2 \phi}{\partial z^2},$$

(F-d) becomes

(F-f)
$$\frac{\partial^2 \phi}{\partial x^2}+\frac{\partial^2 \phi}{\partial y^2}+\frac{\partial^2 \phi}{\partial z^2}=0.$$

Equation F-f (which is known as Laplace's equation) may be transformed by routine substitutions to an expression in *cylindrical* coordinates. The cartesian coordinates x and y are thus replaced by the parameters r (the radial distance from the axis) and θ (the angle of rotation about the axis), while coordinate z is retained as a measure of position along the axis of the cylinder. Note that these parameters are appropriate to a discussion of axially symmetric fields, because the latter are essentially cylindrical in shape. The Laplacian equation, expressed in cylindrical coordinates, is

(F-g)
$$\frac{\partial^2 \phi}{\partial z^2}+\frac{1}{r}\left[\frac{\partial}{\partial r}\left(\frac{r\partial \phi}{\partial r}\right)\right]+\frac{1}{r^2}\left(\frac{\partial^2 \phi}{\partial \theta^2}\right)=0.$$

Since the field considered here is axially symmetric, the potential ϕ does not vary as a function of θ. Thus $\partial^2\phi/\partial\theta^2 = 0$, and (F-g) reduces to

(F-h)
$$\frac{\partial^2 \phi}{\partial z^2}+\frac{1}{r}\left[\frac{\partial}{\partial r}\left(\frac{\partial \phi}{\partial r}\right)\right]=0$$

Further characterization of the axially symmetric field then requires an appropriate *solution* to the general differential equation (F-h).

SOLUTION OF THE DIFFERENTIAL EQUATION CHARACTERIZING THE AXIALLY SYMMETRIC ELECTROSTATIC FIELD

Equation (F-h) may be solved by the technique of assuming a *trial solution* of general form and subsequently particularizing the solution in a manner appropriate to the particular physical case considered. It is found that the appropriate solution is a power series of the form

(F-i) $\phi\,(r,z) = A_0(z)+A_1(z)r+A_2(z)+\ldots+A_n(z)r^n,$

in which $A_0(z), \ldots, A_n(z)$ are a series of (as yet unspecified) functions of z. In obtaining a particular solution of (E-i), these functions of z are evaluated.

For the axially symmetric field, distinctions between $+r$ and $-r$ are purely arbitrary, so that the value of $\phi(r, z)$ cannot vary with the sign of r. For this to be true, all the coefficients of *odd* powers of r must be zero. Thus a modified form of the general solution is

(F-j) $$\phi(r, z) = A_0(z) + A_2(z)r^2 + A_4(z)r^4 + \ldots.$$

Further particularization of the solution may then be achieved by differentiating equation (F-j) to obtain expressions for the terms $(1/r)$ $[(\partial/\partial r)(r\partial\phi/\partial r)]$ and $(\partial^2\phi/\partial z^2)$, which appear in the differential equation (F-h). These expressions are obtained in (F-k) to (F-m) and (F-o), respectively:

(F-k) $$\frac{\partial\phi}{\partial r} = 2A_2(z)r + 4A_4(z)r^3 + 6A_6(z)r^5 + \ldots,$$

(F-l) $$r\left(\frac{\partial\phi}{\partial r}\right) = 2A_2(z)r^2 + 4A_4(z)r^4 + 6A_6(z)r^6 + \ldots,$$

(F-m) $$\frac{\partial}{\partial r}\left(r\frac{\partial\phi}{\partial r}\right) = 4A_2(z)r + 16A_4(z)r^3 + 36A_6(z)r^5 + \ldots,$$

(F-n) $$\frac{1}{r}\left[\frac{\partial}{\partial' r}\left(r\frac{\partial\phi}{\partial r}\right)\right] = 4A_2(z) + 16A_4(z)r^2 + 36A_6(z)r^4 + \ldots,$$

(F-o) $$\frac{\partial^2\phi}{\partial z^2} = A_0''(z) + A_2''(z)r^2 + A_4''(z)r^4 \ldots,$$

where the terms $A_0''(z)$, $A_0''(z)$ etc., are the *second derivatives* of the corresponding functions $A_0(z), A_2(z)$ etc.

Equation F-h states that the *sum* of the two terms evaluated by (F-n) and (F-o) must be *identically equal to zero* for *all* values of both r and z. This can be the case only if the sum of coefficients of *each* power of r is, individually, equal to zero. Thus pairs of the coefficients which appear in (F-n) and (F-o) may be equated, giving

(F-p) $$\begin{aligned} A_0''(z) &= -4A_2(z), \\ A_2''(z) &= -16A_4(z), \\ A_4''(z) &= -36A_6(z), \end{aligned}$$

The function $A_0(z)$ is simply the potential $\phi(0, z)$, which occurs along the axis of the field; that is, it is identical with $\Phi(z)$, the *axial potential*. $\Phi(z)$ is a known quantity in specific cases; its evaluation need not be considered in deriving a general equation. Thus the coefficients of r, that is, the various functions, $A_n(z)$, may be evaluated in terms of the derivatives of the (known) axial potential. That is,

(F-q)
$$A_2(z) = \frac{-\Phi''(z)}{4},$$

where $\Phi''(z)$ is the second derivative of the axial potential;

(F-r)
$$A_2'' = \frac{-\Phi^{IV}(z)}{4} = -16A_4(z); \qquad A_4(z) = \frac{\Phi^{IV}(z)}{64},$$

where $\Phi^{IV}(z)$ is the fourth derivative of the axial potential,

(F-s)
$$A_4''(z) = \frac{\Phi^{VI}(z)}{64} = -36A_6(z); \qquad A_6(z) = \frac{-\Phi^{VI}(z)}{36 \times 64},$$

and so on.

Substitution of the terms so obtained in (F-q), (F-r), and (F-s) into the trial solution (F-j) then yields

(F-t)
$$\phi(r, z) = \Phi(z) - \frac{\Phi''(z)r^2}{4} + \frac{\Phi^{IV}(z)r^4}{64} - \frac{\Phi^{VI}(z)r^6}{2300} + \cdots.$$

Equation F-t, which characterizes the refractive properties of the electrostatic field as a function only of r, the distance from the axis, and of the axial potential and its derivatives, is an *exact* equation. That is, no assumptions have been introduced in order to obtain this equation. The $\Phi(z)$ must, of course, be known in order to evaluate ϕ in any specific case. In order to study the action of a field as a lens, it is necessary to determine the trajectories of electrons within the field characterized by (F-t).

ELECTRON TRAJECTORIES IN ELECTROSTATIC FIELDS

In this section an equation for the trajectories of electrons in electrostatic fields, generally, is developed. The trajectories of electrons in axially symmetric electrostatic fields, specifically, can then be computed (see p. 723 below).

$d(mv)/dt$, the *rate of change of momentum* of an electron which travels through an electrostatic field, may be equated, according to Newton's second law (force = mass × acceleration) with the force acting on the electron. Thus,

(F-u)
$$\frac{d}{dt}\left(m\frac{dr}{dt}\right) = e\left(\frac{\partial\phi}{\partial r}\right),$$

where m is the mass of the electron, e its charge, ϕ is the potential at any point, and r is the distance of the electron from the axis of the field.

Equation F-u may usefully be arranged by noting that

(F-v)
$$\frac{dr}{dt} = \left(\frac{dr}{dz}\right)\left(\frac{dz}{dt}\right) = \left(\frac{dr}{dz}\right) V_{ax},$$

where V_{ax} is the electron velocity in the axial (z) direction, and by noting that, as may be obtained from (16-r),

(F-w)
$$V_{ax} = \sqrt{\frac{2e\phi}{m}}.$$

Upon substitution of (F-v) and (F-w), (F-u) thus becomes

(F-x)
$$\frac{d}{dt}\left(m\frac{dr}{dz}\sqrt{\frac{2e\phi}{m}}\right) = e\left(\frac{\partial\phi}{\partial r}\right).$$

Also, since

(F-y)
$$\frac{d}{dt} = \frac{dz}{dt}\frac{d}{dz} = \sqrt{\frac{2e\phi}{m}}\frac{d}{dz},$$

(F-x) becomes

(F-z)
$$e\frac{\partial\phi}{\partial r} = \sqrt{\frac{2e\phi}{m}}\frac{d}{dz}\left(m\frac{dr}{dz}\sqrt{\frac{2e\phi}{m}}\right)$$

$$= 2e\sqrt{\phi}\frac{d}{dz}\left(\frac{dr}{dz}\phi^{1/2}\right).$$

The derivatives in (F-z) may be evaluated according to the usual rule, $d(uv) = udv + vdu$):

(F-aa)
$$e\frac{\phi}{r} = 2e\phi^{1/2}\left[\frac{1}{2}\frac{dr}{dz}\phi^{-1/2}\frac{d\phi}{dz} + \phi^{1/2}\frac{d^2r}{dz^2}\right],$$

or, combining terms and rearranging,

(F-bb)
$$\frac{d^2r}{dz^2} + \frac{1}{2\phi}\frac{\partial\phi}{\partial z}\frac{dr}{dz} - \frac{1}{2\phi}\frac{\partial\phi}{\partial r} = 0.$$

Equation (F-bb) is an *exact* general equation for the *trajectories of electrons in electrostatic fields*.

TRAJECTORY OF THE ELECTRON IN AXIALLY SYMMETRIC ELECTROSTATIC FIELDS

An expression for the trajectories of electrons in axially symmetric electrostatic fields may be obtained by substituting (F-t) and its derivatives in (F-bb). However, since (F-t) is a *series* equation, an *approximation* to that expression, including only a *finite* number of terms, must be used. The simplest such approximation is one which uses only the *first* terms of the series which give ϕ, $\partial\phi/\partial r$, and $\partial\phi/\partial z$, respectively. These expressions, all obtained by deleting higher order terms from (F-t) or its derivatives, are

(F-cc)

$$\phi \cong \Phi(z),$$

$$\frac{\partial\phi}{\partial r} \cong \left(\frac{-\Phi''(z)}{2}\right)r,$$

$$\frac{\partial\phi}{\partial z} \cong \Phi'(z).$$

Note that the use of these approximate equations is analogous to the use of the approximation $\sin\theta \cong \theta$ in obtaining an equation for the refraction of light by a single spherical surface. In both cases the resulting expressions are valid only for rays (electron trajectories) at small values of r, that is, for paraxial rays. The equation which is so obtained by substituting (F-cc) in (F-bb) is

(F-dd)

$$\boxed{\frac{d^2r}{dz^2} + \frac{\Phi'(z)}{2\Phi(z)}\frac{dr}{dz} + \frac{\Phi''(z)}{4\Phi(z)}r = 0}.$$

Equation F-dd, known as the *paraxial ray equation*, is one of the fundamental expressions of electron optics. The paraxial ray equation is a differential equation in the variable r; specific solutions may, as in the case of (F-h), be found by the method of substituting and particularizing trial solutions. In this case the appropriate trial solution is of the form

(F-ee) $$r = Af(z) + Bg(z),$$

in which A and B are arbitrary constants which must bear no specific numerical relation to each other, while $f(z)$ and $g(z)$ are specific arbitrary functions of z.

To evaluate A and B, note first that, when $z = 0$, $r = r_0$, so that

(F-ff) $$r_0 = Af(0) + Bg(0).$$

f_0 is a *particular* value of r, while $f(0)$ and $g(0)$ are particular values of the respective functions. Thus, if A and B in fact bear no specific numerical relation, it must be true that *either* $f(0)$ or $g(0)$ is zero. One of them, $g(0)$ for example, may be equated to zero to give

(F-gg) $$r_0 = Af(0) \quad \text{or} \quad A = \frac{r_0}{f(0)}.$$

(Note that, if $f(0)$ were equated to zero instead, the final solution obtained would be of slightly different but strictly comparable form.) The general solution of the paraxial ray equation thus becomes

(F-hh)
$$r = \left[\frac{r_0}{f(0)}\right] f(z) + Bg(z).$$

To evaluate B, note that, at $z = 0$, the *slope* dr/dz of the electron trajectory is r'_0, so that

(F-ii)
$$r'_0 = \frac{r_0}{f(0)} f'(0) + Bg'(0).$$

Again, in order to avoid a particular numerical relation between A and B, either $f'(0)$ or $g'(0)$ must be zero. Thus, for $f'(0) = 0$,

(F-jj)
$$r'_0 = Bg'(0) \quad \text{or} \quad B = \frac{r'_0}{g'(0)}.$$

The general solution of the paraxial ray equation is thus

(F-kk)
$$r = \frac{r_0}{f(0)} f(z) + \frac{r'_0}{g'(0)} g(z).$$

The solution is also written in the more concise form

(F-ll)
$$\boxed{r = M(z)r_0 + N(z)r'_0},$$

where $M(z) = f(z)/f(0)$ and $N(z) = g(z)/g'(0)$.

Particular functions $M(z)$ and $N(z)$ can be substituted in (F-ll) in order to determine electron trajectories in a particular lens field. However, in order to show that axially symmetric electrostatic fields, in general, act as lenses, the values of M and N must remain arbitrary.

PROOF OF LENS ACTION

For an electron which initially travels parallel to the axis at some distance r_0, r'_0, the initial slope of the trajectory is zero; thus (F-ll) becomes

(F-mm)
$$r = r_0 M(z).$$

Any other electron trajectory that cuts the first at a distance $z = s$ to the left of the field, and which is of initial slope r'_0 is described by

(F-nn)
$$r = (r_0 + r'_0 s)M(z) + r'_0 N(z).$$

These two trajectories may recross at some position beyond the lens which can be designated as s', and where r is the same for both trajectories. Thus,

(F-oo)
$$r_0 M(s') = (r_0 + r'_0 s)M(s') + r'_0 N(s')$$

or

(F-pp)
$$r'_0 s M(s') + r'_0 N(s') = 0,$$

so that, by cancellation of the common factor r'_0, the equation relating the image and object distances is found to be

(F-qq) $\boxed{sM(s') + N(s') = 0}$.

Thus the image position is found to be independent of the initial slope of the electron trajectories (and, correspondingly, independent of the value of r at which the electrons enter the field), and the axially symmetric electrostatic field is thereby proved to act as a lens.

ACTION OF AXIALLY SYMMETRIC ELECTROMAGNETIC FIELDS AS LENSES

The proof that axially symmetric magnetic fields act as electron lenses is analogous to that just given for the electrostatic case. An equation for the (magnetic) potential throughout the field, and approximations to the solution of this equation, are applied to a general equation which describes the trajectories of electrons in magnetic fields. Thus $\phi_m(r, z)$, the magnetic potential at any point in an axially symmetric field, is given by

(F-rr) $$\phi_m(r, z) = \Phi_m(z) - \frac{\Phi_m(z)r^2}{4} + \frac{\Phi_m{}^{IV}r^4}{64},$$

where $\Phi_m(z)$ is the function describing the axial magnetic potential. Equation F-rr is used to furnish the approximation

(F-ss) $$H_r \cong -\tfrac{1}{2}r\frac{\partial H_z}{\partial z},$$

where H_r are the (scalar) radial and axial components, respectively, of the magnetic field.

Again Newton's second law is used to equate the rate of change of momentum of the electron with the force acting upon it. The result is

(F-tt) $$\frac{d}{dt}\left(m\frac{dr}{dt}\right) = \frac{e}{c}r\theta H_z + m\frac{d\theta}{dr^2}r$$

where θ is the angle through which the electron trajectory is *rotated*, m and e are the mass and charge, respectively, of the electron, and c is the velocity of light. From these equations the magnetic form of the paraxial ray equation is then obtained:

(F-uu) $\boxed{\dfrac{d^2r}{dz^2} + \dfrac{+e}{8mc^2\phi}H_z^2r = 0}$.

[Note that in (F-uu) ϕ is the potential through which the electron beam is accelerated, and that it is an electrostatic, not a magnetic potential.]

The solution of the magnetic paraxial ray equation is of the form

(F-vv) $$r = r_0 M(z) + r_0' N(z),$$

from which it follows, as before, that the field acts as a lens.

OTHER EQUATIONS WHICH CHARACTERIZE THE MAGNETIC LENS

Two important equations which follow from the treatment summarized above specify θ_i, the *angular rotation* of the image formed by the magnetic lens, and f, the focal length. They are

(F-ww) $$\boxed{\theta_i = \left(\frac{e}{8\,mc^2\phi}\right)^{1/2} \int_{z=0}^{z=l} Hz}\,,$$

where l is the linear extent of the magnetic field. As noted in Chapter 17, θ_i specifies the rotation of the image in excess of the 180° rotation normally produced by positive glass or electrostatic lenses.

The focal length is given by

(F-xx)
$$\boxed{\begin{aligned}\frac{1}{f} &= \frac{e}{8\,mc^2\phi} \int_0^l H_z^2\, dz \\[2mm] &= \frac{0.22}{V} \int_0^l H_z^2\, dz\end{aligned}}\,,$$

where V is the accelerating voltage, in practical volts. From (F-xx) it follows that, for any given field strength H_z,

(F-yy) $$\frac{1}{f} = \frac{K}{V}$$

(where K is a constant), and thus that the variation in focal length with accelerating voltage is

(F-zz) $$\boxed{\frac{\Delta f}{f} = C_c \frac{\Delta V}{V}}\,,$$

where C_c is the *chromatic aberration constant* characteristic of the lens.

THE NET CONVERGING ACTION OF ELECTRON LENSES

From the paraxial ray equations (F-dd) and (F-vv) it may be deduced that all electron lenses (i.e., all axially symmetric electrostatic or magnetic fields bounded by field-free space) possess a net convergent action.

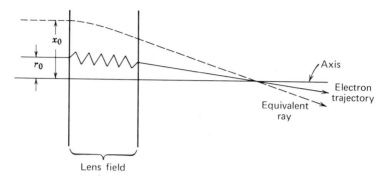

Fig. F-1 Real and equivalent electron trajectories for an electron lens.

In regions of uniform (electrostatic and magnetic) potential, electron trajectories are straight lines. Within a lens field, electron trajectories are in general, curved. While the actual trajectories inside a lens may be complex, the electron path may be represented by an "equivalent ray." The equivalent ray is focused at the same point as the real electron and also travels along straight lines in field-free space but follows a simple and direct course within the lens. This concept is illustrated schematically in Fig. F-1. As shown, the path of the equivalent ray in a *convergent* lens is always *concave* with respect to the lens axis. Mathematically, this means that the second derivative of the function specifying x (the radial position of the "equivalent" electron) must always be negative when x itself is positive (and vice versa).

The convergent action of electron lenses is thus revealed by the fact that expressions which describe the path of the equivalent ray are always found to posess a second derivative of opposite sign. For electrostatic lenses the equivalent ray may be defined to be such that

(F-a*) $$x = r\,\Phi^{1/4}.$$

With respect to this definition, note that, if $r = 0$, then $x = 0$; thus (F-a*) satisfies the specification that the real and equivalent rays must cross the lens axis at the same position. Substitution of (F-aa) in the paraxial ray equation (F-d) leads to an expression in which (d^2x/dx^2) is always opposite in sign to x. To make this substitution the quantities (dr/dz) and (d^2r/dz^2) are first evaluated in terms of x, giving

(F-b*) $$\frac{dr}{dz} = -\frac{x}{4}\,\Phi^{-5/4}\,\frac{d\Phi}{dz} + \Phi^{-1/4}\,\frac{dx}{dz}$$

and

(F-c*)
$$\frac{d^2r}{dz^2} = -\frac{x}{4}\left[\Phi^{-5/4}\frac{d^2\Phi}{dz^2} - \frac{5}{4}\Phi^{-9/4}\left(\frac{d\Phi}{dz}\right)^2\right]$$

$$-\frac{1}{4}\Phi^{-5/4}\frac{d\Phi}{dz}\frac{dx}{dz} + \Phi^{-1/4}\frac{d^2x}{dz^2}$$

$$-\frac{1}{4}\Phi^{-5/4}\frac{dz}{dz}\frac{d\Phi}{dz}.$$

Substitution of these derivatives in (F-dd) then gives an expression which is, after cancellation of terms,

(F-d*)
$$\Phi^{-1/4}\frac{d^2x}{dz^2} + \frac{3}{16}\Phi^{-9/4}\left(\frac{d\Phi}{dz}\right)^2 = 0$$

or, upon rearrangement,

(F-e*)
$$\frac{d^2x}{dz^2} = -\frac{3}{16}\left(\frac{d\Phi/dz}{\Phi}\right)^2 x.$$

On the right side of (F-e*), x is multiplied by the *negative* of a *squared* term, so that the second derivative on the left side of this equation must always be of opposite sign to x. As stated above, this is precisely the condition which reveals net converging action of the lens system.

In the case of magnetic lenses it is unnecessary to define an equivalent ray, since the real trajectories themselves are in fact concave with respect to the lens axis. This is indicated by rearrangement of the magnetic paraxial ray equation (F-uu) to give

(F-f*)
$$\frac{d^2r}{dz^2} = -\frac{e}{8mc^2\phi}H_z^2 r.$$

APPENDIX G

Graphical Integration

Graphical integration can be used to determine the total area under any curve, including those for which the form is not described by any known mathematical function. This procedure may be carried out by (a) direct counting of the number of squares included under the curve; (b) cutting out the graph and comparing its weight with that of a known area of the graph paper; or (c) computer methods. While computer techniques tend to be more accurate, either of the other two methods may give satisfactory results. The determination of the total rate at which quanta of all wavelengths are fluoresced by a solution are described here as an example of graphical integration by means of counting of squares.

Figure G-1a shows a typical fluorescence emission spectrum. When manual readings of fluorescent intensity are made, this curve is approximated by the series of values recorded at discrete wavelength intervals. A crude example of such a plot is given in Fig. G-1b, where emission by the compound considered in Fig. G-1 is measured at 10 mμ intervals. Note that the intensity scale (y axis) of Fig. G-1 is a measure of intensity in terms of the *rate of photon emission* at each wavelength; that is, these spectra have been corrected for phototube response in the manner described in Chapter 25. The number of squares under *either* curve of Fig. G-1 may be counted. The area under the curve given in Fig. G-1b approximates that under the curve given in Fig. G-1a but is less accurate to the extent that the wavelength interval between readings is appreciable. An exact count of squares can be made for Fig. G-1b, but usually *fractions* of squares included under a smooth curve must be *estimated*. (Note that use of graph paper ruled with smaller squares than those shown here aids in making accurate estimates of square count.)

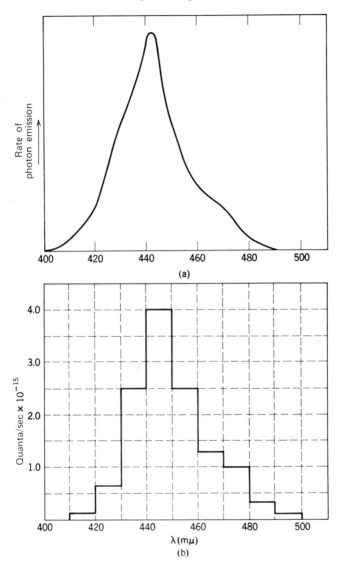

Fig. G-1 (a) Fluorescence emission spectrum; (b) corresponding histogram of emitted intensity.

Each of the complete squares marked in Fig. G-1 represents 5×10^{14} quanta/sec. Counting of squares in Fig. G-1b shows that exactly 25 complete squares are included under this curve. Thus the total emission by this solution is

(G-a) $0.5 \times 10^{14} \times 25 = 1.25 \times 10^{16}$ quanta/sec.

Figure G-2 gives an example of an irregular curve which includes about 14.2 squares.

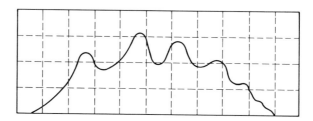

Fig. G-2.

The result obtained in (G-a) can also be obtained by cutting out the graph shown in Fig. G-1 and comparing its weight with that of a piece of the same paper which contains the exact number of squares. While the method eliminates the tedious process of estimating fractions of many small squares, its accuracy is comparable to that obtained by direct counting only if the graph paper is of very high uniformity and if the curves are plotted on a relatively large scale.

Author Index

Subject Index

Numbers in italics refer to sections devoted specifically to the topic cited.